THE BREEDING BIRD ATLAS OF CHESHIRE AND WIRRAL

For singing till his heaven fills,
'Tis love of earth that he instills,
And ever winging up and up,
Our valley is his golden cup
And he the wine which over flows
To lift us with him as he goes . . .

from "THE LARK ASCENDING" by GEORGE MEREDITH, 1829-1909

Copyright by
The Cheshire and Wirral Ornithological Society 1992

ISBN 0 9517301 0 X   The Breeding Bird Atlas of
                       Cheshire and Wirral (hbk)

All rights reserved. No part of this book may be reproduced,
stored in a retrieval system, or transmitted in any form or by
any means, electronic, mechanical, photocopying or otherwise,
without the prior permission of the publisher.

Text set in 9/10pt Plantin.
Printed and bound in Great Britain by the Bath Press, Avon.

THE BREEDING BIRD ATLAS OF
CHESHIRE
AND WIRRAL

J. P. Guest

D. Elphick

J. S. A. Hunter

D. Norman

CHESHIRE AND WIRRAL
ORNITHOLOGICAL SOCIETY

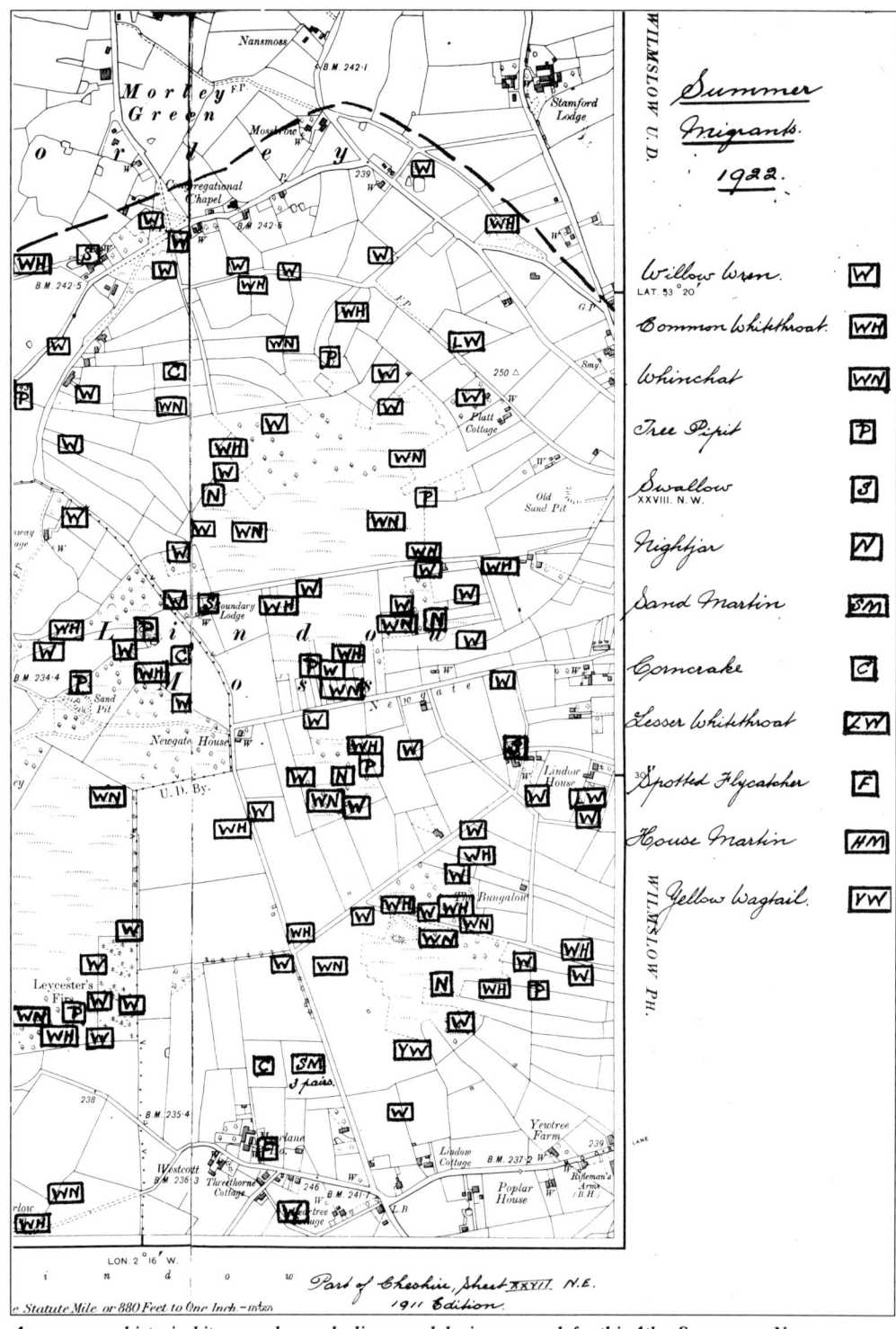

Among many historical items and records discovered during research for this Atlas Survey was Norman Abbott's detailed mapping of his local bird-watching haunts – Lindow Moss shown here with precise locations of singing summer visitors in 1922. (Courtesy of Manchester Ornithological Society)

CONTENTS

1
SPONSORS

2
A BRIEF HISTORY OF THE AREA KNOWN AS CHESHIRE

6
GEOLOGY AND RELIEF

8
BIRD HABITATS IN CHESHIRE

Woodland and plantations; Hedgerows; Heathland and mossland; Moorland; Meres and fresh waters; Towns and suburbia; Salt and chemical industries; Coastal habitats; Farmland

22
CHANGES IN THE CHESHIRE AVIFAUNA: SOME TRENDS OF THE TWENTIETH CENTURY

Recovery of persecuted species; Colonisation by waterfowl and introductions; The rise and fall of industrial habitats; Decline of trans-Saharan migrants; Neglect of mature woodland; Moorland habitat loss; Lowland drainage and succession of marl-pits; Agricultural intensity; Public opinion

26
BIRD-RECORDING IN CHESHIRE AND WIRRAL

Recording areas; The Future?

27
ABOUT THE BREEDING BIRD ATLAS
OF CHESHIRE AND WIRRAL
The Tetrad; Fieldwork; Confidentiality of records

34
ACKNOWLEDGEMENTS AND CONTRIBUTIONS

36
BIBLIOGRAPHY

37
THE BREEDING BIRDS OF CHESHIRE AND WIRRAL

294
APPENDICES

298
GAZETTEER

306
INDEX

The Cheshire and Wirral Ornithological Society gratefully acknowledges the financial assistance of the following towards publication costs of

THE BREEDING BIRD ATLAS OF
CHESHIRE
AND WIRRAL

Altrincham and District Natural History Society
Anonymous
Bayer UK Ltd
British Nuclear Fuels
BTO Research Grants Committee
Cheshire County Council
Cheshire Ornithological Association
Chester and District Ornithological Society
Dundalk Border Services
Dr Paul Griffiths
Hale Ornithologists
Hancock and Wood
Ron Harrison and Steve Barber (Checklist Sales)
ICI Pharmaceuticals
Knutsford Ornithological Society
Laporte Absorbents of Widnes
Manchester Airport plc
Manchester Ship Canal Company
Merseyside Ringing Group
Mitchell Trust
E. A. Noble
Peter Perkins
Clive Richards
RSPB Macclesfield Group
RSPB North Cheshire Group
RSPB Stockport Group
Wilmslow Guild
Wirral Bird Club
Woolston Eyes Conservation Group
Geoff Worthington
Yapp Education and Research Trust

A BRIEF HISTORY
OF THE AREA KNOWN AS CHESHIRE

LYING in Great Britain's southern half and west of the southern Pennine hills, Cheshire might be considered as part of North-West England or of the north-western Midlands.

Historically, Cheshire is neither truly Northern nor Midland. It has ties with the former south of Lancashire, continuing its predominantly low-lying topography, but, above all, it is a Welsh border county. The basically Celtic population had a close affinity with that of northern Wales (Sylvester 1971).

Our county probably came into existence with the shiring of Mercia in the ninth century. A Roman name, Legaceater, meaning "the Shire of the Legions", was recorded a century later in the Alfredian Chronicle. Like a typical English shire, it was centred on a conveniently-sited administrative town. It is believed that, when first split from Mercia, Cheshire extended as far west as Offa's Dyke. This may explain why the city of Chester is at the western edge of the present county.

However, Cheshire became much larger: at the time of the Domesday Book survey (1086), Cheshire stretched north to the River Ribble, encompassing all of southern Lancashire and even parts of the Yorkshire Dales. In the south, Cheshire included Maelor – parts of what later became the counties of Flintshire and Denbighshire.

In the twelfth century, Cheshire contracted as Lancashire became an enlarged county palatine south to the River Mersey. Longdendale, however, remained in Cheshire. As the campaigns of Edward I in Wales concluded (1284), the first Welsh counties were created; Cheshire's western boundary was brought inward to the River Dee. Over on the eastern side, the high land within the shire's boundary was vaguely known as "Lyme", and even towns as far

Fig.1.
Cheshire and surrounding counties before 1st April 1974.

away as Leek (Staffordshire) may have come under Cheshire's jurisdiction at this time.

It was only in 1536, when the union of England and Wales was achieved by Thomas Cromwell under Henry VIII, that Cheshire formally became English.

From then on, Cheshire – made up of "boroughs" (as conferred upon towns from time to time) and Mercian "hundreds" – was relatively stable until successive nineteenth-century reforms and rationalisations created boroughs and urban and rural districts.

Some trimming of Cheshire's boundaries took place earlier this century. The east-facing slopes of the Goyt valley became part of Derbyshire as the boundary was redrawn down the Goyt's western watershed. This excluded from Cheshire its former parishes of Taxal and Yeardsley-cum-Whaley. The northern county boundary was removed from the course of the River Mersey, downstream of Carrington, to be aligned with the Manchester Ship Canal.

Various re-hashings of the ancient counties were mooted in the 1960s and 1970s – it might have come about that Cheshire be eliminated entirely, being absorbed into either a "North-Western Province" or an "extended Staffordshire" division, or divided between both.

Far-reaching changes were enabled by the Local Government Act of 1972. These changes affected many counties in England and Wales, and came into effect on All Fools' Day, 1974. Cheshire did not escape lightly; it had been stated, in respect of Cheshire and Merseyside, that " 'political' (= electoral) objectives appeared to be overriding those of democracy and efficiency" (Wood 1976).

The Act created "metropolitan counties", largely artificial entities meant to streamline the administration of conurbations. With two of these on Cheshire's northern flanks, its human population experienced some

Fig.2.
Cheshire and surrounding counties after 1st April 1974.

traumatic adjustments. Cheshire's birds may have known nothing of it, but some of Cheshire's bird-watchers and Societies experienced confusion, especially with regard to the submission and publication of records. Cheshire "lost" over 380,000 people to Merseyside and Greater Manchester metropolitan counties, and "gained" some 186,000 from a severed Lancashire. (The figures are those of the 1971 census.) This gave a net loss to the county of approximately one eighth of its former total human population.

The following table shows Cheshire's losses and gains as detailed in the Schedules to the Local Government Act 1972, and has particular relevance to study of Cheshire's natural history in an archival context.

Table 1. How Cheshire was affected by boundary changes, subsequent to the Local Government Act, 1972

From the County of Chester
to Greater Manchester Metropolitan County
the parishes of Ringway, Carrington, Dunham Massey, Partington and Warburton (all from Bucklow Rural District); the Urban Districts of Bowdon, Hale, Bredbury & Romiley, Cheadle & Gatley, Hazel Grove & Bramhall, Marple, and Longdendale; the Municipal Boroughs of Sale, Altrincham, Dukinfield, Hyde, and Stalybridge; and the southern part of the County Borough of Stockport (the northern part was within the County of Lancaster)

to Derbyshire Non-metropolitan County
the Rural District (parish) of Tintwistle

to Merseyside Metropolitan County
the Urban Districts of Wirral (which consisted of the areas of Irby, Pensby, Barnston, Gayton, Thurstaston and Heswall) and Hoylake; the Municipal Borough of Bebington; and the County Boroughs of Wallasey and Birkenhead.

To Cheshire Non-metropolitan County
from the County of Lancaster
the wards of Culcheth and Newchurch (both from Golborne Urban District); the parish of Hale and parts of the parish of Bold (all from Whiston Rural District); the Rural District of Warrington; the Municipal Borough of Widnes; and the County Borough of Warrington.

It is worth relating that, prior to the actual date of change – 1st April 1974 –, Ellesmere Port, Runcorn and Widnes were removed from the proposed Merseyside assemblage to the non-metropolitan county of Cheshire.

Highly-vocal public opinion led local councils to oppose the sucking of Wilmslow and the parish of Poynton-with-Worth, both largely rural, into the proposed Greater Manchester county. The House of Lords voted to remove Wilmslow and Poynton from this ignominy and returned them to Cheshire at the last possible moment – less than two weeks before the Act became law. Indeed, Poynton remained in Cheshire by only one vote! (*Hansard*, 16th October 1972).

To simplify presentation, and reading, throughout this book the terms "Cheshire" and "the county" may be presumed to include "the Wirral peninsula". (As with Wilmslow and Poynton, it is true to relate that a majority of Wirral residents regard themselves as "Cheshire".) All the habitat and species maps, however, do show the administrative boundary between Cheshire and the county of Merseyside where it crosses Wirral.

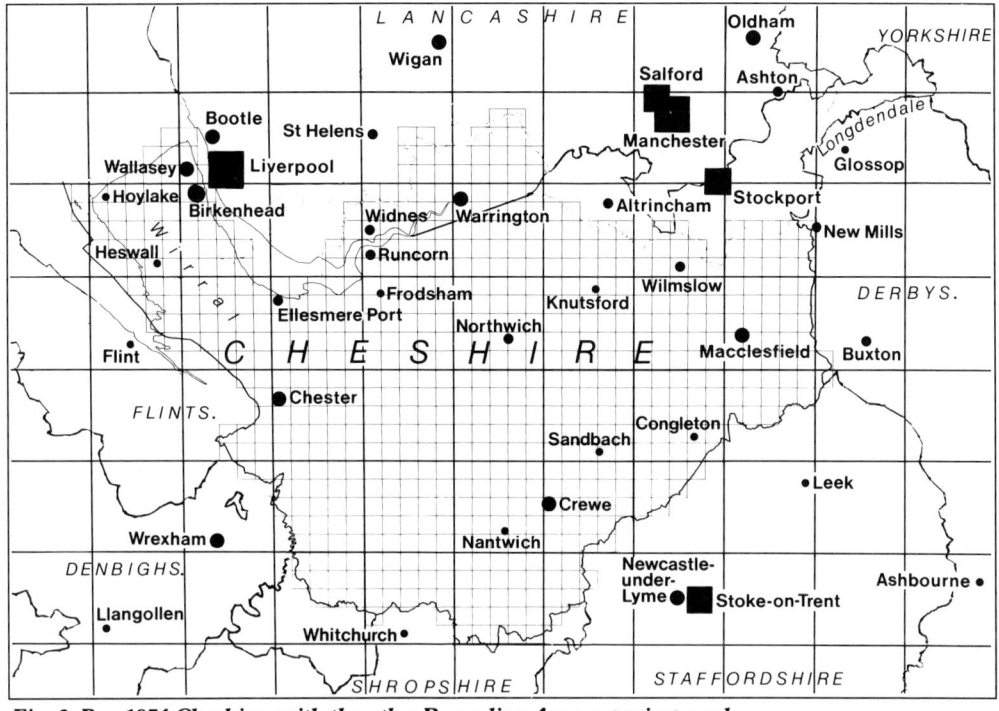

Fig. 3. Pre-1974 Cheshire, with the atlas Recording Area superimposed.

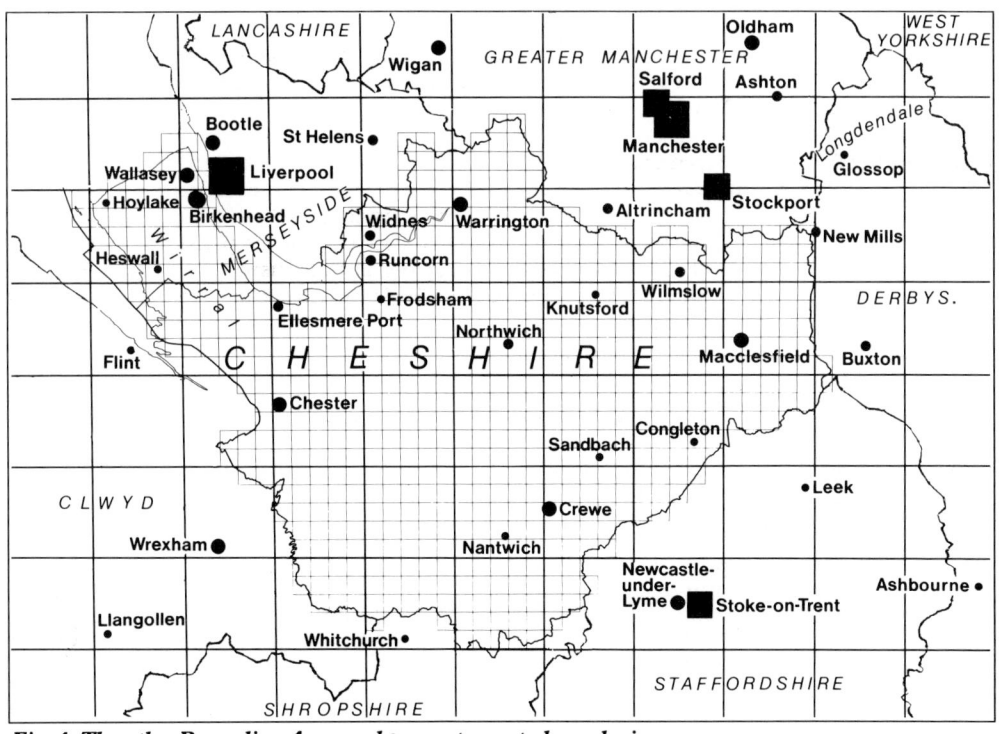

Fig. 4. The atlas Recording Area and present county boundaries.

GEOLOGY AND RELIEF

ALTHOUGH parts of eastern Cheshire at least might be classed as falling within the highland zone of Britain, they contain none of the ancient rocks characteristic of the Lake District or Celtic Britain.

In the east of the county, the land rises towards that part of the Pennines known as the Peak District, reaching its highest point at Shining Tor, 1834 feet above sea-level, on the Derbyshire border. Formed essentially of Carboniferous rocks, the Peak District consisted originally of a dome of limestone overlain by gritstone and shales. Further east this capping has long since been eroded away, leaving an area of limestone at the surface in the Derbyshire Dales, but apart from a tiny outcrop at Astbury, long since quarried away, the surface rocks of the eastern Cheshire hills consist of gritstone towards the county borders, and younger shales at lower altitudes. On the acidic gritstone, podzols have formed, giving rise to a moorland vegetation favoured by a number of specialist breeding birds. The gritstone is relatively resistant to weathering and forms steep edges. The more prominent of these, as at Windgather Rocks for example, are subjected to heavy human disturbance from climbers and others, so are avoided by cliff-nesting birds. However the screes beneath several of the lower edges provide nesting cavities for Wheatears, Ring Ouzels, Little Owls and other birds. Gritstone has long been valued as a building stone, both for houses and drystone walls, and as a raw material for millstones. The resultant quarries and associated screes are home to nesting Kestrels, Jackdaws and Stock Doves.

The shales weather very easily, so form rather rounded, featureless hills. Marine deposits within the shales often result in a less acid grassland developing, generally given over to sheep-grazing. The shale foothills are among the poorest areas of the county ornithologically, although fragments of old woodland enrich the scene locally. In parts of the north-east of the county, around Poynton and Pott Shrigley for example, coal was formerly mined, but the resultant industrial scars have little ornithological impact.

The lowland part of Cheshire is underlain in the north by Permian and elsewhere by Triassic sandstones. The latter outcrop in Wirral, along the Frodsham—Delamere—Bickerton ridge, and at Alderley Edge, being capped in most cases by thin soils on which acidic heathland, or woodland of birch and oak developed. Much of the heathland has been lost to conifer plantations, and the former homes of Nightjars, Tree Pipits and Long-eared Owls are still dwindling in extent.

Sandstone has long been quarried on a small scale for building stone, and at Alderley Edge (and to a small extent at Peckforton) has been mined for copper. At Alderley, Nuthatches have bred in a quarry face, but generally speaking these workings are of little ornithological significance.

The upper, Keuper sandstone forms a syncline in which lie great depths of Keuper marls, principally clays but containing extensive deposits of rock-salt, which were laid down as evaporites in arid conditions. Rock-salt, or "halite", is highly soluble and has been dissolved away over many centuries by underground streams. These reach the surface in places as saline springs which support a characteristic saltmarsh flora, but the major impact on birdlife has been the formation of subterranean cavities which, on collapse, are believed to have formed such meres as Rostherne, Pickmere and Marbury Mere (Great Budworth), and the depression now filled by Wybunbury Moss. Commercial exploitation resulted in an acceleration of subsidence, particularly from the nineteenth century, resulting in the various "flashes" of the plain.

Over much of the county the Keuper marls are overlain by sands, clays and gravels deposited during the last Ice Age. The streamlined, undulating topography of much of the plain was formed by the deposition of boulder clay beneath a southward flowing ice-sheet. During the last readvance of the ice, it reached its limit along the Cheshire–Shropshire border, and across to the Pennine foothills. Debris conveyed within the ice was deposited along the edge of the sheet to form the more varied relief of the Ellesmere moraine, and the hummocky landforms between Congleton and Macclesfield. Meltwater from the ice deposited sands around Delamere, Chelford and Congleton for example, giving rise formerly to heathland and now extensively quarried.

The boulder clay is well suited to the

Fig. 5. Altitude

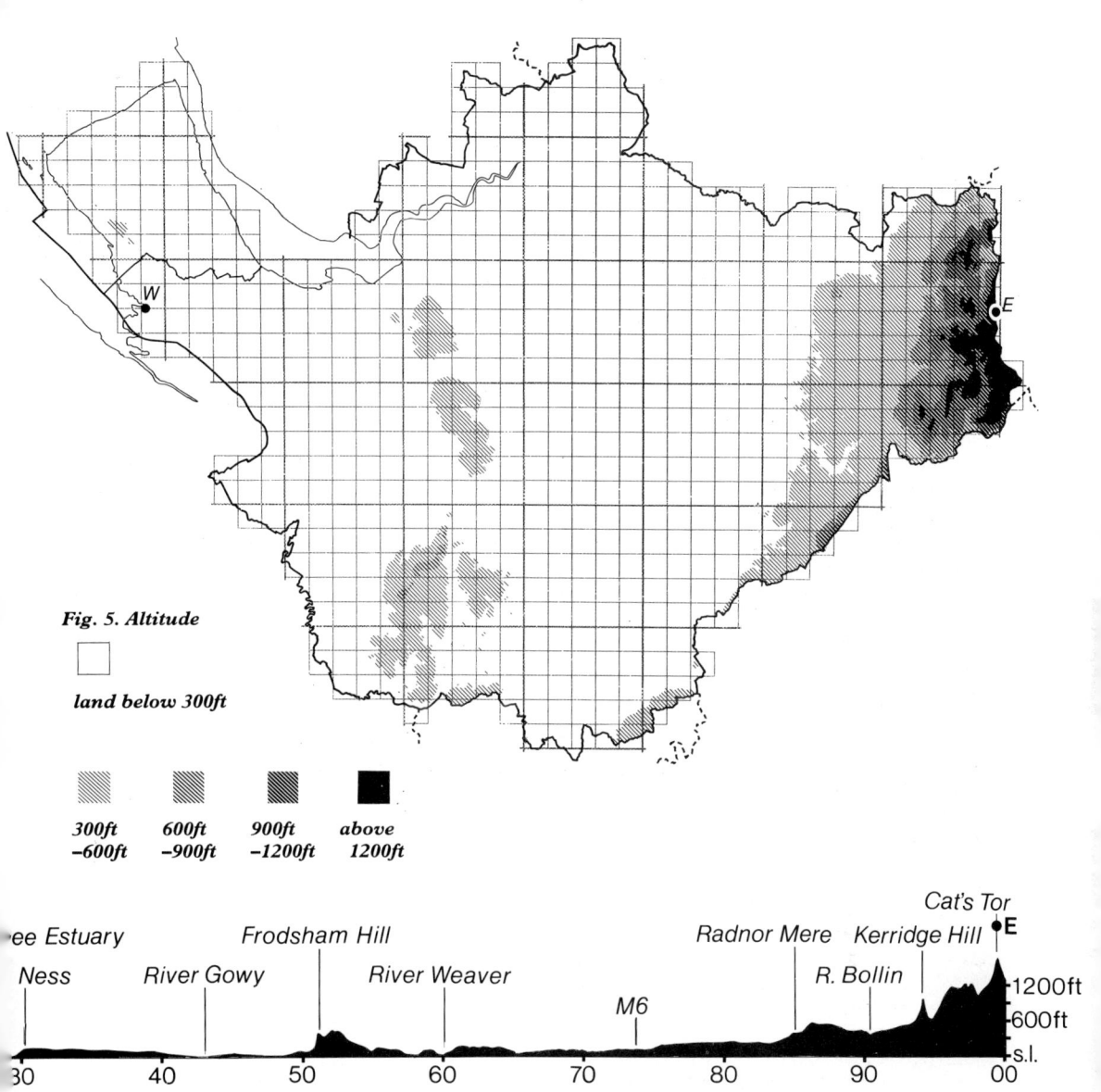

land below 300ft

300ft –600ft | 600ft –900ft | 900ft –1200ft | above 1200ft

Fig. 6. Altitude – cross-section of the county, west to east.

growth of pasture grasses. This clay has been used for several centuries to make bricks for the traditional timber-framed buildings of the lowlands, this practice expanding last century leading to the creation of flooded clay-pits as at Higher Poynton, Rixton and Moreton. These sites tend to develop a cover of scrub vegetation ideal for warblers and other passerines. The area of hummocky glacial clays and sands between Macclesfield and Congleton, and the undulating land of the Ellesmere moraine rising to the Shropshire border are both favoured by Curlews. The former area is characterised by the numerous "mosses" – peaty deposits which developed in waterlogged hollows, now often cloaked in birchwoods or ploughed to show their black, peaty soil.

The sand-dunes which formerly covered the north coast of Wirral have largely been lost to golf-courses and other developments, and their formation has effectively ceased following the construction of concrete sea-walls which prevent the drifting of sand.

BIRD HABITATS IN CHESHIRE

Woodland and plantations

AT the maximum extent of woodland cover in Britain, some 7000 years ago, virtually all of Cheshire, including Wirral, was wooded. By 1086 only some 27% of this native woodland remained (Rackham 1980), and most of this has since been lost. Remnants of this "wildwood", referred to as primary woodland, survive only as tiny fragments usually on ground too steep to cultivate and where pressure from grazing animals has not prevented regeneration.

In the eastern hills, woods of oaks, with a ground flora rather poor in wildflowers but often including cow-wheat, are probably remnants of the native woodland. Other typical components of these woods are rowan, holly and birch, with wych elm, ash and hazel on richer soils along clough-sides.

In the lowlands, many of the primary woods grow on steep land sloping down to rivers or brooks, notably in the valleys of the Weaver, Dane and Bollin. Oak is surprisingly scarce here, wild cherry, sycamore, wych elm and ash being more frequent. Small-leaved lime occurs in a very few woods, for example the Warburton's Wood Reserve of the Cheshire Conservation Trust.

The history of woodlands in the south of the county is obscure, but some appear to be ancient coppices, while others may have been planted more recently as fox-coverts. In medieval times "wychwood" cut from coppices fuelled the Cheshire brine-pans, and it is not difficult to imagine that such woods around Nantwich may have held Nightingales at times during their history. Most of these woods are long gone, but surviving coppices in the south of the plain are typified by derelict hazel stools, often

Parkland at Tatton.
Nineteenth-century timber plantations in Cheshire and Wirral parks now provide nesting sites for many hole-nesting birds such as Stock Doves, Jackdaws, Nuthatches and woodpeckers.

Fig. 7. Woodland

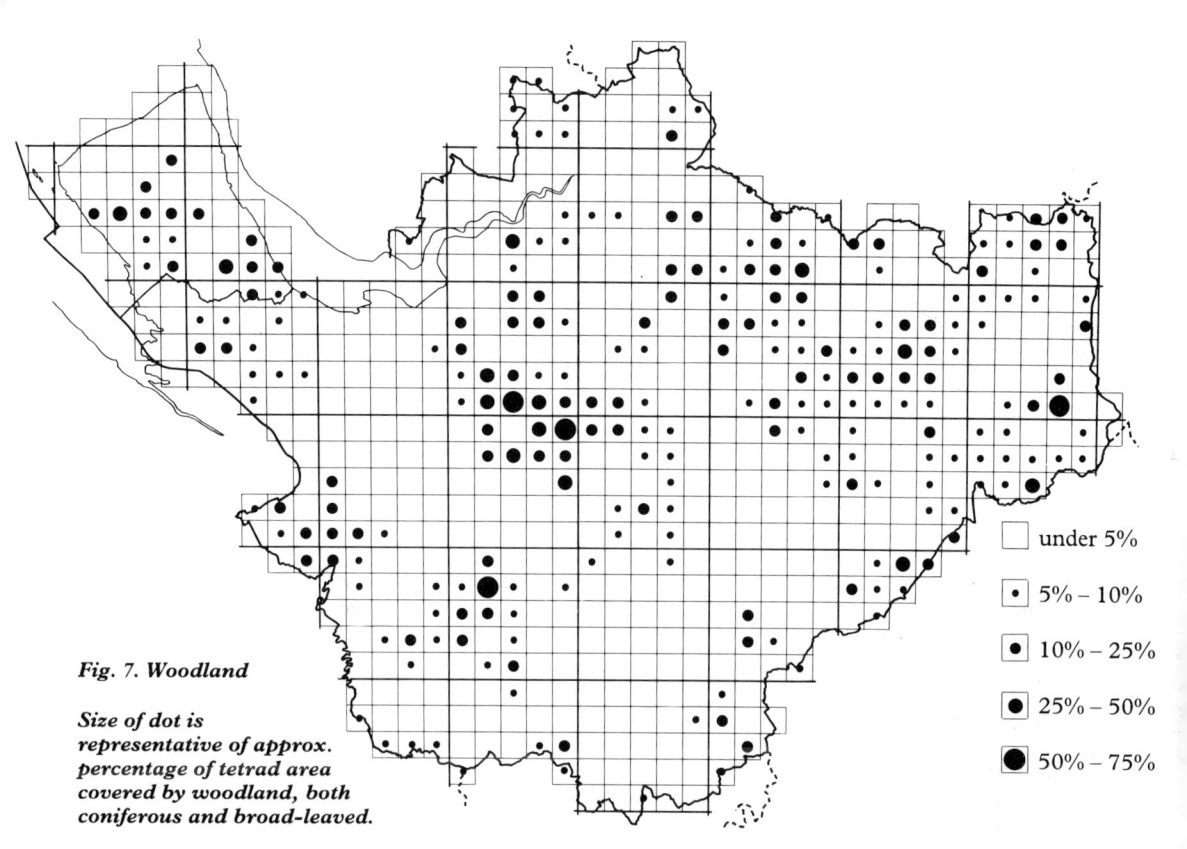

Size of dot is representative of approx. percentage of tetrad area covered by woodland, both coniferous and broad-leaved.

- under 5%
- 5% – 10%
- 10% – 25%
- 25% – 50%
- 50% – 75%

with maple, standard oaks or ashes, and a rich flora including cowslips, goldilocks buttercups and dog's mercury. In some cases hawthorn or blackthorn appear to have been planted in old coppices as cover for foxes, although these woodland edge species may have spread into woods following clear-felling. This scrub attracts roosting thrushes and Starlings which bring in seeds of elder, which may be abundant in such woods.

Whilst there are wildflowers, lichens and perhaps fungi, certainly beetles, hoverflies, molluscs and other invertebrates confined by their weak powers of dispersal to fragments of primary woodland, birds are highly mobile. The only species that shows any association with such woods in Cheshire is the Marsh Tit, which however is also found in mature broad-leaved woodland of more recent origin.

Up to the start of this century, those birds dependent on hollow trees or rotting timber for nest-sites, such as Tawny Owls, Stock Doves, Redstarts, Nuthatches and woodpeckers, had restricted distributions based on one other habitat containing many old trees, namely parkland, where mature trees were allowed to stand for ornamentation. The nineteenth-century change from wood to coal for fuel, and the twentieth-century destruction of tropical forests for "cheap" (but in ecological terms disastrously expensive) hardwoods, has resulted in a plentiful supply of commercially over-mature timber trees, which has been of great benefit to hole-nesting birds.

In the valley bottoms by rivers and brooks, wet woodland is dominated by crack willows, growing where shed branches lodge in developing mud banks, with alder carr on waterlogged ground by rivers and meres. In summer such woods are almost impenetrable, with head-high nettles and butterbur, bound together with goosegrass, adding to the obstacles of fallen trunks and stagnant pools. Such woods, with a distinctive flora and fauna, are fast disappearing as farmers remove fences allowing cattle to graze down to the river edge.

Whilst the trend over the centuries has been towards a reduction in the area of woodland in the county, a countering tendency started from at least the seventeenth century as the gentry of the county took to planting timber trees on their estates. Beech, now a favourite nesting tree for Stock Doves, Jackdaws and Kestrels, is not native to Cheshire. When Lord Stanley planted the Alderley beechwoods around

1640, he had to import mast from the south. The Historical Atlas of Cheshire shows a concentration of Georgian country seats in a belt from Lawton (SJ85) round through Alderley (SJ87) and Tatton (SJ78) to Walton (SJ58). The parks designed around these houses incorporated plantings of timber trees, not just native oak and ash, but also much lime, sycamore, beech, Turkey oak, sweet and horse chestnuts. This parkland belt shows up well in the distribution of many woodland birds.

Coward & Oldham's statement, in 1900, that many of the pheasant-coverts in the county were of recent growth probably refers to the plantations made on former mosslands in the early nineteenth century. Rhododendrons were introduced to many coverts to shelter Pheasants, in some cases spreading so densely as to prevent regeneration of trees. Rhododendron woods are much favoured by Blackcaps and Chiffchaffs, but hold few other breeding birds.

Prior to the nineteenth century there can have been few conifers in Cheshire other than churchyard yews and perhaps a few pines in Delamere. Scots pines (generally referred to in contemporary accounts as Scotch firs) were planted successfully on Rudheath in the late eighteenth century and on the highest points at Alderley Edge from 1745. They were also used in mixed plantings on estates or as shelter belts on mosslands. Larches appear to have crept in somewhat later. They were incorporated in mixed plantings in Delamere from around 1815 onwards, maturing from later that century to the benefit of the Goldcrest. Small stands of larch occur along the towpaths of canals, perhaps planted by the canal companies to provide timber for their own use. Large-scale planting of conifers followed the establishment of the Forestry Commission after the Great War. The Delamere plantations now include much larch, and Corsican and Scots pines. Macclesfield Forest plantations contain a much higher proportion of Sitka and Norway spruce. A few bird species have benefited – the Goldcrest, Coal Tit and Crossbill, but much heathland and rough pasture have been lost and many old coppices have been ripped out and replanted. Since the start of this survey, a number of new conifer plantations have been made in the hills of eastern Cheshire.

The official Board of Trade estimate for 1873 gave 26,374 acres of woodland of all types in Cheshire. This had fallen by 1897 to 24,836 acres, and to 22,100 acres by 1924. A Forestry Commission census in 1947-49 showed that only 20,175 acres (8165 hectares) remained, including 2331 acres of conifers. The latest census, around 1980, estimated 8842 hectares, including 2310 hectares of conifers. These two censuses are not strictly comparable due to boundary changes in 1974, and because the 1980 census includes plots of woodland down to 0.25 hectares whereas the 1947 definition started at a minimum of 2 hectares. In ornithological terms the most significant changes have been the increase in the proportion of predominantly coniferous woodland to 26% of the total as against 11% in 1947, and the corresponding loss of former broad-leaved woodlands with their richer avifauna.

Hedgerows

LARGE parts of Wirral and the Cheshire plain are characterised by single species hedges of hawthorn, although sometimes stretches of holly, elm or blackthorn have been lined out and elder may be invading. Within these hedges, appearing to date from the enclosure movement of the eighteenth and early nineteenth centuries, timber trees were widely incorporated. Typically these are of oak, although ash and sycamore are locally more numerous. The rare native black poplar is also found in some of these recent hedgerows, reflecting a tradition of planting which survived into the last century.

During fieldwork for this survey, considerable difficulty was experienced in finding more than thirty bird species in many of the tetrads of the southern plain. A large area of country around Wettenhall for example is characterised by neatly clipped hedges with typically small oaks, evidently planted within the last 200 years. In a walk of some five kilometres around Baddiley, JPG found only a short stretch of hedge containing hazel. Other hedges were of hawthorn only, or exceptionally holly or blackthorn were present. Of three kilometres of hedges in the Weaver valley south of

*A Lesser Whitethroat –
the archetypal hedgerow bird.*

Nantwich, no hazel was present, only a single holly, and no woodland species whatsoever. Even the most varied hedge here, presumably the oldest, contained such species as maple, suckering elm and crab-apple which are characteristic of hedgerows as much as of woodland. It appears that this area must have been stripped of woodland at an early date and that prior to eighteenth- or nineteenth-century enclosure it may have been even more windswept and birdless than at present.

By comparison, hedges around Marbury in the south-west of the county, for example, often contain six or more shrub species in a 30-yard stretch of hedge, and may be reckoned to be several hundred years older. It is not unusual to find oaks with a girth of 12 feet in this area. Big oaks are also more frequent in the east and west of the county than on the south-central plain. Around Marton and Withington (SJ86/87) is a curious area in which woodland species such as alder, birch, rowan and hazel are found in laid hedges, often in the absence of the usual hedging species. Similar hedges, sometimes also containing gorse, are found in the area of the former Forest of Delamere.

In fox-hunting country, hedges are generally closely clipped, and support very few birds unless timber trees are present for song-posts. Where hedges are left uncut for several years at a time, they form a suitable habitat for various scrub-nesting species, notably the Lesser Whitethroat, but unless cut soon cease to function as hedges. Partridges and Pheasants favour hedges with dense sprays down to ground level.

In the eastern hills drystone walls replace hedges. Indeed some walls are built on earth banks from which the original hedge has died away. Consequently hedgerow species such as the Dunnock, Greenfinch and Yellowhammer decline with altitude.

The 1980 Forestry Commission census of Cheshire, excluding Wirral, suggested a figure of nearly two million non-woodland trees in the county, over 95% of these being broad-leaved. Oak was the commonest species, with some 627,000 trees, followed by sycamore (443,000), ash (283,000), alder (134,000) and lime (105,000). The most abundant conifer was the cypress with some 53,000 trees.

Heathland and mossland

"*THE forest of Delamere, in Eddisbury hundred, still contains about 10,000 acres, in a state of little or no profit.... In Northwich hundred, we meet with another considerable district of waste land, called Rudheath; part of which has, however, been planted within the last few years, chiefly with Scotch fir; the plants appearing vigorous and healthy: while another tract of land upon it has been enclosed, and, at considerable expense, brought into a state of cultivation... Besides these, there is in very many of the townships in the county, some heath, common or moss land; the whole or part of which might be brought into a state of profitable cultivation..... Many extensive mosses and commons, by the aid of draining, marling, &c. have, within the last two or three years, been brought into a state of cultivation*" (Holland 1808).

Although heaths and mosses were classed together as waste land by the improvers of the agrarian revolution, they are quite different as habitats. On the sandy soils of Delamere, the heaths were kept open by rabbits and a now extinct breed of small sheep. The cluster of "Heath" place names around Chester, including Bruera (Norman French for heath) suggests that heathland, cleared from the native woodland, may have been improved for agriculture since ancient times. In the south of the county, around Baddiley for instance, the lack of old hedges suggests that the land was at one time devoted to large grazing ranches. Under such conditions preferential grazing allows the development of scrub in less used areas, and it seems highly probable that areas of gorse developed in such large unhedged areas.

Very little heathland survives in Cheshire today, except in heavily disturbed recreational areas such as Thurstaston

Heathland at Abbot's Moss, Delamere. Formerly home to Nightjars, and grouse, lowland heaths have now almost disappeared under conifers or been excavated for sand. Odd relict patches such as this now hold breeding Yellowhammers and Tree Pipits.

Common, Little Budworth Common (developing into birch woodland), and Lindow Common. Twentieth-century conifer plantations have all but obliterated the remnants of this habitat. The Crown lands in Delamere were enclosed and planted from 1812, initially with broad-leaved trees, these later being replaced by conifers. Cleared plantations here resemble heathland in some respects however, as do the open surrounds of sand-quarries both here and further east. Scrubland birds such as Yellowhammers, Linnets and Tree Pipits are typical of such habitats.

Mosses developed in waterlogged hollows left behind after the retreat of the ice-sheets. Only the wettest of these, for example Abbott's Moss and Wybunbury Moss, have survived into recent times due to difficulty of drainage. Along the Mersey basin in the north a once extensive area of mossland is represented by Risley Moss. At one time contiguous with Chat Moss, Carrington and other mosses, this belt held Red Grouse into the 1930s, and a chapter on Carrington Moss in Coward's *Bird Haunts and Nature Memories* gives the flavour of this all-but-vanished habitat. Nightjars were once characteristic of both heaths and mosses.

Moorland, Burntcliff Top.
Tracts of heathers or of cotton-grasses diminish as improved grazing for sheep and cattle gives an ever-more-uniform regime. Skylarks and Meadow Pipits are most evident where heathers still penetrate grazed moors, the latter birds also inhabiting the lusher grasses of roadside verges. Dry-stone walls, incised cloughs, poorly drained rushy patches allow Pied Wagtails, Linnets, thrushes and Reed Buntings to exist. Wheatears use underground cavities beneath boulders in open grazed situations and on steep slopes as shown above.

Moorland

IT seems likely that the first major clearance of woodland from the hills of eastern Cheshire was carried out by the Danish settlers from the tenth century onwards, with some reversion to woodland following the Norman harrying of the north a century or so later. The flatter hills soon developed a covering of peat, allowing moorland breeders such as Golden Plovers, Red Grouse and Curlews to spread down from the north.

Since local government reorganisation in 1974, the area of moorland in Cheshire has been much reduced by the loss of Longdendale to Derbyshire and Greater Manchester. The remaining moorland in the county has been further reduced in area by agricultural improvement, heavy stocking levels of sheep also being implicated in the conversion of grouse moor to pasture.

A considerable threat to our moorland birds lies in the great increase in number of ramblers who, in the summer months at least, form a steady procession across moors which once held nesting Golden Plovers.

Meres and fresh waters

CHESHIRE is very well off for inland waters. There is a cluster of natural meres in the moraine country close to the Shropshire border, including Combermere, Marbury, Quoisley and Norbury Meres; Oak Mere and Hatchmere in Delamere Forest; and a further cluster in the area between Northwich and Macclesfield including Marbury (Great Budworth), Pickmere and Rostherne. Added to these are a number of ornamental lakes in parks, some natural in origin, but enlarged by damming. These include Tatton, Redesmere, Rode Pool, Tabley and Cholmondeley. The meres and park lakes have long been noted as haunts of Great Crested Grebes and other waterfowl, and the park lakes were probably stocked with Canada Geese soon after their creation.

It is possible that several of the meres, notably Marbury (Great Budworth), Pickmere and Rostherne, formed as a result of subsidence following natural dissolution of underlying rock-salt. Salt has been extracted commercially within the county since Roman times at least, giving rise to increasing rates of subsidence until the earlier part of this century. Changed methods of extraction have reduced the phenomenon more recently.

Very often the depressions resulting from subsidence fill with water to form flashes, those at Sandbach, Winsford, Witton and Billinge Green being particularly well known ornithologically, although Melchett Mere in Tatton Park is also a flash. The Cheshire flashes are typically shallow and as such provide nesting cover for waterfowl, waders and marshland species.

Rostherne Mere. Cheshire's standing fresh water bodies, big and small, as well as sustaining large numbers of wintering wildfowl, also provide breeding sites for grebes, ducks, geese, Coots and Moorhens. Several meres and other expanses of water support reed-beds and their colonies of Reed Warblers – the subject of a special long-term study here at Rostherne.

**Fig. 8.
Waterways and Meres.**

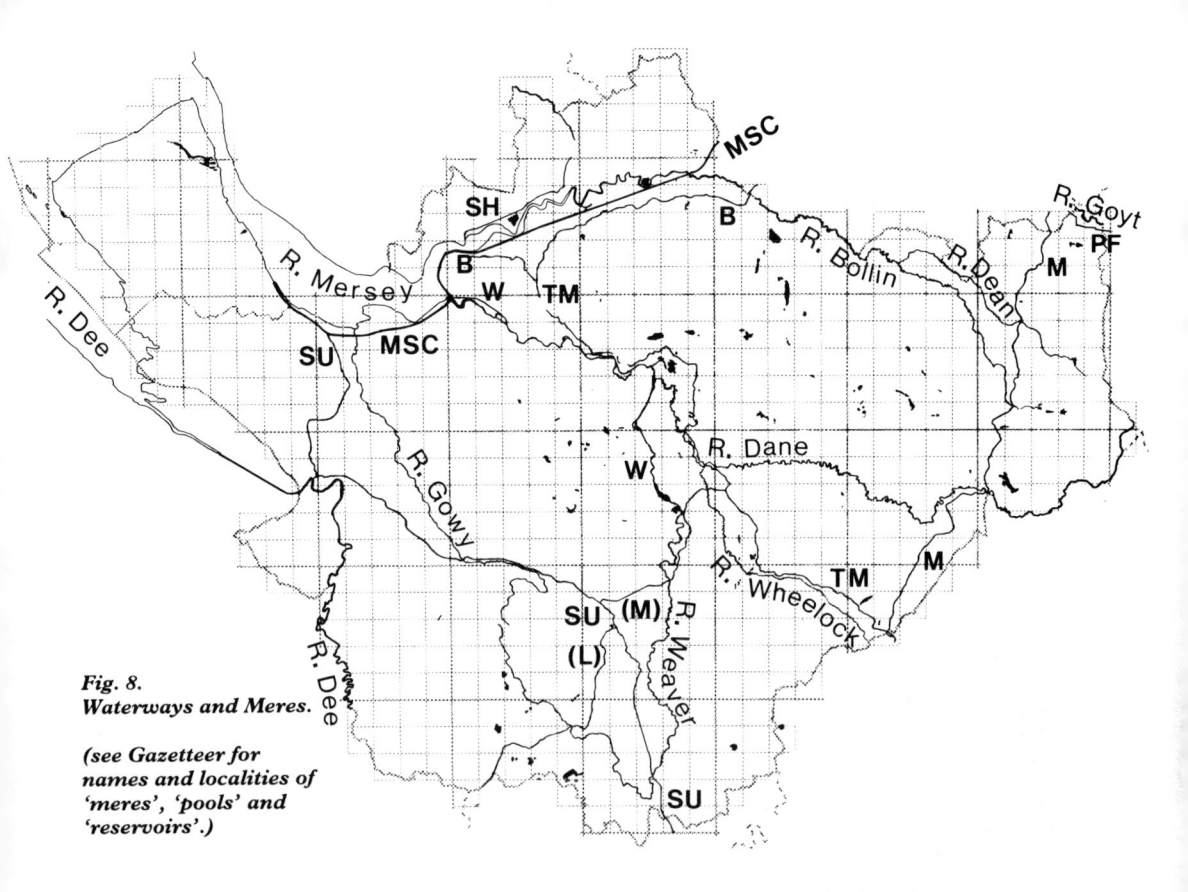

(see Gazetteer for names and localities of 'meres', 'pools' and 'reservoirs'.)

Canals formed the arteries of the Industrial Revolution. Grey Wagtails sometimes breed by the locks of the Trent & Mersey, Shropshire Union and other canals, but the two waterways of greatest significance to birds have been the Weaver Navigation and the Manchester Ship Canal. In both cases the sludge-beds built to accommodate dredgings form sites of enormous ornithological importance. The state of the tanks at Frodsham and Woolston has had a direct bearing on the breeding success of Shovelers, Pochards and other ducks. Huge populations of Sedge Warblers and Reed Buntings also nest. Tanks at Moore (now lost) may have held nesting Bearded Tits for a year or two.

A Bearded Tit – potential colonist?

KEY TO CANALS (above)

B	Bridgewater Canal
M	Macclesfield Canal
MSC	Manchester Ship Canal
PF	Peak Forest Canal
SH	St Helens Canal
SU	Shropshire Union Canal
SU(L)	(Llangollen Branch)
SU(M)	(Middlewich Branch)
TM	Trent & Mersey Canal
W	Weaver Navigation

Canal feeder reservoirs at Appleton, Sutton and Bosley are of limited value to breeding waterfowl since rapid fluctuations in water level prohibit the development of marginal vegetation. Similar conditions apply at Hurleston, where drinking water imported from Wales along the Llangollen canal is stored for treatment, and at the hill reservoirs in the east at Langley, Lamaload and Disley.

Acre Nook Sand Quarry, near Chelford. Waders such as Little Ringed Plovers and Common Sandpipers nest in the flooded beds of such quarries, while large colonies of Sand Martins inhabit vertical banks during excavation.

Motorway building in the 1960s caused a rapid expansion in the extraction rate of sand. The main centres of quarrying are in the Chelford, Congleton and Delamere areas. The Chelford area quarries are worked dry, so that different strata of sand can be extracted separately. The damp quarry bottoms then form suitable habitat for Little Ringed Plovers and Common Sandpipers. All quarries tend to support colonies of Sand Martins. Once quarrying ceases, heathy scrub invades, giving way later to birch woodland. Yellowhammers, Linnets, Whitethroats, Willow and Garden Warblers are typical, with as much likelihood as anywhere in the county of a pair or two of Turtle Doves.

The Moreton clay-pits of north Wirral provide a useful habitat for waterfowl in one of the "drier" parts of the county. Other clay-pits, for example those at Rixton and Higher Poynton, support good populations of scrubland birds.

While the Manchester Ship Canal is almost the only waterway in Europe to have caught fire, and water quality in the Mersey and some of its tributaries leaves much to be desired, most of the county's rivers are tolerably clean from an ornithological point of view. Mallards and Moorhens are the typical breeding species. Fast-flowing hill-streams in the east hold Dippers and Grey Wagtails. Kingfishers nest along most rivers and brooks, even the polluted Mersey and Bollin, flying to marl-pits to feed.

Marl-pits are found, singly or in clusters, in fields throughout the county. Marl, a lime-rich clay, was formerly spread on fields as a fertiliser. The pits filled with rainwater, and form an important habitat for Moorhens and Reed Buntings, although Sedge Warblers, once common at pit-sides, have now all but vanished from this habitat.

Fig. 9. Urban density

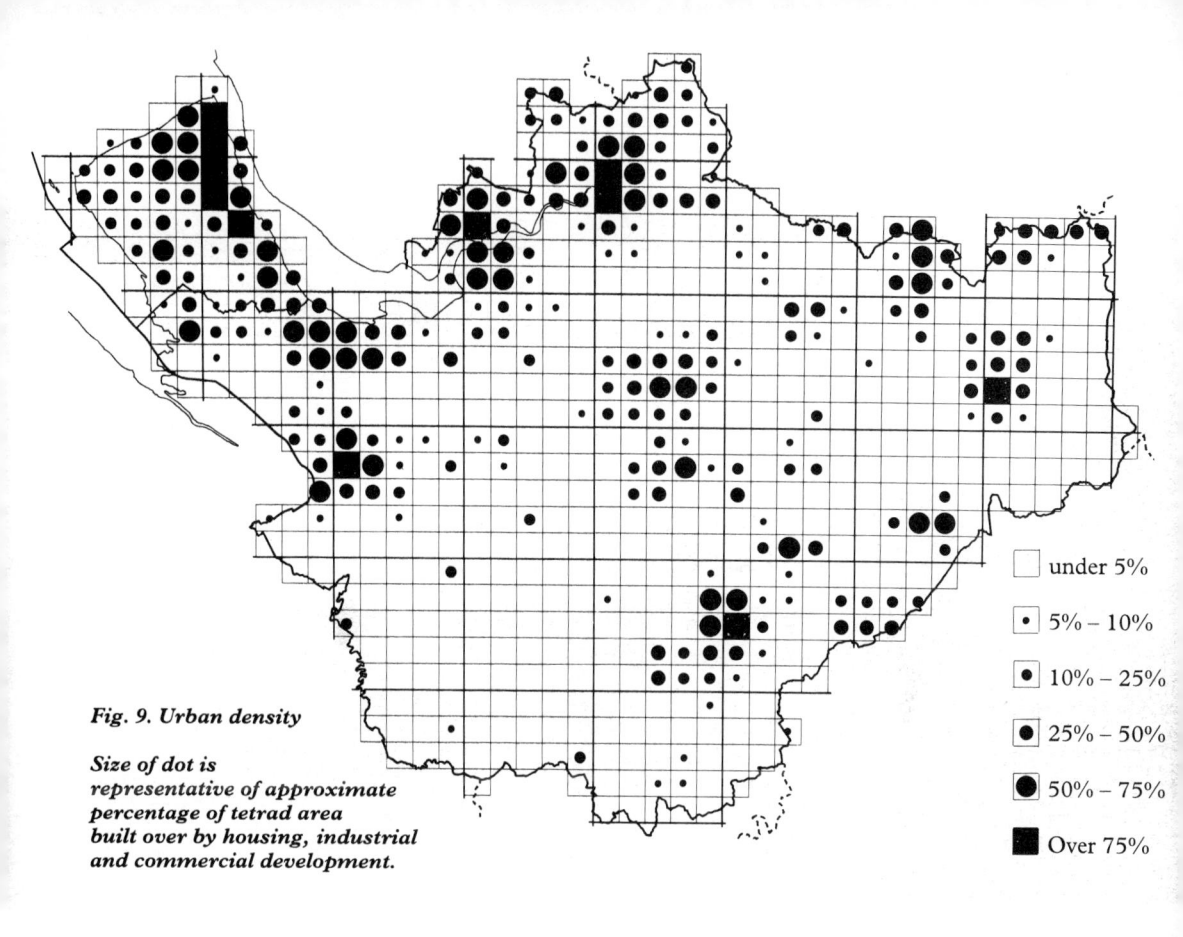

Size of dot is representative of approximate percentage of tetrad area built over by housing, industrial and commercial development.

☐ under 5%
· 5% – 10%
• 10% – 25%
● 25% – 50%
● 50% – 75%
■ Over 75%

Towns and suburbia

AT the time of Domesday Book, Chester was the only sizeable town in the county, with some 500 houses recorded prior to the Conquest. Eastham had a population of over 50, but otherwise only Halton, Tarvin and Nantwich had over 20 residents recorded (Sylvester & Nulty 1958). At the first national census, in 1801, the population had grown to about 191,000, of which 38,000 were employed in agriculture. By 1851, although the total population had increased to 420,000 the agricultural workforce had grown by only 50%, and the towns of Birkenhead and Crewe had come into being as the Industrial Revolution gathered momentum. Of a 1901 population of 816,000 only 125,000 lived in rural areas, with no increase in numbers working on the land over the past fifty years (Sylvester & Nulty 1958). The further increase in population to 1,369,000 by 1961 was made up largely by growth of suburbs to the south of Manchester and Liverpool as people moved out from the crowded cities. Although 209,000 of these people lived in rural districts, relatively few of them were employed in agriculture (Freeman et al. 1966). This trend has continued with a rising proportion of country-dwellers commuting to town for employment.

Into the 1960s the suburban habitat was rather a poor one for birds. Privet was much used for hedging because of its tolerance of a polluted atmosphere. More recently, improved air quality has allowed the growing of a much greater diversity of shrubs. An increase in insect food and nesting cover has allowed the spread of House Martins, Goldfinches and Mistle Thrushes further into town centres, and Collared Doves above all other birds have benefited from the sales of ornamental conifers on which garden centres thrive. Much "Green Belt" farmland around the suburban fringe has been purchased by building companies awaiting a relaxation in planning regulations. Such land, seldom actively farmed, becomes semi-derelict with a vigorous growth of ragwort, thistles and other weeds, and is of considerable benefit to Linnets, Greenfinches and partridges.

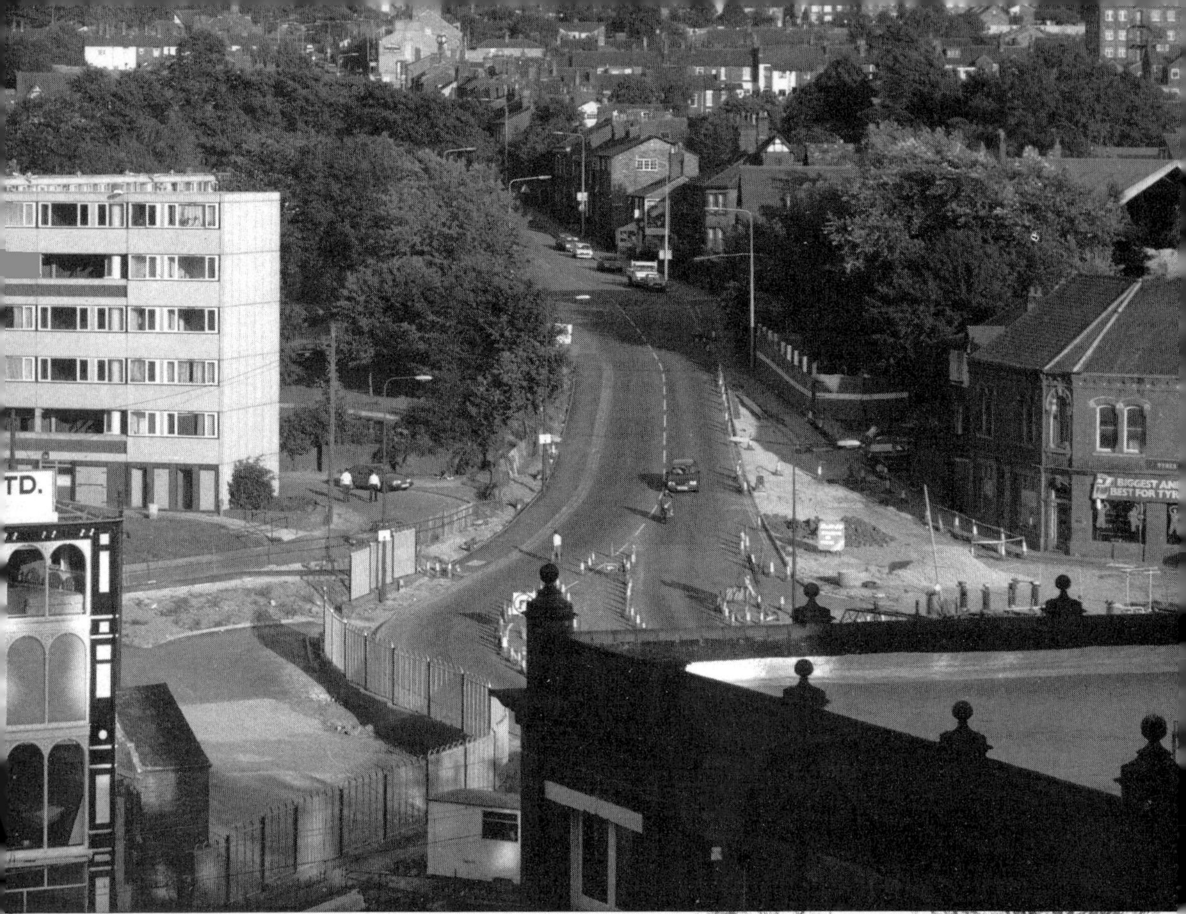
TD.

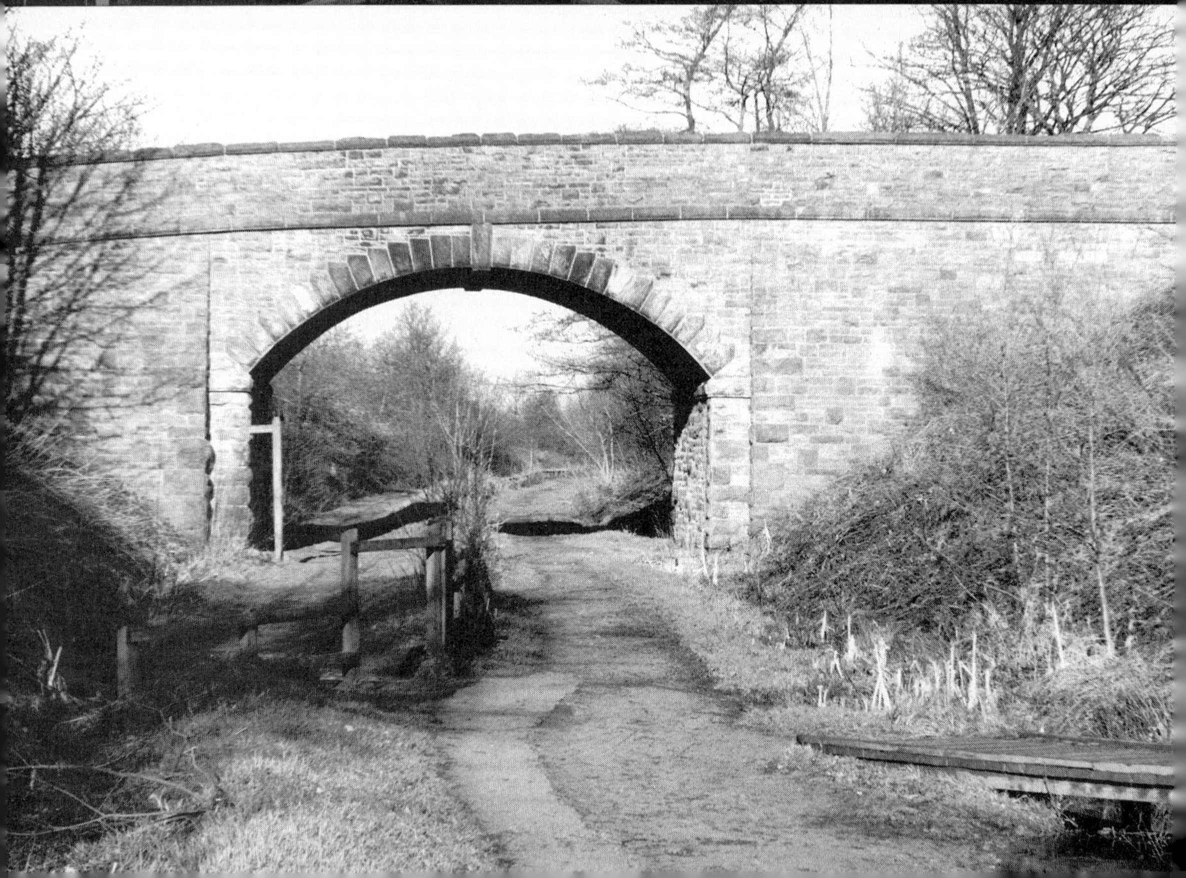

The Cheshire Plain viewed north-westward from Beeston Castle. The parallel curving hedgerows straddling the Shropshire Union Canal mark enclosures from a medieval open field system. Many such agricultural areas of the plain are sparsely wooded, with hedgerow trees instead providing song-posts for Chaffinches and thrushes. Harvesting of silage has caused a decline in numbers of Lapwings and Skylarks.

Left top: Macclesfield. The spread of suburbia has greatly influenced the distribution of many birds. House Sparrows, Starlings Swifts, House Martins and Collared Doves all nest frequently. While farm-land birds have in general been displaced, the belt of semi-derelict fields characteristic of many urban fringes favours Grey Partridges, Goldfinches and others which derive benefit from the proliferation of weeds.

Left below: Middlewood Way, near Whiteley Green. As with various other urban fringe habitats, disused railways soon become overgrown by scrub, forming important habitats for warblers such as Whitethroats, Garden and Willow Warblers and finches such as Redpolls and Bullfinches.

One essential part of the urban infrastructure is the sewage treatment plant. Older parts of these works are generally characterised by willow scrub – home to Redpolls, Reed Buntings and warblers. Similar scrub, plus hawthorns and elders, often develops along railway banks and on the abundant industrial wasteground around Northwich, Warrington and Birkenhead, favouring Bullfinches, Redpolls and Whitethroats.

Salt and chemical industries

THE major environmental impact of the early years of the salt industry was probably associated woodland clearance and coppicing. Large quantities of firewood were needed to boil up the brine to process the salt although coal was used increasingly from the middle of the seventeenth century and by 1806 over 100,000 tons of coal per year were being imported to Northwich and Winsford. As the salt industry and associated chemical works grew, air pollution became chronic. *"In Northwich alone, it was calculated in 1887, the chimney effluent contained 29,000 tons of sulphur dioxide per annum. Hundreds of low-built chimneys belched black poisonous smoke which mingled with the steam to create a perpetual evil-smelling fog. The blackness of the saltmaking districts was legendary. Even in 1698, when Celia Fiennes visited Northwich, she found it... 'full of smoak from the salterns on all sides'. And a writer in the Manchester Guardian in 1888 deplored the effects: 'Every blade of grass in the vicinity is being killed off, and hillsides and fields are reduced to bare surfaces of baked clay'."* (Didsbury 1977).

Coward (*Picturesque Cheshire*, 1903) notes that in 1850 a thriving plantation of "fir" trees extended for about a mile along the top of Keckwick Hill, but by the time of his writing, *"withered trunks stand stark and naked against the sky. Wherever high land catches the vapour-laden breezes from Widnes the trees have suffered, even at High Legh and Dunham the effect may be seen on the western edge of woodlands"*. At Rocksavage, *"Every old tree, fruit and forest, was killed by the chemical gases..."* Evergreen conifers are especially vulnerable to air pollution, and are still scarce along the Mersey valley and through northern parts of Manchester and onto the Pennines. Goldcrests and Coal Tits, birds of the conifers, are correspondingly scarce in this area. Two other species, the Long-tailed Tit and Chaffinch, show similar anomalies in their breeding distributions. Both build their nests from trailing mosses collected from the trunks of trees. Suitable epiphytes are scarce or absent along the Mersey lowlands. Lichens are little used by either species in Cheshire nests but, with pollution levels much reduced by recent industrial decline and pollution abatement strategies, lichens are recolonising rapidly. It may be expected that an increase in moss on tree-trunks will allow Chaffinches back to the Mersey valley in the near future.

A few breeding species have derived benefit from the lime waste sludge-beds associated with the chemical industry. Little Ringed and Ringed Plovers both nest in apparently inhospitable sites at Frodsham and Northwich, and the benefits of derelict land on the urban fringe have been mentioned above.

Coastal habitats

"The sands in the estuary of the Dee... may be reckoned at nearly 10,600 acres; a considerable part of these is now however secured by embankments from the sea, and still more

The coastline of north Wirral is much disturbed nowadays. Red Rocks Marsh, near the north-western tip of Wirral, is better known as a haunt of passage birds than of breeding species. Reed Warblers have bred, with Stonechats and Grasshopper Warblers in the fixed dunes behind. Hilbre Island is beyond, the intervening sands exposed at low tide.

may be, in process of time." (Holland, 1808)

Reclamation of the Sealand area from the middle of the seventeenth century has resulted in further siltation on the seaward side. A large tract of saltmarsh at Gayton is now managed by the RSPB and is notable for its breeding Redshanks. Dunes along the north coast of Wirral which last century were home to Shelducks and Stock Doves have largely been lost to golf-courses or other development. Construction of a concrete sea-wall has limited the mobility of dune sand. The only recent coastal breeding attempts by Ringed Plovers have been on ground disturbed by construction work. The Hilbre Islands, once home to breeding Rock Pipits, are now too popular with trippers to be of any great value to breeding birds. The remainder of the Wirral coast lacks cliffs suitable for nesting seabirds, although Fulmars have been seen inspecting the low sandstone bluffs at Burton Point, now stranded inland behind the growing marsh.

The four tetrads of sheep pasture between the Ship Canal and the Mersey estuary along Frodsham Score proved almost devoid of breeding birds when visited during the survey.

Farmland

THE gradual emergence of pasture following woodland clearance must have helped many birds of open country very greatly. Where could the likes of partridges and Lapwings have lived in a wooded landscape? Sheep and cattle rearing in bygone centuries undoubtedly helped the Jackdaw and Starling (the latter still known as the Shepster in rural areas) and other species to invade the county. Permanent pastures are of great importance to Green Woodpeckers also. Agricultural traditions have played a very great part in shaping the Cheshire landscape. Experience of sheep- and cattle-ranching, from monastic and Tudor times onward, showed that hedged fields allowed the easiest control over grazing animals, thereby helping many woodland birds to spread into farmland. In the Pennine foothills early attempts at hedging pastures sometimes failed, the hedges being replaced and enclosure completed with stone walls. Since many of these have fallen into disrepair, livestock are ranging more widely and areas neglected by their grazing are becoming cloaked with bracken or scrubby woodland.

The growth of urban population from the eighteenth century allowed many Cheshire farmers to specialise in dairying. In parts of the plain this is practised to the virtual exclusion of arable crops, and, since hay crops have been replaced by silage, meadow birds such as Skylarks and Lapwings are being driven away. These intensive dairying areas of the county, with neatly clipped hedges, are the least productive for birdlife.

Cereal and potato growing areas coincide closely with the distribution of the Corn Bunting. This map shows clearly the dairying area in the south-centre of the county. Compare for example the maps for the Grey Partridge or Lapwing. Many birds, such as Yellowhammers and Yellow Wagtails prefer areas of mixed grass and arable farming. Their distribution must have fluctuated over time as patterns of cultivation have altered across the county. For example ridge and furrow formations in fields at Disley (SJ98) and in the dairying areas of the plain indicate not just that arable crops were once grown there, but that their breeding avifaunas were once more diverse.

As well as the change from hay to silage, increased mechanisation has necessitated the introduction of chemical biocides to replace traditional methods of weed and pest control. The collapse of populations of birds of prey and Stock Doves in the 1950s and 1960s, as certain pesticides passed up the food chain, caught the public imagination and steps were taken to remedy the situation. However, the removal of weed species from cornfields, the reduction of floristically rich meadows to single species grass-crops, and the spraying of crops with pesticides has reduced the food supply for less spectacular birds lower down the food chain such as Turtle Doves, partridges and many seed-eating passerines. Despite the expense of storing food-mountains and this degradation of our wildlife, calls for a return to non-chemical farming remain muted. Neglect of "set-aside" land may benefit a few species, but arable land remains inhospitable to its associated wildlife.

CHANGES IN THE CHESHIRE AVIFAUNA: SOME TRENDS OF THE TWENTIETH CENTURY

THE status of individual species of birds is constantly changing, and such changes are discussed in detail under each species heading. However a number of more general trends exist, each affecting a number of species.

Recovery of persecuted species

DURING the nineteenth and early in the present century the populations of many birds were artificially depressed. Goldfinches, Bullfinches and other song-birds were extensively trapped and caged, and became scarce across large parts of the county. Crows, Jays and Magpies were much persecuted by gamekeepers, as were various birds of prey. The decline of gamekeeping during both world wars, and the developing appreciation of wild birds amongst the general public, with consequent legal protection, have helped many of these species to recover to their former prevalence. However, Ravens and Buzzards are still struggling to regain a foothold in the county. Conversely the prevalence of Magpies often causes concern to those who see the nests of Blackbirds and Robins in their gardens predated.

Colonisation by waterfowl and introductions

THE general change in attitudes towards birds has led to a number of foreign species, chiefly waterfowl, becoming established in the wild. Whilst populations of the Greylag Goose have been introduced initially by wildfowlers, its rapid increase, and that of the Canada Goose, reflects the low level of shooting of these species that takes place in Cheshire and Wirral. The Red-legged Partridge, also introduced for shooting, may not yet be permanently established.

The popularity in recent decades of ornamental waterfowl collections, even in fairly small, private gardens, but notably at the centres of the Wildfowl and Wetlands Trust at Martin Mere (Lancashire) and Slimbridge (Gloucestershire), has resulted in the founding of feral populations of several species. The North American Ruddy Duck (ex Slimbridge) has been established in Cheshire since the early 1970s; Gadwalls have visited the county ever more frequently since the opening of the collection at Martin Mere and now breed; and full-winged Barnacle Geese from both centres may yet establish themselves, though hybridisation with Canada Geese dilutes their efforts.

The Eaton estate near Chester, the stronghold of Cheshire Greylags, is also home to a small population of Mandarin Ducks, which is doubtfully truly feral. Snow Geese have been seen there frequently in recent years. Feral populations of this elegant species are now establishing themselves in other English counties, so Cheshire birds should be watched closely.

It is perhaps not generally realised that the Tufted Duck is a recent addition to the region's breeding avifauna. Any bird summering in the county last century was liable to be shot. Pochards too are twentieth-century colonists.

The Little Owl arrived in Cheshire in the 1920s to occupy a niche in farmland that had been vacant since enclosure provided an abundance of hedgerow trees. Whether it could have sustained itself in the days when gunners prowled all but the keepered estates, and the keepers themselves shot anything with a hooked beak, is open to question.

The rise and fall of industrial habitats

INDUSTRIAL sites, by their very nature, break the pastoral traditions of the county and provide new habitats which in turn attract unusual birds. Perhaps the most ornithologically significant industrial sites in the twentieth century have been the immense sludge-beds constructed at Frodsham, Moore and Woolston to accommodate dredgings from the Manchester Ship Canal. Perusal of the species accounts will reveal the extent to which the breeding status of Garganey, Pintail, Teal, Shoveler and Pochard has been influenced by the state of these tanks, which varies from year to year. Those at Woolston currently hold the largest concentrations of Reed Buntings and Sedge Warblers in the county.

Subsidence following salt-mining and brine-pumping led to the formation of the various flashes, again favouring breeding waterfowl, but also waders such as Snipes, Redshanks and Common Sandpipers.

Natural siltation, particularly of those flashes which lie along rivers, notably the Elton Hall Flashes (River Wheelock) and Winsford Bottom Flash (River Weaver) is gradually obliterating these transient features from the landscape.

The Little Ringed Plover took advantage of the numerous sand-workings, shallow flashes and bare, chemical sludge-beds in the county, and in lowland Britain generally, to launch an invasion from the continent. The larger Ringed Plover, long established around but now increasingly displaced from England's coastline, may be trying to usurp the invaders from their industrial breeding-sites.

It should be borne in mind that most, if not all, of these industrial habitats are transitory by nature, and that while their ornithological interest is great, it is essentially short-term.

Decline of trans-Saharan migrants

ENVIRONMENTAL degradation and desertification in Africa is one of the greatest threats to Cheshire's birds. Around the edges of the Sahara the desert is spreading at an estimated 5800 square miles (1.5 million hectares) each year (Clarke 1984). In the Sudan the desert spread southwards by 100 kilometres between 1958 and 1975.

With each passing year the deserts become more formidable obstacles to those of our migrant birds which must pass them on the way to their winter quarters. Indeed some species, such as the Sand Martin and Whitethroat, actually spend the winter months in the sub-Saharan zone known as the Sahel where the effects of desertification are most marked.

Between 1968 and 1973 there was a period of major drought in the Sahel. As the pasture grasses failed, herdsmen cut down trees and scrub to feed their cattle. Compaction of the soil around waterholes by animals' hooves resulted in the destruction of vegetation for up to 30 kilometres around, in effect destroying the winter habitat of Whitethroats. In 1969, Whitethroat numbers returning to Britain were 70% down on the previous year. They have never recovered. By the early 1980s, Cheshire's Sand Martins were reduced in number by more than 90% on the level of twenty years earlier. Sand Martins are less dependent on scrub than are Whitethroats, and have shown some recovery as soon as rain in the Sahel restores insect populations, but the regrowth of scrub, if it occurs, will take longer.

Other species obviously affected are the Whinchat – now approaching extinction as a Cheshire breeding bird –; the Sedge Warbler which, like the Whitethroat, has retreated from eastern Cheshire; the Yellow Wagtail, now seldom occurring in anything like the flocks of 100 or more birds encountered on passage into the 1970s; and the Grasshopper Warbler.

Hazards on migration across the deserts may partly explain the loss of the Corncrake and Nightjar, although other factors are undoubtedly involved, as is the case also with the Common Sandpiper, Turtle Dove, Tree Pipit and Redstart.

Neglect of mature woodland

MANY Georgian broad-leaved plantations and especially hedgerow trees which matured this century have not been felled because hardwoods have begun to be imported cheaply from tropical forests. Thus the ecological disaster afflicting many third-world countries has actually been of benefit to a few British birds. The Great Spotted and Lesser Spotted Woodpeckers were both very scarce in Cheshire at the turn of the century, as was the Nuthatch. Neglect of older woodlands within the county has resulted in a change in their canopy structure, leading to an increase of Blackcaps and Chiffchaffs relative to Garden and Willow Warblers, the latter two species favouring more open scrubby woodland and coppices.

The use of barbed wire has rendered some hedgerow trees hazardous to saws and many are left beyond economically useful age. This has provided an abundance of nest-sites for Tree Sparrows, Stock Doves, Little Owls and other hole-nesting birds.

Pennine foothills looking westward to Kerridge Hill. Lowland hedges peter out to be replaced by dry stone walls. Few species nest in the open hill pastures but Little Owls frequent the scattered trees and barns, and Wheatears may nest in broken walls.

Moorland habitat loss

THE progressive loss of moorland by conversion to grass pastures has led to a reduction in the populations of several dependent species. Early this century the Ring Ouzel, Red Grouse, Black Grouse and Merlin still survived as lowland breeders. Destruction of lowland heaths and mosses drove them back into the hills, where these species, plus the Golden Plover, Twite and Dunlin, retain a precarious toehold in the county.

Lowland drainage and succession of marl-pits

ALTHOUGH Cheshire and Wirral still contain well above the national average of ponds and other waters, many have been lost this century. Marling, the practice of digging out boulder-clay or calcareous marl to spread on fields as a fertiliser, ceased last century as supplies of lime, imported along the canals and railways, became available. Many of the resultant pits have begun to dry out as vegetational succession takes hold, and this natural process is often accelerated by drainage, in-filling or tree-planting.

Natural siltation, mentioned above in the context of flashes, has also accounted for many of the mill-pools which remained in service until the turn of the century.

Underground pumping of water and straightening of rivers have lowered the water-table, this effect being locally accentuated by the digging of deep sand-quarries. Improved field-drainage has also contributed to the loss of breeding habitat for Moorhens, waders, Sedge Warblers and Yellow Wagtails.

Agricultural intensification

DRAINAGE of damp fields is just one effect of the mechanisation of farming this century. Agricultural change has accelerated since the 1940s with increased application of chemical fertilisers and biocides. The replacement of hay crops by silage is displacing Skylarks, Curlews and Lapwings from the Cheshire plain.

Mechanisation means that a farmworker with a day or two to spare can be set to flailing hedges (sometimes the same hedge twice in one year), filling in marl-pits, or spraying weedy hedgebanks – *"all the little, often unconscious vandalisms that hate what is tangled and unpredictable but create nothing"* (Rackham 1986: 28).

Public opinion

IT is now a little over a century since the formation of the Royal Society for the Protection of Birds. Much has been achieved in that century, but more remains to be done. Above all there is a need to find a place for birds within the everyday workings of the economy and not just as a recreational resource. Since agriculture has the greatest spatial impact on the environment, it is of paramount importance that all aspects of farming should once again be brought within the realms of benign husbandry. The next decade may well be crucial.

REFERENCES

Clarke, R. (1984): *The Expanding Deserts*. In Hillary, Sir E. (Ed.): *Ecology 2000*. Michael Joseph.
Coward, T. A.: *Bird Haunts and Nature Memories*.
Didsbury, B. (1977): *Cheshire Saltworkers*. In Samuel, R. (Ed.): *Miners, Quarrymen and Saltworkers*. Routledge & Kegan Paul, London.
Freeman, T. W., Rodgers, H. B. & Kinvig, R. H. (1966): *Lancashire, Cheshire and the Isle of Man*. Nelson.
Holland, H. (1808): General View of the Agriculture of Cheshire. In Marshall, W. (1809): *The Review and Abstract of the County Reports to the Board of Agriculture. Vol. III: Western Department*. 1968 reprint. Augustus M. Kelley, New York.
Perring, F. H. & Walters, S. M. (1962): *Atlas of the British Flora*. London & Edinburgh.
Rackham, O. (1980): *Ancient woodland: its history, vegetation and uses in England*. Edward Arnold, London.
Rackham, O. (1986): *The History of the Countryside*. Dent.
Sylvester, D. (1971): *A History of Cheshire*. Phillimore, London.
Sylvester, D. & Nulty, G. (Eds) (1958): *The Historical Atlas of Cheshire*. Cheshire Community Council.
Wood, B. (1976): *The Process of Local Government Reform, 1966-74*. George Allen & Unwin, London.

BIRD-RECORDING IN CHESHIRE AND WIRRAL

A brief history of recording in Cheshire up until 1961 was given by Bell in his *The Birds of Cheshire* (1962) to which the interested reader is referred. This note summarises the history prior to 1961 and reviews the situation since.

The first comprehensive work on Cheshire birds was Coward & Oldham's *Birds of Cheshire,* published in 1900, which drew on the knowledge of various earlier authors of local relevance – the two most often quoted in this atlas being J. F. Brockholes, who wrote *On birds observed in Wirral, Cheshire,* published in 1874; and Isaac Byerley, author of the *Fauna of Liverpool,* published as an appendix to the Proceedings of the Liverpool Literary and Philosophical Society in 1854. A revised and enlarged account of the county's birds was published in 1910, again by Coward & Oldham, as a part of *The Vertebrate Fauna of Cheshire and Liverpool Bay.*

From 1914 onwards the Lancashire and Cheshire Fauna Committee published annual reports, covering invertebrate and other orders as well as birds, but which evolved over the years into the forerunner of the present *Cheshire Bird Report.* These form one of the main sources of information on the county's avifauna. The journal *Northwestern Naturalist,* which appeared from 1926 until 1955, included a number of invaluable local studies which add significantly to the literature, notably Griffiths & Wilson's *The Birds of North Wirral* (1945), Sibson's *Notes on Birds of Sandbach, S. E. Cheshire* (1945-46) and Bell, Samuels & Pownall's *Notes on the Birds of the Urban District of Wilmslow, Cheshire* (1955).

Two major works to appear in mid-century were Hardy's *Birds of the Liverpool Area* (1941) and Boyd's *Country Diary of a Cheshire Man,* the latter a compilation of his weekly columns in the *Manchester Guardian* from 1933 to 1945. In 1951 Boyd's *A Country Parish* was published as an early contribution to the New Naturalist series. This book contains a wealth of detailed information on the everyday birdlife of the Great Budworth area – too often the commonplace vanishes unrecorded into the mists of time whilst rare occurrences become legendary. In this context, before we move into the modern era, mention must be made of the diaries of Norman Abbott, a correspondent of Coward's, which coincidentally came to light during research for this atlas. Abbott's diaries put almost all present-day watchers to shame by their neatness and content. His chief contribution was the mapping, in 1919, 1921 and 1922, of all singing summer visitors in an area amounting to 22 square kilometres around Wilmslow. His methods would not satisfy the standards of a modern organisation such as the British Trust for Ornithology, but they demonstrate quite clearly just what Coward & Oldham meant when they described species like the Tree Pipit and Whinchat as abundant.

The modern era starts with the publication in 1962 of Bell's *The Birds of Cheshire.* Bell summarised available information on all species then known to have occurred in the county. In 1967 he followed up the main work with a Supplement covering the years up to 1966. Since then county bird reports have appeared each year, giving a deceptive appearance of a period of stability to onlookers. Few years have passed, however, without some substantial change in recording procedure, continuity being provided chiefly by the county recorders who compiled the reports. For much of the period the recorder also acted as editor.

Year	Recorder/Compiler
1964-67	G. A. Williams
1968-77	R. J. Raines (assisted in 1969 by D. J. Bates)
1978-81	J. P. Guest
1982-84	R. Harrison

For the 1964 report the county was divided into four recording areas. The Liverpool Ornithologists' Club, Mid-Cheshire Ornithological Society, Manchester Ornithological Society and South-East Cheshire Ornithological Society each took responsibility for collecting records in their respective areas. This first *Cheshire Bird*

Report was reprinted from the Lancashire and Cheshire Fauna Committee's *Thirty-Fifth Report*. The L&CFC continued to publish bird reports for the county until 1966 in parallel with the new *Cheshire Bird Report* (*CBR*).

A meeting early in 1966 resulted in the formation of the Cheshire Bird Recording Committee, essentially setting the arrangements for the 1964 report on a formal footing, and this CBRC published the 1965 *CBR* in due course. In 1967 CBRC expanded and the county was further partitioned to accommodate the newly-formed Chester and District Ornithological Society. Two years later CBRC mutated into the Cheshire Ornithological Association, then comprising fourteen local societies with an interest in Cheshire ornithology. The five recording areas remained intact. Over the next few years membership of the Association varied. There were 15 member organisations in 1970, gradually increasing to 23 in 1982.

In 1974 local government boundaries were redefined, causing chaos in recording circles. For the 1975 report the Longdendale panhandle was excluded, but areas north of the Mersey, newly added to Cheshire, were somewhat arbitrarily included. Confusion prevailed over the next two years, with no decision reached as to whether to continue to record those areas now in Greater Manchester, and steps were taken towards annexing the whole of Merseyside. At this time, in theory at least, five local societies still retained control over the five recording areas, although the MOS was by now occupied in producing its own Greater Manchester report, and the other societies were finding it difficult to appoint enthusiastic recorders to collate area records.

A fresh start was made in 1978. To avoid duplication with the Greater Manchester report, the new county boundaries were adopted, plus Wirral which was retained because of its geographical continuity with "mainland" Cheshire. Recording areas were redefined along the boundaries of 10-kilometre squares and a sixth recording area, North Cheshire, was created. In 1980 two additional areas, Knutsford and South Cheshire, were divided off from the East and South-east Cheshire areas respectively, with the Knutsford Ornithological Society and Nantwich Natural History Society respectively taking responsibility for recording. The Knutsford and East Cheshire areas continued to be treated as one in the bird reports however. This system remained in practice up to 1983. The recording areas are referred to in some of the species texts in this atlas. Their boundaries are shown in Fig.10.

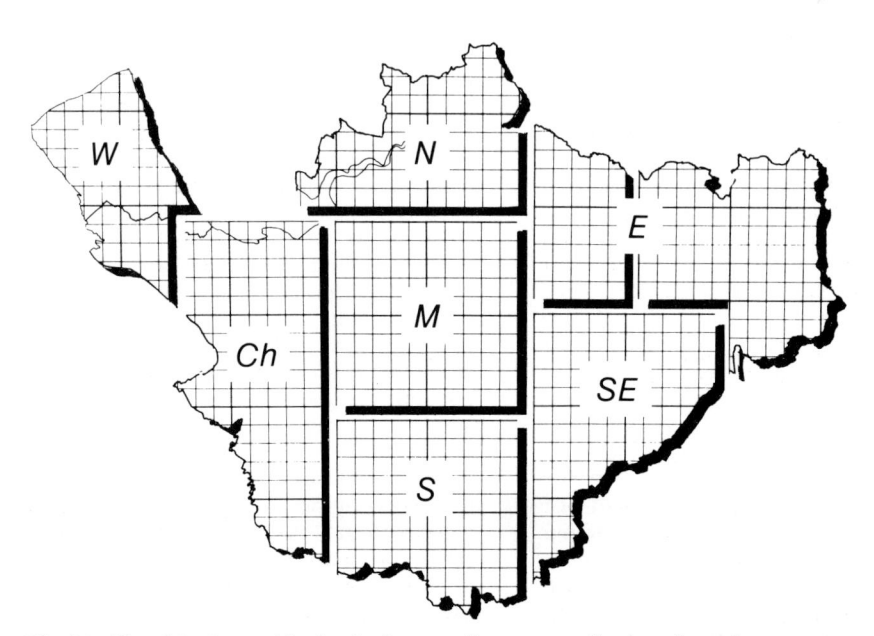

Fig.10. Cheshire's ornithological recording areas during the atlas survey.

From 1981 onwards efforts were made to replace the network of area-recorders with a number of compilers. The expectation was expressed that the compilers would develop a specialist knowledge of the species they covered, allowing more meaningful interpretation of records in the bird report. Unfortunately the COA delegates' meetings opposed this development. Nevertheless a hybrid system was instigated with usually three compilers assisting the recorder each year until 1983. For the 1984 report the switch to compilers was finally put into effect, the systematic list being divided amongst six pairs of compilers and recording areas abolished.

Recording areas

Abolition of recording areas may have been carried out too hastily. On the one hand to ask individual contributors to split their records into several batches was impractical. On the other hand they ensured that the status of each species was noted in all parts of the county, and avoided the frequent confusions which arise between the various Marburys, Strettons, Martons and other places sharing the same name in different parts of the county.

There has been considerable debate as to what form area boundaries should take, generally polarised between protagonists of political and of ecological boundaries. In effect, each species of bird needs to be considered on its own merits, and recording strategies prepared accordingly. Wildfowl and waders congregate within estuaries for example, while the same estuaries form barriers to certain woodland birds.

The Future?

The Cheshire and Wirral Ornithological Society, formed in 1988, operates a similar system of recording to that which originated in 1984. The onus is now on the compilers of the bird report to study the status of those species for which they are responsible and to encourage recording which will add to our knowledge. The breeding distributions demonstrated by this atlas will help define recording strategies. In the case of a declining species such as the Whitethroat, monitoring of the population at sites east of gridline 80 and at strongholds further west is feasible. Never again should a bird like the Whinchat fade away unnoticed. Will Long-tailed Tits soon recolonise the Mersey valley as mosses and insects benefit from cleaner air? Why not census all twenty known Turtle Dove breeding sites? Urban species generally are poorly recorded. Why not extend the Warrington Swift survey to all Cheshire towns? What about rooftop Jackdaw counts? And why not a complete census of town pigeon flocks?

The Common Birds Census, while giving an important guide to the density of our breeding bird populations, needs extending. Even two or three seasons' work on a single plot in the open country between Nantwich and Tarporley would add very greatly to our knowledge. How about a census of Delamere Tree Pipits? The possibilities for easy but instructive fieldwork are endless.

Passage and winter distributions should also be considered. Here information can be gleaned from a variety of sources, not least past bird reports. It is not enough simply to record the largest flocks of Goldfinches for example if such a strategy gives an impression of complete absence away from the estuary fringes in winter.

The content of the county report varies considerably according to editorial policy. It is difficult for the present author to be objective over this matter, but it does seem at times as though certain recorders have found those birds which are not here more interesting than those that are, and have accorded a disproportionate amount of space to lost vagrants of infinitesimal significance, no matter how exciting it may be to come across them in the flesh. The report should first and foremost inform its readers and provide a lasting record of the county's birdlife.

Many of our breeding birds are threatened by developments both at home and abroad. It is the responsibility of all bird-watchers to detect negative trends as early as possible so that public awareness can be heightened and remedial action formulated.

ABOUT THE BREEDING BIRD ATLAS OF CHESHIRE AND WIRRAL

THIS atlas marks the culmination of ten years of work on the most detailed survey yet of the breeding birds of Cheshire and Wirral. The survey was launched in the spring of 1978, fieldwork continuing over the seven seasons up to and including 1984. Fieldwork made use of recording codes and forms, adapted with little modification from those used for *The Atlas of Breeding Birds in Britain and Ireland* (Sharrock, ed. 1976). From 1980 however, the codes were altered slightly to fall in line with the standard codes of the European Atlas Committee (Table 2). These codes remained in use for the last five

---

Table 2. European Ornithological Atlas Committee codes used for the present survey:

i) **Species present during the breeding season and possibly breeding** - one of the following symbols to be entered in the first column of the recording form to indicate the evidence obtained.

O	Species **O**BSERVED in the breeding season.
H	Species observed in the breeding season in possible nesting **H**ABITAT.
S	**S**INGING male(s) present or breeding calls heard in breeding season.

'O' records were in general omitted from the maps as being of unlikely significance. 'H' and 'S' records were plotted as a small dot.

ii) **Probable breeding** - one of the following symbols to be entered in the second column on the recording form. Second column, 'probable', breeding records were entered on the maps as a medium-sized dot.

P	**P**AIR observed in suitable nesting habitat in breeding season.
T	Permanent **T**ERRITORY presumed through registration of territorial behaviour (song etc.) on at least two days, which are a week or more apart, at the same place.
D	**D**ISPLAY or courtship (including copulation).
N	Birds visiting probable **N**EST-SITE (e.g. a Swallow flying into a barn), but see 'ON' below.
A	**A**GITATED behaviour or **A**NXIETY calls from adults.
I	Brood patch on adult examined in the hand, indicating probably **I**NCUBATING.
B	Species observed **B**UILDING a nest or excavating a nest-hole.

iii) **Confirmed breeding or breeding proved beyond reasonable doubt** - one of the following symbols to be entered in the third column of the recording form to indicate the evidence obtained. On the maps the largest-sized dots signify confirmed breeding.

DD	**D**ISTRACTION **D**ISPLAY or injury-feigning.
UN	**U**SED **N**EST or egg-shells found (occupied or laid within period of survey).
FL	Recently **FL**EDGED young of nidicolous species or downy young of nidifugous species.
ON	Adults entering or leaving nest-site in circumstances indicating an **O**CCUPIED **N**EST which is in use (including high nests or nest-holes, the contents of which cannot be verified) or adults seen sitting on the nest.
FY	Adults carrying **F**OOD for **Y**OUNG.
FS	Adults carrying **F**AECAL **S**AC.
NE	**N**EST containing **E**GGS.
NY	**N**EST with **Y**OUNG seen or heard.

years of the survey, data collected during the first two seasons being easily translated to fit the new format.

The symbols in Table 2 are listed in ascending order of significance. Recorders were requested to enter only the highest level of significance obtained onto the recording form at the end of each season, and to attempt to improve on this standard in subsequent years.

Whereas the national atlas had been based on 10-km squares of the national grid, the survey of Cheshire and Wirral took as its recording unit the tetrad of 2 km x 2 km.

The Tetrad

BEFORE the Second World War, a new system of map-referencing was developed for the British Isles. Called the National Grid, it superseded the more arbitrary meridians used by different map-makers prior to its introduction, and has been made familiar by the Ordnance Survey.

From the central meridian are laid out squares with sides (at right-angles and parallel to it respectively) which are multiples of the international metre.

The kilometre reference consists of two letters showing the square with 100-km sides followed by two pairs of numerals showing the square with 1-km sides (and which is bounded by the thinner grid-lines on an Ordnance Survey map). This unit can still further be broken down so that any point on a map may be identified and referred to by a unique combination of two letters and six (or eight) numerals. Almost the whole of Cheshire lies within the 100-km square coded "SJ".

Nationally, biological recorders find a unit of 10 km x 10 km satisfactory, but for the more detailed portrayal of distribution across a county the most suitable recording unit is a square measuring 2 km x 2 km. This unit was first devised by E. S. Edees in 1955 for the Botanical Society of the British Isles and has come to be known as a Tetrad (having four 1-km squares themselves arranged in a square, two by two) and is formed by the even-numbered kilometre grid-lines.

It can be seen that 5 x 5 tetrads will complete one 10-km square which is given an individual number and bounded by the thicker grid-lines on Ordnance Survey maps. Fortuitously, we can allot a different letter of the alphabet to each tetrad in a 10-km square, the letter "O" being omitted as a possible source of confusion with the zero, "0", and it has become the convention, in tetrad recording surveys, to begin the alphabet at the bottom left-hand corner of the 10-km square and work upwards before moving to the right a row at a time (Fig.12).

To make this numbering system clear by stating a few examples: the corner of the Mersey and North Wirral shorelines at New Brighton lies in tetrad SJ39C, Nantwich railway station in SJ65K, and Lamaload Reservoir in SJ97S.

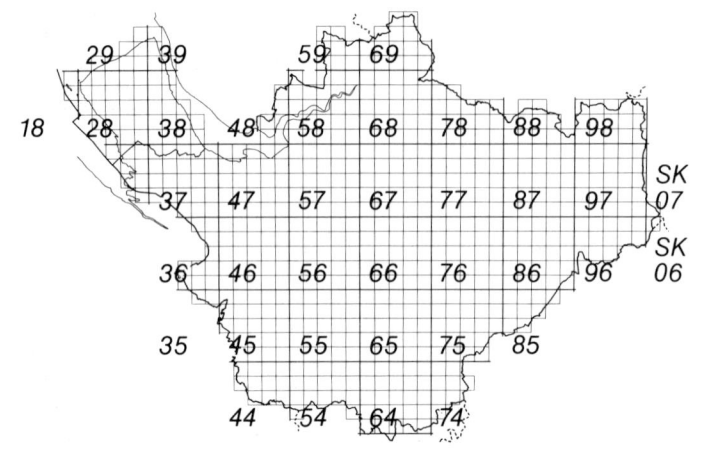

Fig. 11. 10-km square designations for the county. Each of these squares contains 25 tetrads as shown in Fig. 12.

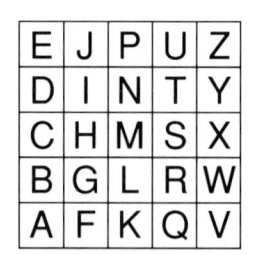

Fig. 12. A 10-km square showing the tetrad letter arrangement

The tetrads forming the Recording Area of the county of Cheshire, with the Wirral peninsula, can be put in three categories:

A)   those of which the entire area of the tetrad is land within Cheshire and Wirral.
In Category A there are 528 tetrads.

B)   those of which some area is land, bounded by the sea or by the Mersey and Dee estuaries. There are 42 such tetrads. In some tetrads in Category B the land area is minimal (e.g. SJ18Y, 18Z, 29X) and the number of breeding species was low; in other cases (27P, 47E, J, P, U) the land area consists almost entirely of tidal mudflats or saltmarsh where no breeding records were obtained. (For convenience, all tetrads in SJ58 through which the River Mersey passes are deemed to be in Category A.)

C)   those of which the land area is shared between Cheshire/Wirral and another county.

There are 154 tetrads in Category C in which the portion of Cheshire land varies between an acre or two and nearly 100 per cent. The Atlas sub-committee agreed to include in the Recording Area each of those tetrads in Category C which contain a Cheshire land area exceeding 25 per cent in total of the whole tetrad. There are 101 tetrads so defined, and records were required to be sought for the whole of the tetrad area, even though up to 75 per cent may lie in another county. In the annual progress maps which follow, results are shown for certain tetrads external to the Recording Area which were visited throughout the period of the survey.

The number of tetrads in the Recording Area defined by the three Categories above is 670. (One tetrad is common to Categories B and C.)

Fieldwork

PROGRESS maps, reproduced overleaf, show how uneven coverage was in the early years. Members of the Chester and District Ornithological Society put in a concerted effort, such that by 1980 the eastern boundary of their recording area was delimited by an abrupt change in the number of species found either side of easting 50. Comparable efforts were made in the east of the county, but holes in coverage were by then appearing down the centre, particularly in the south, where few bird-watchers live. During the closing years of the survey, observers from further afield travelled to assist in covering this area, with the ultimate outcome that all tetrads were visited at least once.

As with the earlier national atlas, those who took part in the survey found their fieldcraft steadily enhanced by the experience. Prior to visiting a tetrad for the first time, the Ordnance Survey map was examined to suggest the best route through the tetrad. In theory, this should have been assessed on purely ornithological criteria, to include sampling of all obviously different habitats. In parts of the county, however, the public footpath network is so poorly maintained that visits were often modified to avoid conflict with landowners.

Keener workers found it useful to have several tetrads on the go at a time. After an hour or two in the field, the flow of new records would begin to dry up. The presence of species could in general be predicted by the habitat, despite the instruction to record what was actually seen, rather than what was expected, and once almost all the expected species had been found, or found absent, further fieldwork was largely wasted effort. Under such circumstances it was far better to spend available time in other tetrads, returning to the original at a later date when the nesting cycle was more advanced and confirmation of breeding more easily obtained.

Seasonality was indeed a major factor. Woodpeckers, especially Lesser Spotteds, are much more conspicuous in early spring. As an April shower died away, Little Owls would begin to yelp from their territories. Nuthatches can be incredibly secretive during incubation, though easily located before eggs are laid and again once the young

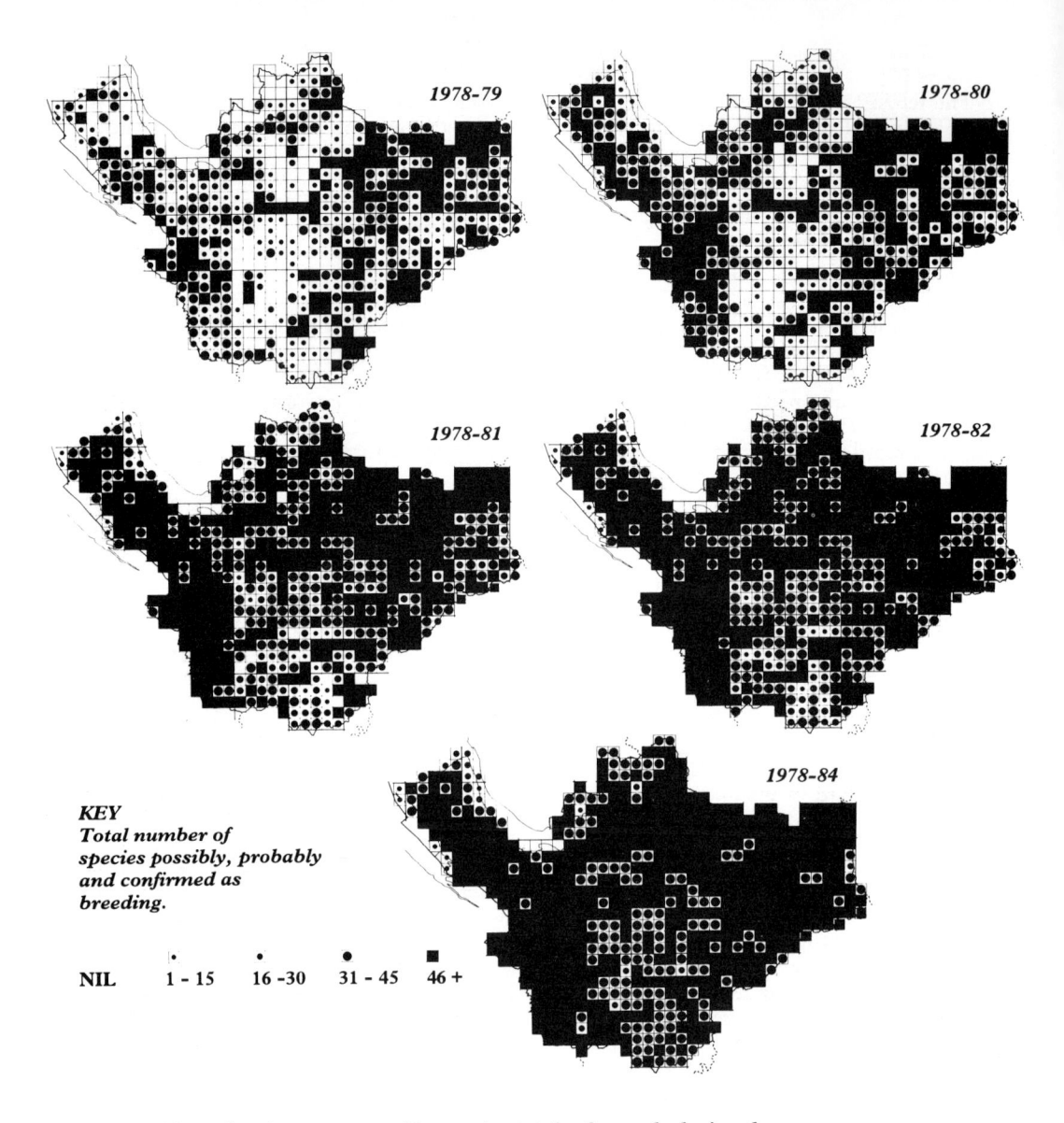

Fig. 13. Maps showing progress of increasing totals of records during the survey.

have hatched. Cuckoos call far more in the early morning, and it was suspected that their apparent absence from the south-centre of the county was due to the arrival of immigrant fieldworkers there later in the day. Spotted Flycatchers must have been missed in many tetrads because of their late arrival and weak song, and the appearance of fledged broods late in the season after much fieldwork had stopped.

Bird song and calls are very much under-used by most bird-watchers, with ever more emphasis being accorded to telescopes and plumage details. Two of the most overlooked species are the Lesser Whitethroat and Little Owl. The rattling song of the former is perhaps familiar to a reasonable proportion of the county's enthusiasts, but how many can claim familiarity with the sharp, ticking anxiety note which drew attention to many a family party being fed on caterpillars from a hedgerow oak? The hoot of the Little Owl, once learned, is easily copied and a good way of provoking birds in unknown tetrads to call

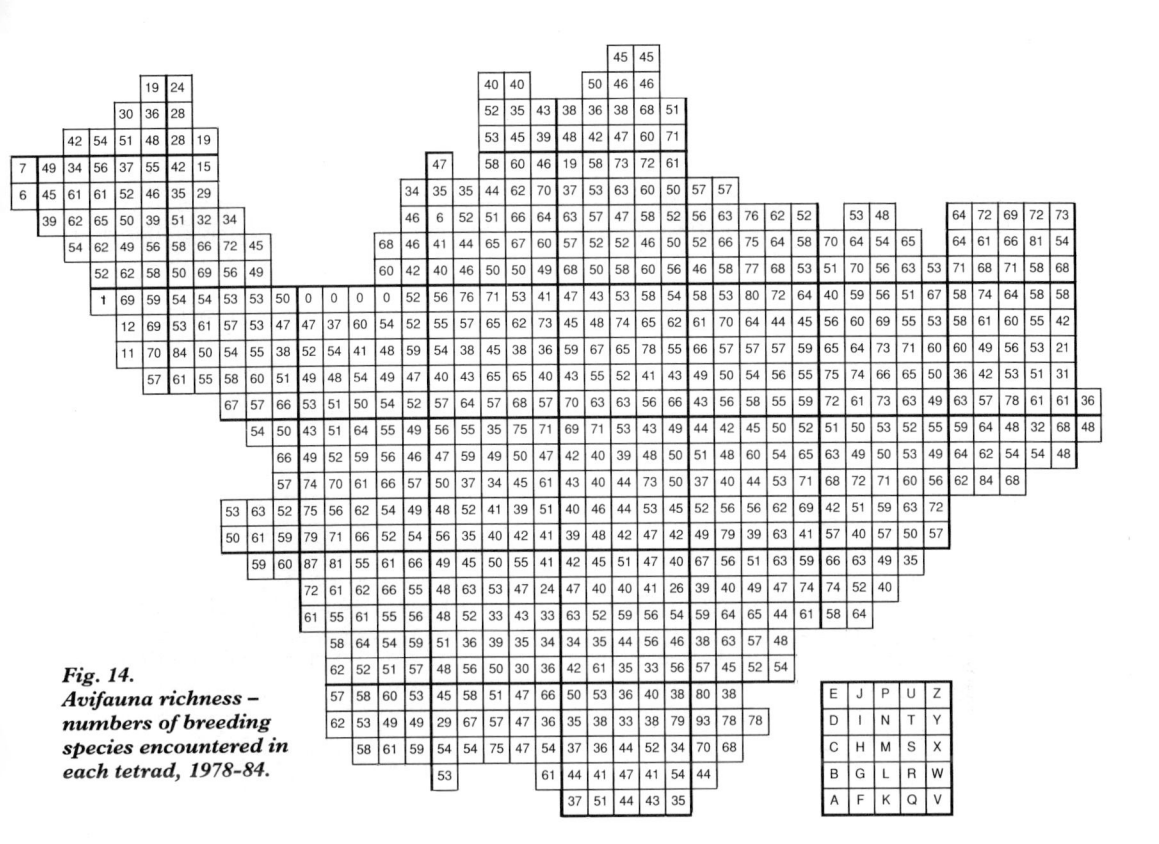

Fig. 14.
Avifauna richness –
numbers of breeding
species encountered in
each tetrad, 1978-84.

back. Athene's bird of wisdom proves to be very gullible!

Many birds have their own particular tricks for concealing their nests. Stock Doves slip off their nest high in a tree while one's back is turned and fly round in large circles after the style of a Mallard. This behaviour alone is almost enough to confirm breeding. Dunnocks often swallow the food intended for their young once they realise they are being watched, but no matter, for 'FY' evidence is good enough. On the other hand, the anxiety behaviour of some species gives away the proximity of a nest and must do so to predators also. One anxious Wood Warbler was heard, through a crash helmet, calling above the noise of a moving motorcycle.

There were many surprises during the survey. Personal highlights included a soaring Honey Buzzard, a family of Water Rails, courtship feeding by Grasshopper Warblers and an anxious Redwing. Cheshire has undoubtedly lost much of its former charm, but still holds many secrets.

Confidentiality of records

WHEN the survey was launched, fears were expressed that scarcer species might suffer if their breeding sites were published. The passage of time has removed many such fears however. There are actually very few species nesting in the county that suffer from human predation. In some of these cases, the nest-sites are already well known, or can be readily deduced by interested parties.

It is sad to report however, that the selfish, undisciplined activity of certain bird-watchers has been the major cause of concern to the editors. Bird-watchers are now the main threat to the survival of Long-eared Owls in the county. The habitat of this species is very limited, only a very few pairs known, and yet birds have been repeatedly flushed from nesting thickets, such behaviour in at least one case precipitating desertion. How do the "bird-racers", beating a wood to kick up a nesting Woodcock, differ from egg-collectors or shooters?

Nevertheless, there have been *very few* cases in which the editors felt it advisable to withhold information.

33

ACKNOWLEDGEMENTS

THIS atlas is remarkable as the first occasion on which a survey of any aspect of the natural history of Cheshire and Wirral has been carried out on so fine a grid. To do so required the efforts of over 350 of the county's bird-watchers. Without their commitment, the survey would have been impossible. Their names are listed opposite. It is hoped that no name has been inadvertently left off - profuse apologies to any who may have been! Special thanks are due to those individuals who covered blocks of tetrads, and particularly to those who travelled many miles during the closing years of the survey to ensure that every tetrad was visited at least once. Jonathan Guest completed forms for no fewer than 70.

A number of individuals undertook to co-ordinate recording activity within particular 10-km squares, thereby reducing duplication of effort to a minimum. They were:

Wirral (SJ28, 29, 37 part,
    38 & 39)   R. A. Eades
    (SJ27)   E. J. Abraham
West Cheshire
    (SJ36, 37 part, 44,   D. Goff &
    45, 46 & 47)   S. W. Holmes
North Cheshire
    (SJ58, 59, 68 & 69) R. A. Smith
Mid Cheshire
    (SJ56, 57, 66 & 67) C. Hansen
South Cheshire
    (SJ54, 55,   1978-79 D. Elphick
    64 and 65) 1980-84 S.J. Gibson
East Cheshire
    (SJ96, 97, SK06 & 07) A. Booth
    (SJ87 and 88)   J. P. Guest
    (SJ98)   S. H. Hind
    (SJ77 and 78)   A. C. Usher
South-east Cheshire
    (SJ74, 75, 76, 85 & 86) D. Elphick

The organisation of the survey was undertaken by Dennis Elphick. This included the mammoth task of collating records onto master sheets - manually, since home computers were not yet commonplace when the survey began; Richard Blindell gave some assistance here. Alan Hunter produced the progress maps which appeared in the county bird reports for 1979, 1980, 1981 and 1982.

Trevor Poyser gave much useful advice on book production when this atlas started its lengthy gestation.

Dr Derek Yalden prepared first draft texts for ten upland species (Merlin, Red Grouse, Black Grouse, Golden Plover, Dunlin, Common Sandpiper, Dipper, Wheatear, Ring Ouzel and Twite), Jonathan Guest working on the remainder, and on the bulk of the introductory sections. Dr David Norman rewrote the text for Sand Martin in its entirety and added a text for Cetti's Warbler to celebrate mist-netting the Frodsham bird! Marilyn Elphick spent many hours deciphering and typing up early, hand-written drafts.

The texts were then discussed by an editorial panel consisting of Dennis Elphick, Jonathan Guest, Alan Hunter and David Norman. Alan Hunter prepared the distribution maps. Several individuals commented on certain texts in the light of local knowledge or personal experience of particular species. In this context the assistance of Malcolm Calvert, Ray Eades, Ron Harrison and Tom Wall is acknow-ledged.

The considerable time between the completion of the survey and publication allowed repeated amendments to be made to the texts as additional information came to light. Steve and Gill Barber, Alan Booth, Stan Dobson, Ron Harrison and the Manchester Ornithological Society contributed valuable references and notes.

During the winter of 1989-90, Jonathan Guest and David Norman sat through a final round of editing. The texts were then transferred to David Jessop who undertook complete typesetting and computer page make-up from Alan Hunters's design for the final production.

Alan Hunter devised the additional maps and text diagrams and acknowledges the considerable co-operation, in the early days of map preparation, of Colin Brown and staff of Fenton & Pattison, and later the assistance of David Evans of Future Graphics. To David Jessop, whose selfless and unstinting commitment to the preparation of this book is deeply appreciated, the authors and the Society express their gratitude.

The Cheshire and Wirral Ornithological Society is indebted to the RSPB for their courtesy in allowing the use of their Charles Tunnicliffe painting of Great Crested Grebes featured on the front of the book-jacket.

Richard Gabb took the excellent photographs which give a flavour of the various bird breeding habitats within the county. The view of Red Rocks reserve and Hilbre was contributed by Mark Turner. Tony Broome, Bill Morton and Ron Plummer produced line drawings which appear throughout the atlas, with other illustrations by Dave Braithwaite, Julian Hodge and Scott Kennedy. Brian Martin and Steve Barber sought sponsorship for publication. Steve Barber and David Steventon contributed greatly in the business of arranging publication schedules and sales.

Next – by no means least – thanks are due to the county's landowners and tenant farmers who allowed access onto their property.

The following contributed the thousands of records which have made this atlas such a thorough portrayal of our county's breeding birds. Future generations of bird-watchers now have an inheritance to emulate, and from which to form comparisons for study and for monitoring – and for enjoying in the process.

To all contributors – thankyou.

K. Ablitt, E. J. Abraham, I. Allcock, Mrs Allman, B. & R. N. Anderson, R. Anderson, B. Andrews, A. Ankers, B. Ankers, C. Antrobus, B. Armitt, A. M. G. Armstrong, C. Armstrong, B. & R. B. Atherton, P. F. Atherton, M. Bailey, Mr Baker, M. J. Ball, N. Bancroft, J. Bannon, S. & G. Barber, M. Barlow, P. B. Barlow, S. Barlow, B. S. Barnacal, I. Bedford, T. H. Bell, G. Bellingham, M. Bellingham, R. Benbow, Mrs Bennion, R. & C. Bertera, M. Billington, A. R. Bircumshaw, T. Bithell, I. Blagden, R. M. Blindell, A. Booth, K. Booth, M. & J. Boswell, J. M. Bowmaker, D. Bowman, H. & I. Bradshaw, A. M. Broome, J. Brotherton, N. M. Brown, D. A. Bullock, Burwardsley Outdoor Education Centre (Mr Jones), C. Butler, A. Butterworth, A. Caldwell, M. Calvert, P. Campbell, Mr & Mrs Carrington, J. Carroll, M. T. Carter, P. Chadwick, Chester and District Ornithological Society, J. Christopher, J. Clare, R. Clarke, T. R. Cleeves, J. Clowes, D. Coan, R. P. Cockbain, P. Coffey, M. Collins, M. Comber, T. Conboy, R. Cook, D. Cookson, M. Cooper, P. Cotterill, Prof. J. D. Craggs, S. Craggs, P. Crockett, D. Cross, P. Cross, W. J. Crowe, B. Crowther, G. Cullingworth, D. Cummings, J. R. Dagley, K. Darwin, S. R. Davidson, J. Davies, A. Davison, B. Dawson, C. Dawson, J. P. Dawson, H. Dean, F. Deardon, B. Devereau, S. Dobson, J. Dolan, M. Donahue, N. Duxfield, R. Eades, W. Edwards, M. Ellis, V. Ellis, C. S. Elphick, D. & M. Elphick, D. J. Evans, P. J. Ewell, C. Farnell, M. Farron, T. Fenwick *et al.*, J. Ferguson, L. Flooks, A. A. Ford, B. Formstone, T. Francis, R. & B. Gabb, Mr & Mrs J. Gibson, S. J. Gibson, J. Gittins, D. Goff, R. Gomes, A. Goode, N. C. Goodier, A. Goodin, A. W. Gradwell, A. Graham, D. B. Green, R. & S. Greenwood, R. D. Gregory, P. J. Grice, A. Griffith, R. & C. Groom, T. Groves, J. P. Guest, A. Guilfoyle, R. Hamer, W. Hamilton, C. Hansen, R. Harrison, Mr & Mrs Harvey, Sir Geoffrey Haworth & P. Haworth, W. Hay, J. G. Heald, P. J. Heald, A. W. P. Hearn, B. A. Heaton, G. Heeson, W. Helliwell, T. H. Helvin, I. Hensby, D. Hewitt, A. Hibbert, M. J. Higginbottom, F. J. Hill, G. Hill, K. Hiller, P. Hilton, H. Hind, S. H. Hind, P. Hines, B. Holmes, S. Holmes, B. Holt, N. Holt, D. Howard, G. Howe, J. Howie, W. J. Hughes, G. Hulme, J. S. A. Hunter, J. Jackson, P. M. Jackson, T. James, Mrs Johnson, V. Johnson, C. M. Jones, D. Jones, E. Jones, F. Jones, J. G. Jones, S. Jones, J. Keeling, A. Kelly, D. E. Kelly, M. Kelly, S. Kemp, H. K. Larsen, D. Lawrence, W. J. Leader, P. Leigh, R. S. Leigh, A. Lewis, P. S. Lewis, C. R. Linfoot, Rev. H. Linn, Dr E. A. Lock, K. Lumb, C. Lythgoe, I. McAlpine, V. L. MacFarland, C. W. Mackley, D. McMaster, G. P. McPherson, J. H. Maddock, M. Margetts, A. W. Martin, B. Martin, D. & D. Mason, L. Mason, K. Massey, T. Mawdesley, Mid Cheshire Ornithological Society, G. Mitchell, Mrs Moore, W. Morton, H. Moss, J. K. Mottershead, M. Mulholland, K. Mullins, E. Mulroy, A. Nall, I. Naylor, Dr D. Norman, B. D. Norris, J. Olerenshaw, A. J. O'Neill, A. Ormond, J. V. Oxenham, R. Palmer, G. R. Parkinson, I. Parsons, Mrs Pass, M. L. Passant, D. Pawlett, A. Pay, C. Pearson, Mr & Mrs K. Penney, C. B. Perkins, Mr & Mrs P. Perkins, Perry Family, G. Phelps, C. M. & S. J. Poole, P. Poole, L. Potter, B. Preece, D. Prince, M. Pritchard, B. Proudlove, C. Rambart, B. M. Rathbone, J. W. Rayner, P. Reay, Mr & Mrs H. Rhodes, C. M. Richards, S. Richardson, S. Y. Ricketts, V. Riggall, D. Roberts, M. Roberts, J. G. Rollason, N. Rydal, E. Samuels, P. Sandle, S. Sankey, C. Saunders, R. P. Schofield, E. A. & J. Seddon, M. J. Self, P. Sellers, A. J. Shaw, B. T. Shaw, B. Sherratt, R. Shewring, Shropshire Ornithological Society (M. F. Wallace and C. E. Wright), W. Simcox, M. D. Simmonds, N. Simmonds, T. Simpson, B. & J. Slack, M. Smith, R. A. Smith, Roger Smith, Ron Smith, W. Smith, W. L. Smith, P. J. R. Smyth, D. Snape, J. V. Soames, P. Solan, South East Cheshire Ornithological Society, South Manchester Ringing Group, P. Spilsbury, T. Spruce, G. Stanley, M. P. Stanyer, Mr & Mrs J. Start, A. Stoddart, N. Sumner, N. Tanner, Roy Taylor Jnr, Roy Taylor Snr, H. Thickens, G. Thomason, S. R. Thomason, J. M. Thompson, Mrs Thompson, A. Thrower, D. W. Todd, L. & B. Tollitt, B. Tomlinson, J. Towers, H. Truesdale, Mrs Turner, W. Turner, W. G. Turner, A. C. Usher, A. Von Dinther, T. Wall, F. R. Walley, A. Walmsley, Mr Ward, J. Warden, D. Weedon, R. West, P. D. Whalley, D. Whitehead, Dr F. Whitwell, C. K. Williams, G. G. Williams, D. Woodward, S. Wooley, S. Woolfall, Woolston Eyes Conservation Group, G. Worthington, Dr D. W. Yalden, J. Young.

BIBLIOGRAPHY

Publications frequently referred to by small numerals in the species accounts.

Principal references
1. Coward, T. A. & Oldham, C. (1900): *The Birds of Cheshire.*
2. Coward, T. A. & Oldham, C. (1910): *The Vertebrate Fauna of Cheshire and Liverpool Bay.* Witherby, London.
3. Lancashire & Cheshire Fauna Committee, Annual reports 1914 to 1966.
4. Hardy, E. (1941): *The Birds of the Liverpool Area.*
5. Boyd, A. W. (1946): *The Country Diary of a Cheshire Man.* Collins, London.
6. Boyd, A. W. (1951): *A Country Parish.* New Naturalist, Collins, London.
7. Bell, T. H. (1962): *The Birds of Cheshire.* Sherratt, Altrincham.
8. *Cheshire Bird Reports* (1964 to 1984): Cheshire Bird Recording Committee and, later, Cheshire Ornithological Association
9. Bell, T. H. (1967): *A Supplement to The Birds of Cheshire.* Sherratt, Altrincham.

Local studies
10. Abbott, N. Unpublished diaries, 1913-1930.
11. Coward, T. A. (1914) Faunal Survey of Rostherne Mere II, Vertebrata. *Mem. & Proc. Manch. Lit. & Phil.. Soc.:* **58,** No.9.
12. Farrar, G. B. (1938): *Feathered Folk of an Estuary.*
13. Griffiths, W. & Wilson, W. (1945): *The Birds of North Wirral.*
14. Sibson, R. B. (1945 & 1946): *Notes on Birds of Sandbach, S.E. Cheshire.*
15. Tunnicliffe, C. F. (1948): *Mereside Chronicle.* Country Life Ltd., London.
16. Boyd, A. W. (1955): Faunal Survey of Rostherne Mere - A revised account of the birds. *Mem. & Proc. Manch. Lit. & Phil. Soc..* **96,** No. 9.
17. Bell, T. H., Samuels, L. P. and Pownall, L. A. (1955): *Notes on the Birds of the Urban Area of Wilmslow.*
18. Raines, R. J. (Ed.) (1960 and 1961): *Birds of the Wirral Peninsula.*
19. Whalley, P. A. D. (1964): *Checklist of the Birds of Sandbach Flashes*
20. Williams, T. S. (1971): *Notes on the Bird Life of Chester.*
21. Annual reports, Wilmslow Guild Ornithological Society.
22. Annual reports, Knutsford Ornithological Society.
23. Annual reports, South-East Cheshire Ornithological Society.
24. Harrison, R. & Rogers, D. A. (1978): *The Birds of Rostherne Mere.* NCC, Banbury.
25. Craggs, J. D. (Ed.) (1982): *Hilbre: The Cheshire Island, Its history and natural history.* Liverpool Univ. Press.
26. Booth, A. & Oxenham, J. V. (1981): The Birds of Danes Moss, Macclesfield. *Cheshire Bird Report,* 1981: 64-69.
27. Goodwin, A. (1983): *Birds of Sandbach Flashes.*
28. The CADOS Committee (1984): *The Breeding Birds of the Chester Area: Results of the Tetrad Survey.*
29. Prince, D. (1984): Bidston Hill (North Wirral) [CBC] Plot - The First Ten Years (1975-84). *Cheshire Bird Report* 1984: 80-84.

National Works
30. Witherby, H. F., Jourdain, F. C. R., Ticehurst, N. F. & Tucker, B. W. (1938): *The Handbook of British Birds.* Witherby, London. (5 vols.).
31. Simms, E. (1971): *Woodland Birds.* New Naturalist, Collins, London.
32. Campbell, B. & Ferguson-Lees, I. J. (1972): *A Field Guide to Birds' Nests.* Constable, London.
33. Newton, I. (1972): *Finches.* New Naturalist, Collins, London.
34. Parslow, J. L. F. (1973): *Breeding Birds of Britain and Ireland.* T. & A. D. Poyser, Berkhamsted.
35. Sharrock, J. T. R. (Ed.) (1976): *The Atlas of Breeding Birds in Britain and Ireland.* BTO, Tring.
36. Cramp, S. & Simmons, K. E. L. (Eds.) (1977, 1979 & 1982): *The Birds of the Western Palearctic,* Vol. I, Vol. II & Vol. III. Cramp, S. (Ed.) (1985 & 1988): *The Birds of the Western Palearctic,* Vol. IV & Vol. V. Oxford Univ. Press, Oxford.
37. Lever, C. (1977): *Naturalised Animals of the British Isles.* Hutchinson, London.
38. Simms, E. (1978): *British Thrushes.* New Naturalist, Collins, London.
39. Perrins, C. (1979): *British Tits.* New Naturalist, Collins, London.
40. Hickling, R. A. O. (1983): *Enjoying Ornithology.* T. & A. D. Poyser, Calton.
41. Lack, P. (Ed.) (1986): *The Atlas of Wintering Birds in Britain and Ireland.* T. & A. D. Poyser, Calton.
42. O'Connor, R. J. & Shrubb, M. (1986): *Farming and Birds.* Cambridge Univ. Press, Cambridge.
43. Spencer, R. and the Rare Breeding Birds Panel (1986): Rare breeding birds in the United Kingdom in 1983. *British Birds* **79**: 53-81.

Neighbouring Counties
44. Annual reports, Derbyshire Ornithological Society.
45. Frost, R. A. (1978): *Birds of Derbyshire.* Moorland, Buxton.
46. Holland, P. K., Spence, I. M. & Sutton, J. T. (1984): *Breeding Birds in Greater Manchester.* Manchester Ornithological Society.
47. Harrison, G. R. (Ed.), Dean, A. R., Richards, A. J. & Smallshire, D. (1982): *The Birds of the West Midlands.* West Midlands Bird Club.

THE BREEDING BIRDS OF
CHESHIRE AND WIRRAL

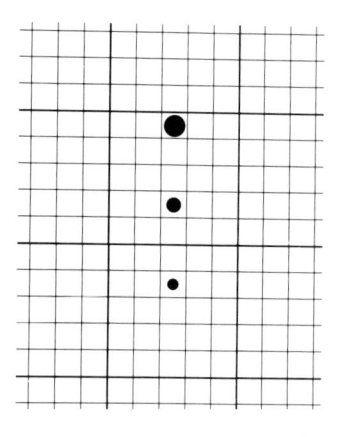

The maps which follow contain dots of three sizes, representing the three categories of breeding evidence obtained during 1978-84.

The largest-sized dot signifies confirmed breeding;

the medium-sized dot shows probable breeding;

the smallest dot indicates possible breeding.

A tetrad lacking a dot is one where the species in question was not encountered, in any breeding circumstances, during the breeding seasons of the survey. (See page 29 for definitions of the above categories.)

Abbreviations used in the Text

'B'	Building (see page 29)	NNR	National Nature Reserve
BTO	British Trust for Ornithology	NNW	north-north-west
CADOS	Chester & District Ornithological Society	NW	north-west
CAWOS	Cheshire & Wirral Ornithological Society	'NY'	Nest with Young (see page 29)
CBC	Common Birds Census	'ON'	On Nest (see page 29)
CBR	*Cheshire Bird Report*	pers. comm.	personal communication (verbal)
COA	Cheshire Ornithological Association	pp	pages
'DD'	Distraction Display (see page 29)	RG	Ringing Group
E	east	RSPB	Royal Society for the Protection of Birds
et al.	*(et alia)* and all	SE	south-east
'FL'	Fledged Young (see page 29)	sp.	species (singular)
'FY'	Food for Young (see page 29)	spp.	species (plural)
GMC	Greater Manchester County	SSSI	Site of Special Scientific Interest
ha.	hectare(s)	SW	south-west
in litt.	in a letter (written communication)	UK	United Kingdom
km	kilometre(s)	W	west
L&CFC	Lancashire & Cheshire Fauna Committee	** WAGBI	Wildfowlers' Association of Great Britain and Ireland
MNA	Merseyside Naturalists' Association		
MRG	Merseyside Ringing Group	WBS	Waterways Bird Survey
'N'	Nest (see page 29)	WECG	Woolston Eyes Conservation Group
* NCC	Nature Conservancy Council	WGOS	Wilmslow Guild Ornithological Society

* The Nature Conservancy Council split into three bodies in April 1991 representing England, Wales and Scotland, and which are co-ordinated by the Joint Nature Conservation Committee (JNCC). ** WAGBI became known as the British Association for Shooting and Conservation (BASC) in 1983. In June 1989 the Wildfowl Trust changed its name to the Wildfowl and Wetlands Trust (WWT).

PLEASE NOTE. Throughout this book, measurements are stated in imperial or metric units – as originally submitted or published. No attempt has been made to standardise by conversion; to do so would have led to figures which look artificial and contrived.

   10 feet = 3.048 metres   100 metres = 328.08 feet
    1 acre = 0.4047 hectares   1 hectare = 2.41 acres
 A Tetrad = 400 hectares
        (or 988.4 acres, or 1.544 square miles)

The sequence of species and their scientific names accord with the proposals in Voous, K. H. (1973 & 1977): List of Recent Holarctic Bird Species, *Ibis* **115 & 119**, followed by Cramp[36] and adopted by the magazine *British Birds*.

Little Grebe
Tachybaptus ruficollis

THE loud trilling call of the Little Grebe may be heard in any month of the year but is usually associated with the breeding season from March, when birds return to nesting areas, until August. During the summer they are often very secretive and prefer sizeable pools, meres and flashes with some dense marginal vegetation in which to nest undisturbed. Their strongholds are in the numerous waters of the Cheshire plain, where pairs also nest in reedy patches along the River Weaver and the canals, and in the sludge-pools and abandoned canals along the Mersey valley in the north. Shallow waters tend to support more plant, and consequently invertebrate, life and so are preferred to the deeper meres and sand-quarries favoured by Great Crested Grebes which feed chiefly on fish[36].

In the east of the county the rivers are too fast flowing and steep-sided to hold grebes, the reservoirs in the hills lack suitable vegetation because of their rapidly fluctuating water levels and indeed few of the eastern meres are sufficiently secluded given the level of recreational disturbance. Disused flooded sand-quarries in central and eastern Cheshire already hold a few pairs and may well support more in future if marginal vegetation becomes established.

There seems to have been some reduction in numbers since Coward's day when Dabchicks nested "in secluded marl-pits and ponds in woods". Boyd, writing in 1936, found Little Grebes were by then absent from marl-pits, but still more abundant and widespread than Great Crested Grebes. Since then many pits have been filled in and others are now largely deserted, some no doubt because they are extensively fished. Many others are now shaded by overhanging trees. Coward & Oldham also mentioned a few pairs breeding in the eastern hills as at Lyme Park and Bosley Reservoir. In some recent years a pair has been present at Bottoms Reservoir, Langley, but otherwise the hills are now deserted.

In Wirral, Brockholes[1] had not seen the species for ten or twelve years, but Coward & Oldham[2] considered it not uncommon on pools at Meols, Heswall and elsewhere. Then, as now, the absence of suitable waters was doubtless the limiting factor.

The floating nest is usually anchored amongst emergent vegetation, but one on a canal at Moss Side in 1973 was still occupied although it drifted with the wind. Proof of breeding often depends on seeing young birds which follow their parents begging noisily to be fed. These chicks generally appear from mid-June onwards, more especially in July and August, and in some years dependent birds are still around as late as October, for example near Sandbach on 25th October 1975. An early chick was reported at Irby, Wirral on on 25th April 1975 by the Birkenhead School Natural History Society.

Accurate assessments of numbers of such an elusive breeding bird are difficult to obtain, but might be best gathered by counts of "trilling" birds in early spring. The 1974 *CBR* contains a report of a pair every 400-500 yards along the canals between Runcorn and Nantwich in July, with two concentrations of ten and 14 pairs plus many young near Nantwich. However, such concentrations did

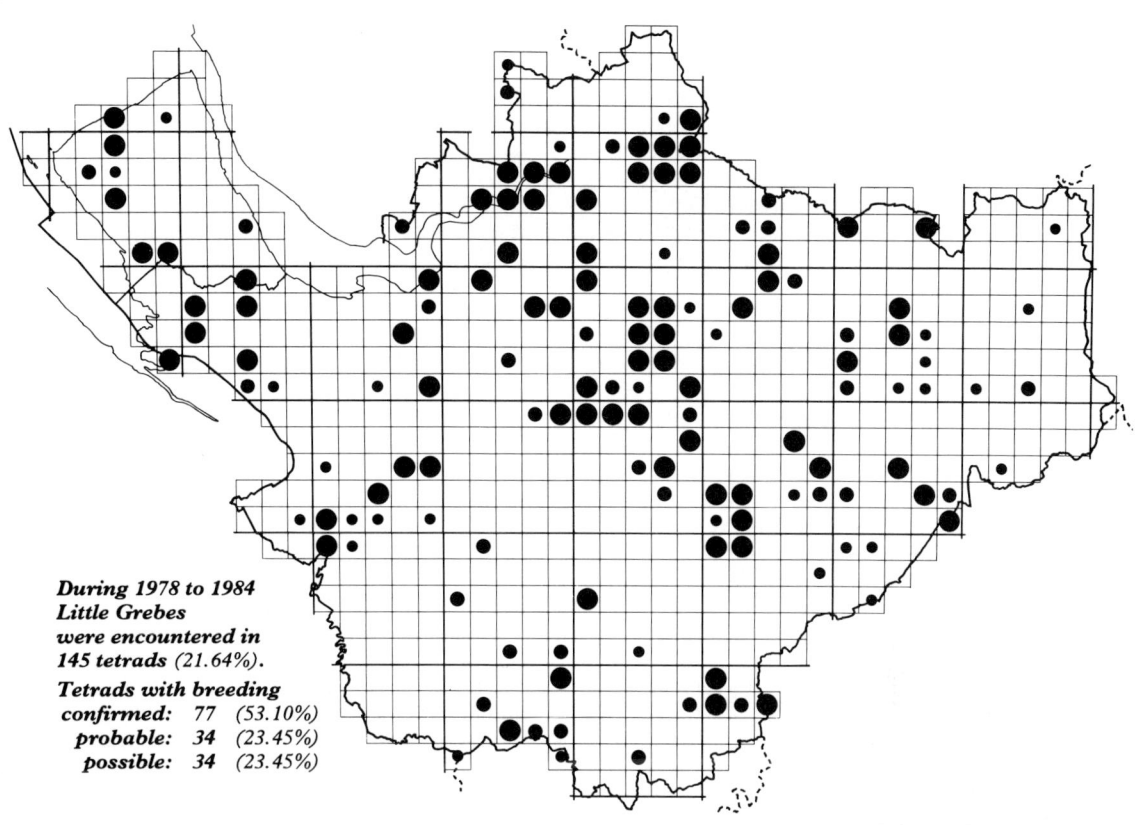

During 1978 to 1984 Little Grebes were encountered in 145 tetrads (21.64%).

Tetrads with breeding
confirmed: 77 (53.10%)
probable: 34 (23.45%)
possible: 34 (23.45%)

not apply by the 1980s and may have been an exaggeration – there are large stretches of canal without any Dabchicks. Without doubt the resurgence of pleasure boating has severely curtailed the value to wildlife of formerly derelict canals. Concentrations do exist in favourable areas, for example eleven pairs or more on the Sandbach Flashes in 1976 and 20 pairs at Woolston in 1982, but such ideal sites are few and far between. With 142 occupied tetrads the county population is probably in the region of 200-250 pairs, but a thorough survey of the mid-Cheshire stronghold might reveal larger numbers.

Great Crested Grebe
Podiceps cristatus

THE natural meres of Cheshire have long been important haunts of the Great Crested Grebe. Harrisson & Hollom (1932) classed twenty English and Welsh lakes as "immemorial" grebe haunts, occupied since at least 1840. Nine of these were in Cheshire and consisted of two main groups: the Delamere group comprising Petty Pool, Little Budworth Pool, Oulton Pool and Tilstone Lodge Pool and the south Cheshire meres comprising Barmere, Cholmondeley Castle, Combermere and Marbury Mere. Rostherne Mere was the ninth. A number of these waters are either man-made or artificially enlarged by damming however.

In the nineteenth century the wearing of "grebe fur" mufflers and hats decorated with grebe skins became fashionable – the species becoming known as the tippet-grebe. Initially the demand was met from continental sources, but from 1857 onwards a steady traffic in birds developed in this country, particularly in Norfolk, the other main stronghold in Britain. The improved efficiency and availability of shotguns, combined with the contemporary demand for stuffed birds and the rising value to collectors of the eggs of such a scarce species, nearly wiped out the British population. Only on the secluded lakes of private estates, where shooting was restricted, did grebes survive in any numbers.

Harrisson & Hollom (1932) gave an estimate of the national population in 1860 as 42 pairs with an approximate maximum of 72 and minimum of 32.

Table 3 shows the figures by county:

Table 3: The probable number of pairs of Great Crested Grebes by county in Britain in 1860, together with the approximate minimum and maximum numbers of pairs.

Cheshire	20	(15-25)
Norfolk	12	(8-26)
Staffordshire	5	(5-12)
Suffolk	3	(2-6)
Lancashire	1	
Yorkshire	1	(1-2)

Legal protection followed, and the Wild Birds' Protection Act of 1880 imposed a close season from March to July. The fashion for grebe "furs" did not die out until the early years of the twentieth century however, helped on its way by groups such as the Society for the Protection of Birds, the forerunner of the RSPB, founded in 1889 in Didsbury, Manchester.

Between 1880 and 1890, 45 new lakes were colonised in England, the species began to spread into Scotland, and big increases were recorded in Norway and Sweden. With the generally more favourable attitude to the bird in Britain in the early years of this century, and this proven potential for recovery, the Great Crested Grebe was able to take advantage of the increased number of lowland waters including reservoirs and flooded gravel pits which were subsequently created.

By 1931, a national census revealed a British population of some 2825 adult birds including 194 in Cheshire with 78 pairs breeding on 27 waters (Harrisson & Hollom 1932). By 1935 numbers had not changed significantly in the county despite prevailing drought conditions (Hollom 1936). Between 1946 and 1955, Boyd co-ordinated an annual census of the species in the county (Hollom 1959). The results, tabulated below, show an initial decrease from the 1931 total, followed by a steady increase to a new maximum in 1953. The reduced number in 1947 may be attributed to the effects of the preceding winter.

In May 1965, 249 birds were counted and 314 adults estimated to be present out of a national total of 4111 and 4651 respectively (Prestt & Mills 1966). However, by 1975, the situation in the county had reversed, showing decreases to 179 counted (down 28%) and 201 estimated present (down 35%). The

Table 4: Annual census totals for Great Crested Grebes in Cheshire 1946-55.

1946	1947	1948	1949	1950	1951	1952	1953	1954	1955
164	126	145	143	163	211	199	253	245	220

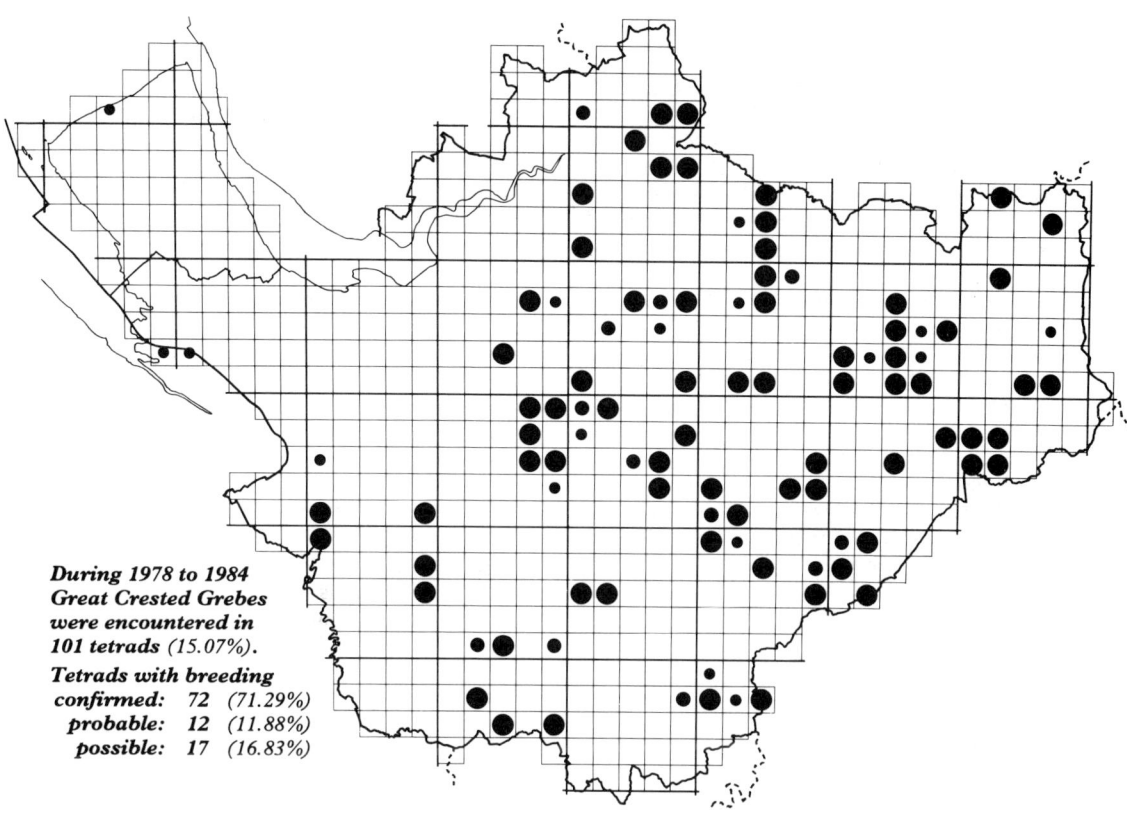

During 1978 to 1984 Great Crested Grebes were encountered in 101 tetrads (15.07%).

Tetrads with breeding
confirmed: 72 (71.29%)
probable: 12 (11.88%)
possible: 17 (16.83%)

national population continued to increase over this period with 6094 birds seen and 6813 estmated, the considerable decrease in Cheshire being unequalled, indeed opposite to the trend elsewhere in the country (Hughes *et al.* 1979). The 1975 *CBR* reported "increasing signs of breeding failure due to disturbance etc.," and increasing pressure from outdoor pursuit groups from the conurbations of Merseyside, Manchester and the Potteries seems the most likely reason for local declines. At Redesmere, where floating and marginal vegetation has been greatly reduced by sailing and angling activites, the species has failed to breed in several recent years. Discarded fishing tackle is a frequent cause of death: of nine young hatched by two pairs at Poynton Pool in 1977 none survived, and several were known to have become entangled. At Marbury Mere, Great Budworth, an oil spillage in the late 1950s severely reduced marginal cover, and nests in the limited reed-beds which remain are often robbed by foxes and children. On larger sheets of water some nests are lost due to wave action in rough weather.

There has been no national census of breeding Great Crested Grebes since 1975 and no county census either. Counts listed in *CBRs* since then give only some indication of the species' success or otherwise. In 1977, breeding success was "generally better" than the previous two years. There were 78 birds on 36 waters in south-east Cheshire in May suggesting a 14% decrease on 1976 but a 32% increase on 1975, whilst at Marbury Mere (Great Budworth) numbers were "well up" with 30 birds and ten young in May-June compared with ten adults and two young in 1976. In the years 1978-1980, minimum counts of 15 pairs (rearing 25 young), 23 pairs (rearing 16+ young) and 21 pairs respectively were recorded in east Cheshire, whilst in south-east Cheshire ten breeding localities were located in 1979 and 17 of the 27 tetrads where birds were present in 1980 held breeding pairs.

Great Crested Grebes breed on most of the larger waters in the county: meres, flashes, ornamental lakes and increasingly on flooded sand-pits (16-19 birds in 1975 as against two in 1965). Birds are generally conspicuous, particularly when displaying, so the map may be taken to represent the breeding distribution very accurately. The scarcity of suitable waters in the west of the county and in Wirral shows up clearly although two summered at Moreton clay-pits in 1983.

Many nests are built in reeds, reedmace or other marginal vegetation, and often the start of nest-building is delayed until such cover is adequately grown. The trailing branches of overhanging willows are a common nest-site at Rostherne and other meres. Other nests are built in the open on emergent mud or, as in the various flashes, on sunken hedges, wrecked canal boats and other relics. Nests on the hill reservoirs tend to be left high and dry as the water level drops in summer. The grebes then find great difficulty in dragging themselves across the mud to and from the nest, and eggs are soon lost or deserted.

The breeding season is prolonged. In 1975, a pair was building at Styperson Pool on 27th February and Boyd[5] had a record of nest-building at Witton on 12th March 1938. Late broods giving their piping hunger cries may be seen following their parents well into October. In recent years such dependent young have been noted at Tatton as late as 14th October 1977, at Radnor Mere on 19th October 1978 and at Farmwood Pool (Chelford) on 14th November 1976.

With 84 tetrads for which confirmed breeding or 'probable' breeding was recorded during the course of the present survey it is unlikely that the county population is more than 100 pairs in any one year.

REFERENCES
Harrisson, T. H. & Hollom, P. A. D. (1932): The Great Crested Grebe Enquiry, 1939. *British Birds* **26**: 62-92.
Hollom, P. A. D. (1936): Report on Great Crested Grebe Sample Count, 1935. *British Birds* **30**: 138-158.
Hollom, P. A. D. (1959): The Great Crested Grebe Sample Census 1946-55. *Bird Study* **6**: 1-7.
Hughes, S. W. M., Bacon, P. & Flegg, J. J. M. (1979): The 1975 census of the Great Crested Grebe in Britain. *Bird Study* **26**: 213-226.
Prestt, I. & Mills, D. H. (1966): A census of the Great Crested Grebe in Britain 1965. *Bird Study* **13**: 163-203.

Slavonian Grebe
Podiceps auritus

THE Slavonian Grebe is restricted as a British breeding bird to Scotland where between 37 and 80 pairs per annum have been recorded since 1971[35, 43]. Birds in breeding plumage have also been recorded on occasions from northern England.

The Slavonian Grebe has very seldom been observed in Cheshire during the summer months. One remained at Marbury Mere, Great Budworth from 6th February until 3rd April 1937[5] by which time it had acquired summer plumage. In 1952 one was seen on Oakmere on 27th April[7], and in 1974 one lingered at Baddiley Mere from 7th March until 29th May.

During the present survey an adult in breeding plumage was found on Ridgegate Reservoir, Langley on 7th April 1980, moving overnight to Trentabank Reservoir where it remained until 12th April. This bird called frequently, which is unusual behaviour outside the breeding season. In 1983, two birds in summer plumage were seen on Marbury Mere, Great Budworth on 23rd April.

Black-necked Grebe
Podiceps nigricollis

THE Black-necked Grebe was unknown as a breeding bird in Britain until early in this century. In September 1928, the hunger cry of a well-grown young bird was heard as it followed an adult on Marbury Mere, Great Budworth. This was in the early days of a general colonisation in Britain as the species extended its range northwestwards across Europe[35]. Black-necked Grebes frequent eutrophic pools with a luxuriant growth of shore plants and submerged and floating aquatic vegetation. It is, therefore, perhaps not too surprising that Cheshire featured in this early colonisation. A pair was again present at Marbury Mere for several weeks in 1937 but without any evidence of breeding, and there were further isolated records here in May 1942 and June 1949[6]. Boyd[5] also referred to single birds, presumably at Marbury, in July 1935 and July 1936, and hoped for the extension of range which seemed to follow shortly after.

In 1938, a pair remained at Rostherne Mere from 15th April until 6th June, one remaining until 12th June, giving rise to suspicion of breeding. In the following year a pair reared two young[24].

In 1941, 1943, 1944 and 1945 a pair nested at Oakmere, and in 1942, a pair, presumably the same birds, nested on another pool in Delamere Forest. In 1953 a pair again nested at Oakmere but the eggs were stolen[7].

The Black-necked Grebe is renowned for temporary colonisation of sites beyond its normal range. A further "invasion" occurred in Cheshire during the present survey with birds being recorded in each breeding season from a total of eight sites throughout the county from 1980 to 1984 when two downy young were seen. A maximum of three adults was seen at any one time and it is probable that the same birds were involved during this period, moving around from site to site. Rostherne Mere NNR and Tatton Mere were particularly favoured after the breeding season.

The Black-necked Grebe is on the list of rare breeding birds in the United Kingdom and is protected under Schedule 1 of the Wildlife & Countryside Act 1981. Until 1980, when a maximum of 21 pairs was recorded in Britain, the population had never exceeded 18 pairs. In 1981, only twelve pairs were recorded but in 1983 there were 32, although this dropped slightly to 28 in 1984. This encouraging position was primarily influenced by one thriving population in northern England. At least two sites, Cheshire included, are threatened by recreational activities.

Cormorant
Phalacrocorax carbo

ALTHOUGH Cormorants rarely breed inland in England, display is often performed in the roost trees at Rostherne Mere in late March and April, and has also been observed at Tatton. Since the 1960s very small numbers of immature birds (up to six in 1984) have spent the summer in the county. In 1974 an adult summered, and in 1981 two adults were seen to display at the Winsford Flashes where they remained into early May.

The origin of the Rostherne birds has aroused controversy over the years. Detailed observations of the roost (JPG) show that prior to spring departure the vast majority of adult birds assume the extensive whiteness on the neck characteristic of the continental race *P. c. sinensis*. Old males of the British race may develop similar markings, but the age-structure of the Rostherne roost would appear to discount this possibility.

Bittern
Botaurus stellaris

BITTERNS nested regularly in the reed-beds fringing the meres until early in the nineteenth century. Coward & Oldham[2] were told by elderly men in many parts of the county that their parents had been familiar with its booming note – hence the dialect names of "Bitter-Bump" or "Bittor". This last name occurs in the Chester plays of the thirteenth and fourteenth centuries. Boyd[6] states that the Bittern was said to have nested in Whitley Reed before it was drained in 1854.

The disappearance of this species in the nineteenth century must presumably have been due to persecution and today there are probably no reed-beds in the county large enough to hold Bitterns.

Grey Heron
Ardea cinerea

HERONS have long been associated with the seclusion of the large private estates. Coward & Oldham[2] noted that "until quite recent times there were heronries on many of the larger estates..." but "owing to... increased fish-preservation and the destruction by prowling gunners..." their numbers had fallen until at the time of writing they knew of only two colonies in the county – at Eaton (near Chester) and at Tabley, a "new" colony founded in 1871. They could, however, list nine colonies which had disappeared for various reasons: tree-felling at Arley and Hooton (the latter was thought to be the oldest in Cheshire and was destroyed to make way for the Manchester Ship Canal); persecution by Rooks and Jackdaws at Arley, Aston (near Frodsham) and at Burton Point; gale damage at Burton Point and Dunham Park (now GMC), where unspecified persecution also took its toll; and complete heronries destroyed at Combermere to protect the fishing interests and at Carden because of the smell of putrid fish refuse. Colonies at Marsh Plantation (Ince) and Oulton Park had been deserted for reasons unknown.

The Eaton and Tabley heronries held 30 and 3 nests respectively in 1874, rising to 78 and 23 in 1907. In 1928 a census of the county showed a minimum of 202 nests, but two heronries were missed. Boyd contributed counts for Cheshire and south Lancashire to national censuses between 1937 and 1953. The precise locations of these colonies are not known, but probably all were within the present recording area. These totals vary between 288 nests in 1937 and 106 in 1947, with other high counts of 282 in 1944 and 279 in 1953 (Lack 1954). Persistent frost locks up the birds' food supply below the ice, resulting in starvation. Tabley held 57 nests in 1939, prior to a severe winter, but only 32 in 1940. Similarly 57 nests in 1946 fell to 29 in 1947[6]. Generally, the population recovers after two or three milder winters. However, after the winter of 1962/63, the worst on record (probably the coldest since 1740), numbers recovered only slowly. In 1962 the county held around 250 pairs, crashing to 93 in 1963.

Birds may occasionally be seen around the colonies during the winter months, but most start to renovate their nests in February. In

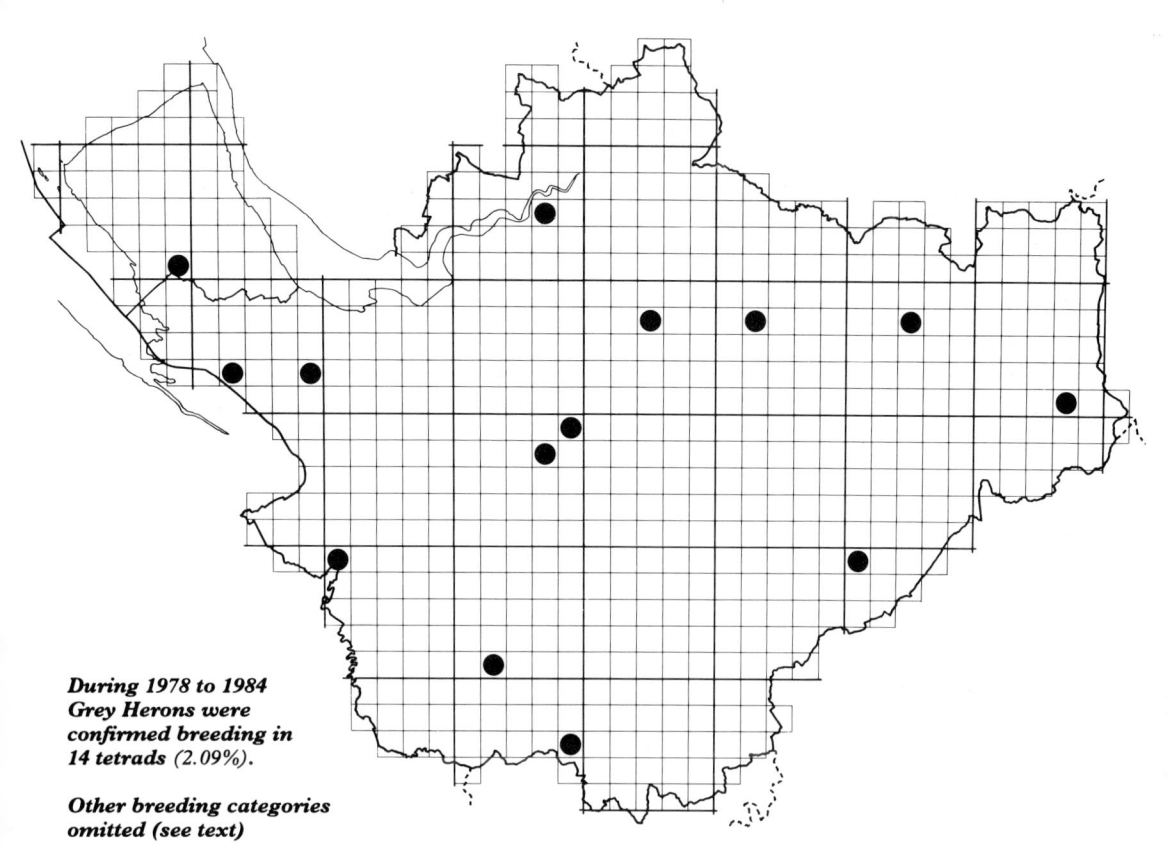

During 1978 to 1984 Grey Herons were confirmed breeding in 14 tetrads (2.09%).

Other breeding categories omitted (see text)

1977, a pair was seen carrying sticks over Mottram St Andrew on 7th March, heading for the heronry at Radnor Mere (Alderley Park) some three miles away. Grey Herons are among the first species to start breeding each year. Some females may be incubating in February in early springs, with the noisy young birds ready to fly by the end of April, although late May is a more usual date in Cheshire. Second broods, or young from late-breeding one-year-old adults, may still be in the nest into August.

During the present survey fourteen heronries were known to exist, with four, including Marbury (Great Budworth – where at least one pair bred in 1928), Nunsmere and Pitts Heath (Runcorn) being founded during the period of the survey. That at Nunsmere seems to have been a temporary offshoot from Oakmere.

Adult birds wander widely from their nests to feed and may be encountered in suitable breeding habitat and in breeding plumage when the nest is in fact miles away. Diurnal roosts are known to be occupied even during the nesting season, as at Pickmere, Witton and Frodsham Marsh. For these reasons a large number of 'possible' breeding records have been omitted from the map. However, pairs will occasionally nest in isolation, as at Elton Hall, Sandbach in 1977 and at Backford, Chester in 1984, and it is possible that one or two such breeding attempts were missed during this survey.

The heronries on the Tabley and Eaton estates remained the major sites in Cheshire. Between 1980 and 1984 they each held between 70 and 80 nests each year. This ranked them in the top ten for England and Wales during this period.

The peak population at Tabley was in 1959 when there were 143 nests. Stafford (1961) described this count as "unusually large for a British heronry – the largest one reported since the annual census began in 1928, and for many years previously". There had also been 111 nests there in 1954, and were again 141 in 1961. During the period 1975-78 inclusive there were just over 100 nests each year and Reynolds (1979) listed it among only six such heronries in England and Wales in 1954 and 1975-77. Tree felling and the cold winter of 1978/79 brought about a significant reduction. By 1979 the heronry had been virtually halved with between 48 and 62 nests[8]. By 1981, there had been a steady increase to 78-80 nests with a slight decline to 74 in 1984. Intermittent tree felling here continued throughout this period probably preventing the number of nests building up to its former high level.

At Eaton, birds moved to the present site about 1920, when the oaks, into which they moved, were about 50 years old. An old story tells of the original heronry being in two huge willows which spanned the Dee at Heronbridge, near Eccleston, in the mid-nineteenth century[2]. By 1941 there were 92 nests and about this time it was apparently the largest heronry in England. However, by 1947, only 27 nests were recorded following the severe winter of 1946/47, with a further decrease to only 19 nests in 1948. Numbers then rose to 67 in 1959, remained more or less stable for four years and then declined again to 22 in 1964 as a result of the 1961/62 and 1962/63 winters. Totals again recovered, reaching 86 by 1971 and maintaining a level of 85-90 until another cold winter brought a drop to 66 in 1979. Since then numbers have stabilised at between 70 and 85 nests.

The population level at some sites varies little from year to year, whilst at others it seems to fluctuate considerably: for example at Combermere the number of nests recorded between 1979 and 1984 were 49, 36, 25, 27 and 30 respectively. On the other hand the heronry at Trentabank Reservoir (Macclesfield Forest) expanded during the course of the survey from a count of two to three nests in 1978 and 1979 to 14 in 1983 and 20-27 in 1985. The overall population in the county in recent years has been fairly stable at between 200 and 250 pairs however, so fluctuations at individual colonies may reflect local redistribution of birds.

REFERENCES
Lack, D, (1954): The stability of the Heron population. *British Birds* **47**: 111-121.
Stafford, J. (1961): The census of heronries 1959. *Bird Study* **8**: 38-42.

Mute Swan
Cygnus olor

THE Mute Swan uses its long neck to browse submerged waterweeds beyond the reach of other surface-feeding waterfowl. Shallow waters with a good growth of aquatic vegetation meet this requirement, such as the shallower of our meres and flashes, ox-bows by the major rivers, large ponds and disused stretches of canals. Swans also need extensive stands of emergent vegetation, typically reeds or reedmace, from which to build their bulky nest. A pair at Radnor Mere felled 100 square metres of reedmace for this purpose. Cheshire

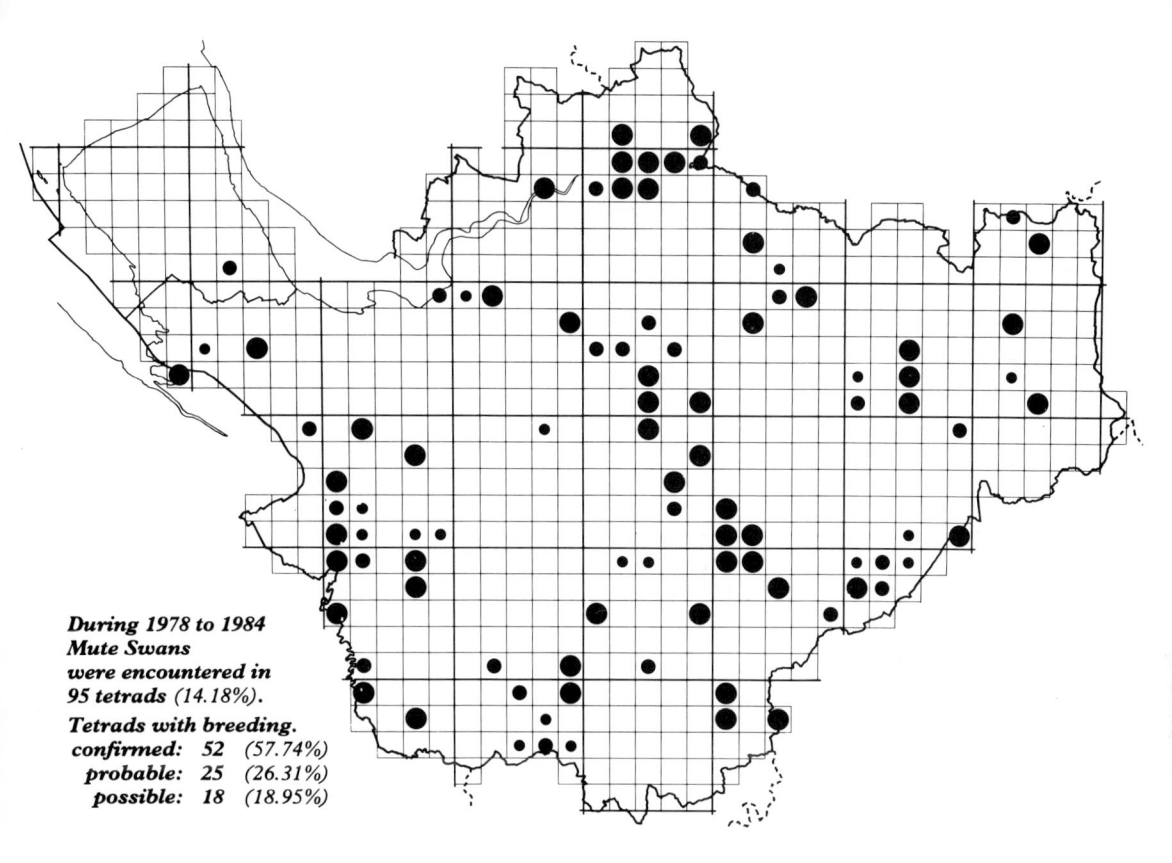

During 1978 to 1984 Mute Swans were encountered in 95 tetrads (14.18%).

Tetrads with breeding.
confirmed: 52 (57.74%)
probable: 25 (26.31%)
possible: 18 (18.95%)

appears particularly well endowed with suitable habitat, although most of the recent sand-quarries have little vegetation around their steep banks and so are of less benefit to the Mute Swan than to some other waterfowl.

Few swans can have been missed during the survey, for this is our largest bird, visible from afar. At a few sites sitting birds were difficult to detect in reed-beds, but territorial behaviour by the off-duty bird aroused suspicions of nesting, and very few nests remained hidden after a brief search. Eggs are laid from the last few days of March, and from the middle of May the appearance of cygnets may provide confirmation, but even if a breeding attempt fails the durable nest may become visible when the vegetation dies down in autumn. A small proportion of records submitted of unconfirmed breeding may have been of "first summer" birds, not yet old enough to breed.

Parslow[31] refers to an increase in numbers almost everywhere in Britain from the end of the nineteenth century, continuing until 1959 when analysis of winter counts first suggested an end to this trend. A census conducted by the BTO in 1978 showed a decline in the national population of between 8% and 15% since a similar survey in 1955. A further census in 1983 suggested a recovery of about 7% since 1978. In Cheshire however the species had been in unrelieved decline for perhaps thirty years up until the end of this survey.

Coward & Oldham[2] found the Mute Swan "in a semi-domesticated condition on most of the meres and ornamental waters", and Boyd noted in his diary for 12th August 1939 that "they seem to be increasing in numbers yearly". He considered that swans were becoming far too plentiful in that they usurped the food and territory of "other and more interesting fowl". Hardy[4] classed them as common on most large fresh waters in the Liverpool area.

The 1955 BTO census revealed 80 breeding pairs and 246 non-breeding birds in the old county of Cheshire, although coverage was estimated to be only 75% and some birds may therefore have been missed. By 1961 a second BTO census showed only 47 pairs. While there is a suggestion that coverage on this occasion was somewhat poorer than in 1955, the drop over six years must have reflected a real trend. The third BTO census in 1978 recorded a continuing decline to a minimum of 20 breeding pairs within the old county boundaries. This survey was admittedly incomplete, but the national organiser's assessment that at least 50 pairs had been missed (Ogilvie 1981) conflicts with the county recorder's consideration that the Cheshire population was unlikely to exceed 30 pairs in that year[8]. With nearly complete coverage during the latest national census in 1983, the county could be stated fairly accurately to hold 29 nesting and a further five territorial pairs. 18 pairs were recorded as raising broods, 66 cygnets having been hatched by 17 of these pairs.

Winter flocks and gatherings of non-breeding birds declined in parallel with the breeding population. On 12th August 1939 Boyd counted flocks of 47 and 113 "adult" birds on just two waters as he travelled, presumably from Northwich, to Staffordshire. A census by the MNA in 1954 showed "150 on the Dee at Chester Weir, 70 in the Burton/Shotton area (partly Flintshire), and 40 in the Weaver estuary"[7], very similar to the total of 246 non-breeders recorded in the 1955 census. Also in 1955 a count of 200 birds was made on the Dee at Chester in August – the largest single gathering on record anywhere in the county[20]. Counts in excess of 30 birds were not infrequent at the Sandbach Flashes into the early 1970s, and 50 were on the Weaver Bend in November 1969, but in the Ringing Report for 1970[8] D. Mirams commented on a reduction in numbers on the Dee at Chester over the past two years. Subsequently numbers fell away rapidly and by the 1980s large gatherings were only likely when frosts forced birds to gather on waters that remained open. Thus up to 46 were counted on Winsford Bottom Flash in January 1982. However, cumulative totals on the various flashes at Sandbach still exceed 20 on occasions in winter.

The decline of the Mute Swan to become one of the rarest breeding birds in Cheshire went against the national trend which shows only a slight decrease since 1955. Reasons for the local decline appear complex. There has been some deterioration of habitat: formerly derelict canal systems have been cleared of waterweeds following the growth in popularity of canal cruising, and the supply of reeds for nest material has been reduced on certain meres by the grazing of moulting Canada Geese, or by anglers making space for a clean cast. These changes alone can account for only a small part of the decline. Three major factors are undoubtedly lead-poisoning, vandalism and collisions by flying swans with overhead power-lines.

Lead-poisoning killed many swans in Cheshire, birds having often been seen showing the symptomatic crooked neck. The sandy bottoms of our rivers, meres and flashes contain relatively little grit, making the ingestion of lead pellets spilt by anglers more likely – swans habitually take up grit to help grind their vegetable food. There is also movement between Cheshire and the River Trent where many swans are known to have died from lead-poisoning.

It seems illogical to assume that power-lines were to blame for a decline in Cheshire when they did not bring about any marked deterioration in other regions where pylons are equally numerous. Nevertheless ringing recoveries showed these collisions to be the commonest cause of death in the county. It may be that birds with a low level of lead-poisoning are more prone to such accidents.

The third major factor is the disturbing increase of vandalism to nests, with reports of eggs being taken, nests burnt, and sitting adults stoned or shot at with air-guns or crossbows. The early disappearance from Wirral and areas adjacent to the suburbs south of Manchester is notable in this respect.

Many swans are accustomed to being fed by people and persistently approach anglers with occasionally disastrous results. Swans and cygnets become entangled in fishing line, in some cases a hook lodging in the bill, and starvation results. Other causes of breeding failure include nests washed out by rising water and the trampling of young birds by cattle. Such losses are inevitable in any population and cannot be classed as causes of the decline.

The map, with confirmed breeding in 52 tetrads, clearly overstates the strength of the species by 1984, although most of these and other sites remained suitable for reoccupation. A full survey in 1985 found 13 pairs breeding in Cheshire (none in Wirral) with non-breeding flocks of 14 at Sandbach in June and 16 at Budworth Mere in August (Elphick 1985).

Footnote: The replacement of lead fishing weights by non-toxic alternatives was followed, in the late 1980s, by a considerable improvement in the status of Cheshire swans.

REFERENCES

Elphick, D. (1985): The Status of the Mute Swan in Cheshire and Wirral. *CBR* 1985, 83-94.

Ogilvie, M. A. (1981): The Mute Swan in Britain, 1978. *Bird Study* **28**: 87-106.

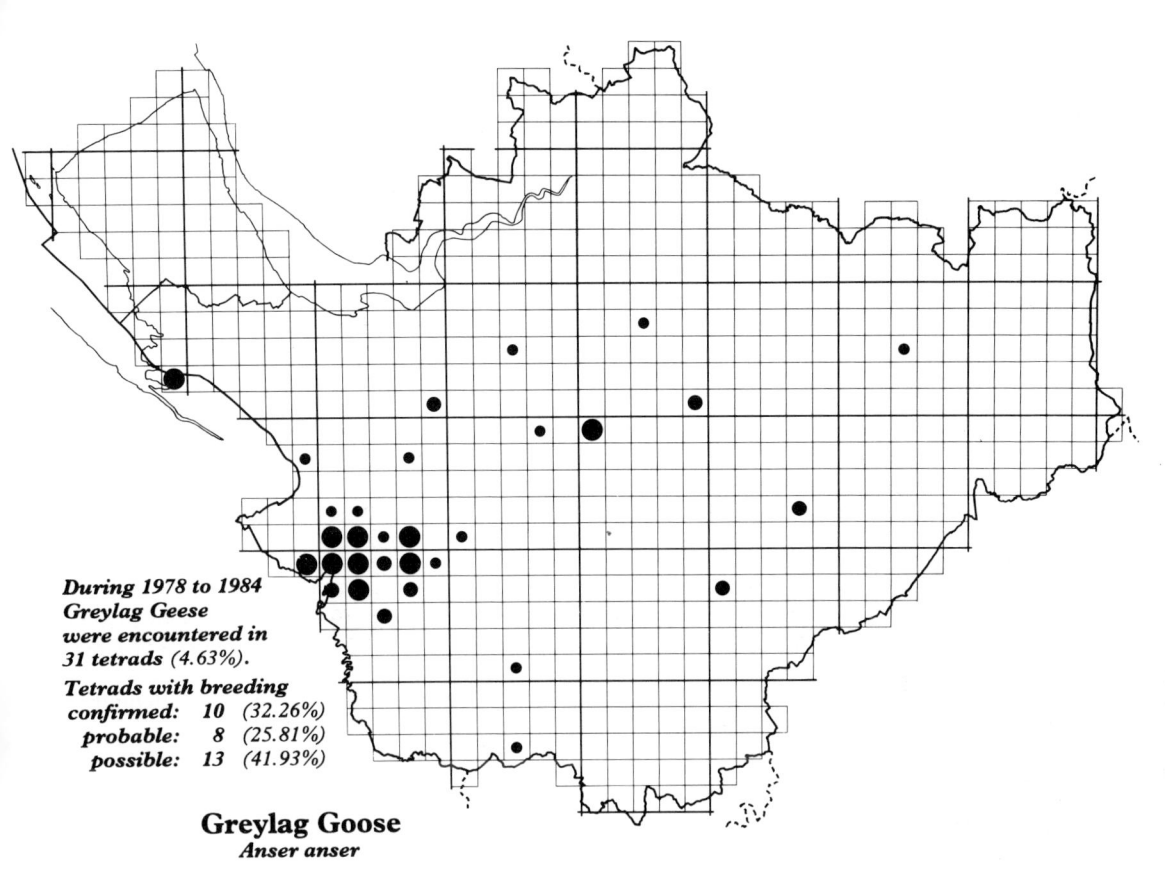

During 1978 to 1984 Greylag Geese were encountered in 31 tetrads (4.63%).

Tetrads with breeding
  confirmed: 10 (32.26%)
  probable: 8 (25.81%)
  possible: 13 (41.93%)

Greylag Goose
Anser anser

WILD British Greylags now breed only in the Outer Hebrides and adjacent parts of north-western Scotland, although before the big drainage schemes of the early nineteenth century birds bred as far south as the fens of Cambridgeshire and Lincolnshire. Indeed, one possible derivation of the name Greylag is the grey goose that lagged behind when other species returned north[32]. Around 1930, eggs and young were transported to estates in south-western Scotland. This new population flourished, and, by 1961, surplus eggs from this area were being used by wildfowlers to found new colonies in England and Wales. By the time of the BTO Atlas (1968-72) feral stocks had become established in many parts of Britain and to some extent in Ireland, such that the breeding range was then wider than at any time in recorded history[32].

In Cheshire, feral Greylags began to be recorded erratically from the 1960s. In 1965, a party of five remained at Huntington from April into June, one was with Canadas in Eaton Park on 12th June and two were at Redesmere in May. No records for the county were entered into the BTO Atlas for the years 1968-72, although in 1970 a pair was at Sandbach from 18th April into May, in 1971 two were at Frodsham from 8th to 15th May, and in 1972 five were at Tatton on 23rd April.

There is a regrettable tendency for birdwatchers to ignore introduced species during the early stages of their colonisations (as with the Canada Goose, Mandarin, and Ruddy Duck), and although the 1975 *CBR* stated that Greylags were then being recorded fairly frequently and might well be breeding locally, no confirmation of breeding was obtained until 1979 when three pairs, out of a total of ten present, were known to have bred on the Eaton estate at Chester. Table 5 (overleaf) shows the maximum counts of birds from the Eaton/Aldford area since 1972; numbers have fluctuated at times but gradually increased to a peak in 1981. In 1980, 13 breeding sites were recorded in this area, all but one of them on the Eaton estate, and, in 1984, there were between ten and 15 breeding pairs.

It is normal for non-breeding geese to move away from the species' breeding areas to moult, and a moulting flock present at Rostherne in recent years has built up in parallel to the Eaton birds. In the summer of 1969 two were occasionally seen and, in

Table 5: Maximum counts of Greylag Geese from the Eaton estate or adjacent area and Rostherne Mere.

	1972	73	74	75	76	77	78	79	80	81	82	83	84
Eaton Estate	8	?	13	28	20	38	15	50	49	55	35	35-40	48
Rostherne Mere	0	0	6	5	22	18	26	29	23	28	28	35	37

1970, two remained to moult.

Table 5 also shows the maximum number of moulting birds at Rostherne since 1972. The remains of a bird ringed at Rostherne in July 1977 were found at Aldford, just outside the Eaton estate, in March 1978. However, another bird ringed at Rostherne in July 1977 and retrapped there the following July, was found shot on Anglesey in January 1984.

In 1980, it transpired that a flock of up to 80 birds was present at Cholmondeley throughout the year with up to 42 in 1981 and 34 in January 1983. Whether these were the same birds as those on the Eaton estate is unclear, but despite the presence of such a flock, no breeding records have yet been reported from the area.

A pair reared two young on a sand-pit at Marton in 1980, and four moulted at Oulton Park Lake in 1981. Although odd birds now occur with Canada Geese throughout the county, it is doubtful whether they will be able to colonise waters further east where there is already severe competition for nest-sites amongst the larger Canada Geese. In recent years, however, odd Greylags have paired with Canadas, and since 1978 when a mixed pair was present at Tabley, hybrids between the two species have occurred with increasing frequency. In 1980, a similar pair reared three young at Tatton and, in 1981, five hybrids were reared at Tabley. Two records, both in eastern Cheshire, of confirmed breeding by Greylags of feral or domestic origin with Canadas have been excluded from the map.

A very few records have been received of birds of the pale, pink-billed eastern race *A. a. rubrirostris*, including a pair at Billinge Green (SJ67V) in April 1983, but there is no record to date of birds of this race attempting to breed in Cheshire or Wirral.

The population of Greylags within the county by 1984 was unlikely greatly to have exceeded the largest single flock reported to date, namely 80 birds, but a thorough survey of breeding pairs and moulting birds is desirable.

Canada Goose
Branta canadensis

THE Canada Goose was introduced to Britain from North America during the seventeenth century as an ornamental bird to grace the parks and estates of wealthy landowners[32]. Coward[11] recorded that "at one time prior to 1914, flocks of 200-300 birds had visited Rostherne Mere, suggesting that such flocks had long been a feature in Cheshire. However, even as late as 1910, when Coward & Oldham wrote that the species had been long naturalised in England, they did not regard Canadas as truly wild birds. It was not until 1938, when Witherby proposed to include the species in his new *Handbook,* that the Canada Goose achieved some respectability.

Boyd[5] referred to large gaggles of Canada Geese occasionally visiting the Cheshire meres for a day or two before disappearing as suddenly as they came. In his country diary for June 1935, he mentioned a large free-flying flock of at least 240 birds based on the lake at Crewe Hall and nesting on a wooded islet there. He added that other birds nested regularly on several of the other meres in the county. Crewe Hall lake was drained in 1945, after which date Tabley became the species' headquarters with a flock of up to 80 birds[6]. Bell[9] considered the principal flocks in the late 1950s/early 1960s to be based at Tabley and Combermere, with maximum counts of between 350 and 400 birds at each locality; there were also smaller flocks at Cholmondeley, Doddington, Tatton and elsewhere. During the early 1960s there was some evidence of an increase in the size of the

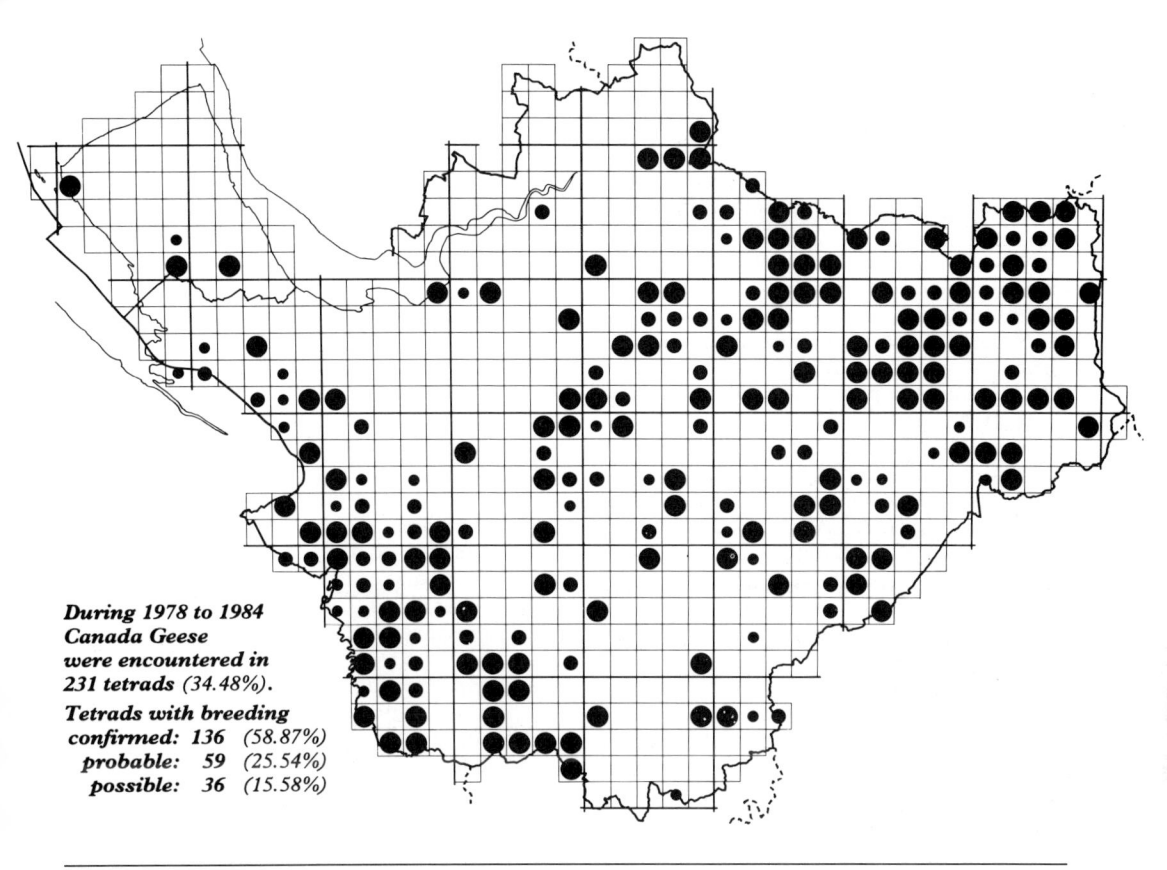

During 1978 to 1984 Canada Geese were encountered in 231 tetrads (34.48%).
Tetrads with breeding
confirmed: 136 *(58.87%)*
probable: 59 *(25.54%)*
possible: 36 *(15.58%)*

Table 6: *Numbers of Canada Geese counted during the National Censuses of 1953, 1967-69 and 1976.*

		1953	1967-9	1976
Shropshire(North) & Cheshire (South)	1	506-598	490	+1589
Cheshire (North)	2	165-219	550	} 666*
Cheshire (West)	3	-	110	

Notes:
1 Includes Combermere, Barmere, etc.
2 Includes Tabley, Tatton, Rostherne, etc.
3 Includes Aldford, Eaton Hall, etc.
* North & West Cheshire combined.

flocks and by 1966 there were at least 500 birds at Tatton and 430 at Bar Mere. The following year up to 600 or more were counted at Tabley.

Full censuses of this species are best conducted in late June or early July when the adults are temporarily flightless during the moult and are accompanied by their growing young. National surveys in 1953, 1967-9 and 1976 suggested that a large number of more or less clearly delimited sub-populations existed scattered throughout Britain (Blurton Jones 1956; Ogilvie 1969, 1977). Table 6 shows figures for the three such sub-populations in Cheshire.

During the 1950s and 1960s farmers throughout Britain began to complain that Canada Geese were increasingly grazing cereal and grass crops and puddling the fields

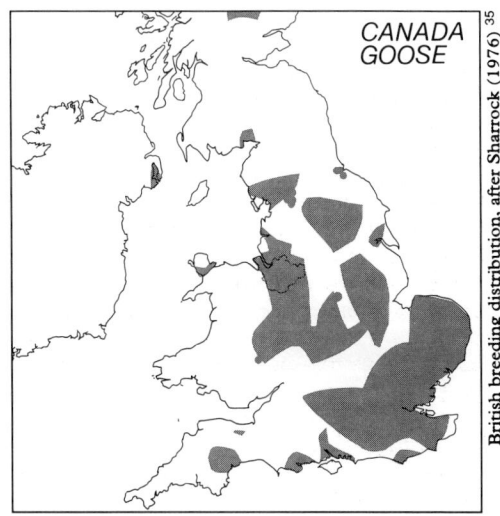

CANADA GOOSE

British breeding distribution, after Sharrock (1976) [35]

in wet weather[32]. In an attempt to reduce such damage birds were transported away from their discrete sub-populations to gooseless areas where landowners were pleased to have them on their estates. Birds from Tabley were moved out of the county at this time to try to reduce numbers there, but there is no evidence that birds were liberated at new Cheshire sites.

Nevertheless by the early 1970s any distinction which might have existed between the Cheshire sub-populations began to blur. Pairs at Nunsmere, Merlewood Pool and Oakmere Hall Farm in mid-Cheshire in 1970[8] and a nesting attempt at Sandbach Flashes in 1972 are further evidence of the species' spread in Cheshire. At Sandbach, which "separated" the "North" and "South" Cheshire sub-populations and from where geese had been absent since the draining of Crewe Hall lake in 1945, birds have nested annually since 1977[19,24].

There is a strong suspicion that the Wildfowl Trust census of 1976 underestimated the number of birds in "North" Cheshire. In 1978, counts by JPG and B. Armitt in the seven 10-km squares (or part-squares) constituting the broadly similar East Cheshire recording area revealed 1110 birds including 335 young. In 1979, a poorer breeding season, 919 (109 young) were counted, and, in 1981, there were 1037 (269 young). Subsequently, no complete moult counts have been made and a comment in the 1982 *CBR* suggesting a 50% decline for east Cheshire in fact referred to a census of the Knutsford 10-km squares (SJ 77 & 78) only (B. Armitt), these being found to contain 527 birds as against 637 (148 young) in 1978 and 594 (165 young) in 1981.

Any decline in the Knutsford area, due partly perhaps to continued public pressure in Tatton Park, has been compensated for by increases elsewhere. In 1981 JPG found 176 (29 young) in mid-Cheshire. In the eastern hills (SJ 96, 97 & 98), where nesting probably dates from the mid-1970s, the summer moult population increased from 65 (27 young) in 1978 to 124 (30 young) in 1981.

No recent moult counts are available from the south Cheshire meres, or from the Eaton estate, although up to 20 pairs may breed there[25]. Breeding now occurs regularly at Thornton Manor Lake in Wirral, with 13 pairs in 1983, and the map also shows a recent spread into north Cheshire.

As early as February pairs split off from the winter flocks to establish territories. Initially, the more favoured sites such as islands in lakes, the marshy fringes of meres, flashes and sand-quarries are occupied. As spring progresses pairs settle in less favourable habitats, for example marl-pits or woodland within walking distance of a mere, and marshy ground by rivers. Undoubtedly, the increasing number of sand-quarries in Cheshire over the past 20 years or so has provided many additional nesting sites.

Some ganders are very aggressive in defence of the nest and will even fly at human intruders, but many eggs are lost to predators, especially foxes. At Radnor Mere seven out of ten nests were robbed in this way in 1981 and eight out of ten in 1982. On hatching the goslings are led to the safety of a mere or other large water body. A number of smaller waters, some less than half an acre in extent, regularly hold broods, but in such cases overhanging trees or shrubby islands provide cover. The average date for the first appearance of goslings at Radnor Mere between 1974 and 1983 was 9th May, the earliest being 2nd May. A late brood of seven young, not yet fully grown, was at Watch Lane Flash, Sandbach on 15th August 1981. During the moult period the geese are reluctant to wander far from water and swim around the shoreline browsing suitable vegetation. It is at this time that damage to reed-beds may occur[21], and such unlikely foods as horse chestnut leaves may be taken.

There is evidence from ringing that some of our non-breeding birds may now be involved in the developing moult-migration from various English counties to the Beauly Firth, Highland, Scotland. A synchronised winter census and thorough count of moulting birds would be useful to assess the county's breeding population, which appears to lie in the range 600-700 pairs.

REFERENCES

Blurton Jones, N. G. (1956): Census of breeding Canada Geese 1953. *Bird Study* 3: 153-70.
Ogilvie, M. A. (1969): The status of the Canada Goose in Britain 1967-69. *Wildfowl* 20: 79-85.
Ogilvie, M. A. (1977): The numbers of Canada Geese in Britain, 1976. *Wildfowl* 28: 27-34.

Barnacle Goose
Branta leucopsis

NO truly wild Barnacle Geese have ever been known to breed in the British Isles. The spate of records in recent years is attributable to the increased popularity of wildfowl collections, including the Wildfowl Trust centre at Martin Mere, from which birds periodically fly free. Since 1971, small numbers have appeared annually in Cheshire amongst flocks of Canada Geese. In April 1982 a party of eight was recorded at Capesthorne and Rostherne Mere. Several of these bore Wildfowl Trust rings.

In 1976, a female Barnacle Goose paired with a male Canada Goose at Tatton Park and laid at least one egg, but this was subsequently found broken in the nest. The following year three mixed Barnacle/Canada pairs were present in the park but did not nest. In 1978, a Barnacle gander stood guard by a sitting female Canada for at least ten days before the eggs disappeared. A pair of Barnacles roaming the eastern meres in April of that year settled at Radnor Mere and successfully reared a gosling. One of these birds bore a Wildfowl Trust ring. In 1980, a Barnacle sitting on an empty nest at Radnor was thought to be paired to a Canada. Subsequent *CBRs* have paid little attention to this species, although the first successful hybrid pairing is known to have taken place at Arclid sand-quarry in 1982. This remained the only such record up until the end of the survey despite a statement in *CBR* for 1983 implying otherwise.

Shelduck
Tadorna tadorna

GENERALLY regarded as a coastal species, the Shelduck has long nested around the estuaries of the Dee and Mersey. During the 1970s a small population established itself in the centre of the county. The two main habitat requirements appear to be extensive mudflats or shallow water for feeding, and holes for nesting such as rabbit burrows, holes under tree-roots, amongst rubble, and so on.

Coward & Oldham[2] noted that in the middle of the nineteenth century Shelducks bred "along the whole range of sandhills on the Cheshire shore". Subsequently, these nesting haunts became almost deserted but, by 1910, Shelducks were increasing at a remarkable rate, perhaps largely due to the Wild Birds Protection Acts of 1880 and 1895, and to protection in the Heligoland Bight (German Waddensee) where most of the north-western European population concentrates for the annual moult. At that time, Shelducks were probably as numerous as they had ever been in the county, despite the development of the former sandhill areas of Wirral. Along the Mersey, Shelducks nested in the embankments of the Manchester Ship Canal throughout the stretch from Eastham to Frodsham, with many pairs in the spoil-heap at Mount Manisty.

Breeding Shelducks are unlikely to have been overlooked during the present survey, but many observers were probably unaware of the bird's status inland. Prospecting pairs at inland waters in spring may have been dismissed as casual wanderers and not reported, although since the early 1970s there has been a tendency for birds to appear by meres and flashes between March and May. Inland breeding is not unprecedented. During the period 1900-10, Coward & Oldham[2] cited several instances of birds being seen in spring away from the estuaries, particularly in the low-lying area between Chester and the River Mersey. In 1907, twelve young were reared at Catton Hall, along the Weaver upstream from Frodsham. Between then and 1910 a few pairs nested on the Helsby and Overton Hills and on several occasions the ducklings were captured as the parents led them down to the estuary. Further inland there was circumstantial evidence of breeding at Oakmere in 1900. In April 1907 Coward saw two pairs displaying there, and Hardy[4] reported nesting in 1909-10. Records of occasional birds elsewhere at this time reflected the increasing numbers at the coast. A pair bred at Walton, Warrington in 1955, and, in 1957, a female and five young were recorded on the Dee at Chester, probably hatched within the city boundary[20]. Apart from these no further records of inland breeding are known until the recent influx.

In the early 1970s birds began to appear inland in spring with increasing frequency. In 1973, a pair remained at Rostherne Mere from late April until 22nd May and in 1974 a pair was present from 9th to 17th May. In 1974 five were at Kingsley on 12th June, and from that year onwards birds lingered around the Sandbach Flashes each spring until, in 1977, a brood of 13 young was hatched in late June, followed by a second brood of eight which flew in early August. In 1978 and 1979, young were reared at both Sandbach and Doddington, and in 1980 a total of 31 young fledged from these two waters. Birds, possibly prospecting for new breeding sites, were seen at Great Budworth and Lostock in May 1978, at Tatton, Rostherne and Redesmere in April 1979 and at Farmwood Pool in 1980. Subsequently, pairs have bred successfully at Combermere, Nantwich, Winsford, Marston, and have attempted to breed at Acton Bridge.

The principal breeding population remains in the Dee estuary however. Here the young birds gather into creches which are supervised by just a handful of adult, or possibly non-breeding birds, which remain to moult on the breeding grounds while the

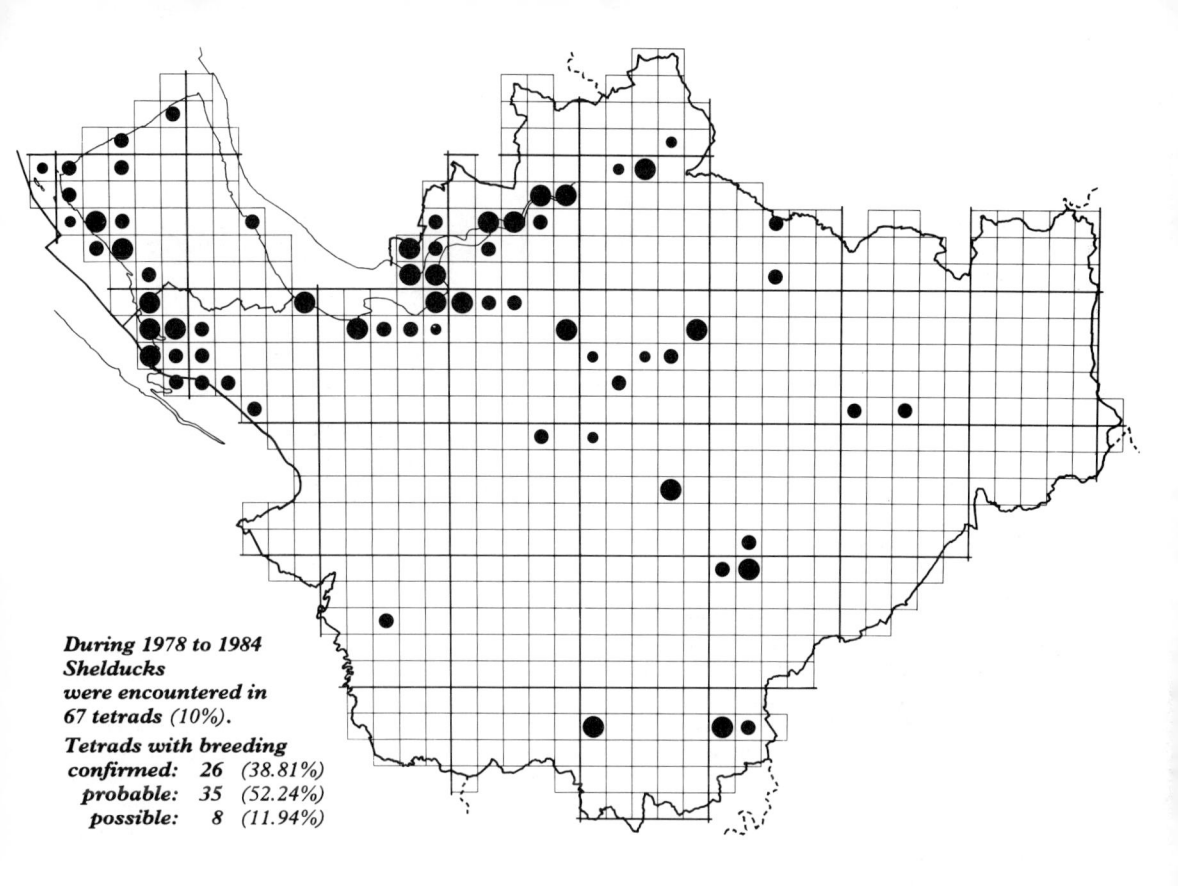

During 1978 to 1984 Shelducks were encountered in 67 tetrads (10%).

Tetrads with breeding
confirmed: 26 *(38.81%)*
probable: 35 *(52.24%)*
possible: 8 *(11.94%)*

majority of adults fly to Heligoland to moult. In July 1973, 120 young were in the Dee off Heswall and the 1979 *CBR* stated that the Shelduck "remains a common breeding species all along the Dee estuary from Red Rocks dunes to Burton Point". In the Mersey estuary 21 young were seen in June 1979 and 33 in August 1981. Breeding occurs regularly upstream as far as Warrington, and in 1978 a pair was seen with four young at the Woolston Eyes.

Where more than an odd pair of Shelducks are present the drakes display noisily and aggressively, and later in the spring the females can be watched inspecting rabbit burrows and other potential nest-sites, sometimes to be followed out by an indignant rabbit! Once the female is incubating, the drake maintains watch over his feeding territory, driving off ducks of any species which attempt to land. Three inland breeding records were first suspected when solitary drakes were seen driving off Mallards, Herons and other water birds, and subsequent watching revealed the female feeding busily while the drake stood guard and later escorted her on the flight back to the nest.

A suggestion that the present inland breeding population originated from a wildfowl collection near Crewe remains unsubstantiated, and whether feral or not they have conformed to the migration pattern of wild birds since before the first recent breeding record. The Wirral and Cheshire breeding population may lie in the range 25-50 pairs.

Mandarin Duck
Aix galericulata

MANDARIN Ducks were first brought to Britain from China or adjacent parts of eastern Asia in 1747 and first bred in captivity here in 1834. In 1971, the species was admitted to the official British and Irish checklist, primarily on the strength of a flourishing population in Surrey and Berkshire, stemming from birds released in 1928 and subsequently. As soon as it became acceptable to regard Mandarins as truly "wild" birds Cheshire bird-watchers took notice and the first mention in *CBR* soon followed (1971), referring to a colony at Eaton Hall as "thriving and increasing". In fact birds were being reared specifically at the request of the then Duke of Westminster.

The history of this colony prior to the 1950s is unknown, although Boyd[5] refers to a bird shot in Delamere in 1943 which "had obviously escaped from captivity – probably from a place on the Welsh border where a number of unpinioned birds are kept". This may have been Walcot Hall, Shropshire, where a population which numbered about a hundred free-flying birds in 1939 was allowed to disperse during the war. Some of these were known to have reached north Wales and Lancashire[37]. In its natural range the species is dispersive and migratory and although most movements in Britain are relatively local there are a few long-distance recoveries of birds originally ringed in captivity.

In the 1950s, a flock of free-flying Mandarins was kept at Bruera, near Eaton Hall, as part of an extensive wildfowl collection (G. Williams pers. comm.). Ten or twelve birds were released each year until 1970, at which time there was a flock of 52 free-flying birds. In the 1950s and 1960s the Eaton gamekeepers reported small numbers of pairs regularly breeding on the estate. The 1975 *CBR* mentioned irregular sightings and suggested a decline. There were very few reports over the next few years.

Pair formation by adults starts in September and continues through the winter, with most birds paired by February. During the breeding season, Mandarins become very secretive, and the discovery of a tree-hole nest is usually by accident. The Eaton estate with its mature, open broad-leaved woodland containing secluded lakes typifies the preferred habitat, although broods have also been seen on quiet streams in adjacent farmland and birds are occasionally reported nesting some way from any of these waters. Oaks, sweet chestnuts and beech trees are particularly important in providing acorns, nuts and mast which are the staple winter foods[35].

During the present survey some upturn in records has become apparent, but this is probably due solely to increased observer activity. In fact, when, in 1981, two broods were seen in the Eaton area, these were the first records confirmed by bird-watchers of Mandarins breeding in the wild in Cheshire. Two broods were also located in the Eaton area in 1982 and three in 1983 although it is thought that the total population may have been between five and ten pairs (D. Goff pers. comm.). It has become apparent however that the population owes its survival to continued releases of captive-reared birds. Groups of 30 and 52 birds were recorded during this period.

Away from the Aldford/Eaton area occurrences are erratic but the records, in 1979, of two males and a female at Petty Pool

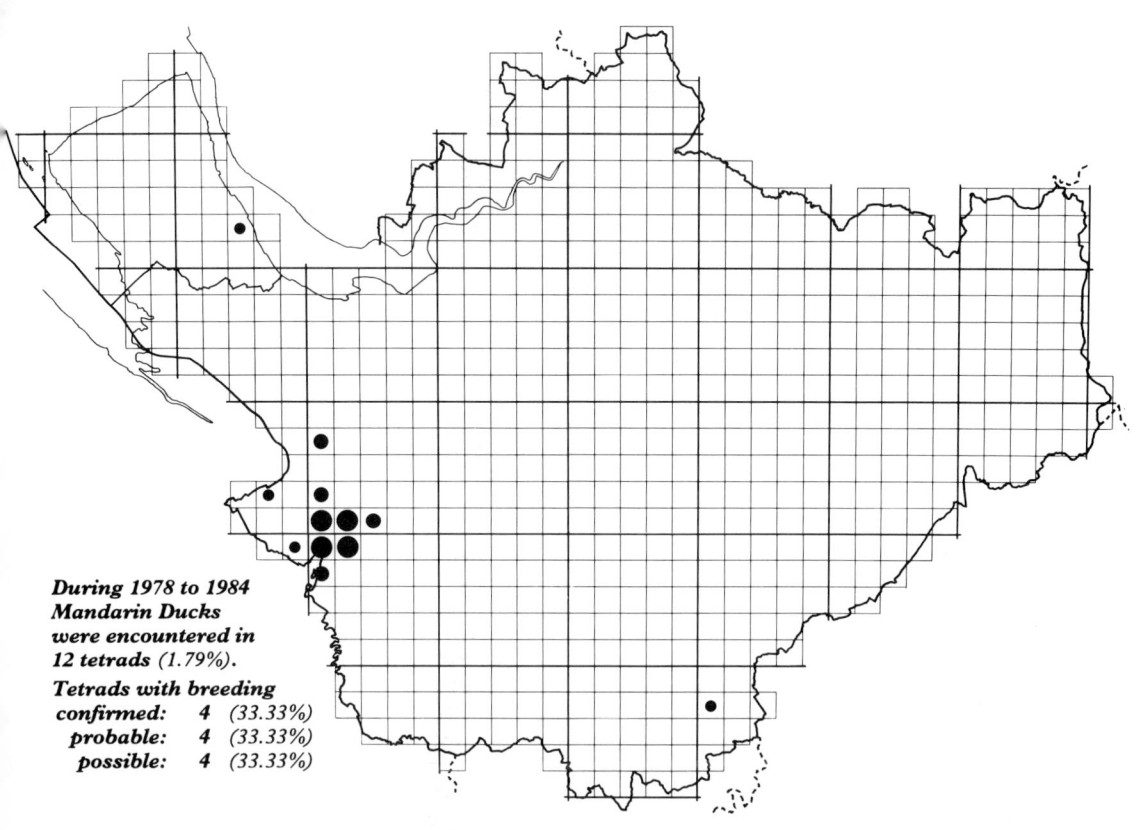

During 1978 to 1984 Mandarin Ducks were encountered in 12 tetrads (1.79%).

Tetrads with breeding
confirmed:	4	*(33.33%)*
probable:	4	*(33.33%)*
possible:	4	*(33.33%)*

in April and a pair reported occasionally at Hale Decoy in April and May, suggest that breeding could occur elsewhere. Also in 1979, it was reported that three or more young birds had flown from a small private collection in Wirral in the previous two years. Elsewhere in Britain breeding birds are generally associated with waterfowl collections and this is the likely source of most sightings of birds in Cheshire away from the Eaton estate. Unlike the Chinese and Japanese populations, British birds are steadily increasing. Indeed they may now form the largest population of Mandarins in the world (Davies 1985). Nevertheless it seems unwise to regard the Mandarin as a permanent addition to the Cheshire avifauna.

REFERENCE

Davies, A. (1985): The British Mandarins - Outstripping the Ancestors? *BTO News* **136**: 12.

Wigeon
Anas penelope

THIS northern duck first bred in Scotland in 1834, that country becoming fairly widely colonised over subsequent decades. By the turn of the century breeding had begun in northern England. Southward expansion did not continue, however. Indeed by 1950 some local reversals began to occur[35].

Brockholes[2] reported two instances of Wigeons rearing young near Puddington, evidently in 1862 and 1863, although Coward & Oldham[2] assumed that these had been escaped birds. They knew of no other breeding records although a drake was present on Redesmere between 30th May and 8th June 1900. Hardy[4] refers to "an old Cheshire nesting report" given in Hartert's *British Birds Magazine* but adds no detail. Similarly, Boyd[6] knew of occasional summer records, notably a pair which remained at Tabley from 10th May to 12th June 1947, but could give no breeding records.

Wigeons prefer a well concealed site for nesting, generally, although not exclusively, in upland areas. Thick cover under overhanging vegetation, grassy tussocks or scrub, and usually near water is preferred[36]. Over the last 20 years or so summer records of very small numbers, mostly drakes, have increased in the county. In 1970, two drakes and one duck spent the summer on the flooded sludge-pools at Frodsham and, in 1971, at least five birds were present there. In 1980, three drakes were present in suitable nesting habitat in a flooded sludge-pool at Woolston although no females were seen and the drakes disappeared at the end of May. Otherwise, the only suspicion of nesting has been on Burton Marsh where, in 1973, three pairs remained until mid-May, when courtship and display-flighting were observed[8], and on Railway Flash, Sandbach in 1976, when a pair stayed into June.

Gadwall
Anas strepera

UNTIL recently the Gadwall, which prefers more southerly and easterly climes than most other species of wildfowl, was a scarce visitor to Cheshire and Wirral. Coward & Oldham knew of only two records prior to 1910 and Boyd[5] could add only one more, in 1914, prior to 1934. All were from the Dee estuary. On 1st March 1934, Boyd[6], who had regularly visited many of the Cheshire meres and flashes over many years, came across a pair on Witton Flashes. He noted that a few years previously a number of birds had arrived in Staffordshire and it was suspected that one pair had bred. In Cheshire, up to three birds were seen in spring at Tabley from 1943 to 1947 and Boyd observed that, like other species of ducks, the Gadwall seemed to be extending its range. Bell[7] traced fewer than 70 reports up to 1961, most of these occurring after 1944, generally in April or the September/October period.

Indeed, the Gadwall did not breed in Britain before about 1850 when a pair of migrants, decoyed in Norfolk, were pinioned and turned down in the Brecks[35]. A substantial population built up there probably aided by continental winter visitors being induced by the feral flock to settle and breed. However, their spread away from Breckland was slow and it was not until the 1950s that the species could be said to be widespread throughout East Anglia. Other releases took place as far apart as the Scilly Isles and the Lake District, and there is little doubt that the present British stock is essentially feral in origin (Fox 1988).

References in *CBRs* from 1964 onwards show a marked increase in occurrences particularly from 1970 onwards. This may have been influenced by releases by the Wildfowler's Association of Great Britain and Ireland (WAGBI) although the documented record shows that the nearest releases to Cheshire were Leicestershire (1963 & 1964), Radnor (1964) and Cumbria (1968-1973) (A. D. Fox *in litt.*). In 1976 the editor of the *CBR* referred to the influence of the Wildfowl Trust's new reserve at Martin Mere, Lancashire, where a feral population of Gadwalls was becoming established.

In 1980, four pairs settled in a reedy corner of Budworth Mere in late April, one pair remained until 11th May and display was noted on a number of occasions. The same year two or three birds remained at Woolston from 9th May until August. In 1981, up to four pairs displayed at this latter site in spring and at least one pair nested, rearing one young – the first occasion on which breeding was confirmed in Cheshire. The following year a brood of five ducklings was seen there; in 1983, no young were seen although a pair summered; and in 1984 two pairs hatched a total of 15 young. In retrospect it seems possible that breeding may have been attempted in 1970 when one or two birds lingered on the flooded sludge-pools at Frodsham throughout the summer.

REFERENCE

Fox, A. D. (1988): Breeding status of the Gadwall in Britain and Ireland. *British Birds* 81: 51-66.

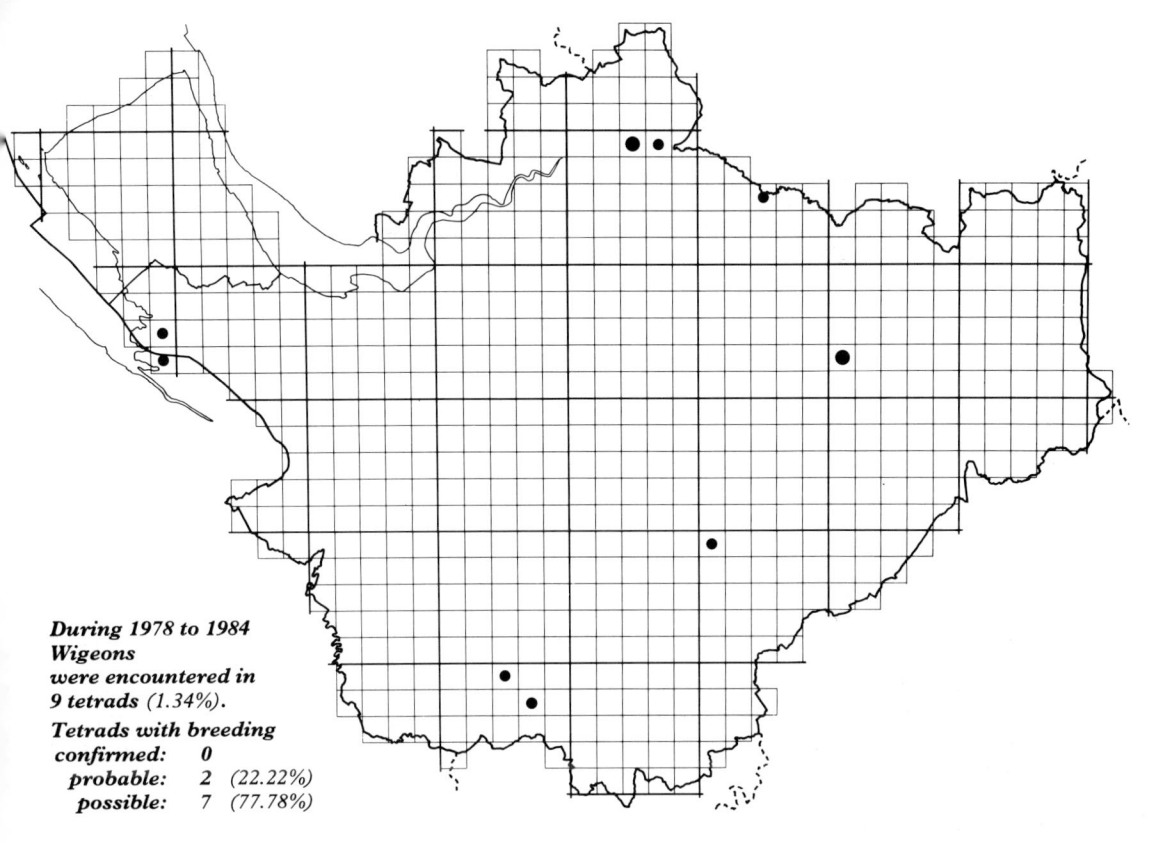

*During 1978 to 1984
Wigeons
were encountered in
9 tetrads (1.34%).*

Tetrads with breeding
confirmed:	0	
probable:	2	(22.22%)
possible:	7	(77.78%)

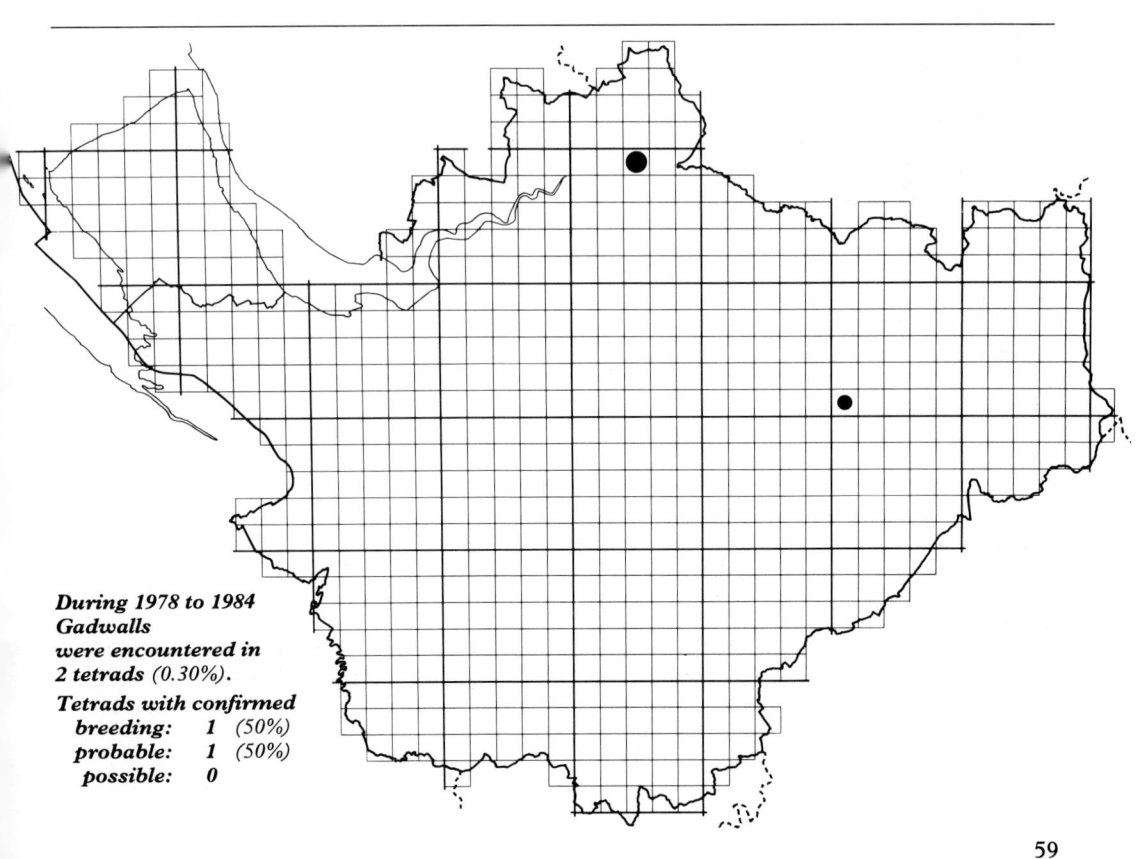

*During 1978 to 1984
Gadwalls
were encountered in
2 tetrads (0.30%).*

Tetrads with confirmed
breeding:	1	(50%)
probable:	1	(50%)
possible:	0	

Teal
Anas crecca

NESTING Teals are very retiring, preferring rushy moorland and heath pools, bogs and peat mosses in upland areas. However, they will also breed in lowland areas where lakes, rivers, streams and shallow marshes (brackish and freshwater) are frequented, but only then if there is plenty of emergent or peripheral vegetation to provide cover[35]. As such, their breeding distribution in Cheshire is limited.

Coward & Oldham[2] knew of few breeding pairs, old mossland and marshland haunts having by and large been drained, but listed the following breeding sites: Duckwood at Eaton (Chester); Bagmere, Brookhouse Moss and elsewhere in the district between Sandbach and Somerford; a few places in the Delamere area; some of the more secluded parks and woods such as Tabley; and the vicinity of Bosley and other places in the hills east of Macclesfield. A few pairs were said to nest on the Gowy marshes near Stanlow Point, and they believed odd pairs bred occasionally on the northern meres such as Marbury (Great Budworth) and Rostherne. A flourishing colony in Tatton Park resulted from captive ducks escaping in the early 1900s. There was no recent breeding record from west Wirral although at least one pair had nested at Leasowe in 1857.

The *L&CFC Report* for 1939-1942 refers to a decided increase in the preceding 15 years both as a winter visitor and as a breeding species. It was then nesting regularly in "Merseyside", Burton Marshes, at Tatton, Arley, Tabley, Delamere and elsewhere. Boyd[6] refers to Teals nesting regularly at Arley and Tabley. Bell[7] had little information on breeding, but in his supplement[9] stressed the scarcity of recent nesting records, a situation which holds true to the present day.

The drainage of mosslands and marshes, lamented by Coward & Oldham, has continued and very few such sites remain suitable. Bagmere, Risley Moss and perhaps Danes Moss are examples of mossland habitats occupied in recent years. The cluster of 'probable' breeding records in SJ75/76 during the present survey corresponds roughly to that group of sites east of Sandbach which Coward & Oldham mentioned. Delamere and the Gowy marshes are no longer suitable due to disturbance and drainage, and there are no recent breeding records from the northern meres, the colony at Tatton having faded out undetected some time before 1960 when the park was opened to the public.

The 1967 *CBR* states that "as in all recent years no breeding records were received", although odd birds had been seen in summer, and it is possible that a few pairs were breeding undetected. From 1968 onwards Teals were present in summer on the Manchester Ship Canal sludge-pools at Frodsham, and similar sites remain the most favoured breeding localities up to the present day. Recent records are from the Weaver Navigation sludge-pools at Acton Bridge and Kingsley, and from the Mersey valley, particularly at Woolston.

Teals breed on the wet bogs of the north Staffordshire hills, but not apparently on the Cheshire moors, although pairs regularly stop off at the eastern hill reservoirs in spring.

Most records of 'possible' and many of 'probable' breeding may in fact refer to non-breeding birds and, in addition, the map exaggerates the true picture in any one year. Broods of ducklings may remain hidden in cover and are seldom seen, but even allowing for their elusive nature it seems unlikely that more than ten pairs now breed in Cheshire.

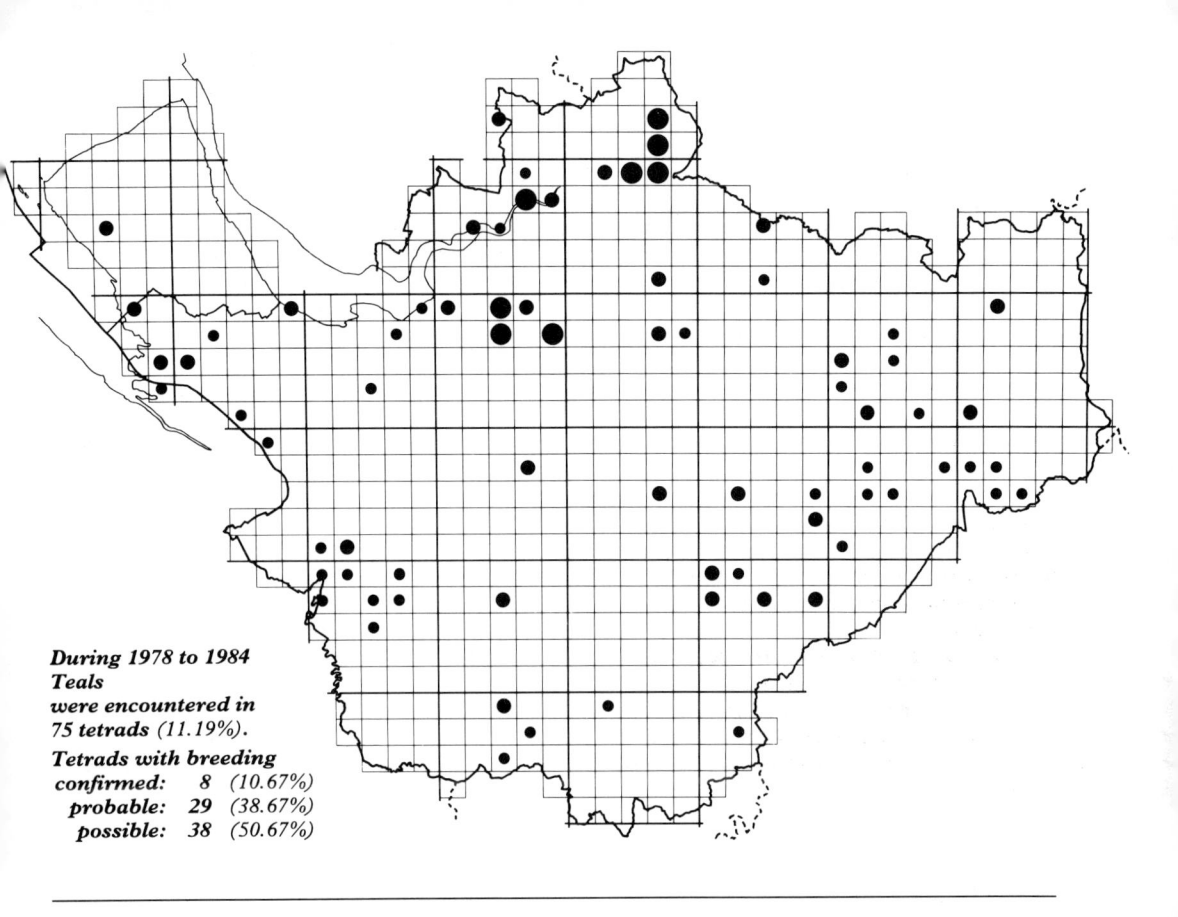

During 1978 to 1984 Teals were encountered in 75 tetrads (11.19%).

Tetrads with breeding
confirmed: 8 (10.67%)
probable: 29 (38.67%)
possible: 38 (50.67%)

Mallard
Anas platyrhynchos

THE Mallard breeds throughout the county wherever there is water, from the saltmarshes of the Dee to the moors of the Pennines. It is our most familiar, adaptable and successful wild duck. Many nesting-sites are in undergrowth near water or in tussocks emerging from pools; many more are found amongst bluebells or under brambles in woods, or even amongst heather or cotton sedge on the eastern hills. Occasionally, cavities or disused nests in trees are occupied: in 1977 it was reported that a duck had hatched a brood 25 feet up in an oak tree in mid-Cheshire, and, on several occasions, birds have been seen flying into or out from Crow nests 40 feet up in the tops of conifers in Tatton Park. In 1984, a duckling was successfully hatched on an unoccupied Heron's nest at Radnor Mere, only a yard or so from an occupied nest, and a duck was sitting on an old Crow's nest 30 feet above Snape Brook.

Display begins in October. For example, on 5th October 1974 two pairs were displaying and a third pair copulating on Rostherne Mere. From then on throughout the winter, pairs can be seen head-pumping and mating even when the meres are largely frozen over. Eggs are laid from March onwards, exceptionally earlier, although these may well become chilled and be deserted. It was considered that a duck with young at Mobberley in 1974 must have started laying in the first half of February. In 1983, a female was on eggs on 3rd March in a Chester park[8].

Many eggs are taken by crows or foxes, and in many tetrads predated eggs were the only confirmation of nesting. Ducklings appear from April to July, not only on the larger waters, but also on smaller brooks and marl-pits. Exceptionally, a duck with young, estimated to be about two weeks old, was seen at Pettypool on 4th August 1969[8] and, in 1977, a day-old chick was reported at Tatton on 8th August.

Once eggs are laid, drakes gather in flocks by meres, flashes and other waters prior to moulting. For example 200 were at Rostherne in May 1976 and 500 at Gayton Sands in June 1979. Many an unfortunate female that wanders too near such a gathering finds herself the object of ardent pursuit by five, six or more relentless drakes.

Large numbers of Mallards are reared and released by the Withington Estate at Chelford. A large count of 735 in 1976 was considered to include many released birds. Similarly, 1000 were reportedly released in 1977 and a count of 811 in September 1976 was thought to be partly composed of such birds[8] as were 1500 reported at Dairyhouse sand-quarry on 27th July 1984. Releases also occur at Adlington and doubtless elsewhere in the county.

The sparsity of breeding records in central and southern parts of the county may be due in part to below-average coverage, particularly in February and March when early spring display-chases are conspicuous. However, intensive dairy farming with associated silage fields, tidy hedgerows and little woodland characterises this landscape, and there is a lack of rough undergrowth in which the ducks could nest. Along the Delamere–Peckforton ridge the lack of surface water due to the porous nature of the underlying sandstone is probably a further inhibiting factor.

In 1976, seven broods totalling more than 60 young were counted along half a mile of the River Dean at Newton. At Woolston 20 broods were reported in 1981, 18 in 1982, 17 in 1983 and 16 broods totalling 128 young in 1984[8] although it is likely that many more broods were missed. In 1978 between 25 and 30 broods were seen in Tatton Park. Several other waters may produce ten or more broods each year, and sightings of five or more broods in a day are common at many additional sites. With 404 tetrads in all holding birds known to be breeding, and a further 105 tetrads with 'probable' breeding records, the county population probably lies at around 1500 pairs.

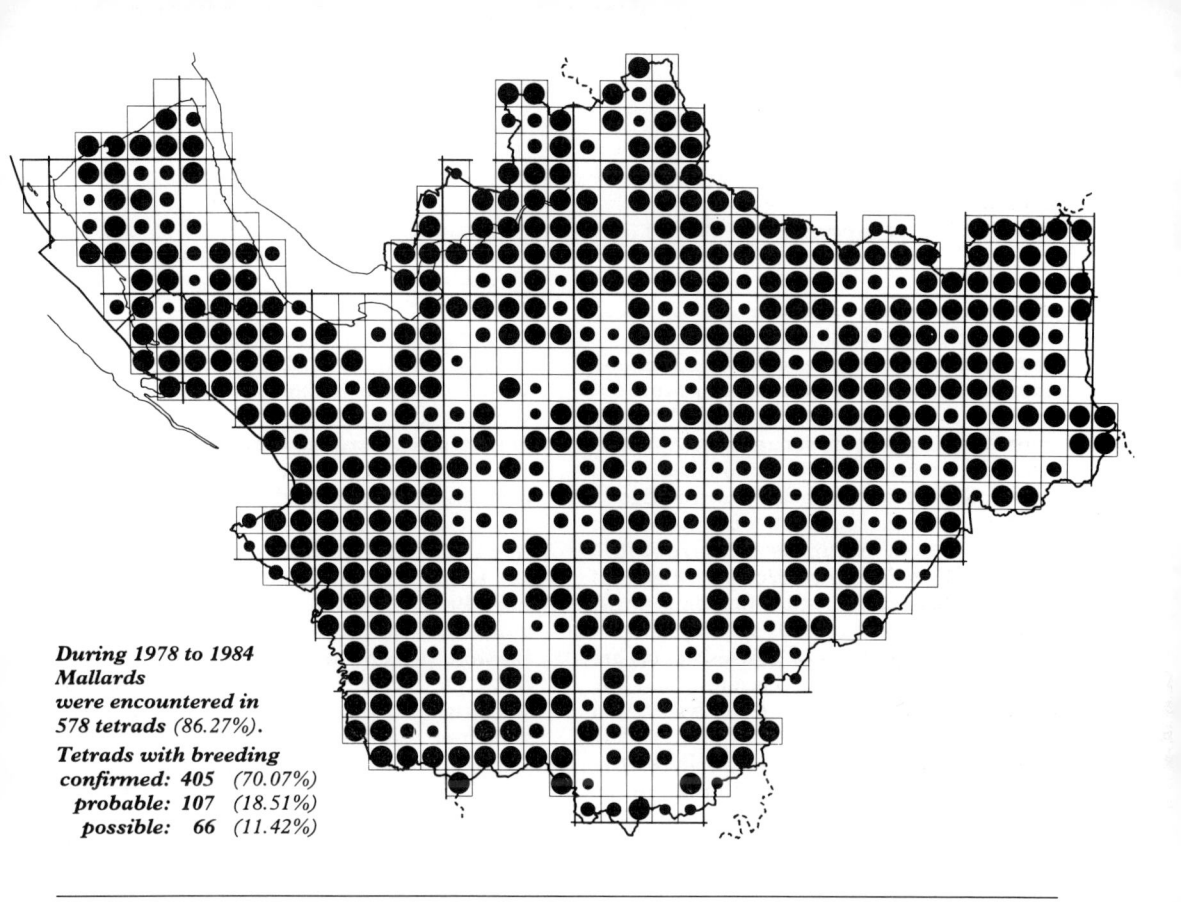

During 1978 to 1984 Mallards were encountered in 578 tetrads (86.27%).

Tetrads with breeding
- **confirmed:** 405 (70.07%)
- **probable:** 107 (18.51%)
- **possible:** 66 (11.42%)

Pintail
Anas acuta

THE breeding habitat of the Pintail is diverse and includes moorland pools, lowland lakes and freshwater marshes although nests, which tend to be more exposed than those of most other ducks, may be as much as 200 metres from water[32]. As a breeding bird, the Pintail, like several other duck species, is a relatively recent addition to the British list, first proved breeding in Scotland in 1869, and in England in 1910. However, it remains one of the rarest of all British breeding ducks with perhaps as few as 50 breeding pairs[32], and its distribution in western and central Europe is very patchy[33].

Towards the end of the 1890s a drake paired with a Mallard duck at Toft Park near Knutsford. The eggs were taken and hatched under a hen and the drake was shot by a gamekeeper[2].

Since the 1950s, numbers wintering in the estuaries have increased and there has been a corresponding increase in summer records. One was at Bromborough on 7th June 1957[7] and a pair at Burton Marsh on 2nd July 1961[18]. Since 1967 birds have occurred annually in summer.

Between 1968 and 1971 breeding was probably attempted annually at the Frodsham sludge-pools, then an extensive area of shallow marsh with emergent vegetation. On 6th July 1969 a pair was seen with five young barely able to fly and, in 1971, breeding was again attempted but one bird was found shot and another found dead[8].

In 1973, a pair remained on Burton Marsh from late April into July, and, since the establishment of the Gayton Sands RSPB Reserve, small parties of non-breeding birds have occurred there in summer, notably six males and four females on 16th June 1979 and four males and nine females on 10th June 1983. In 1977, 1978, and 1981, birds were seen in summer at Rostherne Mere and in 1980, 1981, and 1984, at Woolston, but there was no evidence of breeding at either site.

Garganey
Anas querquedula

"THE Garganey is unique among British waterfowl in being exclusively a summer migrant, arriving in late March and April and departing during September" although immigration may extend into May or early June, and return movements sometimes start as early as mid-June[35]. In Britain, the species is on the western edge of its breeding range and there are considerable fluctuations in numbers with sporadic nesting. Many Cheshire records will be of passage or summering, non-breeding birds. However, nesting Garganeys are very difficult to locate. Shallow fresh or brackish pools are required for feeding and breeding and the nest is usually within 50 metres of water, well concealed amidst tall grass[35].

Coward & Oldham[2] described the Garganey as "a rare visitor in autumn to the Dee Estuary". Between 1910, when there were five birds on Marbury Mere on 9th and 10th April, and 1951, birds were recorded in at least thirteen years, usually in April[6]. The first summer record was of a pair on 8th July 1928 at Rostherne[24]. In 1932, a pair remained at Arley Pool into June but disappeared when the pool dried out. In 1948, a pair settled at Tabley but again was not proved to breed. Hardy[4] referred to nesting in Cheshire but without any supporting detail, whilst Williams[20] records a nest found in the Aldford area about 1950 from which the eggs were taken and hatched under a broody hen. However, the first substantiated record for the county came in 1957 when a pair reared three young at Halton Moss near Warrington[3]. Breeding was again proved at this site in 1959 and 1960 but the site was drained in 1961[9]. Since then there have been regular pre-breeding and post-breeding records.

As with several of our scarcer duck species, the state of the county's various sludge-pools is crucial to the Garganey. Between 1969 and 1973 small numbers were regularly present during summer at Frodsham, with larger autumn counts possibly suggesting breeding success: up to 24 in 1969, six in 1970 and seven in 1972 and 1973. Additionally, there were four birds in August 1968, although there was no evidence of nesting, and there are summer records for 1975, 1976 and 1979 suggesting that breeding may have been attempted.

The Sandbach Flashes are the other main source of records in the county, although here records are more erratic. Two birds of unknown sex were recorded in June and July 1968. Nesting was suspected in 1969, when a pair was recorded from 26th April to 3rd June and nine were present in August; in 1972, when at least two birds summered, and in 1974, when a male was regularly seen into June. Breeding was first proved in 1975, when a duck was flushed off eggs on 2nd June and six were present in August and September. Subsequently, breeding may have occurred in 1976, when a pair summered, and in 1978, when one was present from the 9th to 22nd July.

Despite the increased coverage of suitable sites during this survey, no confirmed breeding records were forthcoming. Nationally, the Garganey has declined in recent years and was included in the national Rare Breeding Birds list for the first time in 1980, when only three pairs were proved breeding in the whole of the United Kingdom[43].

In 1981, display was observed at an undisclosed site and a pair was present on one day in June at a second site. In 1982, birds were recorded at five sites and two immatures were at Tabley on 15th and 18th July. In 1983, three sites featured, including a pair at an undisclosed site on the 29th April, followed by a male there on 12th May and, subsequently, by two males in eclipse together with a female and four immatures on 16th July – successful breeding at this site must have been a distinct possibility. In 1984, three sites featured again, including summer records of a male at Woolston on 13th May and from 8th to 10th June, and a pair at another site on 11th May together with five immatures on a nearby river from 4th to 11th September. These records suggest that one or two pairs may have bred or attempted to breed in most years during this survey. Since 1982 all Cheshire records, including probable spring and autumn passage birds, have been submitted to the Rare Breeding Birds panel and are listed in their annual reports[43].

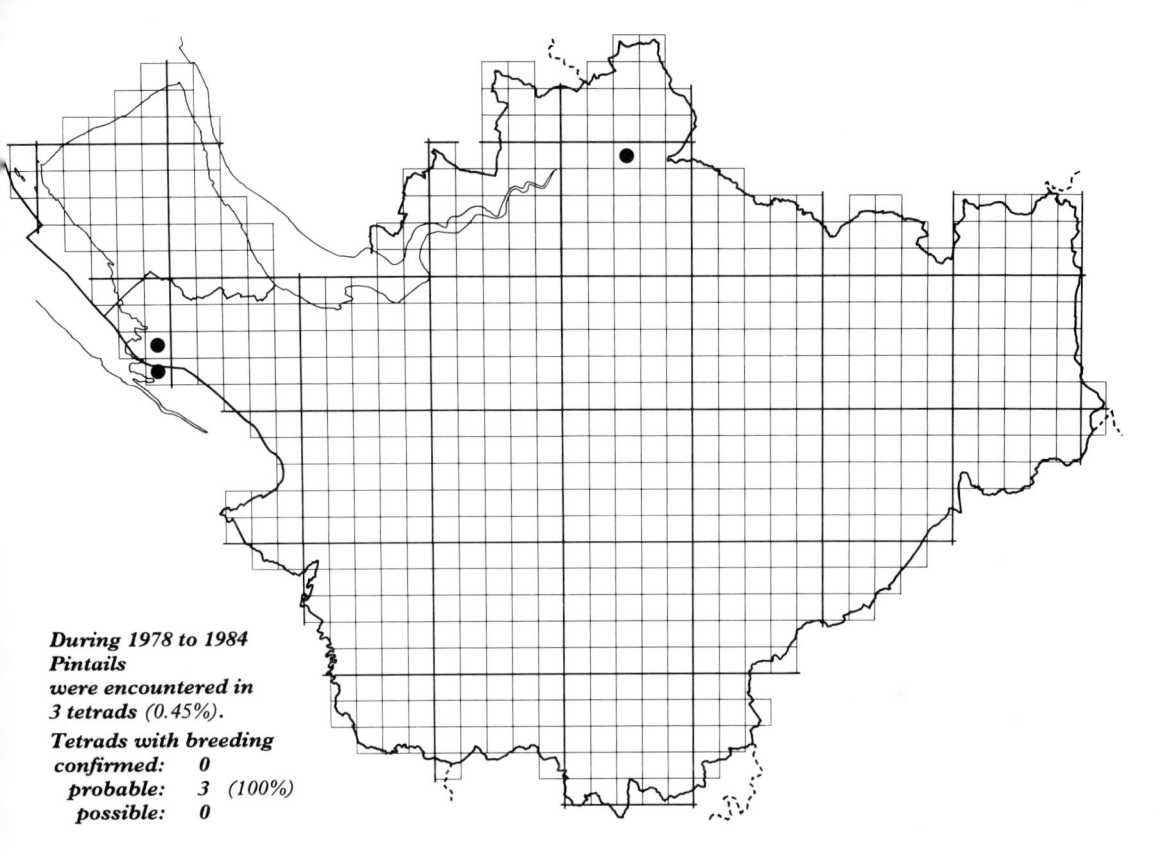

During 1978 to 1984
Pintails
were encountered in
3 tetrads (0.45%).

Tetrads with breeding
  confirmed: 0
    probable: 3 (100%)
    possible: 0

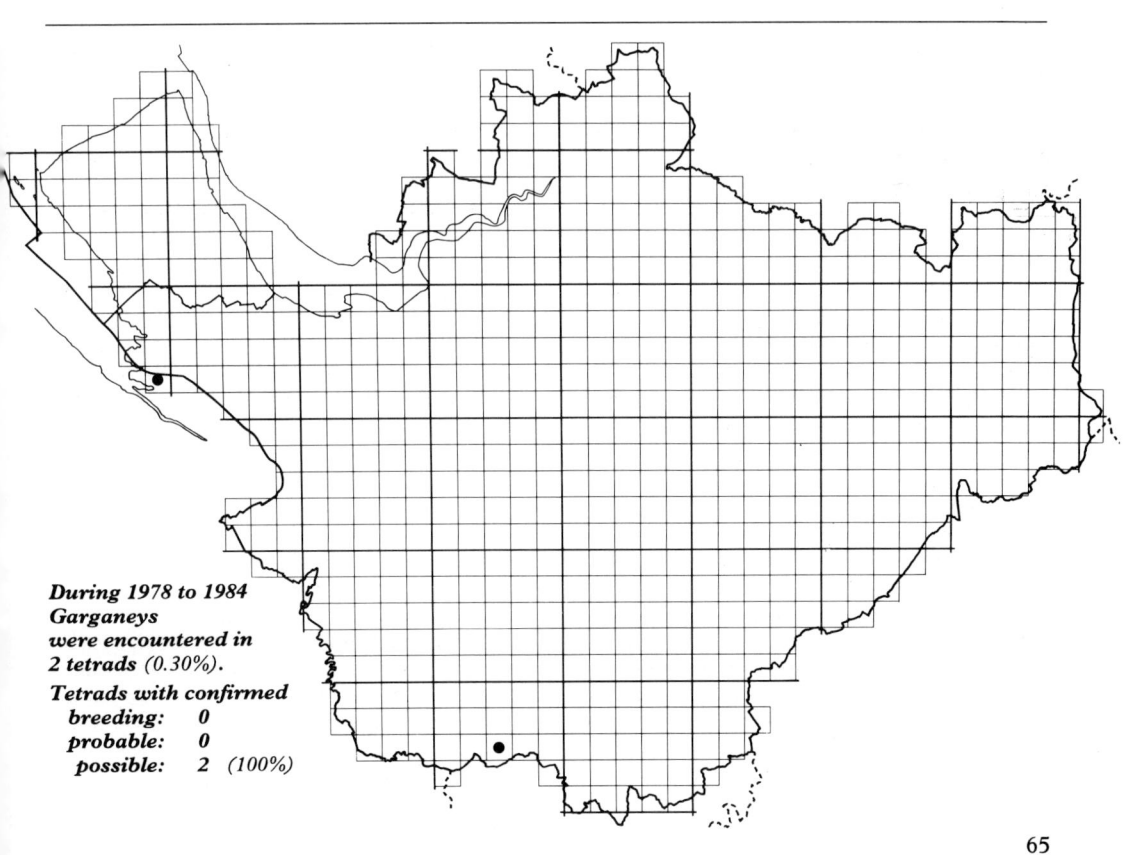

During 1978 to 1984
Garganeys
were encountered in
2 tetrads (0.30%).

Tetrads with confirmed
  breeding: 0
  probable: 0
  possible: 2 (100%)

65

Shoveler
Anas clypeata

BREEDING Shovelers favour shallow waters with extensive reedy fringes and abundant aquatic vegetation. Cheshire has only a limited number of such sites, and Shovelers have never been more than a scarce nesting species here, although large numbers occur on autumn passage and, to a lesser extent, in winter. The brightly coloured drake is one of the most striking of the duck species and the spatulate bill, which gives the species its English name (and the old Mersey wildfowler's name of Spoonbill), is diagnostic for identification. The mandibles of this specially adapted bill are edged with fine, intermeshing, comb-like lamellae. This distinguishes the Shoveler as a filter-feeder and restricts it, almost exclusively, to shallow, eutrophic waters at all seasons.

The Shoveler is a breeding bird of long standing in Britain, albeit a rare one, and, like many of our other native duck species, it extended its range north-westwards in Europe only during the early part of this century, perhaps in response to climatic amelioration[35]. As with the Tufted Duck and Pochard, this resulted in a major increase in the Shoveler's British breeding population.

Coward & Oldham[2] considered it possible that a few pairs bred in Cheshire, but their only "presumptive evidence" of nesting success came in 1908 when three young birds, incapable of flight, were shot at Tatton in August. Two years later a pair certainly bred at Tatton and in 1912 a pair nested on the Dee marshes. This was followed by some six pairs at Burton in 1922 "and for some years previously" [7]. Hardy[4] had also "occasionally found nests on the saltings" there, and in 1941 he found three pairs on the Frodsham marshes. At least one of these was nesting. He also reported birds at Oakmere during the summer of that year. Boyd[5] saw birds on various pools in the summer of 1936 and records putting a duck off a nest of nine eggs in Staffordshire that year, adding the comment that this was another duck species which was spreading as a breeder. Bell[7,9] also listed breeding at Tatton, Rostherne and Sandbach Flashes in 1939, and probable breeding at Frodsham in 1953 and at Sandbach Flashes in 1966. To this may be added Boyd's[6] reference to a few pairs nesting regularly at Tabley, Williams' report[20] of possible breeding at Huntington Reservoir, Chester where the species was present from early March to late July 1961, and a reference in the 1964 *CBR* to a pair rearing two young at Sandbach Flashes.

From the publication of Bell's supplement[9] up to the start of this survey, breeding was confirmed in only three years. In 1971, seven young were reared at Frodsham, in 1974 birds bred at Woolston and in 1976 young were seen at Burton Marsh and Sandbach Flashes. During this survey, however, breeding was attempted in each year with young seen in all seasons except 1978.

Initial location, especially of the drakes, is easy and evidence of breeding intent often follows when a pair is seen in display, pumping their heads up and down in unison. Few actual nests were found during this survey, although one with ten eggs at Woolston in 1984 was more than a quarter of a mile from the nearest water.

It is clear from the map that the breeding population is concentrated along the valleys of the Mersey and lower Weaver. The Woolston sludge-pools currently hold the vast majority of our birds – up to 63 were counted there in June 1982, and five broods in 1983 totalled 29 young. It is unlikely that the county population is greater than ten breeding pairs.

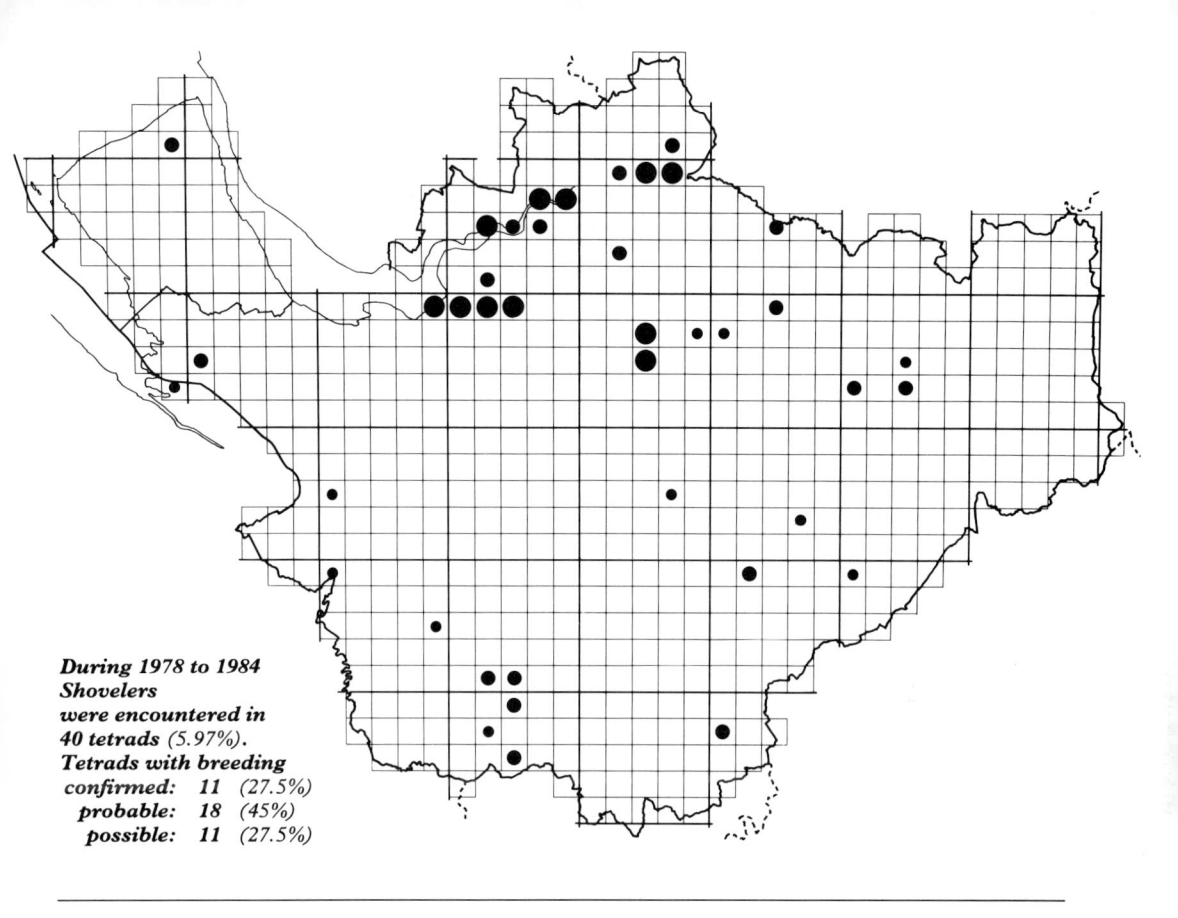

During 1978 to 1984 Shovelers were encountered in 40 tetrads (5.97%).
Tetrads with breeding
confirmed: 11 (27.5%)
probable: 18 (45%)
possible: 11 (27.5%)

Red-crested Pochard
Netta rufina

SMALL numbers of feral Red-crested Pochards have bred in Britain but to date there are no such records for Cheshire. Birds have been seen in most years since the mid-1970s however, including some in the breeding season. Most birds are thought to originate from the small collection of waterfowl in the grounds of Tatton Hall. In 1978 a pair reared nine young here which were allowed to fly free. Full-grown young birds have also been seen on several occasions amongst the milling flocks of newly-released Mallards at Withington in September.

Pochard
Aythya ferina

ALTHOUGH a breeding species of long standing in Britain, until the mid-nineteenth century the Pochard was restricted to East Anglia. In common with other duck species it has spread slowly northwards within the last hundred years, mainly through eastern Britain. It remains generally scarce, but with local concentrations. Further increase and spread is unlikely unless the species can adapt to less specialised feeding requirements[35, 36]. For nesting, Pochards prefer large pools, lakes, marshy fleets and slow-flowing streams edged with dense emergent vegetation. Unlike Tufted Ducks, they do not use the many man-made gravel pits and reservoirs for breeding, probably because, being primarily vegetarians, they require shallower waters with abundant plant-life.

The Pochard started to breed regularly in Cheshire only during the period of this survey. Coward & Oldham[2] regarded it as a common winter visitor but could list only two summer records: a young bird at Redesmere on 19th July 1906, and two on Tatton Mere on 13th May 1908. By 1915 it was becoming more frequent as a casual visitor in June and July. This was perhaps due to moulting birds wandering from distant breeding grounds, as occurs to a greater extent with parties of drakes nowadays. It was not until 10th July 1932 that Boyd encountered a duck with five young at a locality, believed to be Baddiley, in the south of the county. About 20 Pochards had spent the previous summer there. Very small numbers continued to linger on the meres during summer for the next thirty years and then, with increasing frequency, during the early 1960s[7, 9].

In 1973 a duck was watched sitting in a Coot's old nest at Rostherne, arranging twigs around herself[24]. Then, in 1974, came the second and third breeding records: a brood of eight ducklings at Rode Pool on 7th July, of which only one survived, and a brood of four seen on a canal near Northwich in late July. Four pairs summered at Rode Pool in 1975.

During the present survey, the emergence of the colony on the Manchester Ship Canal sludge-pools at Woolston has been remarkable. In 1979, a few birds stayed into early May and during the following five years birds bred with varying success:

Table 7: Pochard breeding success at Woolston, 1980-84

Year	1980	81	82	83	84
No. of Broods	7	17	8	9	8
No. of Young	43	83	28	46	50

In 1982, 43 drakes and nine ducks were recorded as late as 29th May although only eight broods were raised. The success of this colony is largely dependent on the state of the sludge-pools. In 1980 and 1981 there was considerable dense, emergent vegetation on No. 2 bed making it very suitable for breeding but since then silt dredged from the canal has been pumped onto the bed making it progressively less attractive.

Elsewhere, a brood of four well-grown young was on the mere at Mere in 1978, three young were seen at Doddington in 1979, and single broods were noted at two sites in 1980[8].

Several suitable sites are not yet occupied, and in due course the right conditions may develop at certain disused sand-quarries. At present however, the Pochard's breeding status in Cheshire depends mainly on the state of the Woolston sludge-beds.

Australian Pochard
Aythya australis

FOUR drakes and two ducks of this species – undoubtedly birds escaped or released from captivity – frequented the reed-bed at the southern end of Tatton Mere from November 1978 into 1979. Three drakes and two ducks survived into March of that year, and a duck was seen later with four ducklings. Three drakes and one duck were last reported in November 1979.

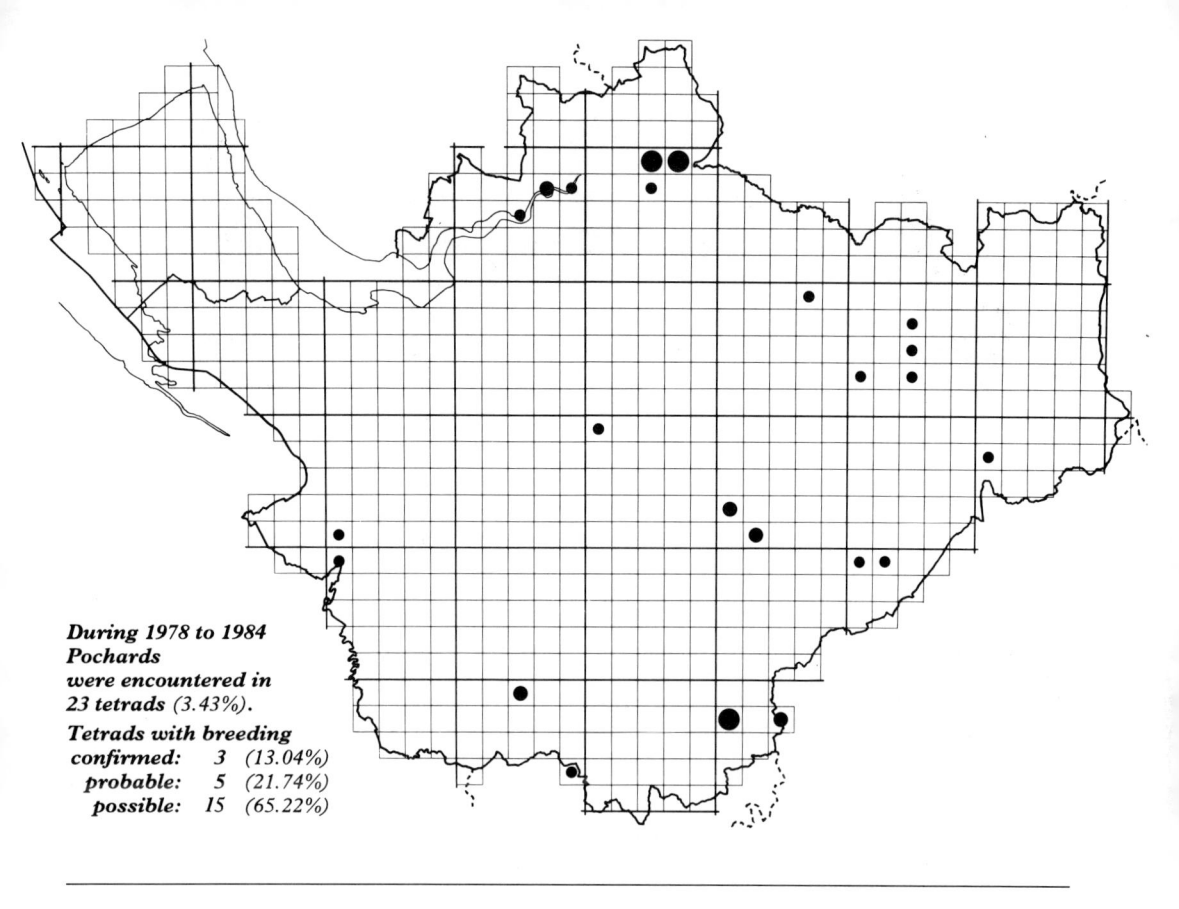

During 1978 to 1984 Pochards were encountered in 23 tetrads (3.43%).

Tetrads with breeding
confirmed: 3 (13.04%)
probable: 5 (21.74%)
possible: 15 (65.22%)

Tufted Duck
Aythya fuligula

AS a breeding species the Tufted Duck was unknown in Britain until the middle of the nineteenth century, but by the 1930s it had occupied most suitable waters. The initial spread was probably associated with climatic amelioration and reduced persecution in the breeding season. Whatever the causes, the species has continued to spread, colonising artificial freshwater sites such as new reservoirs and flooded sand-pits.

Coward & Oldham[2] observed that prior to the passing of the Wild Birds Protection Act of 1880 the Tufted Duck was apparently rare in Cheshire at all times of the year. However, by the turn of the century birds were beginning to stay on certain meres throughout the summer and the first recorded instance of breeding was in 1908 when a brood of three young was seen on Redesmere. Bell[7] records the spread of breeding pairs to seven sites in eastern and central Cheshire by 1915, and since then the species has bred at some time or other at most suitable waters in the county. Boyd[5,6] saw at least seven broods on one water in 1933, six broods totalling 29 ducklings at Tabley in 1944, and three broods of six young each at Arley in 1946 – this latter pool was subsequently drained. Nesting occurred at Frodsham in 1952[7] and probably on the Eaton estate in 1967[20]. In 1984, breeding occurred at Moreton clay-pits – the first breeding record for north Wirral.

Suitable waters are scarce in the west of the county but there seems to have been some colonisation there since the BTO Atlas survey in 1968-72. During the BTO survey there were no confirmed breeding records compared with the five during the present survey. However, the breeding pool at Hapsford was being filled in during 1983.

Tufted Ducks nest amongst rank vegetation within a few feet of the water's edge and prefer larger pools, meres, flashes and the like. The most favoured sites are shallower waters with extensive fringing reed-beds or alder carr. Flooded sand-quarries may be occupied even before work has ceased so long as at least some clumps of vegetation are available in which to nest. In Cheshire, other sites include the small boggy pool at Danes Moss and an oxbow of the River Bollin near Styal, where pairs bred in 1981 and 1982 respectively. Flooded marl-pits at Astle, Withington and Siddington held pairs in the spring of 1983.

The number of birds present in Cheshire during the summer far exceeds the total of broods seen, and, as display is seldom witnessed even on breeding waters, it is difficult to tell how many of these birds intend nesting. Breeding seldom starts before the middle of May and nests are generally well hidden and difficult to find without potentially harmful searching. Most records of confirmed breeding relate to sightings of the chocolate-brown ducklings between late June and August. The number of broods reported fluctuates widely from year to year. A record of eleven young at Woolston on 17th June 1981 was unusually early. Losses of young can be heavy and may reduce brood sizes considerably within a short time – a brood of ten young at Woolston in 1980 was reduced to two in a little over a week. It was thought that minks were responsible. Similar rapid losses at Redesmere have been attributed to pikes.

Tatton Mere was undoubtedly the major stronghold of the species at the time of this survey. There birds nest in the Knutsford Moor reed-bed and adjacent alder carrs. In 1964 eleven broods totalled 71 young; in 1966 14 broods totalled 84 young; and more recently 12 broods on 8th July 1978 totalled 61 young with several further broods on other dates. 20 broods were seen in 1979.

In 1965 *CBR* contained details of 21 broods totalling 80 young including five broods on Combermere, four on Tabley Mere and four on Redesmere. In 1979, by far the most successful year on record, 74 broods were reported in the county, 60 of them, totalling 180 young, being in the East Cheshire recording area. In 1983, 30-32 broods located in the SJ87 10-km square contained some 165+ young (JPG). In such good years a thorough census might reveal in excess of 100 broods in Cheshire and Wirral.

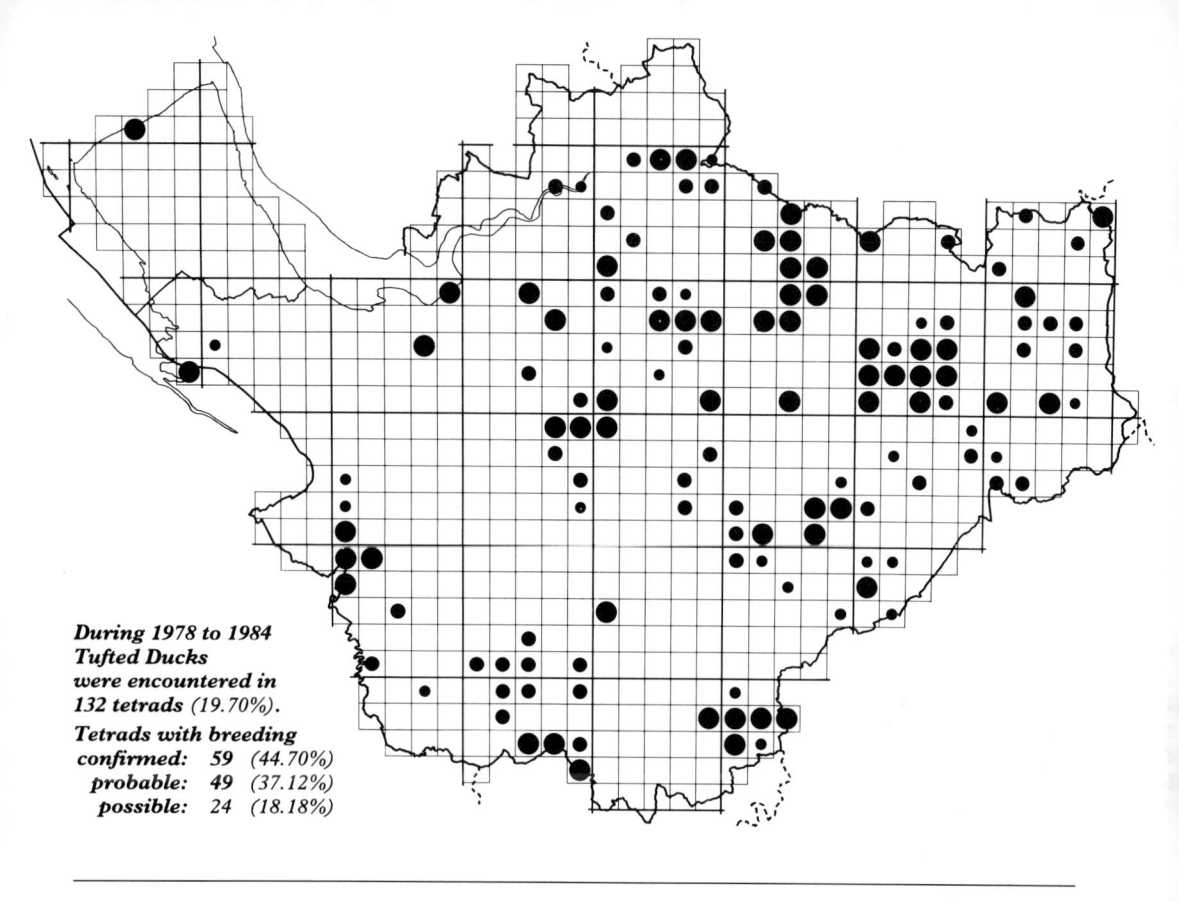

During 1978 to 1984 Tufted Ducks were encountered in 132 tetrads (19.70%).

Tetrads with breeding
confirmed: 59 *(44.70%)*
probable: 49 *(37.12%)*
possible: 24 *(18.18%)*

Goldeneye
Bucephala clangula

THE Goldeneye has bred on at least two occasions in Cheshire and summered on several others. A clutch of seven eggs, complete with full down, was taken by a "well known" ornithologist from a nest in a Cheshire marsh on 28th May 1931. The nest was in the entrance of an old rabbit burrow adjacent to saltmarsh at Burton Marsh on the Dee Estuary[35]. The duck was seen on the nest and a 'keeper reported seeing drakes in the area. He also reported that nests had been found in previous years and expressed no astonishment at the species' presence despite the lack of woodland. He was aware that it was not a regular breeder there and it was thought that the birds involved were "winged" and unable to fly properly ([3], Taylor 1938). Another nest with eight eggs was found in the same marsh in 1932. This was the first reported breeding of the Goldeneye in Britain and there was no other instance until 1970. Since then a small population has become firmly established in Scotland. This has partly been due to the provision of nest-boxes to which the species readily adapts in the absence of hollow trees[35].

In 1905 and 1908, single birds with damaged wings remained on Marbury Mere, Great Budworth, throughout the summer. A number of full-winged birds, which were considered possibly to have been attracted by wounded birds, were seen on the same mere in the summers of 1924 and 1925[6].

An adult drake spent the summer of 1967 on the Sandbach Flashes, and since 1974 *CBRs* have featured summer records annually. Almost invariably these refer to single birds although two females were on Tatton Mere in August 1978, two females were seen at Rostherne on 8th June 1980 and three birds of unknown sex were reported at Sandbach Flashes on 5th August 1984[8]. On a number of occasions birds have taken up residence for several weeks during the summer particularly at Rostherne, Doddington Pool, Tatton Mere and Appleton Reservoir.

This increase in summer records and a tendency nowadays for the species to delay departure in spring – birds often remain well into May – coincide with the rapid increase of breeding numbers in Scotland. The Goldeneye has been on the list of the Rare Breeding Birds Panel since the panel's inception in 1973 and is on Schedule 1 of the Wildlife and Countryside Act 1981. From 1978 the number of confirmed breeding pairs in Britain entered double figures and by 1984 there were 53 such records. Birds have summered in parts of Wales, the Lake District and several other widely separated localities and it is thought that these, together with the Cheshire records, could be the precursors of a spread in the British breeding population.

REFERENCE

Taylor, F. (1938): Untitled note on Goldeneyes nesting in Cheshire. *Bull. Brit. Ool. Assoc.* 5(59): 5-7.

Smew
Mergus albellus

BELL'S latest date for Smews was 6th April in both 1929 and 1950[7, 9]. Since 1969 birds have stayed later than this in eleven years. In 1969 a female remained at Tatton until 9th May, and in 1979 a drake stayed at Rostherne Mere until 16th June, when it was found recently dead. The British Birds Rare Breeding Birds Panel's report noted that such records (a female also summered in Suffolk) were not unexpected following the influx during the preceding winter. The Panel reports only six other records of birds summering in Britain since 1973 – one in 1975, two in 1980 and three in 1981 – with the comment that "there has never been any suspicion of breeding, but this is the sort of build-up which could be a prelude to colonisation"[43].

Red-breasted Merganser
Mergus serrator

THE Red-breasted Merganser has long been a resident of Britain and by 1870 had a wide breeding range in several parts of Scotland. A major expansion of this range was apparent by the 1880s and has continued. Expansion also took place in Ireland at the beginning of this century. In England, breeding was first proved in 1950, at Ravenglass (Cumbria), while the first Welsh breeding occurred on Anglesey in 1953. During the next 10-15 years the Lake District was colonised and subsequently contiguous parts of north-western Yorkshire and also north-western Wales[35]. Breeding Mergansers are commonest in coastal districts but will nest far inland on the larger river systems and the bigger sheets of water such as reservoirs.

In 1962 a female Merganser built a nest at Oakmere, but no eggs were laid[9]. The early spring of 1965 saw the largest concentration ever recorded on the Dee. There were some 250 birds on 5th March, and on 21st a drake was seen displaying to Goosanders inland, at Doddington. Numbers in the estuaries have remained high (although never again approaching the level of 1965). This perhaps reflects the southward spread of Mergansers as breeding birds into the northern Pennines, and the colonisation of Wales. The 1979 *CBR* noted that the species had become more frequent on the flood pools of Burton Marsh where up to 13 were present and displaying during March of that year.

On 17th April 1967 a drake was seen at Rostherne Mere. Every year since 1971, except for 1973 and 1977, such April or early May sightings of single birds have become more regular on inland waters. It seems likely that these birds are moving between the Dee and waters in north Derbyshire where a small population is becoming established, with confirmed breeding in 1973[45], 1978 and 1982-84[44]. Birds have been seen on several hill reservoirs and meres in eastern Cheshire. In early May 1971, three drakes and a duck were present at Lamaload and a male was present there in late March/early April 1983. On other waters single birds, some staying for several days, have been recorded in spring in 1979, 1980, 1982 and 1983. On 29th March 1980 a drake at Lamaload was displaying to a Mallard duck. Future colonisation of such sites seems a distinct possibility.

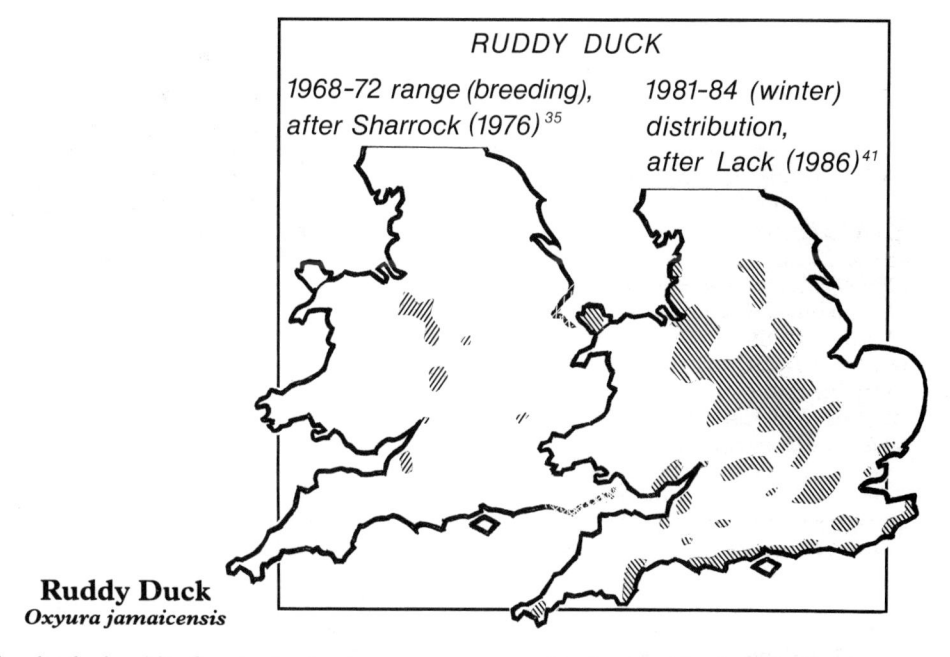

RUDDY DUCK

1968-72 range (breeding), after Sharrock (1976)[35]

1981-84 (winter) distribution, after Lack (1986)[41]

Ruddy Duck
Oxyura jamaicensis

THE blue-beaked, white-faced, chestnut-backed drakes of this North American species are not easily missed. Even when hidden by an inlet into the reeds, their peculiar chuffing display may be heard. With tails erect, they rear out of the water then subside in a series of jerks as the head is pumped rapidly up and down, while bubbles, trapped under the breast feathers, escape to the surface of the water. At times the appearance of a person at the water's edge seems to prompt nervous displaying. The female is rather drab, vaguely recalling a small duck Pochard in shape and colour. On two occasions JPG has seen Ruddy drakes displaying to duck Pochards.

The Ruddy Duck is ungainly out of water because its legs are set very far back, and consequently the nest is usually floating, anchored amongst reeds, reedmace, sedges or similar emergent vegetation. Most sizeable pools or lakes appear suitable, provided such vegetation is present. In many cases even a narrow fringe of reeds may be occupied provided the shoreline is little disturbed or sheltered by woodland, scrub or even a crop – one nesting attempt at the Sandbach Flashes failed when the hay was cut only feet from the nest. Deeper waters, such as certain sand-quarries, have not yet held successful breeding birds, possibly because the insect life in such sites is inadequate to feed the ducklings.

Few nests were reported during this survey, and generally confirmation of breeding came from seeing ducklings. Broods appear from June to September, but unlike our other breeding ducks the family structure is ill-defined. Ducklings are almost as likely to follow drakes as ducks, and broods often appear to amalgamate or intermingle. On several occasions, both at Tatton and Woolston, broods of Pochards and Tufted Ducks have included a Ruddy duckling, possibly the result of a female laying eggs in the other species' nests. Indeed it is difficult to define the breeding population in terms of pairs, and counts of adults or drakes tend to be the best figures available.

Feral breeding in Britain began in Somerset in 1960; young, unpinioned birds having escaped from the Wildfowl Trust's captive breeding population at Slimbridge. The first Cheshire breeding record took place as recently as 1968, at Quoisley Mere, near the Shropshire border, although a male had been seen here on 25th July 1965 (G. A. Williams pers. comm.). There is no mention of the species in *CBRs* until 1971 when the Quoisley colony was said to be flourishing and birds were seen at Barmere where the species had previously bred. In 1972, breeding occurred at Barmere and Cholmondeley, and in 1973 young were seen for the first time in the east of the county with single ducklings at Redesmere and Capesthorne.

Subsequently, it has proved difficult to assess the county population from year to year, the original scepticism over recording "foreign" birds having given way to blasé acceptance of the species' presence. Limited records from *CBRs* prior to the start of this survey include:

1974 three broods Redesmere; two pairs Rode Pool.
1975 12-15 pairs in the county, sites including Rode Pool, Redesmere, Capesthorne, Radnor Mere, Tatton, Tilstone Fearnall, Oakmere and Quoisley.
1976 At least six southern meres occupied; two attempts (by one female) at Rostherne and one at Sandbach.
1977 Successful breeding on at least nine waters with three broods at Warmingham Flash, Sandbach and two broods at Tatton.

Hudson (1976) considered that Rostherne and Tatton meres were the most northerly breeding sites in the country at that time. A pair was seen on the Eaton estate in May 1979 and birds were again present in spring from 1982 to 1984. In 1980, two broods were seen at Woolston and, in 1981, breeding occurred at both Abram and Pennington in Greater Manchester[46].

In 1974 the largest winter count was of just 35 birds on Barmere. A count of 152 at Farmwood Pool just ten years later gives some indication of the phenomenal increase of the species. As the map shows, birds are now present on most suitable waters.

The total number of broods reported in any one year includes 22 in 1978 and 31 in 1983. However, these almost certainly understate the true figures. JPG estimated that there were some 75 adults in east Cheshire alone during the summer of 1979 and, with 59 tetrads occupied during the survey, and concentrations of up to twelve drakes at Woolston in April and nine to ten pairs at Sandbach in 1982, the county population may have been about 100 pairs in that year as against some 50-60 pairs nationally in 1975[35].

There is no doubt that most of the Cheshire meres provide ideal breeding sites for the Ruddy Duck and have played an important part in its colonisation of Britain.

REFERENCE

Hudson, R. (1976): Ruddy Ducks in Britain. *British Birds* **69**: 132-143.

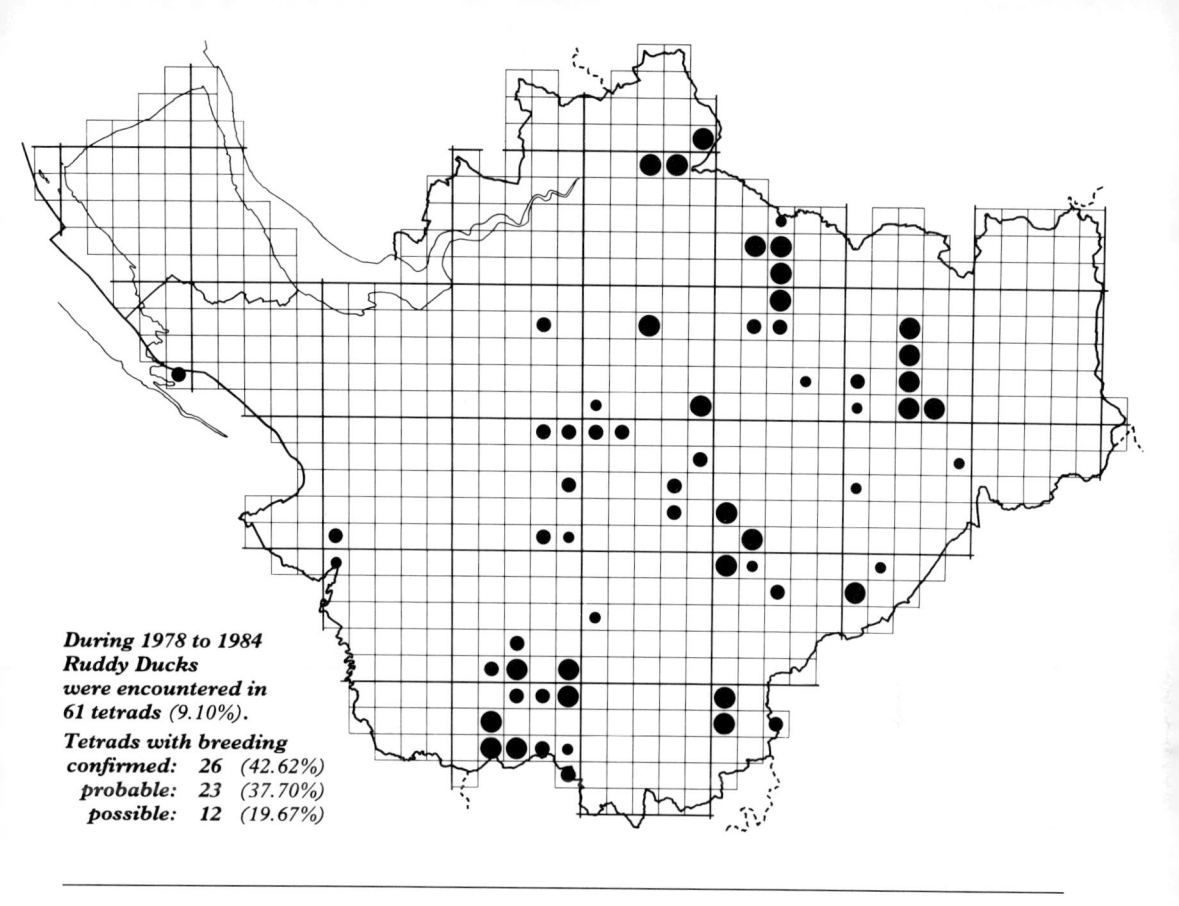

During 1978 to 1984 Ruddy Ducks were encountered in 61 tetrads (9.10%).

Tetrads with breeding
confirmed: 26 (42.62%)
probable: 23 (37.70%)
possible: 12 (19.67%)

Honey Buzzard
Pernis apivorus

THERE is no record of this species ever having bred in the county but in late May 1980 a Honey Buzzard was watched soaring over a wooded area suitable for breeding.

Red Kite
Milvus milvus

THE Kite was formerly resident in Cheshire, where it was known as the Gled or Glead Hawk (but see Buzzard). Daniel King in his "Vale Royall" of 1656 described the Kite as occurring in the "...forests of Delamere and Maxfield...". Coward & Oldham[2] took a letter written by the first Lord Stanley of Alderley in 1791 as evidence that the species was still plentiful in the county at the end of the eighteenth century: "On the other side of this [Radnor] mere the eye rests on a thick venerable wood of beech trees above 140 years old... The silence that reigns there is only broken by the shrieks of the large kites, which constantly build their nests in the neighbourhood....". However they state that by the time of their writing, "The Kite has long since vanished..."

Hardy[4] refers to records of payments for Kite heads in old parish registers at Woodchurch, Neston and Eastham, and to a mention under the heading of vermin in a Wallasey schoolmaster's diary of 1720. The memory of the Kite is recalled in the name of Gleadsmoss near Withington.

Marsh Harrier
Circus aeruginosus

COWARD & Oldham[2] state, "There can be little doubt that until the early part of the last century the Marsh Harrier nested regularly on the mosses which then covered thousands of acres in the broad valley of the Mersey" and suggest that, like some other larger birds of prey, this species was probably exterminated as a breeding bird during the second half of that century.

There has been an increase in sightings in recent years, particularly in 1980 when there were some 25 records between May and October, although some of these undoubtedly related to the same individuals. Most of these recent records are from the Dee estuary whilst Frodsham Marsh, Woolston and Sandbach also feature regularly. Most British breeding records come from extensive reed-beds although there have been some cases of nesting in arable crops. Given the scarcity of reed-beds in modern Cheshire, and the threats to the few in existence, future breeding seems unlikely.

Hen Harrier
Circus cyaneus

HEN Harriers probably bred on the heather-clad hills of eastern Cheshire in bygone centuries, although there is no written record of their having done so[2]. They now breed regularly in north Wales and sporadically in the Peak District, not far from the Cheshire boundaries, and it is perhaps surprising that there are no summer records in the county.

Montagu's Harrier
Circus pygargus

THERE are very few satisfactory records of this species in Cheshire or Wirral. The best documented occurrence, and the only one in any way suggestive of breeding, was of a sub-adult male which took up residence on the Dee Marshes between 8th May and 2nd August 1980, and was seen to display with nesting material. Its arrival coincided with a large movement of Marsh Harriers during persistent south-easterly winds. On 9th June of the same year a male Montagu's Harrier which was thought to be a full adult flew north at Hale Decoy.

Goshawk
Accipiter gentilis

SHARROCK[32] describes Goshawks as frequenting extensive woodland, whether coniferous, mixed or broad-leaved, although beech and pine are preferred. The nest is usually placed high up in a fork, close to the main trunk and birds are most easily located in March and April when display flights high over the large territory may be seen, although this may not be a precursor to successful nesting (Marquiss & Newton 1982).

Up until the 1880s birds are known to have bred in the pine forests of the Scottish Highlands. However, the species had declined in England somewhat earlier, prior to most other raptors, and its history is neither well documented nor fully understood ([32], Hollom 1957). Thereafter, until the mid-1960s there were very few breeding records. Fieldwork for the BTO Atlas[32] revealed a healthier British population than for at least 90 years previously, including a small concentration in the Peak District. Since then there have been up to 40 pairs annually nesting in Britain (Marquiss & Newton 1982).

The Goshawk was unknown to Coward & Oldham[2] or to Boyd[5, 6] either as a breeding bird or casual visitor to Cheshire, and Bell[7] could give only one record, that of a bird over Caldy Hill on Christmas Day 1955. The next two records from Cheshire, as now defined for recording purposes, came in 1968. A female which had escaped from captivity at High Lane (now GMC) on 5th January was shot at Disley on 15th January[8], and a female at Rostherne on 16th October "...appeared to have something on its legs, probably bells as used in falconry" [21].

The fortunes of the Goshawk seem to be intrinsically linked with falconry over many centuries. Sub-fossil records from Derbyshire show the Goshawk to be an early colonist after the last Ice Age and beaks of "hawks" were found in four "tumuli" dug out in the nineteenth century, perhaps suggesting falconry to be even older than suspected[42]. 26 hawk's eyries were listed in the eleventh-century Domesday Book for Cheshire (Morgan 1972). Most birds in Britain today originated from falconry, either by escaping or being released from captivity (Marquiss & Newton 1982). The two 1968 Cheshire records bear this out, and a bird seen around Macclesfield in 1980 and 1981 bore a blue ring suggesting it too had escaped.

Shooter & Anthony (1981) referred to some 20 or more pairs breeding within the Peak District National Park since 1958, although this must be an over-estimate (Marquiss & Newton 1982), and more recently this population has suffered regular persecution from egg-collectors and falconers[41]. During the early 1970s birds were reported in most years from various parts of Cheshire and breeding was first confirmed soon after, with breeding attempts at the same location in most years during the survey. However, the nest was robbed in many years, and it seems doubtful whether the species will sustain its breeding attempts without recruitment of young birds, whether reared in the wild or in captivity. Breeding was confirmed at two other sites during this survey.

REFERENCES
Hollom, P. A. D. (1957): The rarer birds of prey. Their present status in the British Isles, Goshawk. *British Birds* 50: 135-136.
Marquiss, M. & Newton, I. (1982): The Goshawk in Britain, *British Birds* 75: 243-60.
Morgan P. (1987, ed): *Domesday Book: Cheshire*. Phillimore, Chichester.
Shooter, P. & Anthony, S. (undated, 1981 approx): *A Checklist of the Birds of the Peak District*.

Sparrowhawk
Accipiter nisus

FOR many years the Sparrowhawk's fortunes in Cheshire were inversely proportional to the extent of game-rearing. Prior to the First World War pairs were thinly scattered throughout the lowlands and in the eastern hills despite the attentions of keepers – a pair even built a nest on top of a rusted, unsprung gin-trap at Plumley![2]

The scarcity of cartridges during both major wars and the accompanying lapse of game-keeping and associated persecution allowed numbers to increase considerably as Boyd noted in his diary in 1943. By 1951 he found Sparrowhawks "probably not much less numerous [than Kestrels] except at Arley where the keepers discourage nesting pairs". However, due to their low, fast hunting flight they were seen much less often than the hovering Kestrel – remarks that apply equally well today.

In the late 1950s a decline set in, linked not to the activities of gamekeepers but to the much greater danger of pesticides. Toxic agricultural chemicals were absorbed by the small passerines on which the hawks feed, and accumulated in the predators causing reduced fertility and egg-shell thinning. The decline, which continued into the 1960s until the use of the offending pesticides was restricted, was most marked in the arable areas of eastern England. Although numbers in Cheshire were much reduced, the use of chemical sprays in such a pastoral county was not so extensive and even in the early 1960s a few pairs continued to breed in widespread localities[9].

By the early 1970s a recovery was well under way and the number of birds reported in the county has increased annually[8]. At the present time this may actually be our most numerous bird of prey in certain parts of eastern Cheshire, although the recent decline of Kestrels has helped the Sparrowhawk to achieve this status. It is now regularly reported from suburban gardens and a high percentage of ringing recoveries are of birds killed striking windows.

Many records of 'probable' breeding refer to displaying birds in spring. From March onwards hawks can be seen soaring high over their nesting woods, and not infrequently pairs indulge in a spectacular display, the male stooping shallowly, flicking his wings at the bottom of the dive, and bouncing back skywards while the female circles calling. Egg-laying does not usually start until May; thus the young hatch when there is an abundance of young fledged passerines on which to be fed. During incubation and brooding of the young, whilst the female remains at or near the nest, the male does all the hunting and can be watched repeatedly carrying back food. The female resumes hunting only when the

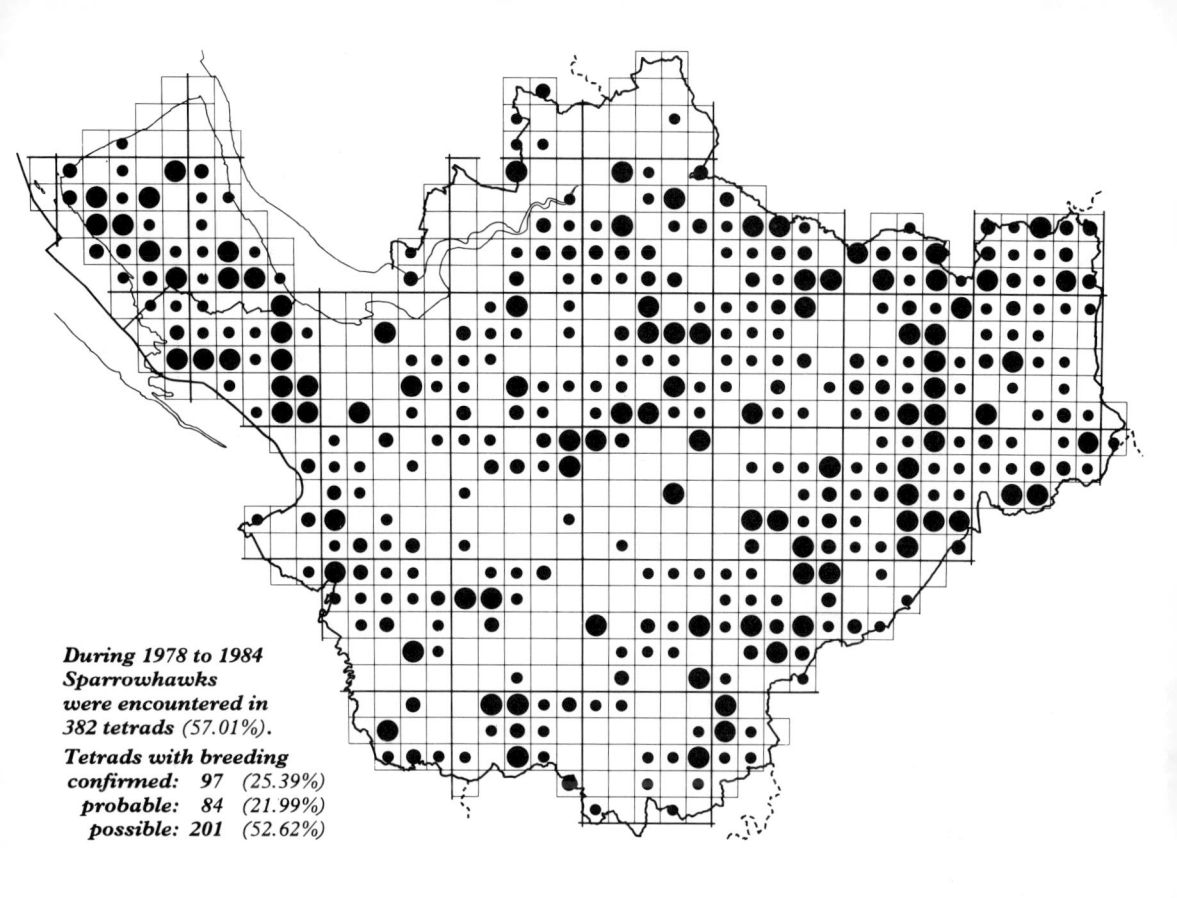

During 1978 to 1984 Sparrowhawks were encountered in 382 tetrads (57.01%).

Tetrads with breeding
confirmed: 97 (25.39%)
probable: 84 (21.99%)
possible: 201 (52.62%)

young are about three weeks old. Fledgling males, which are little more than half the weight of their sisters, often escape bullying by moving along branches away from the nest. After fledging the young remain dependent on the parents and continue to whistle noisily well into August.

The large difference in size between the adult male and female means that a wide range of prey can be taken. In the breeding season males concentrate on finches, sparrows and tits, and may take up to a quarter of fledgling Blue and Great Tits in some woods (Newton 1986). Females tend to go for thrushes and Starlings but can manage to kill Collared Doves and even Woodpigeons twice their own weight. More unusual prey has included a Black Tern, a Kingfisher, a Teal, a bat and a rabbit[8] and MRG records from Sparrowhawk nests or plucking posts in north Wirral include rings from Goldcrests, Willow Warblers and House Martins.

In the absence of any significant natural predators, the main factor inhibiting further spread of the Sparrowhawk is disturbance by man. They are still persecuted in some areas, but a bigger threat locally may now be from would-be falconers who annually pillage a number of suburban nests. However, the species is now reasonably widespread and the county population in any one survey year was estimated to be about 150 breeding pairs.

The high proportion of 'possible' breeding records may be attributed to birds hunting in adjacent tetrads or wandering non-breeding birds. However, lack of coverage in late summer when the young are noisy may have led to pairs being overlooked particularly in southern parts of the plain. The lack of woodland in this area shows up well on the map however. Concentrations of confirmed breeding records in Wirral and in the east of the county may indicate more effective coverage than the average, although these are both well-wooded areas. Conversely the absence from parts of the Delamere—Bickerton ridge may be due to inadequate coverage.

REFERENCE

Newton, I. (1986): *The Sparrowhawk*. T. & A. D. Poyser, Calton.

Buzzard
Buteo buteo

THE first apparent reference to Buzzards in Cheshire is in Daniel King's "Vale Royall" where he mentions "Ravens, Crows, Choughs [i.e. Jackdaws], Kites, Gleads and such like" as being in the Delamere and Macclesfield forests in 1617. The old word "Glead" could mean either a Kite or Buzzard but in this context the latter is perhaps intended. Mapping the Buzzard's likely breeding distribution in 1800, Moore included virtually the whole of Britain in its range, but he showed it to be absent from Cheshire by 1865 although still present in Shropshire and the northern Pennines (Moore 1957). Tubbs (1974), summarising historical "vermin campaigns", states that "From about 1670 there are abundant references to payments" for Buzzards' heads "in churchwardens' accounts from Cornwall to Lincolnshire", these entries tailing away towards the middle of the seventeenth century presumably because there were few birds left. It is possible that the Cheshire population was also severely depleted by this time although this is not evident from Moore's 1800 map.

Perhaps surprisingly, Coward & Oldham[2] make no mention of Buzzards ever having nested in the county although they described the species as "still present in the more secluded districts of the mountains of North Wales" and that occasionally birds were seen in Cheshire and Wirral. Boyd[5] in his country diary for 1st September 1939 could see no reason why Buzzards, "if unmolested", should not nest in the Pennine foothills or in the woods at Delamere or Peckforton, but knew of only two summer records between 1924 and 1936. Later, he added a sighting in May 1940[6]. In his survey of 1954, Moore (1957) mapped Shropshire, Staffordshire and Derbyshire as having low-density breeding populations but showed none in Cheshire.

1955 marked the start of a long struggle to recolonise the county. A pair nested in Eaton Park, Chester and an attempt at Rostherne aborted when one of the adults was shot[21]. In 1957, a bird spent the summer near Shutlingsloe and, in 1962, a pair summered in Eaton Park, where one was also present in 1964. In 1965, a pair was seen around Peckforton, where single birds were seen in the spring of 1964 and again in 1966, although there was no evidence of nesting.

Buzzards have been regularly reported since the late 1960s/early 1970s. A spate of records from the Tatton and Rostherne areas occurred at this time, with display observed from 1971 and nest-building reported. In 1972, a party of six was seen nearby in October and in 1973 the frequency of sightings, including display, indicated that three or four birds were resident in the area. It seemed that a small population was becoming established and there were reports of attempted breeding in Tatton Park that year. Then, in March 1976, an adult male and a first-year female were found lying dead beside a rabbit carcass in Tatton Park; post mortems proved poisoning. Although three or four birds were seen later that year sightings have been infrequent since and almost none have been reported since 1980 when a further dead bird was found in the Park.

At Peckforton one or two were present regularly from 1973 to early 1975, in which year a pair was said to have bred near Congleton and at another unnamed site. Subsequently, the period of this survey brought the best run of breeding records for over a century, with the following reports:

1978	singles at Tatton in summer.
1980	one pair bred; one at a second site in June.
1981	one pair bred.
1982	one pair bred; display observed at a second site.
1983	one pair bred.
1984	one pair bred; display observed in east Cheshire.

Perhaps this heralds an eastward expansion from the Welsh hills where there is a thriving breeding population.

REFERENCES

Moore, N. W. (1957): The past and present status of the Buzzard in the British Isles, *British Birds* **50**: 173-197.

Tubbs, C. R. (1974): *The Buzzard,* David & Charles, Newton Abbot.

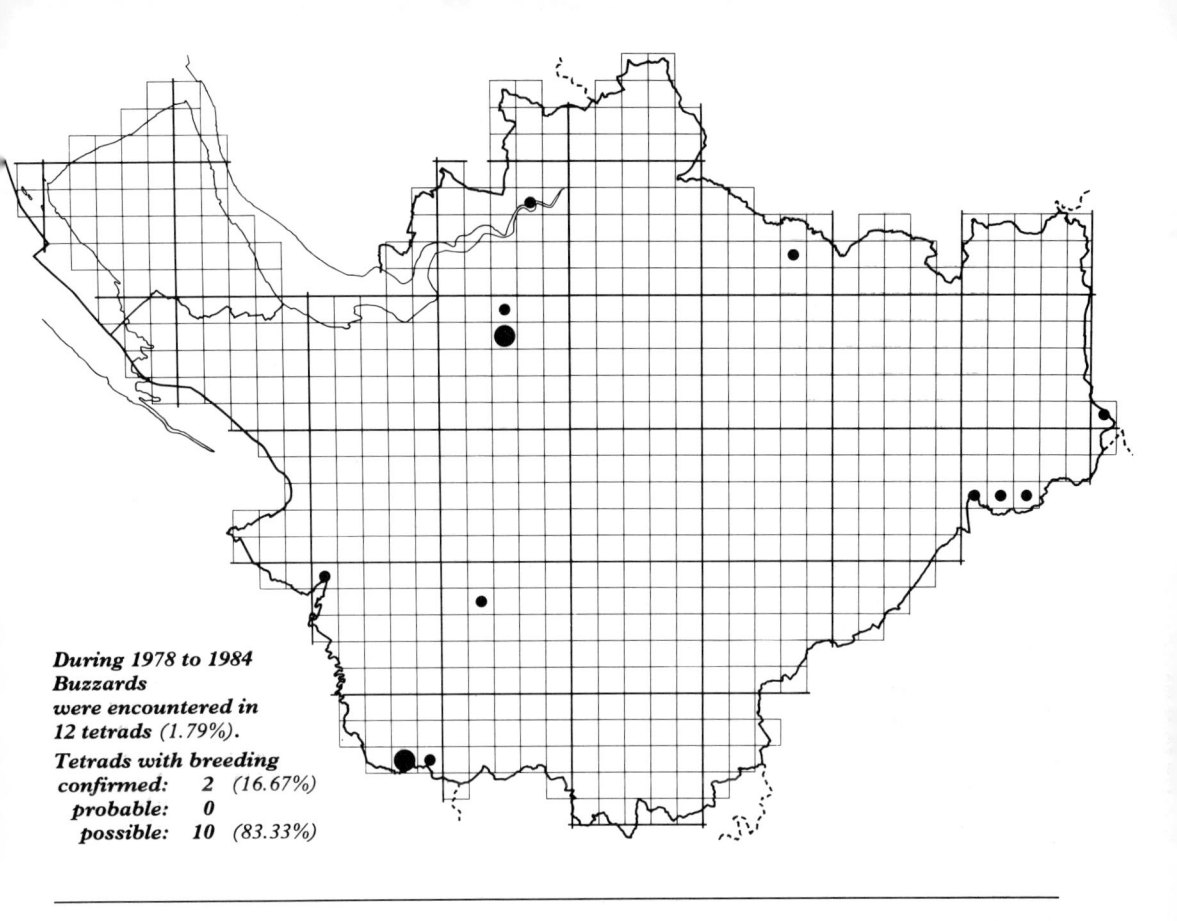

During 1978 to 1984 *Buzzards* were encountered in 12 tetrads (1.79%).

Tetrads with breeding
- confirmed: 2   (16.67%)
- probable: 0
- possible: 10   (83.33%)

Kestrel
Falco tinnunculus

As in Coward & Oldham's day, when there usually two or three Kestrels to every Sparrowhawk on a gamekeeper's gibbet, this remains the most familiar and widespread bird of prey in the county. Hardy[4] noted it as a "...common nester...often seen in the suburbs..." and that on the Gowy marshes and around Ellesmere Port numbers fluctuated according to the local vole populations. Craggs[22] recorded birds breeding on Hilbre in the three years 1945-47 at a time when short-tailed voles were particularly common. In 1943 Boyd[5] wrote that there were more Kestrels than ever, possibly owing to the lack of cartridges during the war years. Bell[7] also described it as "...much the commonest hawk..." and *CBRs* in the 1960s/early 1970s suggested this continued to be the case. However, it may now be outnumbered locally by the Sparrowhawk and there is some evidence that it has declined in certain areas.

From early spring the pair displays near the nest-site, circling with tails spread and rapidly flickering wings, the male sometimes stooping repeatedly at the female and all the while calling "kek-kek-kek...". The nest is typically in a hollow tree, tall beeches being a particularly favoured site, but old nests of Crows, Magpies, pigeons or other large birds, or ledges on bridges, chimneys, pylons, buildings or a quarry-face are used, as are suitably large nest-boxes. Brockholes[1] knew of eggs being laid in the old nest of a Sparrowhawk. Young birds call loudly and give away the location of many nests, occasionally attracting the attentions of young would-be falconers, particularly in suburban areas where some nests are robbed.

Harrison & Rogers[21] describe how birds attempting to breed at Rostherne were persecuted by Jackdaws unless the latter had already begun to breed. In 1981, at Watch Lane Flash, Sandbach, a pair of Kestrels was watched chasing a resident pair of Carrion Crows from their nest. This harassment went on for several days with a second pair of Crows often becoming involved. The Kestrels won the dispute and successfully raised three young in the Crows' nest.

Hunting birds hover conspicuously over patches of rough grassland such as the headlands of fields, motorway verges or industrial wasteground, and, as they will hunt at a considerable distance from the nest, many records of 'possible' breeding will refer to nothing more than wandering birds on the look-out for food. Small mammals form the bulk of the Kestrel's diet with birds of secondary importance. Boyd[5] had seven pellets from a nest analysed and they contained the remains of at least two wood mice and a common shrew. Several of the pellets contained only fur. More recently, pellets collected from beneath a roost at Prestbury Sewage Works showed that 96% by weight of that bird's diet consisted of field voles. Urban Kestrels tend to take more birds than rural ones, and remains reported at nests include pigeons and Starlings.

The increasing scarcity of this falcon in some rural areas is probably attributable in part to more intensive land-use, rough fallow areas being increasingly pressed into agricultural use. The much publicised association with motorway verges may not be entirely beneficial: as well as hunting voles in the verges, Kestrels drop onto the carriageway to take struggling birds and insects maimed by the traffic and themselves fall victim to oncoming traffic.

In 1972 at least six pairs bred within three miles of Wilmslow station, but only two tetrads in the same area held breeding pairs during any of the seven years of the present survey. By 1977 a similar decline was reported in south-east Cheshire and the 1978 *CBR* stated that it had declined considerably in the east of the county. Indeed the map confirms the relative scarcity of the species away from the Mersey valley or Wirral. Kestrels are little affected by urbanisation, indeed they often benefit from the rough farmland associated with green-belts, other rough ground around sewage works and railway embankments for example, and from the availability of nesting ledges on factory chimneys and other industrial structures.

In 1981 three occupied nests were found in one tetrad west of Northwich, but most occupied tetrads hold only single pairs, and only records of 'probable' or confirmed breeding should be taken to signify nesting pairs. With 339 such tetrads the county population is probably about 350 pairs. The BTO Atlas estimate of 37.5 pairs per 10-km square would give a figure of 1000 pairs and this was clearly too high by 1984, although just how much the Cheshire population may have fallen in the years since that publication is unknown.

Merlin
Falco columbarius

As a breeding bird, the Merlin has had only a marginal status in Cheshire throughout this century. Coward & Oldham[2] mention one breeding record from Longdendale (now GMC/Derbyshire) in 1894, and thought that the odd pair might breed on the moors east of Macclesfield. Bell[7] reported a female being flushed from three eggs in a tree nest on the Longdendale Moors in 1952 and that the same nest had been used in 1951. Since then the only indication of possible breeding is isolated records from the moors east of Macclesfield in 1966, 1970 and 1972[8].

Coward & Oldham also thought it possible that the species used to nest on the extensive mosses of the plain and mention a freshly-killed bird gibbetted on Carrington Moss (now GMC) in 1883. Hardy[4] considered a pair, which used an old Crow's nest in an oak at Oulton Park in 1911, were more likely to have been Kestrels. However, Boyd (1942), reviewing Hardy's book, commented that, "...the hen (which was shot) was identified by the late T. A. Coward, and... Merlins nested and reared a brood some years later in the same place, as the present writer can testify".

In a review of the species' status and recent history, Newton *et al.* (1981) collated evidence of 45 pairs nesting in the Peak District during the 1950s, only one of which was in Cheshire. They documented the

During 1978 to 1984 Kestrels were encountered in 564 tetrads (84.18%).

Tetrads with breeding
confirmed: 187 *(33.16%)*
probable: 151 *(26.77%)*
possible: 226 *(40.07%)*

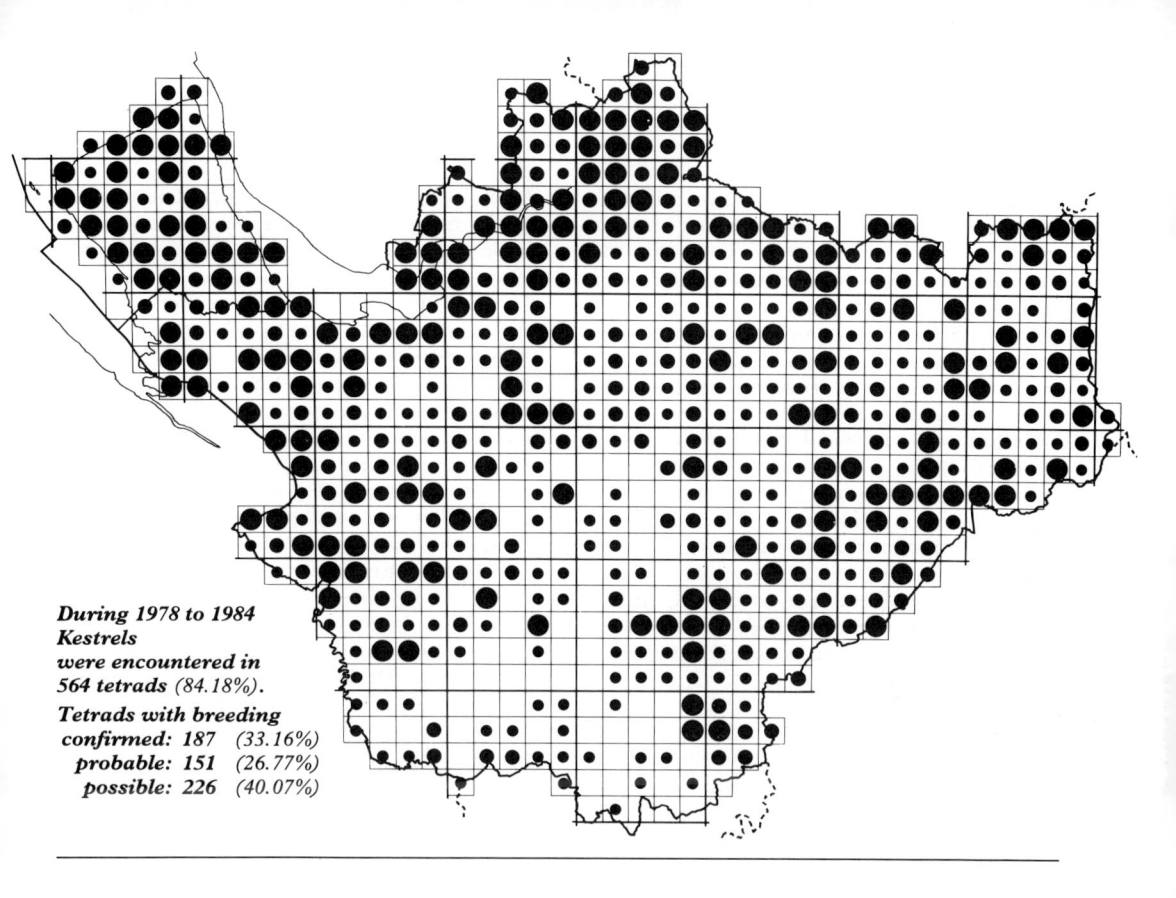

catastrophic decline in the Peak District population that occurred through the 1960s and 1970s, until, in 1980, the species was probably extinct as a breeding bird. In these circumstances it is not surprising that there were no proven or 'probable' breeding records submitted during this survey; indeed there was only a single record of 'possible' breeding, i.e. of a bird seen in suitable habitat.

Newton *et al.* (1981) discussed the probable causes of the decline. Although there has been some loss of habitat, much suitable terrain remains available and the food supply (Meadow Pipits, especially) seems still abundant. Though recreational disturbance and accidents, such as moorland fires, have also cost some nest-sites, there are still many former sites that seem unaffected. Their conclusion was that organo-chlorine pesticides were primarily to blame, and very serious contamination of Peak District eggs was recorded. However, there are at least hints that the bird's status in neighbouring counties is starting to improve, and it is reasonable to hope for a recovery of the population over the next decade or two. On a well studied moor in Clwyd however numbers, having increased from four or five pairs in the mid-1960s to eight pairs in 1975 and 1976, had decreased to only one pair in 1982. The decline was attributed to moorland fires, the taking of both eggs and chicks, and high pesticide residues causing addling, egg-breakages and chick deaths (Roberts & Green 1983).

DWY

REFERENCES

Bibby, C. J. & Nattrass, M. (1986): Breeding Status of the Merlin in Britain. *British Birds* **79**: 170-185.
Boyd, A. W. (1942): Hardy on the Birds of the Liverpool Area. *Ibis* pp290-291.
Newton, I., Robson, J. E. & Yalden, D. W. (1981): Decline of the Merlin in the Peak District. *Bird Study* **28**: 225-234.
Roberts, J. L. & Green, D. (1983): Breeding Failure and Decline of Merlins on a North Wales Moor. *Bird Study* **30**: 193-200.

Hobby
Falco subbuteo

THE Hobby is a trans-Saharan migrant present in its restricted breeding area for only five months of the year. Typically, the species is found on dry heaths and downland in southern England with areas of open country for hunting, nesting in isolated clumps, shelter-belts or tall hedgerow trees[32].

In his claim to retain the master-forestership of Mara and Mondrem (now Delamere) in the mid-fourteenth century, Richard Done "...claymeth to have all sparhawkes, marlens and hobbys found within the said forest..." (Husain 1973). No doubt the chicks were intended for falconry. Coward & Oldham[2] quoted two instances of Hobbies nesting around Delamere, at Vale Royal in 1895 and at Oakmere in 1898, adding that a Mr F. Nicholson wrote in 1875 that it had occasionally done so. In 1894, a pair bred near Buxton, Derbyshire[42].

Only three birds were recorded within the present county boundary in the first half of this century, all of them being shot, the last at Alderley Edge in 1934. In 1938, a pair almost certainly bred in the Goyt Valley, which then formed the Cheshire/Derbyshire boundary[42]. There has been a marked resurgence since 1966 when one was watched over Tatton Mere on 25th June, and since 1970 birds have been recorded annually, with at least nine birds in 1979 and 1984. Similar increases have been recorded in the West Midlands since the 1940s[44] and, to a lesser extent in Derbyshire since 1955[42]. However, most of these records will relate to either passage birds "over-shooting" their more southerly breeding areas in spring or, as Sharrock[32] suggests, roaming immatures summering in Britain but still too young to breed.

The breeding range of the Hobby has been spreading slowly northwards in recent years. There was a slight suspicion of nesting in south Wirral in 1973 and 1975 and, in 1982, two birds were seen together in Cheshire for the first time this century. In 1984, two birds were regularly seen in the eastern hills on the Cheshire/Staffordshire border between 28th July and 1st September. That same year a pair nested successfully in Derbyshire and by the end of this survey birds had probably bred in Shropshire.

Reports of birds breeding in open farmland are increasing[40]. The 1984 Derbyshire nest was in a small wood amidst pasture and arable farmland[41]. Given the abundance of such habitat in Cheshire it is conceivable that a pair or two may already be nesting undetected.

REFERENCE
Husain, B. M. C. (1973): *Cheshire under the Norman Earls.* Cheshire Community Council.

Peregrine
Falco peregrinus

COWARD & Oldham[2] indicated that "...although some of the rugged escarpments in Longdendale afford suitable nesting sites, the constant vigilance of the gamekeepers makes it impossible for the birds to establish themselves..." and they found a report of nesting for two or three seasons (possibly prior to 1880) at Manley in Delamere to be implausible given the appearance of Simmonds Hill, the supposed nesting site.

There is no obvious site in the eastern hills where the species might nest, Longdendale now no longer being part of Cheshire. However, the use of buildings or industrial structures cannot be ruled out, especially given a record of two young found dead around 1947 at a site on a Liverpool warehouse (Ratcliffe 1980).

REFERENCE
Ratcliffe, D. (1980): *The Peregrine Falcon.* T. & A. D. Poyser, Calton

During 1978 to 1984 Red Grouse were encountered in 11 tetrads (1.64%).

Tetrads with breeding
confirmed: 6   (54.55%)
probable: 3   (27.27%)
possible: 2   (18.18%)

Red Grouse
Lagopus lagopus

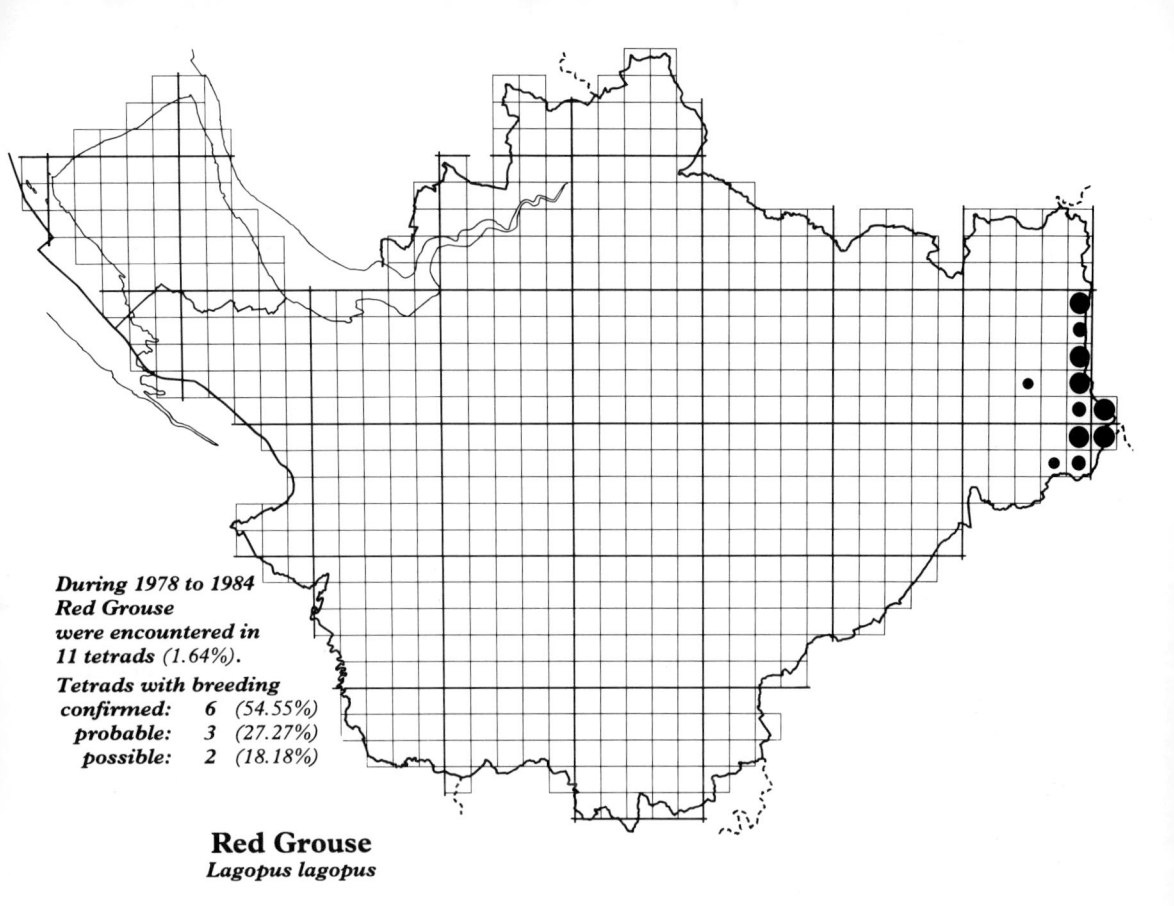

THE Red Grouse occupies an equivocal position: on the one hand, as a game bird, it is ignored by some bird-watchers, being regarded not much more highly than are feral pigeons; but, on the other, it is a major sporting bird, subject of an important tourist industry, and a major research animal for many zoologists. Its ecology has been studied possibly more thoroughly than that of any other British bird.

Red Grouse are conspicuous enough when holding territory and their typical moorland habitat is so clearly demarcated that predicting their likely presence (or absence) is straightforward. "Clocker droppings", (large, globular droppings deposited at specific locations) left by incubating hens which leave their nests to defecate infrequently, are a good indication of breeding, but the nests themselves are obvious enough, especially after use when the eggshells, split around the circumference, are left in the nest. Parents are generally very attentive, giving a spectacular distraction display, while the flying young remain in family parties for several weeks. This not only provides ready confirmation of breeding for an atlas like this, but also allows a biologist to assess breeding success in a population by counting the proportion of young to adults in early August.

Despite its economic importance, the Red Grouse population, in Cheshire as elsewhere, has shown a steady decline. Formerly it bred on the lowland mosses of both Cheshire and Lancashire including Carrington Moss and Chat Moss (both now GMC) and Holcroft, Rixton and Padgate Mosses near Warrington. Birds even wandered to Wirral on occasion[2]. As with the Black Grouse, Lord Delamere attempted unsuccessfully, around 1902, to reintroduce birds to Newchurch Common. Carrington Moss was a well-stocked grouse moor until 1886, but it was then bought and converted to agricultural land: according to Coward & Oldham[2] the last sighting was of five on 25th July 1894. Hardy[4] indicated that there were still a few nests on the adjacent Chat Moss in 1935 and that wandering birds were to be found on the Warrington mosses up until that time, but they seem to have died out soon after. Thus the Grouse became confined to the hills.

In 1910 Coward & Oldham were able to write "From the hills of Lyme southward to Bosley Minn and east to the Derbyshire border, the bird abounds wherever the moorlands are uncultivated. In the upper part of the Goyt Valley and elsewhere in Macclesfield Forest, the hillsides for miles are clothed with ling, heather, bilberry and cranberry...".

Bell[7] mentioned a few records of stragglers from the lowlands, but otherwise dismissed it as still common on the heather-clad hills. However, during a survey in 1969 based on 1-km squares, birds were found in only 53 squares (Yalden 1970), showing that, even before the loss of Longdendale, the species occurred in only 2% of Cheshire's area; within the present boundary it occurred in only 14 squares. A patch of heather on Bosley Minn was overlooked and although it held a pair in 1973[8] it has since been limed and heavily grazed, and grouse have not been recorded there in recent years. Yalden (1979) suggested that the population within the present county boundary might have been around 185 pairs in the early 1970s. There are few records in *CBRs* until 1983 by which time every sighting was of importance. Fieldwork for this atlas recorded the species as confirmed or 'probably' breeding in only nine tetrads, with a further two 'possible' breeding records; these records apply, in fact, to about 18 one-kilometre squares. However, various lines of evidence suggest that the 1984 population of the area was only about half what it was in the early 1970s, perhaps less. 15 males were noted in a survey of most of the Cheshire moorland on 14th May 1984. Normally one would expect to see only about 30% of the population at that time of the year, thus from "abounding" in Coward & Oldham's time, the population appears to have declined to only about 50 pairs.

The decline of the species reflects directly the decline of its heather-dominated habitat (Yalden 1981). A small amount has been lost to forestry, but the greatest losses have been to pastoral farming. In some cases this has been an active process: with substantial subsidies on offer, farmers have drained, ploughed, limed, fertilised and reseeded moorland converting it to good pasture. Just as significant has been the more passive change due to overgrazing. Sheep prefer to eat grass, but in a grass/heather mosaic eat away the edge of the heather so that it slowly retreats. The decline of bilberry is perhaps even more important, because the new spring shoots of bilberry are more nutritious than heather and enable the hen grouse to form her eggs. However, sheep also find bilberry very nutritious, especially over winter. The present economics of hill use are weighted very much against the grouse and in favour of farming: the sheep themselves are subsidised, and no rates are levied on farmland, whereas grouse moors are heavily rated. It remains to be seen whether changes in the Common Agricultural Policy are in time to affect the fortunes of the Red Grouse as a Cheshire bird.   DWY

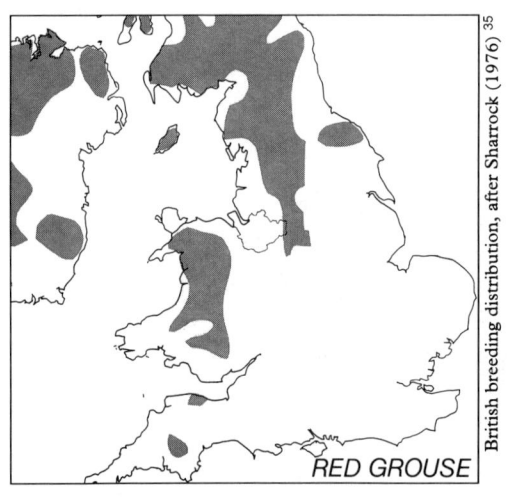

British breeding distribution, after Sharrock (1976) [35]

RED GROUSE

REFERENCES

Yalden, D. W. (1970): The Distribution and Status of the Red Grouse in Cheshire. *Cheshire Bird Report* 1969: 22-26.
Yalden, D. W. (1972): The Red Grouse (*Lagopus lagopus scoticus*) in the Peak District. *Naturalist* 922: 89-102.
Yalden, D. W. (1979): An Estimate of the Number of the Red Grouse in the Peak District. *Naturalist* 104: 5-8.
Yalden, D. W. (1981): Loss of Grouse Moors; pp 200-202 in Phillips, J., Yalden, D. W. & Tallis, J. - *Peak District Moorland Erosion Study: Phase 1 Report*. Peak Park, Bakewell.

Black Grouse
Tetrao tetrix

THE Black Grouse has been declining in both numbers and range in England for at least 200 years: the scrubby moorland-edge habitat which it prefers is all too easily "tidied up" by farmers or foresters or degraded by over-grazing. In Cheshire, Coward & Oldham[2] referred to the species breeding in the plantations and wooded cloughs between the Goyt and the Dane. They specifically mention Lyme, Bakestonedale Moor, the Goyt between Whaley Bridge and Goyt's Moss, wooded cloughs between Sutton and Bosley, on Bosley Cloud, and the Dane Valley between Bosley and Wincle. They also mention its former occurrence on the wooded heaths of the Cheshire plain: at Delamere Forest up to the 1860s/70s, at Rudheath, Allostock in the early part of the nineteenth century, and at Roe Park, Congleton. In 1901 Lord Delamere attempted to re-establish the species at Vale Royal by releasing birds on Newchurch Common and around Abbots Moss but the attempt failed because of the abundance of foxes; few birds survived the first year and the last was seen in 1905[2]. Hardy[4] referred to the species being "...common at one time near Ince and Frodsham...". This was probably long before "...a cock was shot at Ince Hall, November/December 1888". Boyd[5] in his country diary for 31st October 1936 commented that there was some evidence that Blackcocks were more numerous in the eastern hills. This was considered to be good news as the species had declined in many of its previous haunts. However, Bell[7] regarded the upper Goyt (now Derbyshire) as the only reliable locality for the species, and could give no recent breeding records. The *L&CFC Report* for 1963 recorded successful breeding at Danes Moss in 1960, and that appears to be the last breeding record for the county, although birds were present here until 1962.

The sharp decline in the species' fortunes in Derbyshire between 1960 and 1975 was documented by Lovenbury *et al.* (1978), but the species was then still reasonably numerous in the adjacent parts of north Staffordshire (55 cocks was their estimate of the population). They were able to quote a few sightings for Cheshire which seemed likely to be wandering birds from the Staffordshire population, and the scatter of occasional records continued up to 1980 when one was seen near Macclesfield Forest[8]. However, the population in Staffordshire has declined sharply since 1978, and is believed now to include no more than 20 breeding males (Yalden 1986). A report of lekking at Wincle in 1975[8] was probably the last occasion birds displayed in Cheshire.   DWY

REFERENCES

Lovenbury, G. A., Waterhouse, M. & Yalden, D. W. (1978): The Status of the Black Grouse in the Peak District. *Naturalist* **103**: 3-14.

Yalden, D. W. (1986): The Further Decline of Black Grouse in the Peak District 1975-85. *Naturalist* **111**: 3-8.

Red-legged Partridge
Alectoris rufa

THE Red-legged Partridge is a bird of sunnier climes, introduced into Britain from south-western Europe for sporting purposes. As such it prefers drier soils, and for this reason is restricted to southern and eastern England, avoiding the wetter north and west. As the distribution map for the UK shows, Cheshire lies on the edge of the bird's range.

Coward & Oldham[2] knew of birds being reared occasionally from introduced eggs, but as the species had never succeeded in establishing itself in the wild they did not consider it a Cheshire bird. Hardy[4] noted that "shortly after...[1935]...others were introduced to the Leverhulme estate in Wirral...". Bell[7,9] listed records of breeding at Storeton, Wirral in 1939, Delamere in 1954 (where a pair was believed to have bred in 1956 also), Audlem in 1963, Puddington, Wirral in 1964 and at least four pairs in the Peckforton area in 1965 - all were believed to refer to introduced birds. Birds were also liberated at Tatton in 1957[24].

Since 1970 records have become more frequent. Coveys of ten at Doddington in 1974 and 21 at Wybunbury in 1975 marked the start of a major programme of releases on the Checkley estate (Betley) where 400 were "turned down" in 1976. By 1979 large numbers were also being released on the Withington estate, a figure of 1600 being quoted for 1980. Subsequently birds have wandered out from these centres and in addition other minor releases have been made, as at Adlington and in the Comberbach area. Birds were also released in the Dibbinsdale area in 1983 and were subsequently seen frequently around the Dibbinsdale and Heswall area, although these birds were not reported for this survey, perhaps reflecting observers' traditional scepticism of introduced species. In fact the species was considerably more widespread and numerous by the end of the survey than at any time previously. Coveys will wander considerable distances, and the sprinkling of records in the eastern hills is thought to relate to birds travelling from Adlington or elsewhere into the higher country. Chalk downs and rocky slopes are favoured habitats elsewhere in the species' range, and birds calling on top of dry-stone walls at the edge of peat moors seem somewhat incongruous. A small population has persisted in the vicinity of Dale Top since 1978 or earlier.

The lowering of the water-table due to more efficient drainage and consequently drier topsoil may have favoured the species' present success in the county, and it is perhaps significant that the Withington population is based on an area of glacial deposits where continuing sand-quarrying activities have lowered the water-table even further. In 1972 it became known that game-farm stocks included the closely related Chukar *A. chukar* and many of the eastern Cheshire birds appear to be Chukar/Red-legged hybrids *A. chukar* x *rufa* which may also be more vigorous under our climatic conditions (Hudson 1972).

Where the species is numerous, pairs are frequently encountered in arable fields before the crops have grown up in the spring, and even isolated cocks can be located by their loud, clucking calls, rather resembling small

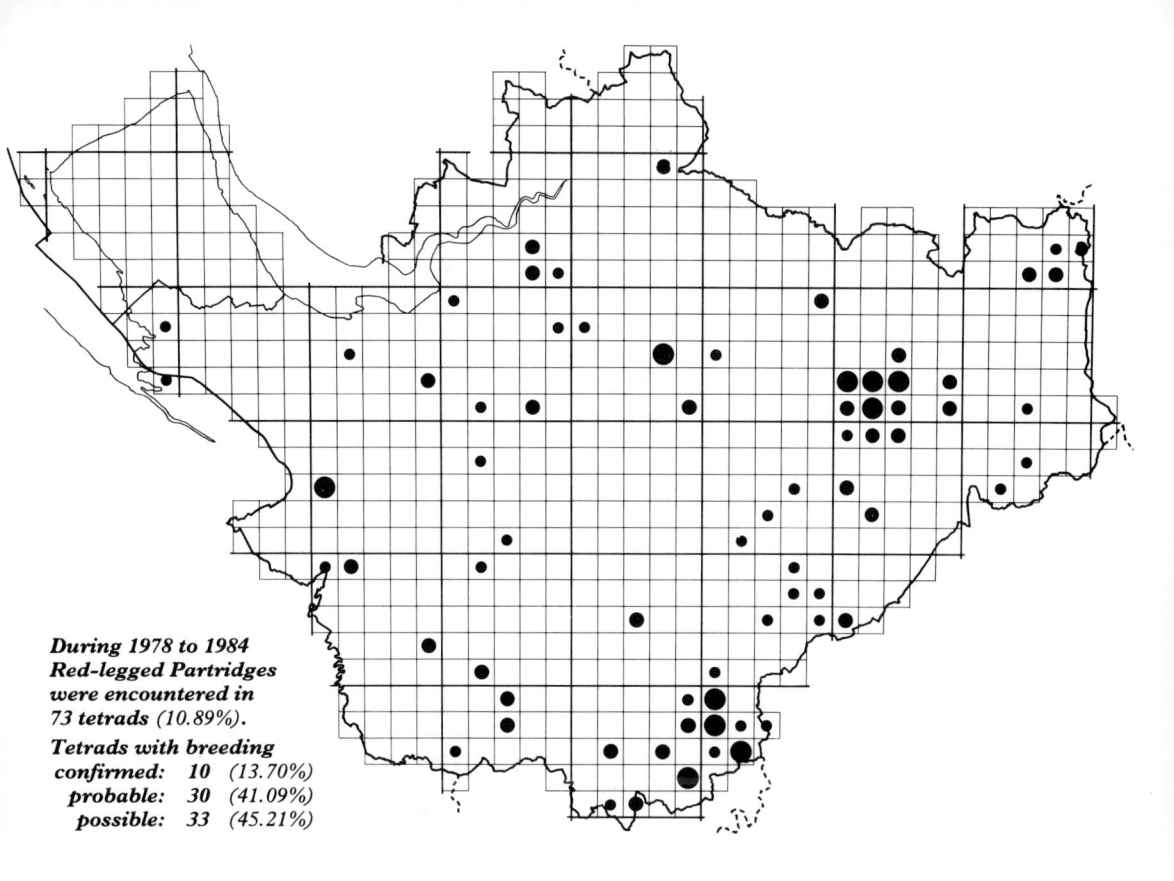

During 1978 to 1984 Red-legged Partridges were encountered in 73 tetrads (10.89%).

Tetrads with breeding
confirmed: 10 *(13.70%)*
probable: 30 *(41.09%)*
possible: 33 *(45.21%)*

steam-engines. However, as with other birds of arable land, and particularly game-birds, proof of breeding can be difficult to obtain; few observers found nests. Clutches are often laid in two nests with the male and female each raising a brood[35] and many sightings of individual birds as well as pairs may have indicated breeding attempts.

With large numbers of hand-reared birds, it is difficult to assess the feral breeding population, which may have been between 50 and 100 pairs at the time of the survey.

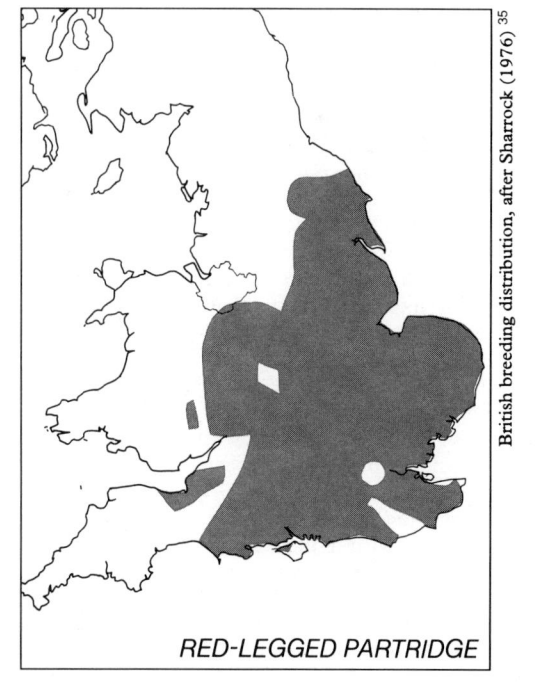

British breeding distribution, after Sharrock (1976) [35]

RED-LEGGED PARTRIDGE

REFERENCE

Hudson, R. (1972): Gamebird introductions, News and Comment. *British Birds* **65**: 404-405.

Grey Partridge
Perdix perdix

THE rasping calls of the Partridge were once a typical sound of the spring twilight throughout rural Cheshire, but the decline of game preservation and the intensification of agriculture have resulted in a considerable contraction of the species' range within the county. Coward & Oldham[2] acknowledged that its abundance in lowland Cheshire and its presence in numbers on higher ground was due to protection. Boyd[6], examining the game-books of the Arley estate, showed that in ninety seasons from 1853 to 1948 (1914-19 omitted) birds were most plentiful between 1865 and 1877 with a good period from 1893 to 1899, after which bags were small. Average bags for the three 30-season periods from 1853/4 to 1882/3, 1883/4 to 1912/13 and 1913/14 to 1947/8 were 146, 100 and 21.5 birds respectively although the last did include the period of the Great War for which no figures were recorded. He also noted an increase during and after the 1939-45 war which he attributed to the increased acreage under corn.

By the 1960s a further decline was noticed throughout Britain which Potts (1970) considered to be "probably the most spectacular status change of any common bird species in recent years". He pointed out that it was the only farmland species to have declined throughout the decade and that the decline was unprecedented. The *L&CFC Report* for 1966 referred to a "very great decrease in numbers" in Wirral, and the 1968 *CBR* recorded a general impression that the species was declining throughout Cheshire, although there was no objective evidence to confirm this. This resulted in a considerable number of records being submitted for the 1969 *CBR*, and a request that more attention should be paid to this species resulted in even more records for 1970. The 1970 *CBR* suggested that there were few birds left in Wirral, whilst in central, north and east Cheshire there were still plenty of birds although numbers were not very large and, in many instances, were probably maintained by releases of hand-reared birds. By 1975, it was reported that the species was apparently recovering from the very thin distribution of a few years previously. In 1976, something of a resurgence was noted in eastern Cheshire following the hot summer (JPG), but the trend has been downward ever since.

A Partridge shoot is still held in most winters on the Frodsham Marshes, where 74 were seen on 3rd February 1980 and 60-70 on 27th December 1982. The only other sizeable coveys in recent years have all been in the north or west of the county, for example 49 at Moss Side on 17th February 1980; 32 at Parkgate on 15th October and 43 in the Fiddler's Ferry/Moss Side area on 30th November, both in 1983; 47 at Woolston Eyes on 27th October, 38 at Dungeon, Heswall on 27th December and, exceptionally, 185 in the Great Sankey area on 19th November, all in 1984.

Partridges often nest beneath hedges, particularly under a low horizontal spray, or similar cover in mixed farmland areas. They lay large clutches, for example two hens laid 36 eggs in one nest at Norton Priory[4], a clutch of 19 at Smallwood and a brood of 20 at Malpas in 1974, and 18 eggs at Danes Moss in 1977. Records prior to 1975 suggest that pairs in "keepered areas often reared up to ten or more young whilst birds nesting on waste ground, an increasingly common habit, seldom rear more than two or three" (JPG/[21]).

The noisy, cheeping young, which fly long before they are fully grown, feed on invertebrates, particularly ants on grassland and aphids and tenthredinid sawfly larvae amongst corn and other arable crops. Increased use of chemical herbicides and the decline in undersowing of cereals has had the effect of reducing the availability of such food. Potts (1970) has also shown that late, damp springs, which have been more frequent in recent years, have further reduced the food supply for newly hatched chicks. This is exacerbated by the fact that, unlike most other species which can delay their breeding cycle, in some cases by several weeks, the Grey Partridge can only delay by a maximum of three days. Consequently chicks hatch before sufficient prey items are available.

The map suggests that Grey Partridges are absent from large areas of south central Cheshire, where the emphasis is very much on dairy-farming, grass being cut early for silage, leaving a lack of cover for the young. However, despite intensive survey work, few

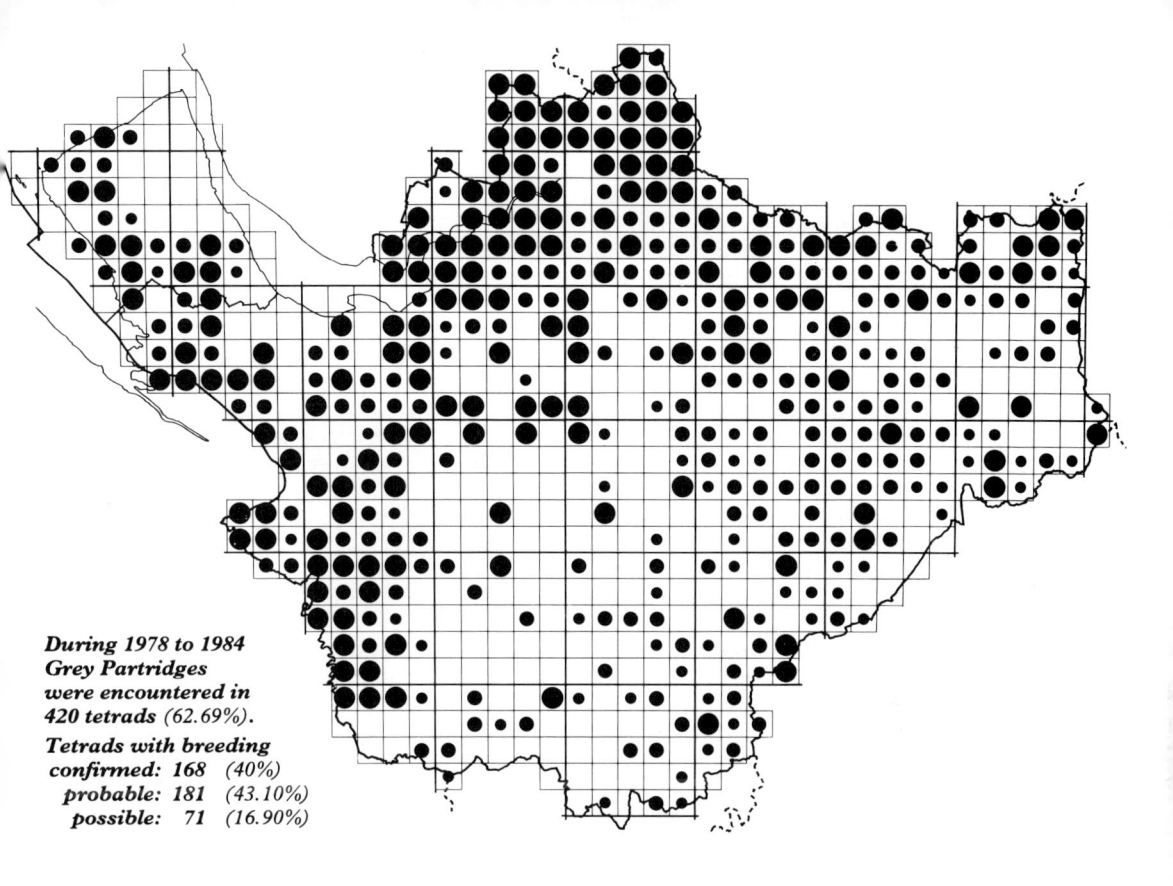

During 1978 to 1984 Grey Partridges were encountered in 420 tetrads (62.69%).

Tetrads with breeding
confirmed: 168 (40%)
probable: 181 (43.10%)
possible: 71 (16.90%)

pairs were proved to breed in east Cheshire where there is a considerable amount of arable farming. This indicates a scarce if widely scattered population in that area. In the hills the population remains small, although Coward & Oldham[2] alluded to its once having been more numerous there "...when, owing to the high price obtainable for British wheat, more land was under corn". The Grey Partridge in Britain is essentially a sedentary species although one bird ringed near Frodsham during the breeding season became a road casualty at Ellesmere Port, 13 kilometres west, two years later and this may indicate that some spread from traditional areas could occur.

Grey Partridges are most easily located at dawn or dusk when territorial birds are calling and displaying, thus providing many of the 'probable' breeding records. A lack of evening visits may therefore have led to some under-recording in south and central Cheshire.

Confirmation of breeding is difficult, however, except where the species is numerous and the chance of encountering broods of chicks is relatively high; this was the most frequent method of confirming breeding. Clearly the present strongholds are along the Mersey valley in the north, where cereal-growing is prevalent and low hedges remain numerous, in the game-rearing country south of Chester, and in Wirral. Few Grey Partridges are reared today, although some are still released on the Eaton estate, around Widnes and elsewhere. A breeding population of one pair per three acres is achieved through game management on Frodsham Marsh, showing what can be sustained in good habitat. Densities elsewhere are much lower however, and an assumption of five pairs for each tetrad with confirmed and probable breeding and two for all other occupied tetrads would give a county population of 1100 pairs.

REFERENCE

Potts, G. R. (1970): Recent changes in the farmland fauna with special reference to the decline of the Grey Partridge. *Bird Study* 17: 145-166.

Quail
Coturnix coturnix

THE Quail appears never to have been more than an erratic summer visitor to Britain although various authors refer to the species having been common up to the end of the nineteenth century (e.g.[35]). However, Moreau (1951), researching the history of the Quail, considered that "there was undoubtedly a marked and progressive decline of Quail in Britain in the first half of the 19th century" and that by the time of More's 1865 assessment of the distribution of birds during the nesting season, the species "had already reached nearly the low ebb at which it was to continue on into the twentieth century". Moreau further suggested that human predation, particularly in the Mediterranean countries, rather than changes in agricultural practices, caused this "long downward tendency". He went on to record a slight upturn in numbers from 1942 suggesting that protection was the main cause.

In Cheshire, the Quail was classed as "occasional" by More in 1865 and in 1910 Coward & Oldham said it was "...an occasional summer visitor...noticed at irregular intervals". Boyd[5] referred to them as "...most erratic in their appearance" and Bell[7] did not consider its status had changed. It is recorded more frequently now than at any time since records began, although this probably reflects the abundance of birdwatchers (and the extent of the recording network) rather than any real increase in numbers of birds. The species' arable habitat, increasingly devoid of weeds and invertebrates, must be far less favourable now than in centuries or even decades past.

Irruptions of Quails into Britain have been documented on a number of occasions and are often associated with warm dry springs and southerly or south-easterly winds which cause over-shooting of the normal breeding range in western Europe. Notable irruptions occurred in 1870 and 1893, but there were no more good "Quail years" until 1947. Few years have passed since the 1964 *CBR* without at least one or two Quails, usually cock birds uttering the distinct trisyllabic call "wet-my-lips", being recorded in the county. 1964 was an outstanding year, with at least 32 birds noted within the present county boundary. During the present survey smaller invasions were detected in 1979, when up to eight were calling around Doddington, and in 1982. It is possible that these last two irruptions were detected mainly because observers taking part in fieldwork for this survey were active in rural areas which are otherwise seldom visited.

Cheshire Quails occur primarily in grass fields, from which they are often evicted by silage cutting, or in cornfields. Undulating fields seem to be preferred, and a few areas seem to be particularly favoured, notably around Marton and Spurstow. In 1983, a bird was calling near Haslington as early as 24th April, but normally none is heard until late May or June, with records continuing into early August. A late bird was seen at Doddington on 27th October 1979.

Confirmation of breeding is difficult to obtain because of the species' habitat and secretive behaviour. The only such record during the present survey, at Doddington in 1979, came from a farmer who saw young whilst cutting barley in late August.

REFERENCES
Moreau, R. E. (1951): The British Status of the Quail and some Problems of its Biology. *British Birds* **44**: 257-276.
Moreau, R. E. (1956): Quail in the British Isles, 1950-53. *British Birds* **49**: 161-166.

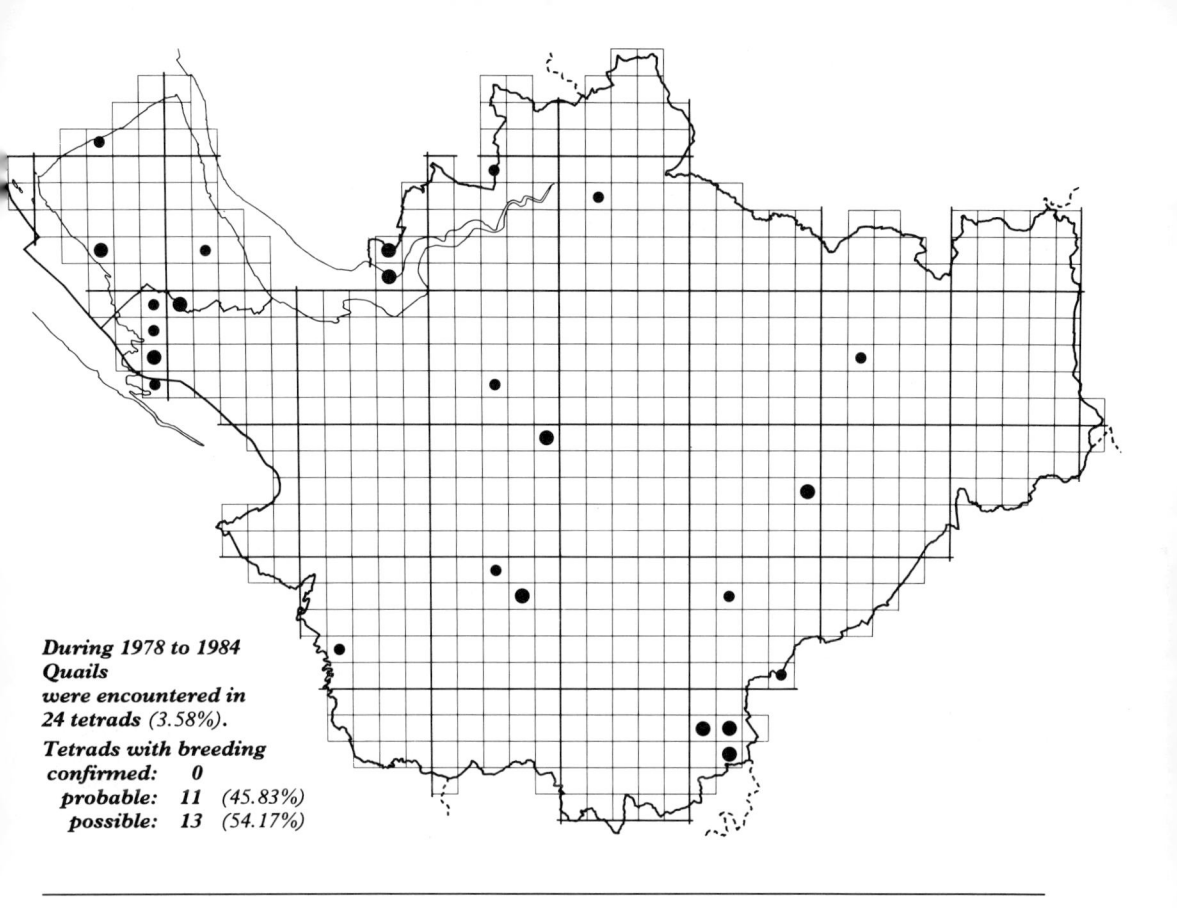

During 1978 to 1984 Quails were encountered in 24 tetrads (3.58%).

Tetrads with breeding
confirmed:	0	
probable:	11	(45.83%)
possible:	13	(54.17%)

Pheasant
Phasianus colchicus

THE Pheasant was introduced to England in the mid-twelfth century, possibly earlier, but large-scale rearing did not occur until the late eighteenth century (Rackham 1986) and the species sustained itself in the wild during the intervening period. A wide range of subspecies breeds across Asia from Turkey to Japan, showing a great diversity of plumage patterns. The original stock released in England was of the Caucasian type *P. c. colchicus*, but, by 1900, birds with a white neck-ring, showing descent from the Chinese type *P. c. torquatus*, had become dominant[2,35]. The Japanese Pheasant *P. versicolor* was also introduced in the nineteenth century, and most birds at that time were either pure-bred *P. c. torquatus* or were affected by the crossing of either that subspecies or *P. versicolor* with *P. c. colchicus*[2]. A dark melanistic type of doubtful origin has since occurred around the Tatton and Capesthorne estates and doubtless elsewhere.

"Except...in those parts of the hills east of Macclesfield where there is no suitable cover, the Pheasant is extensively preserved and exists everywhere in a semi-domestic condition. Hand-reared birds are turned down in their thousands in woods and coverts, which were maintained and often planted solely for their benefit." So wrote Coward & Oldham in 1910, but the onset of the Great War four years later saw an end to much of the 'keepering, and although this practice was revived somewhat between the wars, the Second World War proved a further set-back. Boyd commented in 1942, "There are far fewer Pheasants, for none are hand-reared now...". The gradual breaking up of the big estates has had a further adverse effect on game preservation, but at the present time there are still a few large estates in the county releasing several hundred birds annually. On the other hand, as the map shows, Pheasants are absent from suburban north Wirral, scarce on the mosslands north of Warrington, and largely absent from a belt of open farmland in the centre of the county.

Indeed, although very much a traditional feature of the countryside, Pheasants do not naturally belong in the farmland habitat, preferring the security of game-coverts, and the present distribution reflects the availability of woodland for shelter (see the map for Partridge, an open country species). On Frodsham Marsh released birds frequent the reed-beds and reedy ditches akin to their native habitat in Asia. Presence in farmland areas may depend on continued releases of captive reared birds. Many of those that survive the winter's shoots are recaptured for breeding, but a few remain free to breed in the wild.

In addition, some birds wander away from the point of release. Birds released at Tatton have been shot subsequently at Toft (5-6 kms), and within weeks of birds being liberated at Adlington, in 1979, cocks had moved four kilometres down the Dean valley and taken up territories near Handforth. Some of these wandering birds settle in areas of rough grassland or marsh and establish more or less isolated feral populations. For example a small population has existed for many years around Foden's Flash near Sandbach, one or two females nest annually in rank undergrowth at Prestbury Sewage Farm, and the Woolston population appears to be stable at eight to nine pairs. Whether such populations could maintain themselves indefinitely in the absence of replenishment by wandering released birds is unknown, but good numbers survived in a wild state around Arley for years in the absence of artificial rearing[6].

Cock birds reveal their presence by crowing noisily, but as the nest is generally well hidden under a hedge or amongst undergrowth, most records of confirmed breeding resulted from seeing young. There may be from two to five pairs in most occupied tetrads, with much denser concentrations in some areas, giving a Cheshire feral population perhaps in the range 800-2000 pairs.

REFERENCE
Rackham, O. (1986): *The History of the Countryside*. Dent.

Water Rail
Rallus aquaticus

WATER Rails are best known as winter visitors to the county. Much smaller numbers remain through the summer however, breeding in the concealing vegetation of reed-beds, or in dense alder carr beside the meres.

Coward & Oldham[2] knew the Water Rail as "...a not uncommon resident...but owing to its shy and retiring habits it has been overlooked in many parts...". They also reported that it occurred throughout the year at Holmes Chapel and nested at Knutsford Moor, by the Serpentine at Tabley, in an alder swamp at Redesmere, at Rostherne and "...doubtless in many other similar situations in other localities". Hardy[4] mentioned nests on the Dee and Mersey (Cheshire) marshes, at Delamere (Oakmere) and elsewhere. Tunnicliffe[15] referred to a bird at Sutton Reservoir in April and another with two chicks at Capesthorne on 30th June 1946. Boyd[16] thought that it probably still bred at Rostherne and Bell[7] considered that Coward & Oldham's general comments regarding its

During 1978 to 1984 Pheasants were encountered in 427 tetrads (63.73%).

Tetrads with breeding
confirmed: 168 (39.34%)
probable: 130 (30.44%)
possible: 129 (30.21%)

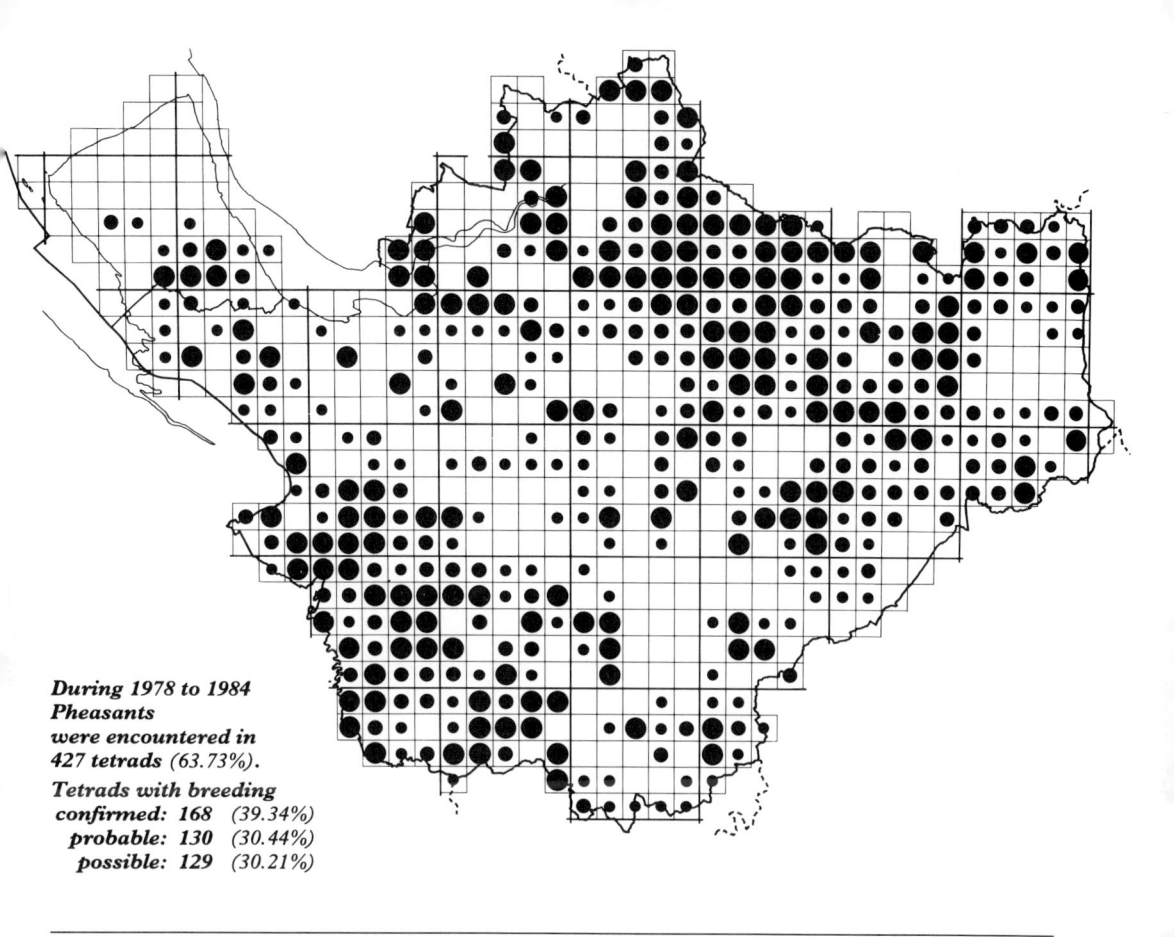

status remained true. In his Supplement[9] Bell reported that "Little evidence of breeding is available..." and could only cite birds heard calling in May 1966 at two of the Sandbach Flashes.

The BTO Atlas survey (1968-72) sparked off interest in the species' status and a run of records of 'probable' or confirmed breeding came from the Chester area: at Poulton in 1969, 1970 and 1972, then again in 1976; at Huntington in 1972, 1973, 1974 and 1975; and at Eaton Hall in 1975. However, there were only seven 'possible' breeding records and two 'probables' in west Cheshire during the course of the present survey and it was considered to be extremely scarce during the last six years[28].

In July 1968 up to four birds were seen at Rostherne and birds heard calling after dusk at Knutsford Moor[21] and at Rostherne[24] on 16th and 23rd May 1976 respectively were considered to be summering individuals and probably breeding. In the following year "song" was heard at Dog Wood, Tatton on 26th March and, in 1979, a bird was present at Rostherne until 15th April. Young were seen at Kingsley sludge-pool in 1972, a site again found to be occupied during the present survey, and a bird in the Frodsham area on 3rd May 1975 may have summered. In Wirral, breeding was suspected near the Wirral Way at Thurstaston in 1973 and single birds were at Red Rocks in July 1979 and at Neston in June 1980. At Sandbach Flashes, birds were recorded "summering" in 1973 and 1976 although these probably refer to early immigrants as breeding almost certainly did not take place. In 1977, a bird was seen in late March, apparently with nesting material, but all birds had gone by 9th April.

Many of the records of 'possible' breeding during this survey refer to birds calling unseen. There were only two cases of confirmed breeding, both of which occurred when observers were fortunate enough to see young birds crossing gaps in the reeds – at Knutsford Moor in 1978 (when it was thought two pairs summered) and at Bagmere in 1983. Nesting is undoubtedly more frequent than these records suggest and as many as four to six pairs were present at Woolston in 1981 although there was no confirmation of breeding there. Juveniles recorded at Sandbach Flashes in August, in both 1983 and 1984, were possibly passage birds.

The extensive marshes which have developed in certain disused sludge-pools form ideal habitat. Most disused sand-quarries are too steep-sided to develop sufficient marsh vegetation for this species, with the notable exception of Shakerley Wood (Allostock) quarry where presence was noted in 1983. Sympathetic landscaping of such quarries could provide additional breeding sites for the future.

Casual daytime visits are unlikely to detect isolated pairs of rails, and presence is most easily confirmed after dusk when birds indulge in a variety of grunts, squeals, trilling and whistles. More nocturnal visits by bird-watchers to suitable habitats might well have revealed presence in additional tetrads. The current breeding population is estimated to be 15 pairs.

Spotted Crake
Porzana porzana

THE Spotted Crake is currently an extreme rarity as a breeding species in Britain which is on the very north-western edge of its range. Breeding is very sporadic and numbers reported fluctuate from year to year. Even in the past, few nests were found – even by the most skilled of searchers – and nowadays this species is listed under Schedule 1 of the Wildlife and Countryside Act 1981 making it an offence to disturb nesting birds or their young. The "whiplash" call, generally made at night, is probably the best way to determine the species' presence.

Spotted Crakes probably bred locally but fairly commonly in the U.K. until the drainage of their former haunts during the eighteenth and first half of the nineteenth century, although there was something of a resurgence between 1926 and 1937[35]. Most birds seen in Britain today are on either spring or autumn passage although the occasional bird may overwinter.

One was "obtained" near Warrington in June 1892[2] and Boyd in his country diary for the 14th October 1936 states that "a pair or two bred...a few years ago"; this probably refers to two pairs which nested on Puddington marsh in 1934. A further nest with ten eggs was found there on 20th June 1938[7]. Raines[18] reported two nests in the Burton area in 1935, one nest in 1937 and an indication that birds probably bred in other years. A freshly dead bird picked up at Thurstaston on 20th July 1957 proved on dissection to be in breeding condition and on 9th June 1959 a bird was flushed several times in the Hoylake fields. A bird was calling at Pickmere on 29th July 1963[9] and during late May 1971 one called regularly in the evenings at Red Rocks marsh[8]. Two years later one was present at Pulford withy-bed from 21st April until 5th May and, in 1974, one was calling nearby at Plowley Brook, Aldford on 1st June[8].

Two birds recorded during this survey may both have been on passage: one found dead, caught on a fence, at Hale Marsh on 5th May 1978, and another at Bagmere on 9th April in the same year. Spotted Crakes like thick riparian and marsh vegetation, from lowland swamps and fens to highland sedge and rush bogs[32]. Some suitably wet, boggy habitat, such as at Bagmere, still exists in Cheshire and may provide the opportunity for a nesting attempt by this elusive species.

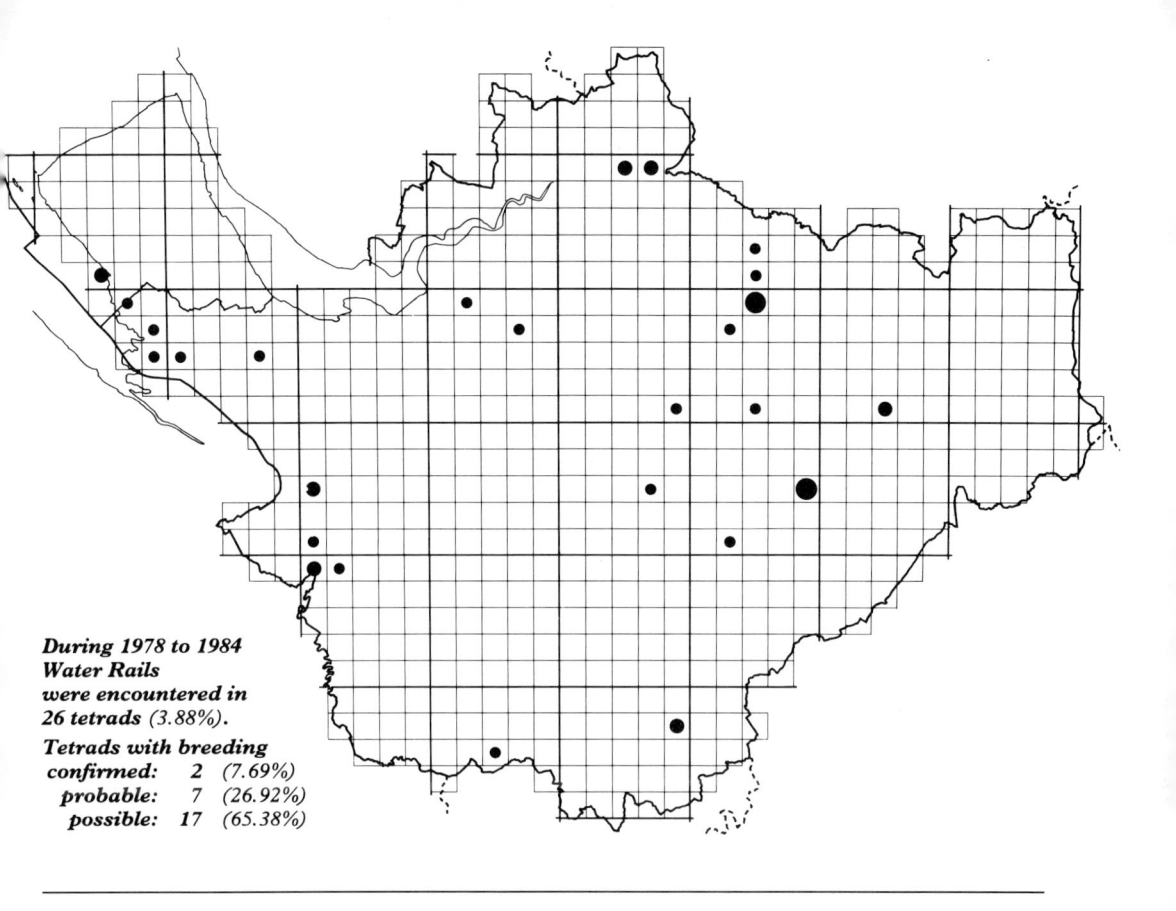

During 1978 to 1984 Water Rails were encountered in 26 tetrads (3.88%).

Tetrads with breeding
confirmed:	2	(7.69%)
probable:	7	(26.92%)
possible:	17	(65.38%)

Corncrake
Crex crex

THE fortunes of the Corncrake have declined steadily since 1900 throughout western Europe, primarily with the mechanisation and intensification of agriculture. By 1979, the British and Irish population represented a substantial proportion of the species' overall population in western Europe but the total continues to decrease.

The decline in Britain was first noticed during the second half of the nineteenth century in the more intensively farmed areas of south-east England, and the species' range has continued to contract steadily to the north and west[32]. The introduction of mowing machines to cut hay, and earlier mowing as a result of mechanisation had taken a very heavy toll of eggs and young (Norris 1947). A continuing decline, even in the agriculturally backward Hebrides, has been tentatively ascribed to the spread of the Sahara resulting in greater mortality on migration from wintering quarters in central Africa (Swann 1986). The proliferation of overhead wires during this century has undoubtedly also taken its toll of migrant birds and must have contributed to the species' decline[32].

Coward & Oldham[2] knew the species as a common summer resident in Cheshire: "During the last week in April or the first few days of May the familiar call of the Land Rail...may be heard in the lowlands and on the hill-pastures up to the edge of the moors... If the grass be short when the bird first arrives it seldom calls by day, but later, during May, June and July, its presence is advertised day and night by its monotonous and incessant *crake*... Towards the end of July the note becomes less frequent, and in August it is seldom heard. When, however, partridge-shooting begins in September, Land Rails are often flushed and shot. Most of the birds leave in September, but individuals, wounded or from some reason unable to migrate, have occasionally been obtained in Cheshire during the winter months".

Norris (1945, 1947) recorded a notable decrease in Cheshire after 1910, and subsequently local fluctuations were especially marked. In 1938, he recorded the Corncrake as still breeding in small numbers, especially in the east of the county. By the following year however there had been a 50% decrease. After 1930, Boyd[6] knew of only one nest (1947) in the Frandley area, when one of the adults was killed by a mowing machine in July whilst, at Lower Peover, Sir Geoffrey Howarth (pers. comm.) noted that the last bird was heard on 1st May 1944 and that "our early cutting of grass for artificial drying and silage would be enough to discourage breeding".

Griffiths & Wilson[13] stated that it had been a common summer visitor in Wirral until some twenty years before 1945, but had gradually decreased since and was then rarely seen. Bell[7] refers to a gradual disappearance throughout the county from the 1920s. He reported isolated instances of nesting in 1954, 1956 and 1959.

Subsequently, no records of confirmed breeding have appeared in *CBRs*, although there were scattered breeding season records throughout the 1960s and to a lesser extent in the early 1970s. The BTO Atlas shows four 10-km squares in Cheshire where breeding was 'possible' over the years 1968-72 and two where it was 'probable'[32].

There was only one record during the current survey, a bird seen and then heard calling briefly at Mobberley on 29th April 1984. The demise of the species in Cheshire ran parallel to that in neighbouring counties. The last breeding record in Greater Manchester was at Ringway in 1971[43] and in the West Midlands area at Chesterton, Worcestershire in 1969[44].

Apart from the records listed by Boyd and Norris, there is little indication of how numerous Corncrakes might have been prior to their decline. However, Abbott[10] mapped twelve territories over an area of roughly 22 square kilometres in the Wilmslow/Alderley Edge area in the period 1919-1922. If the replacement of the scythe by the mowing machine started the Corncrake on its decline, then the modern fashion for cutting grass for silage offers no hope for its revival. The preferred habitat of hayfields is increasingly scarce.

REFERENCES

Cadbury, C. J. (1980): The status and habitats of the Corncrake in Britain 1978-79. *Bird Study* 27: 203-218.
Norris, C. A. (1945): Summary of a report on the distribution and status of the Corn-crake (*Crex crex*). *British Birds* 38: 142-148, 162-168.
Norris, C. A. (1947): Report on the distribution and status of the Corn-crake. *British Birds* 40: 226-244.
Swann, R. L. (1986): The recent decline of the Corncrake *Crex crex* on the Isle of Canna. *Bird Study* 33: 201-205.

Footnote: A record of breeding at Tatton in 1987 (*CBR*) should be regarded with caution. It rests on the finding of a dead, fully-grown juvenile bird in the latter half of June, long before any youngster is likely to have been fully grown. JPG.

Moorhen
Gallinula chloropus

MOORHENS breed by marl-pits, meres, canals, rivers and streams throughout most of the county. The map shows that this abundant and widespread species avoids the higher land of the eastern hills and the dry sandstone areas of the Delamere ridge and Bidston Hill. It is also noticeably absent from the centres of towns such as Birkenhead, Runcorn, Widnes, Warrington and Crewe. Lack of coverage may explain the few remaining gaps in the south of the county.

The nest is built typically amongst rushes at the water's edge or in branches overhanging or emerging from the water, but various other sites are recorded. Occasionally birds will be found nesting high in trees or bushes, either building their own or using nests of other species. Coward & Oldham[2] described one eleven feet above the ground in a fir, and another 18 feet up in a spruce at

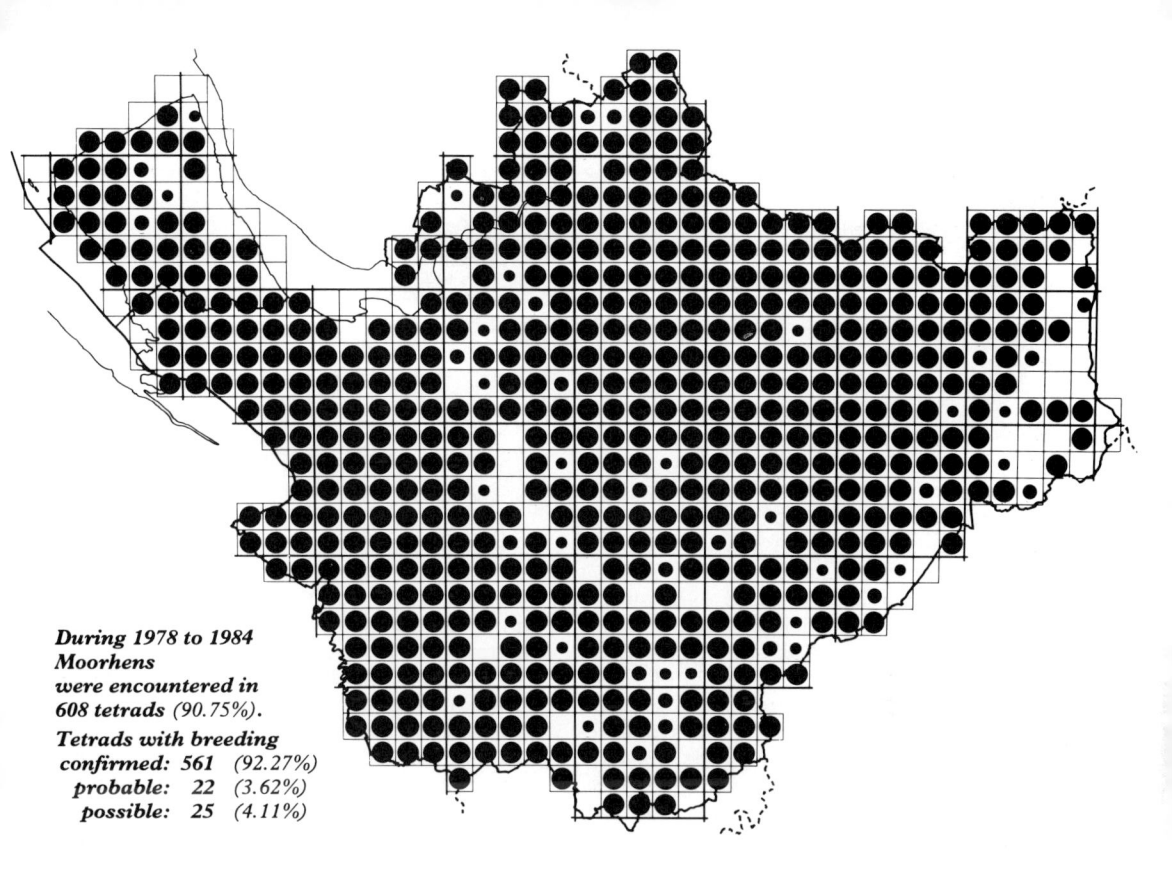

During 1978 to 1984 Moorhens were encountered in 608 tetrads (90.75%).

Tetrads with breeding
  confirmed: 561 *(92.27%)*
  probable: 22 *(3.62%)*
  possible: 25 *(4.11%)*

Astle. In another instance the old nest of a Woodpigeon was used. Boyd[5, 6] reported a bird sitting on four eggs in a Jay's nest 20 feet up in a hawthorn. In 1966, a pair nested 18 feet up in a Magpie's nest at Sandbach and another pair, at Churton, used a Crow's nest 15 feet up in an apple tree. There are several recent records of birds vacating nests in trees and the tops of thorn bushes, although in some cases the birds may simply have been roosting rather than nesting. One was sitting in a disused Heron's nest at Sandbach in 1977 and may have had eggs, and Boyd[6] mentions six eggs in a thrush's nest lined with rushes in a hawthorn hedge.

With the frequent use of marl-pits as farm tips the centres of discarded rubber tyres have been recorded as nest-sites. Further records are of a nest on top of a wall emerging from water, others on bare mud in the flashes at Witton and Sandbach, and on exposed shingle banks in rivers. Three separate nests in marl-pits near Prestbury found during this survey had polythene built into the lining but only one held eggs. Polythene is also frequently used to line nests at Rostherne where nests were also built in Dutch nesting-baskets set out for Mallards (T. Wall pers. comm.).

From late February onwards birds are to be seen displaying, their tails erect and expanded showing off the white under-tail. On 17th April 1980, eleven were displaying together by Railway Flash, Sandbach. In some years the first eggs have been laid by the end of March, and chicks may be seen from mid-April. Huxley & Wood (1976) give the egg-laying period as March to July with breeding slightly later in the north than in the south. Hardy[4] records an early nest found with a week-old brood at Rookery Wood, Delamere on 2nd April 1933. The *L&CFC Report* for 1948 records "late" nests with eggs on 2nd and 15th August. However, more recently, records of young have spanned the period from 14th April to 15th September suggesting a later date for nests with eggs. Moorhens not infrequently lay in Coots' nests. This has often been noted at Rostherne but at Sandbach in 1964 a Coot's egg was found with six Moorhen's eggs in a nest of the latter.

Coward & Oldham[2] reported that pairs bred on the hill-streams in the east of the county where the species is now scarce, but otherwise there has probably been little change in the Moorhen's status. Eggs used to be collected for human consumption and at

Rostherne, between 1938 and 1960, were taken and boiled to feed Pheasant chicks on the Tatton estate[21]. While such activities appear to have ceased, the filling-in of marl-pits, particularly on the urban fringe, has undoubtedly already reduced the available habitat in many areas and gaps may be expected to appear on future maps if this practice continues. Improved field drainage and predation by minks are probably also factors causing the decline of Moorhens in some areas.

Whalley[19] estimated 30-35 pairs breeding around Sandbach Flashes. In July 1976, 13 pairs were nesting along a two-mile stretch of the River Dean between Dean Row and Adlington, rearing some 30 young. Such a concentration was probably due to several nests on the shingle banks in mid-stream which had been exposed by that summer's drought and because many nearby pits and normally damp hollows were too dry for nesting.

Counts at Rostherne gave figures of 26-29 pairs in 1980 and 23-24 pairs in 1982. The breeding population of Woolston's No. 3 bed was estimated at 40-50 pairs in 1983, although few other tetrads in Cheshire hold such high totals. A full survey of SJ45J (Aldford) revealed 27 pairs in 1983 and a farmland CBC in west Cheshire gave an average breeding density of 15.2 pairs per tetrad. Sharrock[32], quoting national, farmland CBC figures for 1972, gives an average density of 3.8 pairs per square kilometre. Such detailed quantitative information is unfortunately scarce, but there may well be an average of 10-15 pairs per occupied tetrad, giving an estimated county population of at least 6000 breeding pairs.

REFERENCE

Huxley, C. R. & Wood, N. A. (1976): Aspects of the Breeding of the Moorhen in Britain. *Bird Study* **23**: 1-10.

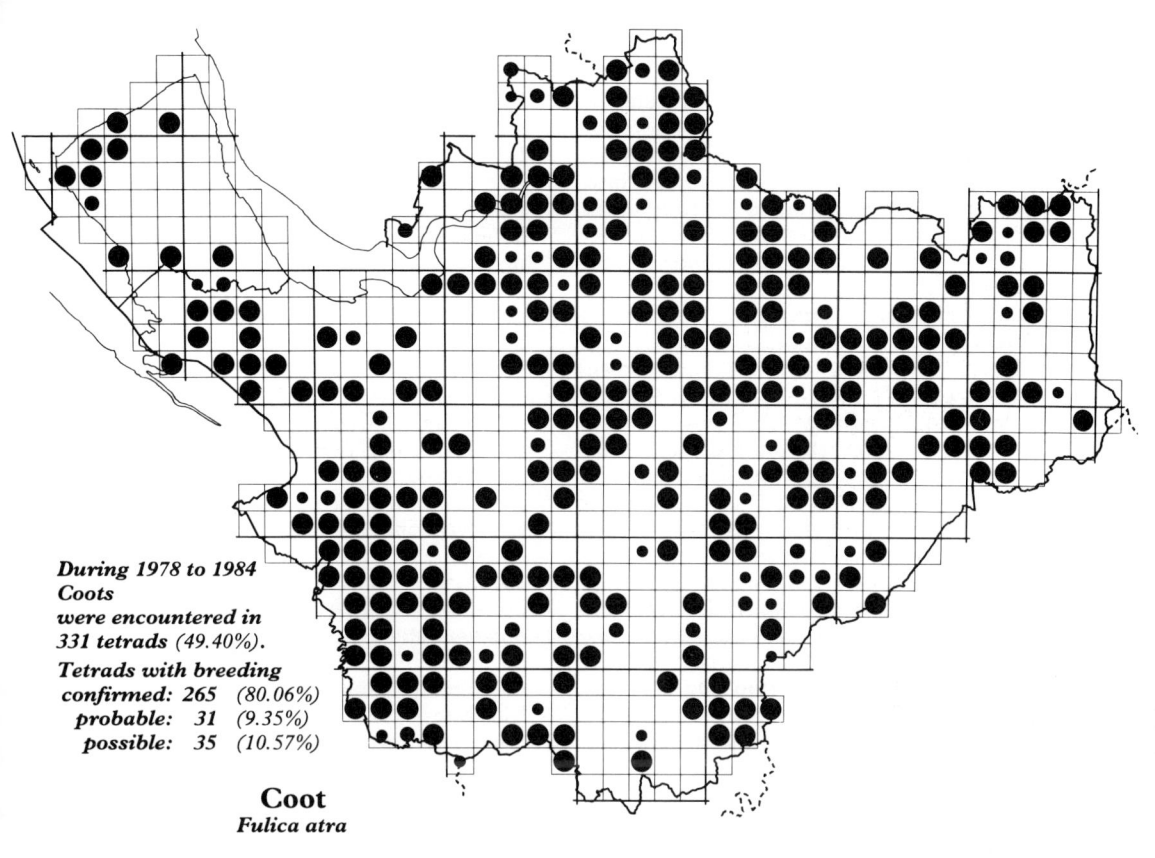

During 1978 to 1984 Coots were encountered in 331 tetrads (49.40%).

Tetrads with breeding
confirmed: 265 (80.06%)
probable: 31 (9.35%)
possible: 35 (10.57%)

Coot
Fulica atra

A TYPICAL bird of all but the smallest waters in the county, the Coot breeds on meres, flashes, flooded sand-quarries and clay-pits, slow-flowing rivers and canal "wides", and in recent decades larger marl-pits have been occupied in some areas. Even the hill reservoirs hold a pair or two, with the exception of those at Higher Disley and Lamaload where fluctuating water levels and lack of vegetation prohibit nesting.

Coward & Oldham[2] described the Coot as "a common resident" but "confined to the meres and larger waters" where it was "locally abundant". In Wirral, it was rare in the breeding season because of the lack of such stretches of water. It bred in some numbers on all the meres on the plain and was particularly abundant where there were extensive reed-beds. They considered the lack of cover was sufficient reason for the absence of Coots on the hill reservoirs although they nested at Lyme, Adlington, Poynton and Bosley.

The nest is often built amongst fringing reeds or sedges, on overhanging branches trailing in or branches emerging from water. Disused grebe nests are sometimes renovated. Many nests are exposed to the weather and become flooded during summer squalls. At Sandbach Flashes nests are frequently found on emergent mud-banks and at Elton Hall Flash even on mud temporarily exposed while the River Wheelock is at a low level during dry weather. Such nests are very vulnerable to flooding and eggs are often taken by Herons or other predators.

Breeding Coots generally require at least 0.5 hectares of open water[32]. Both Coward & Oldham[2] and Boyd[6] stated that Coots did not breed on marl-pits, but by 1956 a pair or two began to nest on pits at Arley. Such nesting has since been observed at Bradwall, Lach Dennis, Siddington and elsewhere on the Cheshire plain. Birds have also bred on a small boggy pool at Danes Moss. Breeding in such habitats may indicate the high level of the population; territorial birds are very aggressive and excess pairs are forced into marginally suitable habitat. In 1981, up to six pairs were breeding in a shallowly flooded rushy field near Pickmere village, but in 1982 only two pairs were present, presumably as a result of mortalities during the intervening severe winter. No comparable decrease was evident on the larger, more favoured waters. Also, in 1981, a pair arrived in a freshly flooded field (caused by a blocked drain) at Dean Row, where the species had not previously been noted, and hatched young.

Egg-laying begins in April or occasionally in late March. Unusually, in the mild spring of 1975, a bird was on eggs at Elton Hall Flash, Sandbach on 9th March. Breeding success of this species is generally low with several British studies showing some two-thirds of eggs failing to hatch – many of these being lost to predation, flooding and the clumsiness of adults in knocking eggs out of the nest or burying them in the structure of the nest[33]. At Rostherne, where there is little shallow water and a consequent lack of aquatic vegetation for food, an average 27 pairs breed, 45% of eggs are lost before hatching and 90% of chicks die before fledging[21]. At Sandbach Flashes, where between 60 and 70 pairs held territory annually in the late 1970s, a maximum of 35 broods was seen in 1979 (JPG) although further details of success rates were not noted. At Marbury Mere (Great Budworth) ten breeding pairs were present in May 1982 but no young were seen.

The aggression of territorial Coots contrasts markedly with the sociable behaviour of non-breeding birds which may be encountered in flocks throughout the summer. These non-breeding flocks are irregular in occurrence, indicating the mobility of our Coot population – surprising since birds are seldom seen to fly between waters. For instance up to 50 were present at Rostherne in 1970 and a similar number was on Railway Flash, Sandbach in 1978. In 1982, 50+ on Railway Flash on 8th June increased to 75+ by 30th June as failed breeders abandoned their territories to join the flock.

A compilation of counts of pairs on East Cheshire waters over several years suggests some 285 territorial pairs in 60 tetrads with confirmed breeding (JPG), including concentrations at several meres and sand-quarries. A similar density across all 265 tetrads with confirmed breeding would give a county population approaching 1250 pairs.

Oystercatcher
Haematopus ostralegus

WHILE the coastal sandhills of Wirral must have furnished ideal nesting habitat for Oystercatchers in bygone centuries, Coward & Oldham[2] could find no evidence of breeding on the Cheshire shore, although they referred to Brockholes' belief that birds had done so at Hoylake. At the turn of the century Oystercatchers were an abundant visitor to the estuaries from autumn to spring, but only a few non-breeding birds remained between late May and early August.

The first recorded instances of nesting took place in the 1930s. In 1936 Hardy[4] found a nest by the Mersey at Mount Manisty and in the following year Boyd found a pair nesting on Frodsham Score, where a pair also bred in 1941. Since 1936 nesting has occurred sporadically on Burton Marsh. In 1939, three or four pairs were present, and three nests were found there in 1940. These early nests on the marshes may actually have been in Flintshire however. In 1941 a pair also nested at Leasowe and another on the wall of the Manchester Ship Canal between Norton and Runcorn. Bell[7] listed breeding records at Thelwall Eye in 1947 and subsequently, and at Weston Marsh in 1958. In his Supplement[9] he refers to eight pairs nesting on Burton Marsh in the early 1960s. Three pairs bred there in 1968[8]. Buxton (1962) gave additional records of nesting at Ince in 1950 and on Hale Marsh in 1951.

Nesting has continued by the Mersey estuary up to the present day, although only a handful of pairs seems to be involved. In 1968 two pairs nested on the Frodsham Marshes,

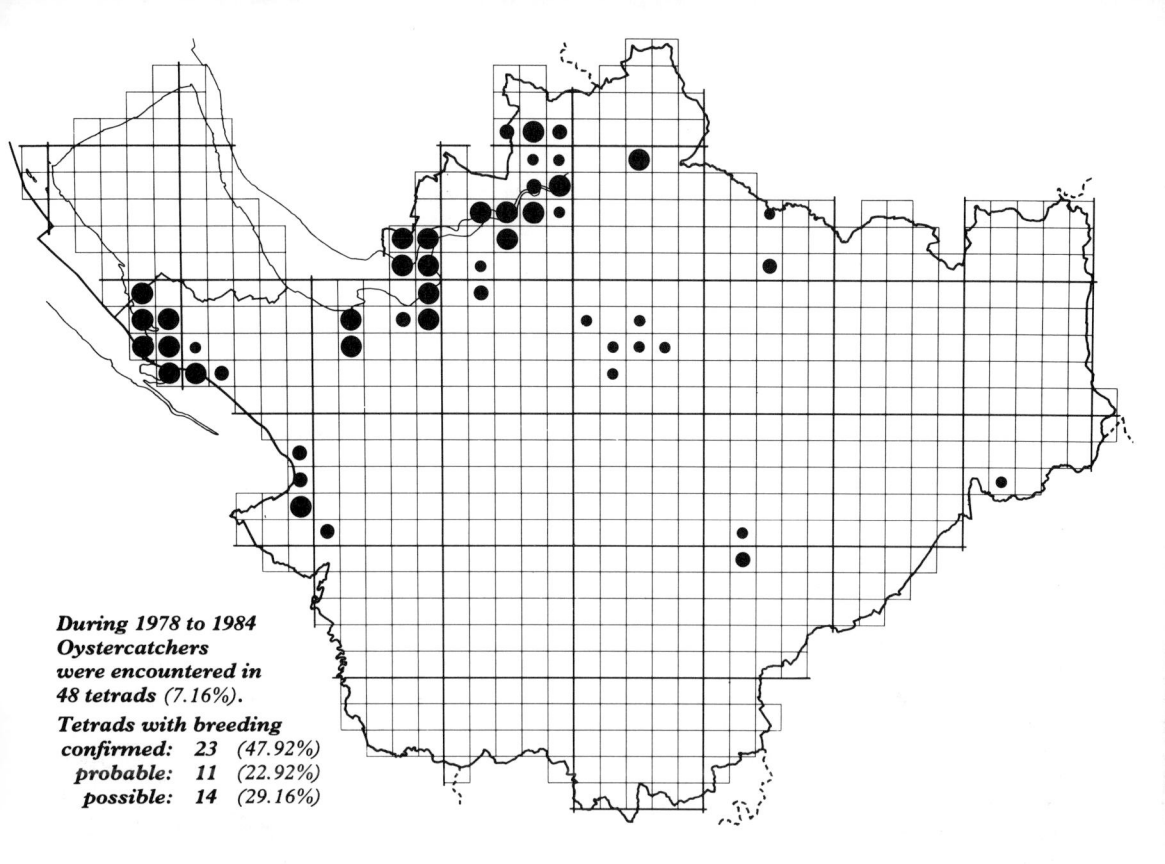

During 1978 to 1984 Oystercatchers were encountered in 48 tetrads (7.16%).

Tetrads with breeding
confirmed: 23 (47.92%)
probable: 11 (22.92%)
possible: 14 (29.16%)

and *CBRs* since 1972 refer to a pair or two in most years in the Moss Side/Moore area. The only reported instance of successful breeding was in 1973 when young were seen at Moss Side. Of four nests detailed in the reports three held three eggs each, and the other held four. Nests around the estuary are often in cereal fields. Elsewhere nests have been reported from rough, wet farmland close to the Dee estuary and one on a gravel ridge in industrial wasteland.

The estuarine distribution, implicit in the above records, holds true to the present day. However, further north in Britain, from the Ribble northwards, Oystercatchers regularly nest on shingle banks by rivers or on adjacent fields, and there is evidence to suggest that birds have been trying to colonise similar inland habitats in Cheshire. In June 1956 birds were seen in the Goyt Valley (presumably in Derbyshire) and nesting was considered a future possibility although none was present during a brief visit in 1960 (Buxton 1962). Between 1971 and 1974 nesting was reported to have been attempted near Ringway, now Greater Manchester[8]. In 1972 up to four birds were reported in June from Tatton Park and the 1974 *CBR* lists breeding at Ashley and Tabley. Subsequent enquiries have failed to substantiate these records and the Ringway records are not mentioned in the Greater Manchester Atlas[46] although they are included in the BTO Atlas[35].

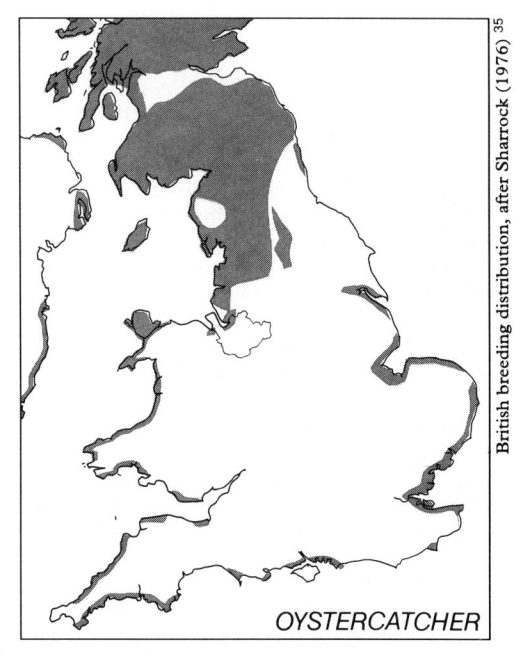

British breeding distribution, after Sharrock (1976)[35]

OYSTERCATCHER

103

Table 8: Totals of Oystercatchers recorded away from the coast and estuaries by month for the years 1978-80 and 1983[8].

Jan	Feb	Mar	Apr	May	Jun	Jul	Aug	Sep	Oct	Nov	Dec
1	2	13	30	38	9	21	29	2	7	1	1

Bell[7] stated that Oystercatchers seldom occurred inland even on passage. Recent records however show definite peaks in March-May and July-August, the spring peak largely of what appear to be prospecting birds, often in pairs, that linger at such sites as the Sandbach Flashes or Withington sand-quarries for days or even weeks.

The food of the Oystercatcher is mainly bivalve molluscs, especially cockles and mussels, although they will also feed on annelid worms, particularly when inland. Unlike most waders, the chicks are fed by their parents, and may be dependent on them for anything from six weeks to ten months depending on the type of prey being fed on. If worms are the main prey the chicks will be almost as efficient as their parents in probing for worms when six weeks old. However, if bivalves are the predominant prey items chicks will be dependent on their parents for anything from three to ten months as they are inefficient at opening bivalves, a technique which has to be acquired and learnt over a period of time. Inland breeding and feeding may therefore be beneficial to the species.

During this survey twelve pairs bred with limited success at Gayton Sands RSPB reserve in 1980. 1500 non-breeding birds were counted here in June of the same year. Ten pairs nested in Wirral in 1984, but success was low because of flooding on the high spring tides[8]. Flooding has long been a hazard to coastal nesters (e.g.[12]).

Such a conspicuous bird, with its black-and-white plumage and elaborate, piping display, is unlikely to have been overlooked in any square during this survey, although prospecting birds at inland sites may have been dismissed out of hand. Confirmation of breeding by Oystercatchers came from 23 tetrads during the seven years, but it seems unlikely that the county population will have exceeded 20 pairs in any one year.

REFERENCE
Buxton, E. J. M. (1962): The Inland Breeding of the Oystercatcher in Great Britain, 1958-59. *Bird Study* **8**: 194-209.

Little Ringed Plover
Charadrius dubius

THE Little Ringed Plover, a recent addition to our avifauna, is very much a bird of industrial waste ground, nesting in Cheshire primarily on settling-beds of silt or chemical waste, in the bottoms of active sand-quarries, or on demolition sites. It was restricted as a breeding bird to shingle banks beside continental rivers until the middle of this century when adaptation to man-made habitats enabled it to spread rapidly across England. Riverside nests are unknown in Cheshire, although at Sandbach Flashes pairs have occasionally nested on the muddy shores of the flashes when these are exposed by summer drought. Also at Sandbach, eggs have been laid in scrapes on a car park of crushed brick, a newly-created lorry park at Fodens' motor works, and the concrete and cinder-strewn site of a demolished factory. A pair occupied the scrap-yard of the aluminium factory at Warrington in almost every summer of this survey. The absence of water within a quarter of a mile of one of these sites did not deter the pair from nesting, although usually the territory incorporates an area of damp mud for feeding.

Where there are several pairs breeding close together, Little Ringed Plovers can be noisy and demonstrative, but while many sitting birds can be spotted from a considerable distance others are very wary and employ various ruses to disguise the whereabouts of their nest. For example, the bird will settle down in a hollow as if settling onto eggs or, if suddenly aware of the presence of an observer when approaching the

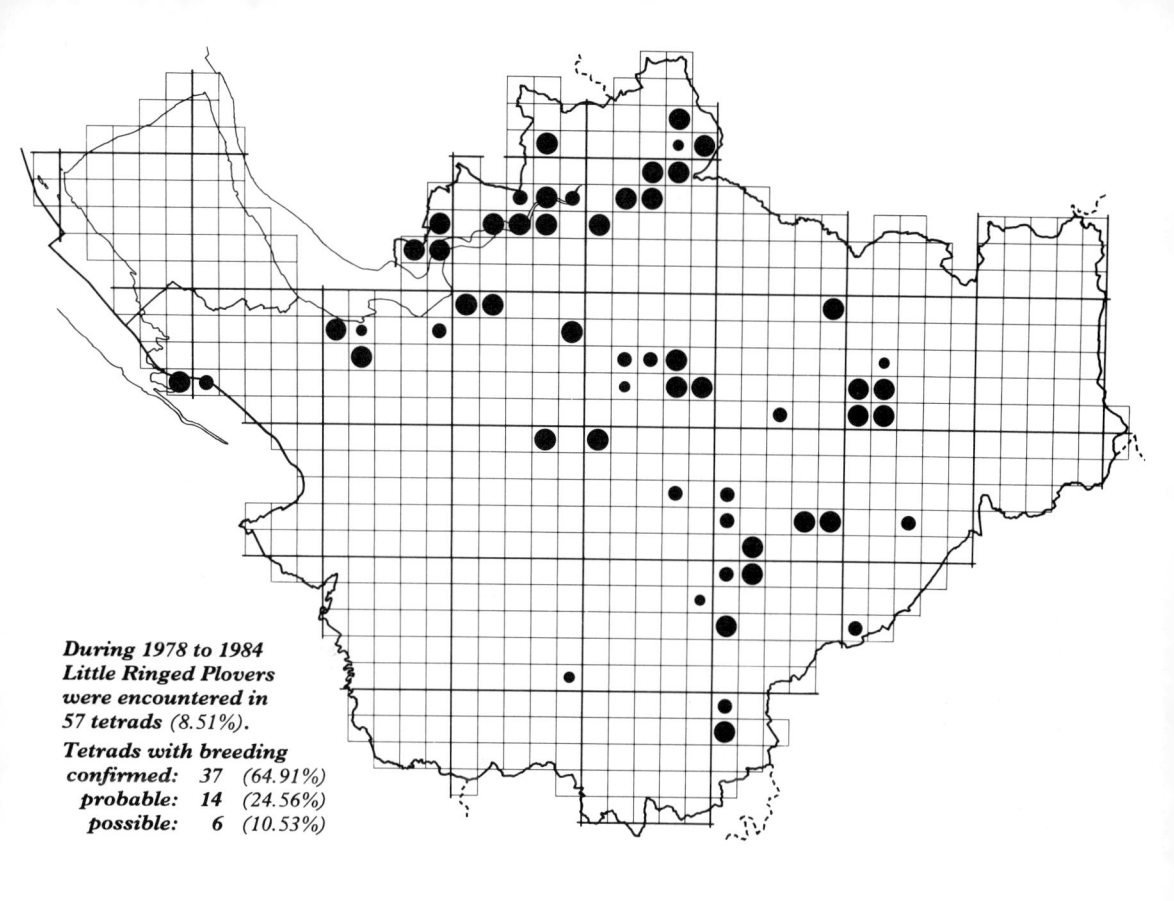

During 1978 to 1984 Little Ringed Plovers were encountered in 57 tetrads (8.51%).

Tetrads with breeding
confirmed: 37 (64.91%)
probable: 14 (24.56%)
possible: 6 (10.53%)

nest, will peck at the ground as if feeding. In some cases observers were able to see the fluffy chicks running for cover as the adults gave their loud alarm calls. However, as with most waders, once the young birds have frozen motionless they are so well camouflaged that there is a real risk of their being trampled if the observer unwisely ignores the adults' frantic displays and continues to walk through the territory. This is a Schedule 1 species under the 1981 Wildlife and Countryside Act, so such behaviour is illegal as well as irresponsible. Fortunately it is easy to confirm Little Ringed Plovers breeding without closely approaching the nest or young. Patient watching from a concealed position will be rewarded by views of chicks picking insects from the surface as their parents stand guard.

The first British breeding record was at Tring, Hertfordshire in 1938 and the excitement of their early colonisation was dramatised in a novel (Allsop 1949). Sixteen years later, in 1954, a pair nested at a site in central Cheshire but lost their eggs to a predator (Arnold 1955). A single bird was again present here in May 1955 and in June 1957 one was seen in a sand-pit at Sandiway.

On 12th July 1961 a clutch of four eggs was found at Plumley and two young successfully reached the flying stage – the first successful breeding record for Cheshire[7]. In August of that same year a pair and a young bird were present at Sandbach Flashes.

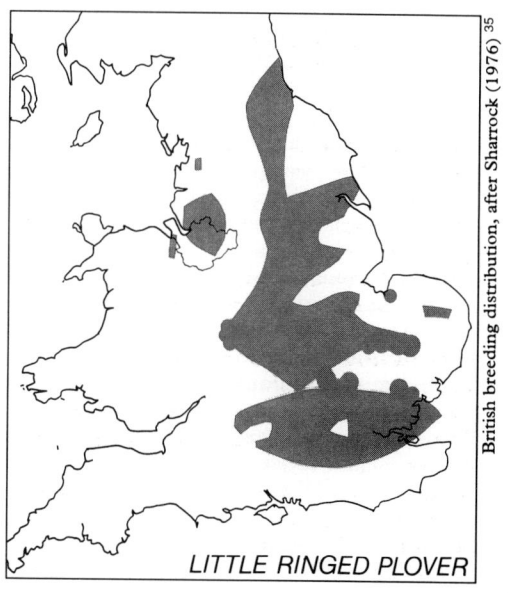

British breeding distribution, after Sharrock (1976)[35]

LITTLE RINGED PLOVER

In 1962 and 1963 three and two pairs respectively bred at Sandbach. In 1963, there were also single pairs at both Plumley and Chelford and by 1965 nine pairs were known – six at four sites around Northwich and single pairs at Chelford, Plumley and Sandbach. By 1970 at least 20 pairs were nesting in Cheshire[8] and it seems that this level was maintained until 1982 when the county report suggested between 25 and 29 pairs. By the spring of 1984 a national survey organised by the BTO revealed a slight decline to 21-24 pairs hatching a minimum of 28 young.

Breeding success varies from year to year but is often reported as "low". Of twelve pairs recorded in 1969 three failed to breed and twelve young reached the flying stage. At Sandbach, in 1973, five pairs reared eight young to flying, in 1974 seven pairs hatched 36 young of which 12-15 fledged and in 1975 seven pairs fledged only nine young. In 1974, 14 pairs in mid-Cheshire were thought to have reared only six young. The earliest recorded date for eggs was 15th April in 1980, and birds may be seen sitting on second or repeat clutches until late July. Of nine hatching broods at Sandbach in 1974, one was of five chicks, seven of four, and one of three. Hatching dates for those clutches, of which at least one was a repeat clutch, ranged from 17th May to 5th August (JPG). Subsequent post-breeding assemblies at Sandbach reached a maximum of 25 between mid-July and mid-August. Most adults had left by the end of August, but a few immatures remained into September, the last being recorded on 28th. In 1975, the last bird recorded at Sandbach was on 13th October.

The seven-year survey map clearly overstates the Cheshire distribution in any one year, new sites being occupied as some older ones are deserted when, inevitably, they become overgrown, flooded or built upon. In the 1980s, however, some sites, particularly in the Mersey valley, were abandoned whilst still suitable for breeding, because this species was driven off by the increasing numbers of Ringed Plovers. Although they take different prey items and do not compete for food, several observers noted interspecific aggression, with the smaller Little Ringed Plovers always losing out.

REFERENCES

Allsop, K. (1949): *Adventure Lit Their Star*. Latimer House, London.
Arnold, E. L. (1955): Little Ringed Plover breeding in Cheshire. *British Birds* **48**: 176.

Ringed Plover
Charadrius hiaticula

IN Cheshire the Ringed Plover is most familiar as a passage migrant, on northward passage from Africa to its Arctic breeding grounds. These birds mainly frequent the Dee estuary, with some on the Mersey and a few occurring regularly at inland sites, particularly in May. Smaller numbers of the larger British race winter on the Dee. In recent years single birds or isolated pairs have started to appear at sand-quarries, industrial settling-beds and other suitable inland sites during March or April. Some of these have established territories and stayed to nest.

Coward & Oldham[2] indicated that by the middle of the nineteenth century the species had ceased to nest regularly on Wirral shores, although it had formerly bred in some numbers, presumably along the strand-line below the sand hills. They could list nesting records only from Hilbre in 1876, Heswall in

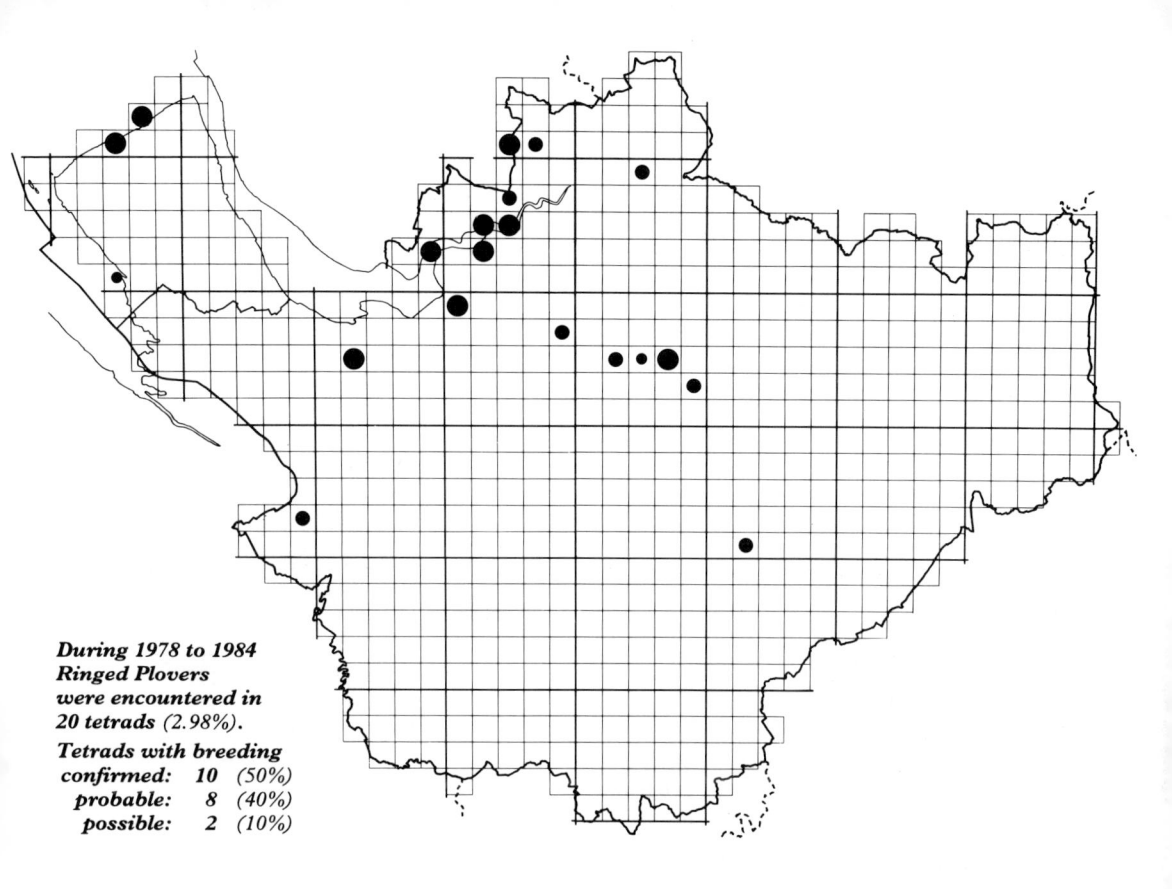

During 1978 to 1984 Ringed Plovers were encountered in 20 tetrads (2.98%).

Tetrads with breeding
confirmed:	10	(50%)
probable:	8	(40%)
possible:	2	(10%)

1904, Caldy in 1907 and at around the start of the century on Chester Golf Links. Bell[7] added a single pair which nested behind the Wallasey Promenade in 1939, 1940 and 1941, and an unsuccessful attempt at Hilbre in 1954. Since then only two coastal nesting attempts have been recorded – a pair bred at Leasowe in 1978 and a pair in 1983.

In 1975 breeding was recorded on an industrial sludge-bed near Northwich. Birds bred again there in 1976, and possibly at a second inland site. In 1977 a territorial pair displayed at Sandbach Flashes and may have had eggs, but deserted at the start of the angling season. During the present survey nesting occurred in all but one year and, as Table 9 shows, Ringed Plovers were becoming well established at inland sites.

Sharrock[35] stated that the Ringed Plover was primarily a coastal nesting species which during the 20 years up to 1975 had increasingly nested inland, particularly in northern England and Scotland. The spread of this habit may well prove to be the species' salvation in Cheshire, as its coastal nesting sites in Wirral and elsewhere have been devastated by development during the last century, or have become excessively trampled by human visitors. It would appear that this species may yet oust its smaller relative, the Little Ringed Plover, from a number of the latter's nesting sites, not only because of its larger size, but also because it tends to arrive from the coast earlier in the spring, establishing territory during March, before the Little Ringed Plover arrives from Africa.

Ringed Plovers display conspicuously, and, as many of their nesting sites are regularly visited by bird-watchers, it is unlikely that they have been overlooked, although it is possible that they escaped detection in some inaccessible industrial sites. With 13 pairs in 1984, the Cheshire breeding population was at its highest level for well over one hundred years.

Table 9: Total number of pairs of Ringed Plovers breeding inland in Cheshire 1975-84.

	1975	1976	1977	1978	1979	1980	1981	1982	1983	1984
No. of nesting pairs	1	2-3	2+	1	-	2	1	5	8	13

Golden Plover
Pluvialis apricaria

THE plaintive "peeoo" of an anxious Golden Plover is for a bird-watcher the characteristic sound of the blanket bog in the Peak District during May and June. However, there is now very little of this habitat left in Cheshire.

The adults are extremely demonstrative, often starting to express their anxiety when an intruder is 400, even 800, metres away. Locating either the eggs or the young that they are guarding is, however, very difficult, and they are unlikely to allow themselves to be "watched back". Three of the confirmed breeding records were of adults giving the distraction display ('DD') and four were sightings of small, downy chicks ('FL'). This matches the national results for 1968-72[35] when most of the confirmed breeding records were 'DD' or 'FL'.

Birds return to the moors in late March and early April. One was heard in song-flight over High Moor on 15th March 1981 and, exceptionally, a pair was observed on Birchenough Hill during mild weather conditions in January 1983. Both birds of the pair incubate, the off-duty bird flying off to feed on nearby pasture (Parr 1980). This suggests that some of the 'possible' breeding records may be of feeding birds, rather than of birds on territory. Much of the food of adults seems to be earthworms, but in Scandinavia, Byrkjedal (1980) found their main prey to be weevils, ground beetles and tipulid larvae, with crowberries an important additional food. Family parties usually remain together in or near the territory for a while after the young have fledged. During this period the parents share responsibilities for the young which feed themselves. Parr (1980) documents how the parents lead their chicks to small grassy patches within the territory, although the parents themselves may still fly off to feed on nearby pastures leaving the young on their own. Golden Plovers favour short vegetation, often colonising newly burnt patches of heather, presumably because these are easier areas to work for food.

The history of this species in Cheshire is not well documented. Coward & Oldham[1] wrote that "a few pairs...breed annually on the higher moors of Longdendale and of the country east of Macclesfield", which would be a reasonable description of the present status. A survey of the species in the Peak District from 1972 to 1974 produced an estimate of 380 breeding pairs of which only 15 were in Cheshire (Yalden 1974). In 1978, 17 pairs and two single males were recorded during a single day's coverage of most of the available habitat within Cheshire (JPG), and there was probably one additional pair to the south of the area covered. A partial survey of the Peak District moorlands for the RSPB in 1981 (Campbell 1982) recorded a similar population overall to the 1972-74 survey (159 territories compared with 183), but did not cover the Cheshire area. However, the RSPB surveyors discovered that there had been a substantial shift of the population, with 50% fewer pairs in the southern part (Kinder–Bleaklow) partly offset by a 30% increase in the population further north. The Cheshire part of the Peak District seems to have suffered a decline similar to the rest of the southern part. Atlas surveyors tried to cover all the available habitat but could find evidence of only seven or eight territories.

In part this halving of the population may reflect deterioration of the habitat; the high moors seem more heavily grazed, more grassy, than they were in the early 1970s and one former site (Wood Moss) has been converted to pasture. However, there is considerable evidence that this species is very susceptible to

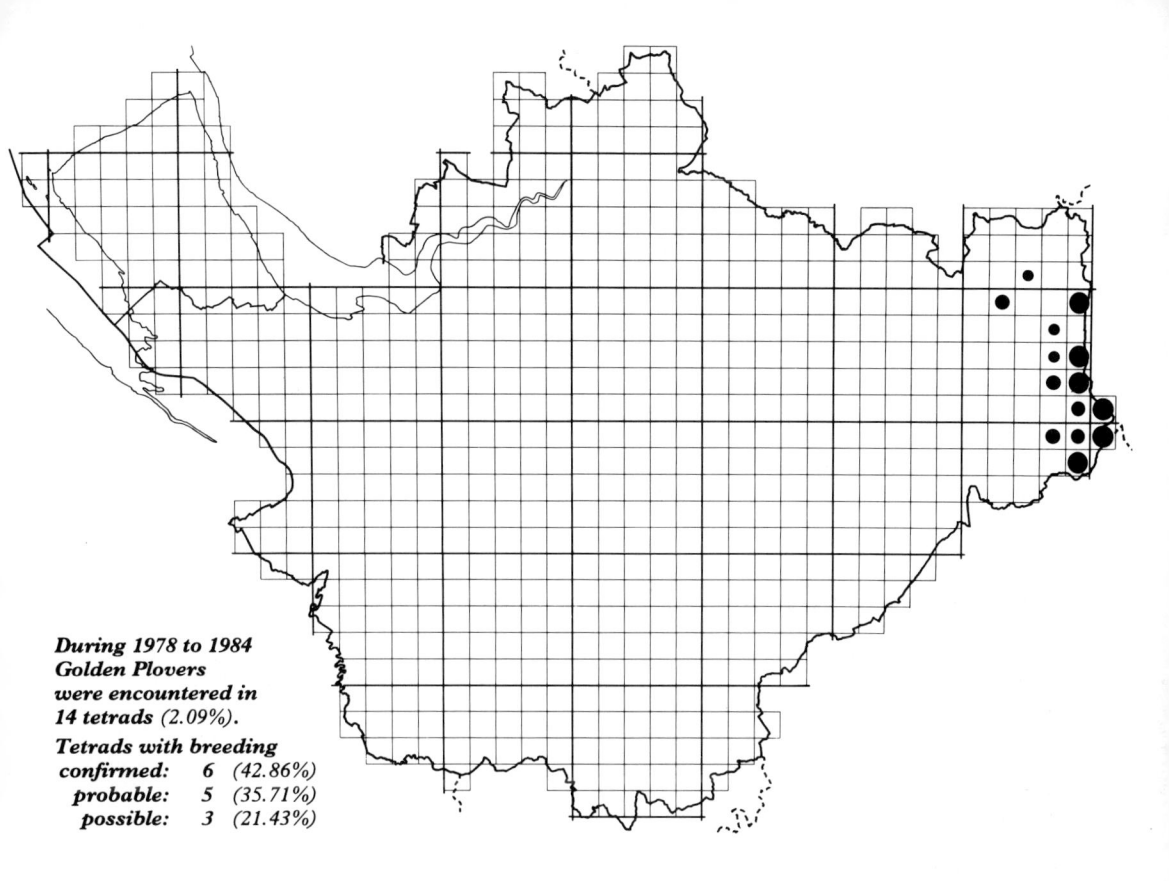

During 1978 to 1984 Golden Plovers were encountered in 14 tetrads (2.09%).

Tetrads with breeding
confirmed:	6	(42.86%)
probable:	5	(35.71%)
possible:	3	(21.43%)

disturbance. In particular, Parr (1980) mentioned that birds would stay off the nest for several hours if someone was working in the vicinity or if a vehicle was parked nearby. Thus the cumulative effect of a succession of ramblers passing through an area of good habitat (such as along the Pennine Way) could be harmful (Yalden & Yalden 1989). The Cheshire moors do not themselves suffer heavy recreational pressure, but they are part of a larger area which certainly does. Given that the population is already so small, this species is clearly endangered as a breeding bird in the county.   DWY

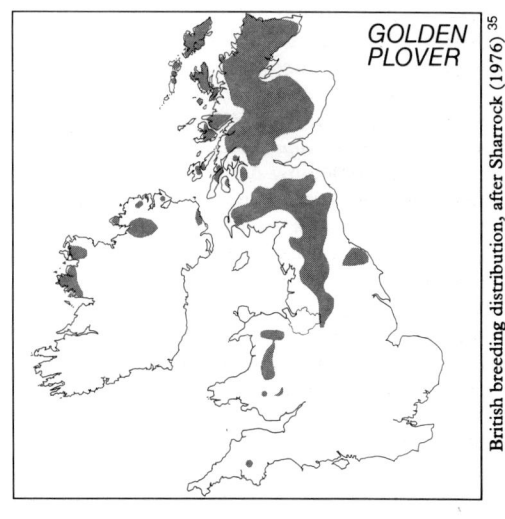

GOLDEN PLOVER

British breeding distribution, after Sharrock (1976)[35]

REFERENCES

Byrkjedal, I. (1980): Summer Food of the Golden Plover (*Pluvialis apricaria*) at Hardangervidda, Southern Norway. *Holarctic Ecology*, **3**: 40-49.

Campbell, L. H. (1982): Peak District 1981: Report of the Survey for the RSPB Conservation Planning Dept. (Unpub.).

Parr, R. (1980): Population Study of Golden Plovers (*Pluvialis apricaria*) using Marked Birds. *Ornis Scandinavica* **11**: 179-189.

Yalden, D. W. (1974): The Status of Golden Plover (*Pluvialis apricaria*) and Dunlin (*Calidris alpina*) in the Peak District. *Naturalist* **390**: 81-91.

Yalden, D.W. & Yalden, P.E. (1989): The sensitivity of breeding Golden Plovers (*Pluvialis apricaria*) to human intruders. *Bird Study* **36**: 49-55.

Lapwing
Vanellus vanellus

THE wild display flight and jubilant song of the Lapwing are welcome signs of spring. As the season progresses displaying birds can even be heard on mild, moonlit nights. Pairs may split off from the winter flocks as early as mid-February, or more usually in March, to establish territories on their breeding grounds. Display usually starts in the first week or two of March and may then continue until mid-May. For some days before the first eggs are laid, birds may be seen rocking and rotating on the ground as they prepare the nesting scrape. Eggs may be laid in late March but more usually in April. Chicks hatch a month later and by the middle of June young birds from early nests are independent, although a small number may still be seen in the custody of their parents in late July. On hatching, chicks are often led by their parents to some damp hollow where it is likely to be easier for the young birds to find insects and larvae in the soft ground, and where clumps of rushes or tall grasses provide shelter from predators. Boyd[6] recorded an adult leading a four-day-old chick across a railway to a meadow on the other side, and there are recent records of chicks attempting to cross busy roads under the guidance of their parents – a very hazardous operation since the chicks instinctively crouch at their parents' alarm signals instead of running from the traffic.

Dry springs may cause feeding problems for the young birds. On 12th May 1984 a female was brooding ten one-day-old chicks in a damp ditch on Frodsham Marsh (DN). It was suspected that the adults, which defend feeding areas from other Lapwings, had taken over chicks from other broods, probably the result of confusion when several families competed for scarce feeding areas. A full brood normally consists of four young, although there may occasionally be three, rarely two or five. The Lapwing is single-brooded but may lay up to four replacement clutches[36].

Coward & Oldham[2] stated that almost every field in rural districts of Wirral and the lowlands supported at least one pair of birds, despite systematic egg gathering for consumption by farm labourers and others, a practice seldom, if ever, occurring today. In the eastern hills they considered Lapwings were abundant and "equally at home in the pastures and amongst the heather...". However, during the BTO Lapwing Habitat Inquiry of 1937 the return for the Cheshire moorlands, including areas of heather and ling, was negative (Nicholson 1939). Today, Lapwings still seem as common in the hill-pastures as anywhere in the county, and may even be seen displaying over the cotton-sedge moors, but, as in 1937, there are no recent reports of nests in the heather.

Of 46 nests that Boyd[6] located in the Great Budworth area, 25 (54%) were in meadow and other grass, eleven (24%) in cornfields and potatoes, eight (17%) in marsh and two (4%) on cinders. Surveys carried out by E. Cohen in north-east Cheshire from 1934 to 1937 inclusive showed that of a total of 77 nests 38 (49%) were in "arable under green crops", 19 (25%) were in "permanent pasture", ten (13%) were in "rushy fields", six (8%) were in "arable down to grass" and four (5%) were in "rough lowland pasture".

In April 1983, JPG surveyed the SJ87 (Alderley Edge) 10-km square. Of 134 pairs or territorial birds observed (no attempt was made to find nests) 54% were on ploughed or tilled land, 22% in cereal or newly-seeded grass-crops, 19% on pasture and 5% in the bottoms of sand-quarries. Fields from which the turf has been removed for sale also often

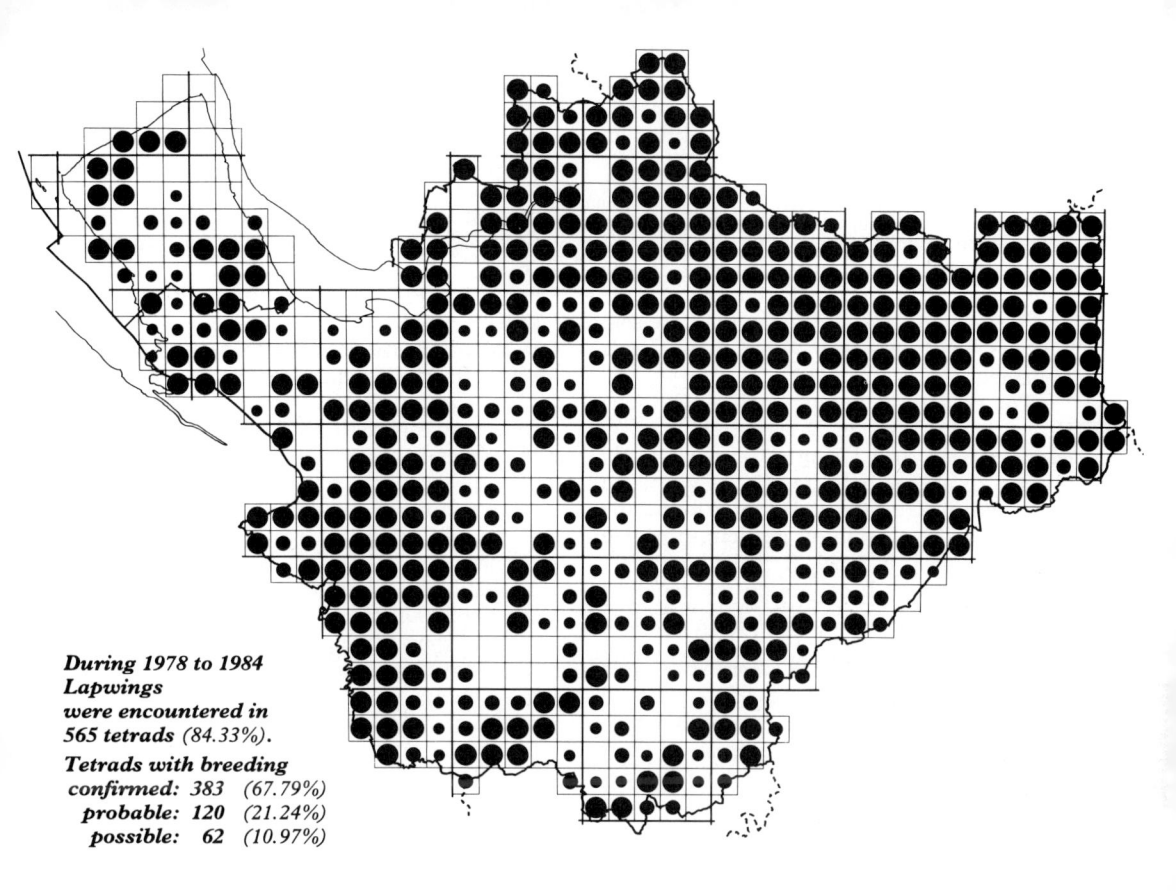

During 1978 to 1984 Lapwings were encountered in 565 tetrads (84.33%).
Tetrads with breeding
  confirmed: 383 (67.79%)
  probable: 120 (21.24%)
  possible: 62 (10.97%)

support nesting birds. It appears that modern pastures are too intensively grazed to permit safe nesting, more efficient drainage having led to the disappearance of the once typical clumps of rushes. At one time all marshy land in Cheshire was known as "peewit land", whether or not Lapwings were present. The 1967 *CBR* commented that in parts of mid-Cheshire "...this species appears to have largely ceased nesting on arable land in favour of marginal land with short grass, and disused lime-beds...". Such industrial sites are certainly much favoured, as is shown by the record, in the same report, of seven pairs nesting on one lime-bed. Whalley[19] referred to five nests within 200 yards of each other on the sludge-pool by Elton Hall Flash, Sandbach in May 1963. This tendency for Lapwings to nest in groups had not been mentioned by earlier Cheshire authors, although elsewhere it has been linked to a plentiful food supply[36]. During the 1983 census of SJ87, colonial nesting was found to be the norm with up to nine pairs close together in a single field (SJ87U), and relatively few isolated pairs. An intruding predator or bird-watcher crossing a nesting field may be mobbed by up to a dozen or more Lapwings, and there is a clear defensive advantage to nesting in groups, particularly given the enormous increase in Crows and Magpies in recent decades. The pair or two of Lapwings which nest annually by the Prestbury Sewage Works invariably lose their eggs to Crows.

Lapwings are unlikely to have been overlooked in well-covered tetrads during this survey, displaying birds being highly conspicuous in early spring. In undulating country, nests are usually situated towards the top of the field, and in any open site sitting birds can easily be detected by scanning through binoculars. The anxious calls of adult birds with eggs or young often draw the observer's attention – no doubt it also gives a clue to approaching predators who must then risk an aerial buffeting. Tetrads not visited in the early part of the recording season, however, may appear blank. The 1983 survey of SJ87 revealed territories and nests in tetrads where no birds had been located during intensive summer fieldwork in previous years.

The apparent absence from areas in the south and centre of the county is more than a quirk of coverage however. Even in the better-covered squares the species is notably scarce. Lapwings are known to have declined in recent years from Egerton Green, SJ55 (F. R. Walley pers. comm.). This area consists primarily of grazing land or fields cut for silage, such operations allowing the birds no time to nest between cuts. Scarcity on the Peckforton Hills and in Delamere may be attributed to the dry, sandy nature of the soil. The absence from south Wirral is not easily explained.

The centres of towns, such as Chester, Runcorn and Widnes, Warrington, Crewe, Macclesfield and Congleton are clearly avoided, as are the built-up parts of Wirral. In north Wirral, Griffiths & Wilson[13] found the Lapwing to be a common resident, breeding all over the area. In 1959, A. A. Bell found 94 pairs and 34 non-breeders on the Hoylake Fields, but a repeat census in 1961 revealed only 28 pairs and eight non-breeders.

The 1983 census of SJ87, carried out in a wet spring, revealed some 134 occupied territories (1.34 pairs per square kilometre). Compared with Boyd's 1930 estimate of 13 or 14 pairs in 750 acres (around 4.5 pairs per square kilometre) of arable and pasture at Frandley, Great Budworth, this does suggest some decrease. Furthermore CBC counts from south-eastern Cheshire in 1978 gave a total of only ten pairs in 393 hectares (2.5 pairs per square kilometre) of farmland. An exceptionally high density of breeding pairs was recorded from the No. 4 sludge pool at Frodsham however, where in 1984 some 50 pairs were estimated to be nesting in an area of roughly one square kilometre, and 131 chicks were ringed (DN). At Rostherne Mere NNR, Harrison & Rogers[24] considered the species less numerous than in 1914 when Coward & Oldham reported "many pairs" nesting in fields around the mere. By 1976 there were probably not more than 12 pairs. In 1983 there was for the first time no confirmation of breeding on the reserve (NCC).

In some years southward passage of birds across the county is evident by early June, and such wandering birds may have given rise to some 'possible' breeding records. Furthermore there is a substantial population of non-breeders, which could account for some records. Assuming a population of five pairs per tetrad in which breeding was confirmed and one pair for each tetrad with 'probable' breeding, the county may hold in excess of 2000 pairs.

The species seems particularly susceptible to hard winters, and spectacular cold-weather movements are frequently seen. Cheshire-ringed nestlings have been found in Cornwall, Eire and France. Many birds find their way back to their natal areas to breed, but one Frodsham-bred bird had evidently changed breeding areas, its ringed leg being found eight years later in a Scottish Peregrine eyrie. The national CBC index has been almost constant since 1974, and in particular showed no changes from 1978 to 1979 or from 1981 to 1982, spanning severe winters, showing the success of their strategy of cold-weather movements. There is, however, no quantitative information on any recent changes in the Cheshire population.

REFERENCE

Nicholson, E.M. (1939): Report on the Lapwing Habitat Inquiry, 1937. *British Birds* 32: 170-191, 207-229 & 255-259.

Dunlin
Calidris alpina

ALTHOUGH familiar enough on passage and as a wintering bird, occurring in large flocks on the mudflats of estuaries such as the Dee and Mersey, as a breeding bird the Dunlin is poorly known in Britain, where it is at the extreme south-western edge of its range in the Palaearctic. Most of the wintering birds are of course visitors from the tundra, and that is where the breeding biology of the species has been principally studied.

There is nevertheless a substantial Pennine breeding population[35]; all are of the smaller, southern race *C. a. schinzii* – the southern-most section of which breeds in the Peak District. A survey in the early 1970s suggested a population there of 140-150 pairs associated especially with the wetter areas of cotton-grass bog (Yalden 1974). At that time the Cheshire population was estimated at around 26 pairs but almost all of these were in

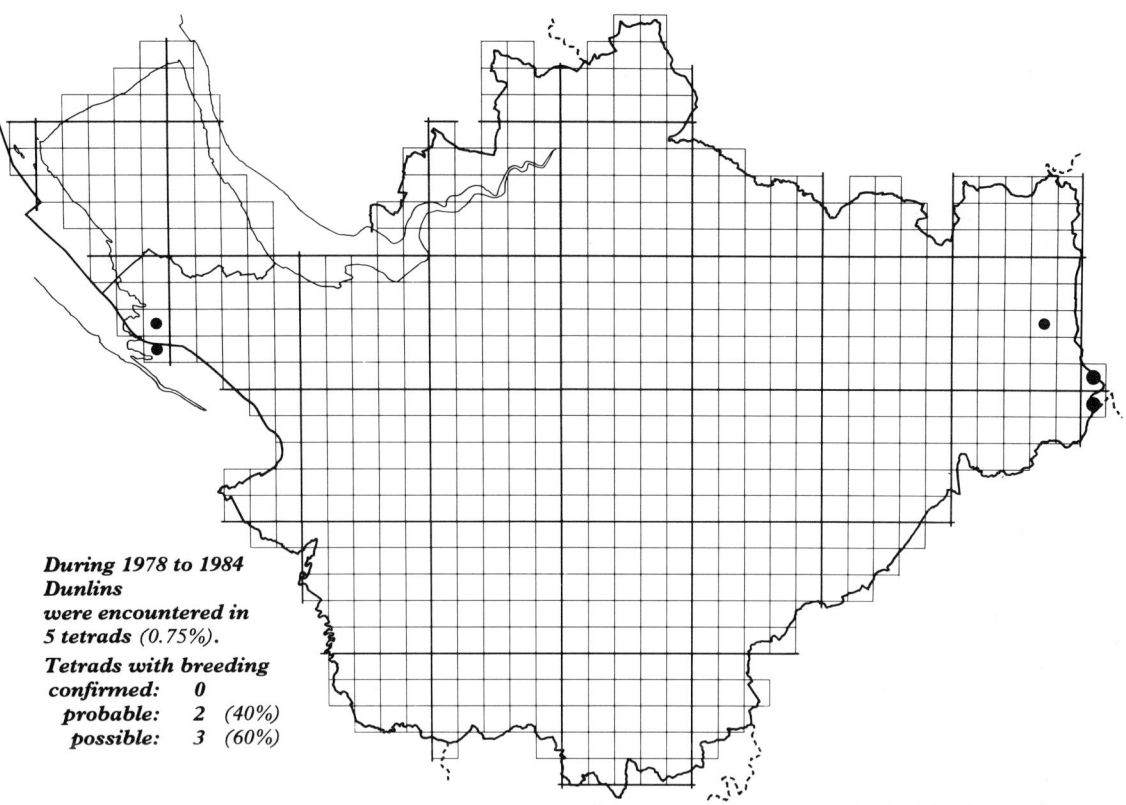

During 1978 to 1984 Dunlins were encountered in 5 tetrads (0.75%).

Tetrads with breeding
confirmed: 0
probable: 2 (40%)
possible: 3 (60%)

Longdendale. There were at most three sites holding perhaps four pairs in eastern Cheshire, but the population in any one year was no more than two pairs. A partial re-survey of the Peak District moors by the RSPB in 1981 (Campbell 1982) found the total population essentially unchanged but they did not cover the Cheshire moorlands. Survey work for this atlas found evidence of no more than one pair in any one year, and the Dunlin now has the slenderest of toe-holds as a breeding species in the county. Very little of the wet cotton-grass habitat still exists in the reduced area of moorland, and in drier years even that seems to have been unsuitable.

Coward & Oldham[2] stated that Dunlins bred in several places on the Pennine range but they had no evidence of birds breeding on the Cheshire moorlands. Brockholes[2] noted that "...a few breed in suitable places in Wirral..." and cited the receipt of eleven eggs in 1871 which had been taken from the Dee marshes near Puddington and Shotwick. However, by Coward & Oldham's time most of these marshes were under cultivation and there was no evidence of breeding on the Cheshire side of the Mersey, although it had recently bred at Cuerdley Marsh, a few miles below Warrington on the Lancashire side of the river (Jackson 1906). Hardy[4] cites the record of a bird flushed on Burton Marsh in the summer of 1939. It ran in circles and feigned a broken wing but no attempt was made to locate a nest. Bell[7] also gave one record of nesting in the Frodsham area in 1955 and another[9] from the Dee in 1961.

The species does still breed, albeit sparingly, on saltmarshes along the Lancashire coast and further north. In May 1974, birds were seen repeatedly giving the song-flight over a patch of rushes at Elton Hall Flash, Sandbach. This trilling song is sometimes heard from birds on spring passage and is presumably the origin of the 'possible' breeding records on the Dee. Given that the species is difficult to prove breeding, it is still possible that it may nest occasionally in coastal habitats in Cheshire.   DWY

REFERENCES

Campbell, L. H. (1982): Peak District 1981: Report of the Survey for the RSPB Conservation Planning Dept. (Unpub.).
Jackson, A. (1906): *Zoologist* pp21-25.
Yalden, D. W. (1974): The Status of Golden Plover (*Pluvialis apricaria*) and Dunlin (*Calidris alpina*) in the Peak District. *Naturalist* **390**: 81-91.

Ruff
Philomachus pugnax

COWARD & Oldham[2] considered it possible that the Ruff had bred in Cheshire at some time, but had no proof. However, birds regularly occurred in spring and autumn when on passage, which is true to this day. This century birds have been noted in breeding plumage on various occasions. Up to three males remained at Witton Flashes from 8th May to 30th June 1949. The first bird, which arrived on 8th May, was joined by a second bird on the 18th and during the next three days displayed to the former, although there were no females (reeves) present[6]. A third bird arrived on 23rd May and remained until 29th after which only the first arrival stayed on. This bird, which never attained full breeding plumage, remained until 30th June.

On 30th May 1963 a male displayed in the presence of a female at Elton Hall Flash, Sandbach. 1966 was considered to be "remarkable for the number of records of birds in full summer plumage": at Sandbach Flashes there were three, two Ruffs and a reeve, on 19th May, a single Ruff on 1st June with two on 2nd June, and at Frodsham there were up to three different Ruffs during June[8]. Birds attaining full summer plumage were again reported at Frodsham throughout June in 1970, 1971, 1973 and 1975 (when birds were proved to be breeding in south Lancashire and it was considered that they might be breeding in Cheshire)[8]. Elsewhere, three males displayed regularly to up to three females at Watch Lane Flash, Sandbach, in April 1973; a male displayed to two females at Burton Marsh on 5th May 1974; and up to three birds were present on various dates in June at Sandbach in 1980 and 1981[8]. No other birds were recorded during the present survey apart from those on spring and autumn passage which may, respectively, extend into early June and begin in late June.

Jack Snipe
Lymnocryptes minimus

ON 31st May 1983, a Jack Snipe was seen to ascend in an apparent display flight at an upland site in the east of the county. No vocal sounds were heard.

Snipe
Gallinago gallinago

THE peculiar bleating noise made by the wind rushing through the Snipe's tail feathers as it dives in display is becoming increasingly unfamiliar as damp lowland meadows are drained. Both sexes indulge in this "drumming", particularly in the early spring. The bird towers upward on rapidly beating wings followed by a dive in which the specially adapted tail is fanned out, only to climb again to repeat the performance, all the while circling to remain over the breeding marsh. Drumming in Cheshire has been noted between 15th March (exceptionally as early as 28th February and once, briefly, on 8th January) and 14th July. Breeding birds also utter a repeated "chipper-chipper-chipper" call, often from the top of a post or in flight, though spring migrants especially will call from cover; dates between 1st April (but once on 26th February) and 3rd July have been recorded (JPG). Birds ringed as nestlings near Macclesfield in July 1939 and in Wirral in June 1940 were recovered in Ireland in January and September

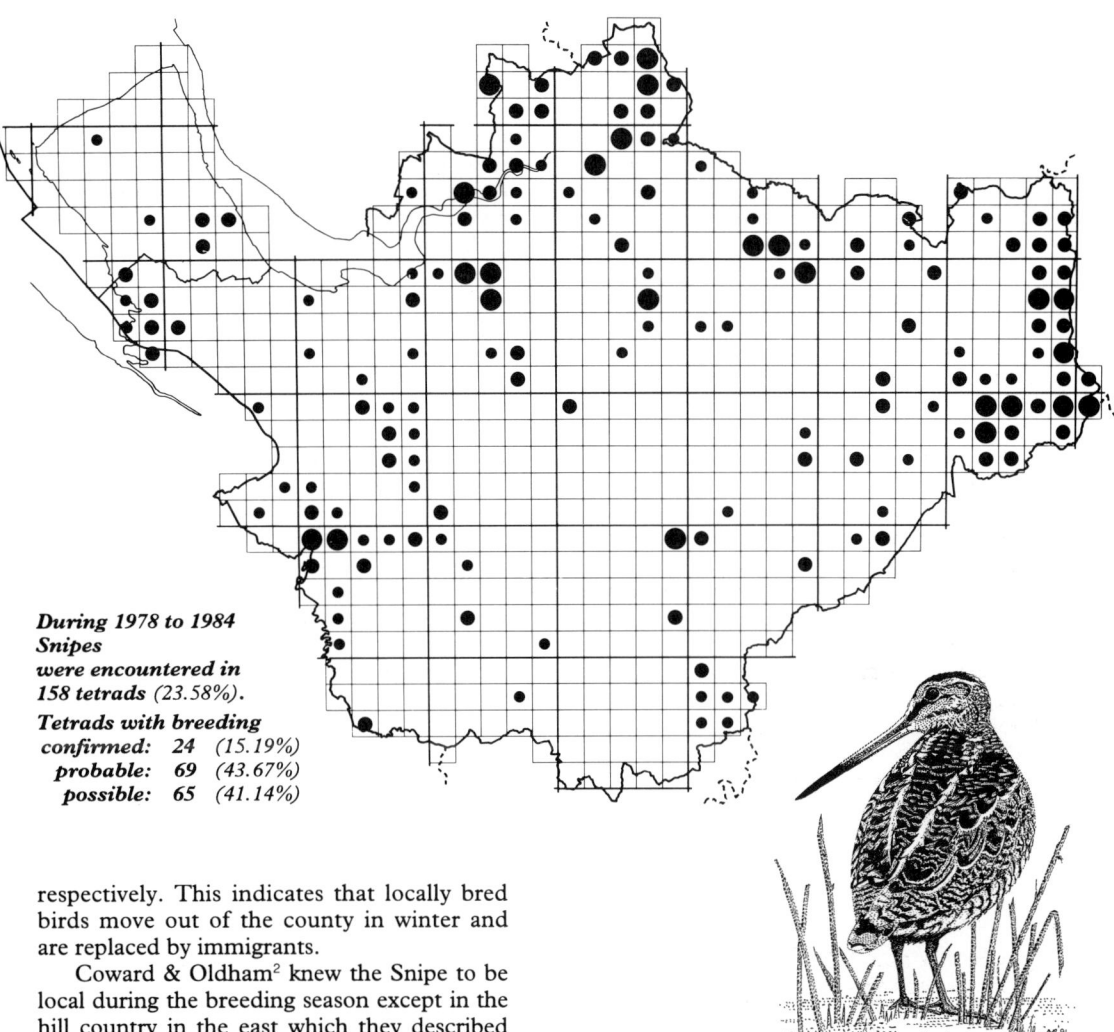

During 1978 to 1984 Snipes were encountered in 158 tetrads (23.58%).

Tetrads with breeding
confirmed: 24 (15.19%)
probable: 69 (43.67%)
possible: 65 (41.14%)

respectively. This indicates that locally bred birds move out of the county in winter and are replaced by immigrants.

Coward & Oldham[2] knew the Snipe to be local during the breeding season except in the hill country in the east which they described as "the great stronghold of the Snipe in Cheshire in the spring". They could quote no authentic instance of nesting in Wirral, although birds bred at Aldford and in the marshes at Frodsham, Helsby and Thornton-le-Moors. It nested at several sites in Delamere Forest and elsewhere on the plain at Knutsford, Mobberley, Cheadle, Handforth and Siddington. At Rostherne it nested "in some numbers in the Gale Bog and the withy beds"[11] Considerable numbers bred on the hills east of Macclesfield.

Griffiths & Wilson[13] described the Snipe in Wirral as typical of a zone of low-lying land, running from New Brighton to West Kirby, protected from the sea by embankments, and they specifically mentioned a few pairs nesting. In his country diary for April 1941, Boyd also referred to a bird "chippering" in wet meadows by the sea. A. Booth (*in litt.*) states that drumming birds were present throughout the summer in fields at Meols in 1956, and these were regarded as nothing out of the ordinary. The map shows only a single possible breeding record from Meols during this survey.

Abbott[10] recorded birds at Mobberley, Lindow Common, Bosley Minn, near Tabley Park, Redesmere, near Astle and near Monk's Heath between 1913 and 1930. Boyd[6] knew of nests in Marbury Park, at Arley, Frandley, Billinge Green and Aston-by-Budworth. Stretton Moss and Whitley Reed were also mentioned, but both sites have since been drained. The latter was originally enclosed and drained in 1854. Subsequently the drains were neglected and a new reed-bed was formed allowing an increasing number of Snipes to breed, but it was then redrained because of the need to increase farm production during the 1939-45 war. At Rostherne, where nesting was last recorded in 1961, Gale Bog and the withy-beds (where

Coward found them) are now for the most part too overgrown. Grazing meadows there are now less suitable as indicated by Harrison and Rogers' statement[24] that "the many clumps of rushes (*Juncus* spp.) still apparently present in permanent pasture fields in the early 1960s, have now nearly all gone". Bell[9], referring to the county as a whole, stated "in recent times there have been no definite breeding records and...the species is now rare in the breeding season...".

Records point to a gradual and continuing decline associated with agricultural drainage: sites around Handforth which held drumming birds into the early years of this survey have since been deserted, and even in the hills, still a stronghold of the species, former nesting sites at Higher Disley and Saltersford have recently been drained.

Damp, rushy flushes in pastureland and on the moors form the typical nesting habitat in the hills. In the lowlands, Snipes breed mainly in low-lying damp meadows, by sludge-pools, sewage works or other boggy areas along the valleys of the Dee, Weaver and Mersey, with scattered pairs in remnants of suitable habitat elsewhere. Most of the confirmed breeding records refer to nests and the majority of 'probable' breeding records were of displaying birds. A cluster of records refers to drumming birds over the saltmarshes of the Dee.

In 1974, there were six pairs at Bagmere, three at Sandbach and two at Tatton Park. From 1978 to 1981 no more than ten territorial or breeding pairs were recorded at lowland sites in any one year[8]. During the BTO/RSPB/NCC Survey of the Breeding Waders of Wet Meadows in 1982 only six pairs held territory in that habitat type, at Woolston (2), Tatton Park (1), Weaver Meadows at Aston (2) and Astmoor (1). However, it was a particularly dry spring and at least one other lowland site, which normally hosts breeding Snipes, was bone dry.

In 1983, six pairs bred at Risley Moss and four birds were drumming at Woolston during May. In 1984, the breeding population in the Chester recording area was estimated to be less than ten pairs with perhaps eight or ten pairs in the Mersey/Weaver valley[8]. Most occupied tetrads held only one or two pairs, although numbers may be marginally higher in the eastern hills. Many sites were occupied in only one or two years of the survey however, depending on the wetness of the ground in spring, and the map undoubtedly overstates the present distribution.

The county population in any one year of this survey probably did not exceed 75 pairs and, of those, no more than 20% bred in the lowlands.

Woodcock
Scolopax rusticola

IN the breeding season the Woodcock inhabits a variety of woodlands – mature broad-leaved and mixed woods or conifer plantations – nesting amongst ferns, brambles or similar undergrowth on relatively dry ground. Tree-lined streams and alder carrs are often incorporated into territories since the birds often feed by night in boggy patches.

Coward & Oldham[2] knew of breeding records at Somerford, Tatton Park, Holmes Chapel, Astbury and Rostherne, and rather more frequently in the hills to the east of Bosley and Macclesfield. Because all known occurrences were listed it is thought that breeding on the plain was uncommon at that time. They also quoted Brockholes' belief that the bird bred occasionally in Wirral. Hardy[4] records nesting at Caldy Heath in 1930, Eaton Park in 1938, Kelsall in 1939, and,

"...in recent years..." in Row's Wood and Appleton Firs. There were also records from Delamere of a nest with four eggs in 1945 and roding birds in 1949[3]. Boyd[6] knew of no breeding records in the Great Budworth area, though he did refer to birds being occasionally present in the Arley woods into April. Bell[7] noted, however, that in the 25 years up to 1962 it was widespread as a nesting species with most records relating to the east and north of the county.

The easterly distribution with a few isolated records from the centre and west of the county are therefore traditional, and not easily explained. However this pattern does fit in with the regional distribution given in the BTO Atlas[35], the species being widespread in neighbouring Derbyshire and Staffordshire but scarce in north Shropshire and Clwyd.

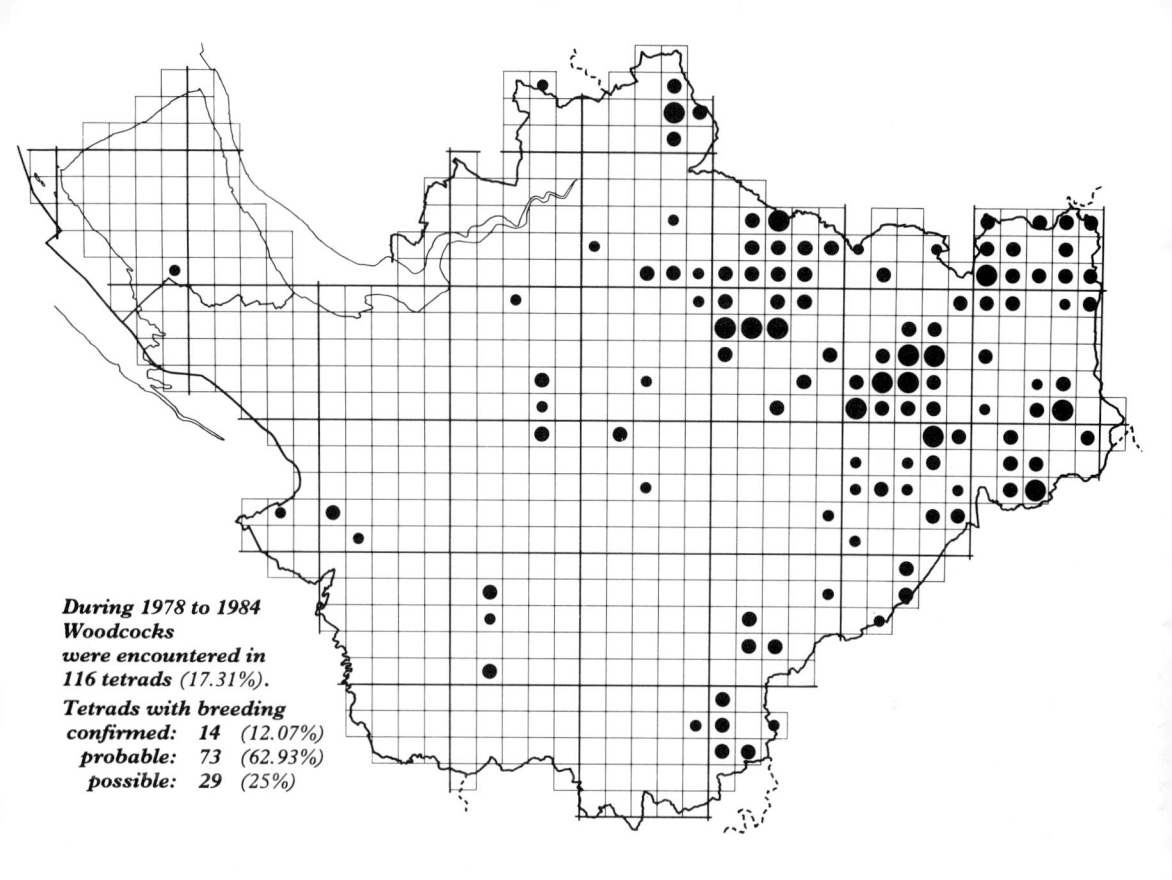

During 1978 to 1984 Woodcocks were encountered in 116 tetrads (17.31%).
Tetrads with breeding
confirmed: 14 (12.07%)
probable: 73 (62.93%)
possible: 29 (25%)

Woodcocks may be found in woods and coverts throughout the county in winter, and there is no obvious reason why they should be absent from large areas in summer. During the last fifteen years or so, records of roding birds have come from Gorstage, Delamere Forest, Abbots Moss and Peckforton in the centre of the county, from the Eaton estate near Chester, Walton Golf Course and one other, unidentified site in the north. In Wirral there have been only late spring records, from Heswall and Thurstaston. The picture is undoubtedly clouded to some extent by the reluctance of observers to check woods away from known localities: outings from certain local bird clubs, for example, annually drive past several occupied woods on their way to traditional rodes.

Most records during the survey in fact referred to roding birds. This display flight, with its distinctive, slow, halting wing beats and the associated characteristic grunts and squeaks of the flight-call, is performed throughout the spring and early summer. Roding has been noted as early as 16th February at Macclesfield Forest in 1981, and in most years probably starts by the end of that month. At this time of year, however, the performance generally starts well after dark, and it is most easily witnessed in May or June when the display begins shortly after sunset. At times roding birds become involved in aggressive chases, accompanied by much squeaking and rough "turrip" calls, during which they may wander over open country, away from their usual flight lines.

An assessment of the number of breeding Woodcocks is made difficult by their nocturnal habits and their far-ranging display flights often encompassing parts of more than one tetrad (which habit may have resulted in some over-statement of their true distribution). Futhermore, the fact that males can be successively polygynous (i.e. mating with several females in turn) or promiscuous adds to the problem. Females will sometimes accompany the males on the rode, and records of three and sometimes four birds flying around together give no indication of how many nesting females may be present in the area. Nonetheless a conservative estimate of two nesting females per tetrad in which breeding was either 'probable' or confirmed would give a county total of rather more than 150 nesting females. Young have been recorded as early as 22nd April in east Cheshire, in 1933, and 19th April at Rostherne Mere NNR in 1957.

Black-tailed Godwit
Limosa limosa

A pair remained at the Sandbach Flashes from Mid-March to 13th May 1976. Display flight occurred over an area of marshy ground on several dates in April, and over Elton Hall Flash on 25th April

---

Curlew
Numenius arquata

IN the early years of this century, Curlews were restricted in the breeding season to the hill areas in the east of the county where concentrations of several pairs could be found in favoured areas such as Shutlingsloe and the Cat & Fiddle Moors[2]. Previously birds had bred on the lowland mosses such as Carrington Moss (now GMC), Danes Moss, and probably Lindow Moss and elsewhere, but, in general, this practice had by then ceased. Hardy[4] cites nesting haunts at Risley Moss (up until wartime building), Holcroft Moss (before the war), Newchurch Common (from 1927 to 1938) and Frodsham Marsh in 1941. Then, as now, flocks of one-year-old, non-breeding birds spent the summer on the estuaries.

Boyd[5] in his country diary for 5th March 1933 commented that the sight of three Curlews flying over "the mere" uttering their beautiful bubbling notes was a comparatively unusual occurrence in the Cheshire plain. In an entry for 11th May 1940, he reported a pair nesting in pastures near Nantwich, adding, "for some years curlews have been extending their breeding range from the moors to the foothills and have now invaded the plain". In 1942 a pair bred successfully within three miles of Crewe and in a diary entry for the following year, Boyd indicated that the species was continuing to spread, nesting on grassland in the plain, although not yet in his own area near Great Budworth. J. E. Guest found a nest in damp pasture by the Alderley sewage farm in 1947. Further spread in the lowlands aroused such interest that the unfamiliar newcomer was dubbed "French Curlew". The *L&CFC Report* for 1949 catalogues further evidence of this spread into the lowlands between 1943 and 1949.

During the two decades up to 1960 breeding was reported from various sites in the east and south of the county, and also at Oakmere and on Frodsham Marsh[7]. Tatton Park was first colonised in 1957 when two or three pairs bred[3], and, by 1972, 15-20 pairs were present in this area in spring. Numbers there have since declined because of increased pressure from public access and agricultural "improvements". Birds were first reported from meadows south of Chester in the early 1960s, and by 1966 the *CBR* detailed 15 occupied sites. There remains a distinct possibility that some of these sites had been occupied much earlier, particularly in view of the above mentioned records from the Crewe and Nantwich area in the early 1940s. Until this survey no further breeding records came from the farmland areas of the south, although with the increase in intensive silage production and up to three mowings during the season, any further spread in the lowlands has been halted and numbers now may be decreasing.

In late February or March the winter flocks on the Dee and Mersey begin to split up, although at Northwich, where a flock is resident each winter, there is a notable exodus in early February followed by an influx, possibly of continental birds or local non-breeders, in late February/early March (Elphick 1979). Smaller sub-roosts form near to some breeding areas, as at Prestbury Sewage Works and possibly Tatton and Bagmere. In the hills a small flock of up to 20

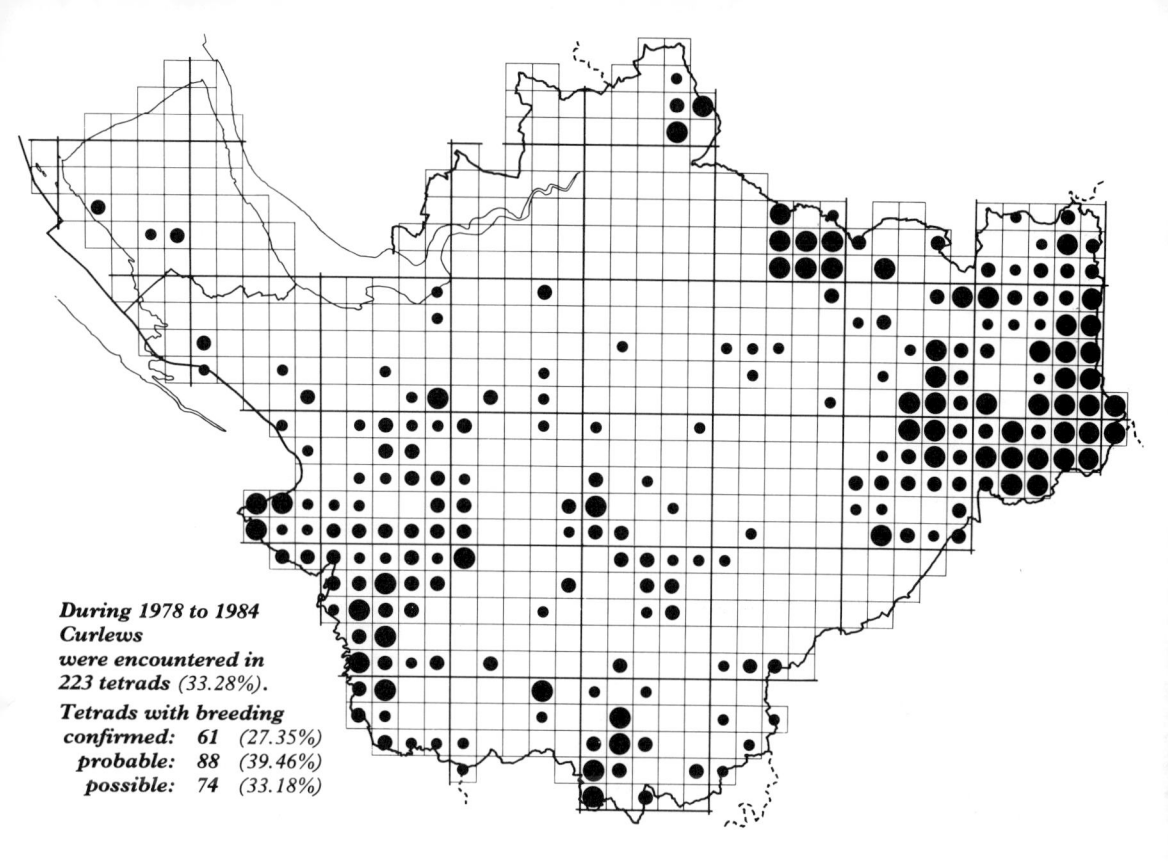

During 1978 to 1984 Curlews were encountered in 223 tetrads (33.28%).

Tetrads with breeding
confirmed: 61 (27.35%)
probable: 88 (39.46%)
possible: 74 (33.18%)

birds is regularly present near Crag Hall, Wildboarclough, at this time. Pairs fly out by day to establish territories, but once the breeding cycle starts they remain to roost by the nest.

On the moors, pairs nest among the cotton-grass or heather, feeding in pastures at the moor edge. In areas around Congleton and west of Macclesfield undulating country is favoured with alternate high vantage points and low-lying damp hollows for feeding. Elsewhere permanent grassland is the key factor, being presumably richer than tilled land in invertebrates such as larvae and earthworms on which the chicks feed. At Tatton the parkland areas are preferred, as are the low-lying hayfields around Wettenhall, pastures around Audlem and low-lying damp meadows south of Chester. Breeding also occurs on some of the few remaining lowland mosses, for example Risley and Danes Moss. There is no record of breeding in Wirral to date, although there were two 'probable' records during this survey.

The song period is extended – birds will sing at any time and may even be heard on the shore in winter. Where pairs are thought to be nesting it is often possible to see the sitting bird run rapidly, half-crouching, from the nest while the observer is still some distance away, but even if the nest is not found, the far-carrying bubbling song of breeding birds is unlikely to be missed. Once eggs are laid the pair is very aggressive, persistently driving off Crows, Kestrels and even dogs which intrude into their territory. On two occasions they have been known to mob Goshawks and once a bird stooped at a Heron. The strident alarm call, "wik-ik-ik", repeated by a circling bird is also a sign that eggs or young are nearby. The adults are very wary, and few observers were fortunate enough to see the beautifully camouflaged chicks, showing a slightly curved bill even when very small.

Many occupied tetrads hold only a single pair, but numbers may be considerably higher on the eastern hills, and in favourable sites on the plain. With 147 tetrads probably or certainly containing breeding pairs, the county population probably lies in the region of 250-350 pairs.

REFERENCE

Elphick, D. (1979): An Inland Flock of Curlews *Numenius arquata* in Mid-Cheshire, England. *Wader Study Group Bulletin* No. 26 pp31-35.

Redshank
Tringa totanus

THE loud, fluty song of the Redshank, uttered from a post, the ground, or in flight, makes this species difficult to miss. The map probably shows all breeding occurrences during this survey. The birds' limited habitat of coastal saltmarsh or damp, rushy meadows inland further restricts the possibility of their being overlooked. Once eggs are laid and particularly when the young have hatched, the adults will fly close overhead mobbing the intruder, or perch on a nearby bush or post calling anxiously. At Sandbach Flashes railway pylons often served as perches.

Breeding Redshanks were first recorded in Cheshire during the second half of the nineteenth century and must have increased rapidly (Hale 1980). Coward & Oldham[2] noted that a few pairs nested annually on the Dee marshes, and, although the "colony" was, at one time, much reduced because of persecution, they considered it was then in a satisfactory condition. In May 1894, they estimated that seven or eight pairs nested but by 1907 there were not only many more birds but the area over which they were nesting was greatly extended. They also referred to at least 50 pairs nesting on the Gowy Marshes between Stoke and Thornton-le-Moors in 1907 and eight to ten pairs on a strip of marshland at Norton near Runcorn. They attributed this increase to a lapse in persecution and the protection afforded by the Wild Birds Protection Acts of 1880 and 1896 and subsequent Orders, although the saltmarsh was also growing rapidly at this time.

Hardy[4] listed nesting at Hale Marsh, Weaver marshes, Frodsham, Mersey marshes, Gowy estuary (Ince and Thornton in 1933 and 1938), Burton and Neston marshes on the Dee, Meols sewage farm, occasionally on Cuerdley Marsh and Norton Marsh above Runcorn. Additionally, he cited Woolston Moss, Warrington and Hulme near Winwick as sites where they had bred. An essentially estuarine distribution extending up the Mersey valley is thus depicted. A. Booth (*in litt.*) recorded birds at Meols Fields from 1954 to 1958 and considered that they were breeding. Bell[9] reported some 25 pairs on the Burton Marshes in 1965 (although the 1966 *CBR* gives 100 pairs here); and breeding recorded or suspected at Puddington and the Weaver marshes. The 1966 *CBR* lists at least three pairs on Frodsham Marsh, plus single pairs at Moore and Puddington.

Away from the estuaries, Abbott[10] recorded nesting at Bosley Reservoir in 1915 and each of the years 1926 to 1929. This extended to Bosley Minn in 1916. Tunnicliffe[15] also recorded birds breeding at Bosley in the 1940s. Boyd[6] detailed the colonisation of the Great Budworth area from 1926 or earlier, with five or more sites holding nesting pairs at some time in the next 25 years. This extension of range was not limited to mid-Cheshire, and in his country diary for 29th May 1936 Boyd notes that "Redshanks have spread over the whole county where once they were seldom seen, and nest in scattered pairs over hill and plain alike."

Bell[7] considered the Redshank to be "common and widespread" by 1962, noting that there were many records since 1910 of birds nesting inland by the meres, flashes and sewage farms. However, in his Supplement[9] he listed only a few breeding sites: regular sightings in summer at Sandbach Flashes, Holford Moss and Tatton Park. The 1966 *CBR* mentions a pair with young on 27th May at Plumley and breeding at Sandbach Flashes. In 1969 a pair bred in Tatton Park and in April 1973 breeding was attempted at Rostherne[24].

Sharrock[35] states that since the peak of their expansion in the 1930s, the numbers breeding inland in southern England decreased, and it appears that this happened in Cheshire also. Drainage of agricultural land and the lowering of the water-table must have contributed enormously, with the total loss, for example, of the Gowy marshes as breeding habitat. The decline continued into the 1970s: the 1970 and 1971 *CBRs* listed only Tatton, Plumley, Lostock and Kingsley as sites where Redshanks either bred or were suspected of doing so. In 1975, there were five pairs around the Sandbach Flashes and successful breeding was again recorded in 1977. The Redshank was one of the two key species identified for the BTO Wet Meadows Survey carried out in 1982. Although that spring was unusually dry, only two territorial birds were recorded in that habitat type for Cheshire: one in the Weaver meadows at Aston and one at Astmoor, Moore.

During the tetrad breeding survey, several pairs bred each year along the lower Weaver and Mersey valleys. The only confirmed inland breeding records came from Pickmere

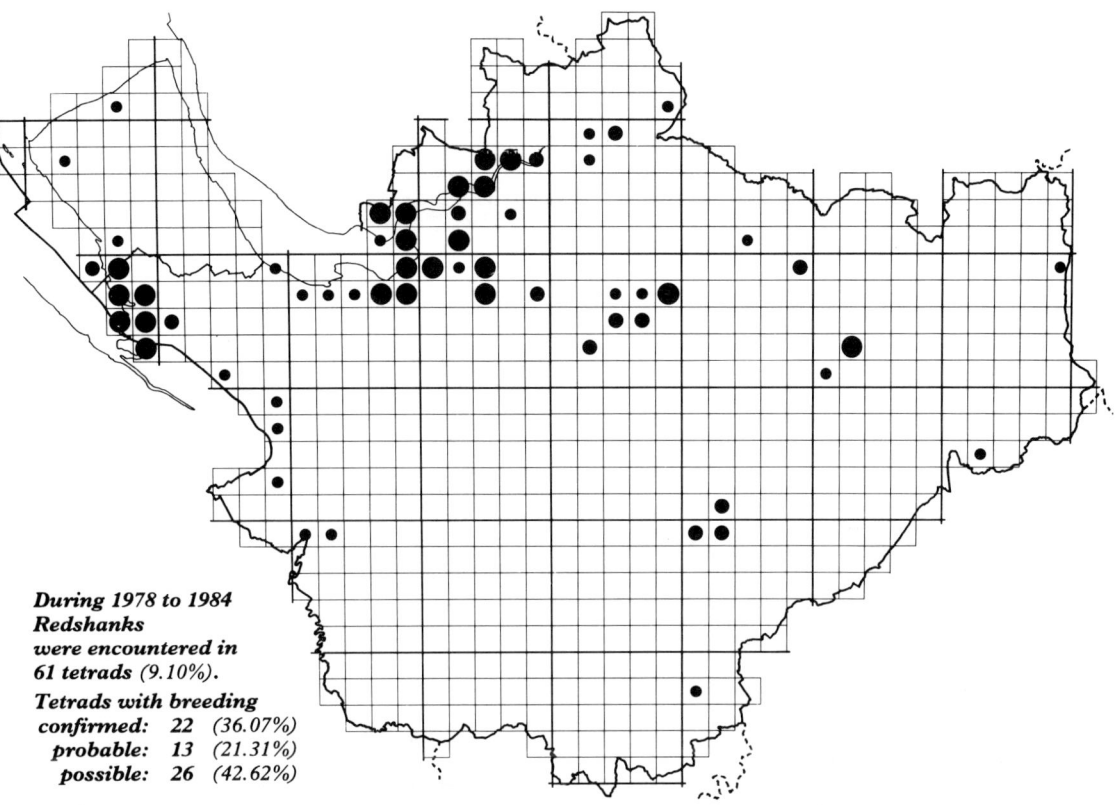

During 1978 to 1984 Redshanks were encountered in 61 tetrads (9.10%).

Tetrads with breeding
confirmed: 22 (36.07%)
probable: 13 (21.31%)
possible: 26 (42.62%)

in 1981 and probably 1982, and, in 1984, from the marshy bottom of Acre Nook sand quarry, Chelford. Mapped records of 'probable' breeding probably relate to non-breeding or passage birds displaying. Allowing for non-breeding birds wandering between the flashes and sludge-beds of the Northwich/Middlewich/Sandbach area, the total county population away from the Dee is unlikely to exceed 10-15 pairs.

The Dee remains the stronghold for breeding Redshanks. In 1972 about 20 pairs were recorded on Burton Marsh from the "Fisherman's Path", a track some 6600 yards in length across the saltmarsh. In 1973, R. A. Eades carried out a similar line transect observing 15 pairs from a one-mile transect giving a total estimate of 200 pairs on the Cheshire saltmarshes[8], but no more recent figures are available. The breeding success of the birds on the Dee saltmarsh is often poor, with nests being flooded by high tides in May or June. However, they can be long-lived birds, with the oldest known Cheshire bird still alive 14 years after ringing, so a low reproductivity rate may be tolerable.

This species seems more susceptible to hard winters than most other waders but the Cheshire birds probably avoid this problem since many, perhaps most, English Redshanks winter further south – a Frodsham-bred chick was shot in its first autumn on 11th October 1981 in Finistère, France. The birds which winter in large numbers in the estuaries are from the Scottish population and the larger Icelandic race. These sometimes display on bright days in spring before departure to their breeding grounds.

REDSHANK

British breeding distribution, after Sharrock (1976) [35]

REFERENCE
Hale, W.G. (1980): *Waders*. New Naturalist, Collins, London.

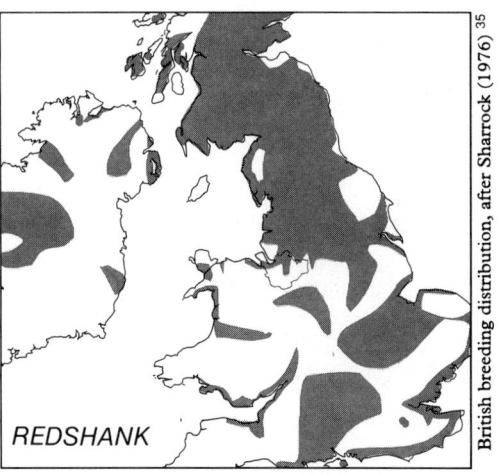

Common Sandpiper
Actitis hypoleucos

NOWADAYS the Common Sandpiper is, as a breeding bird, largely restricted to hill-streams and reservoirs which have shingly shores (the former because of occasional spates following heavy rain, the latter because of the "draw-down" in most summers). This restriction is all the more surprising because outside the breeding season the species occurs in a wide range of waterside habitats, including saltmarshes, sewage farms, lakes, estuaries and rivers. Moreover its stay in the breeding areas is very brief, birds leaving almost as soon as the young can fly. A ringed bird from the study population in the Peak District, seen on territory in May, was recovered in Morocco on 15th June – presumably this was a failed breeding attempt, but it emphasises the quick departure of some birds (Holland et al. 1982a). Few are seen in their breeding areas after mid-July.

The Common Sandpiper is very vocal in display and during territorial establishment, but can be secretive and easily overlooked when incubation takes place in late May. Nests are quite difficult to find, but, after hatching, the young are defended by extremely strident adults, and this makes confirmation of breeding relatively easy. One or usually both parents fly around in an agitated manner, calling loudly, and they can lead a dog away very successfully. True distraction display, with wings often drooped and tail fanned in a "rodent-run" may also be seen, though it is more often given when the bird is displaced from eggs.

Coward & Oldham[1,2] mention a number of lowland breeding sites where the species no longer occurs, particularly on the saltmarshes at Thornton-le-Moors, Ince and Helsby, on the Dee between Farndon and Chester, on the Bollin between Wilmslow and Warburton and on the Dane at Holmes Chapel. They also mention a few sites in Wirral (Burton, Leasowe Embankment and Woodchurch) where, however, the species was scarce and apparently irregular, but they regarded it as not uncommon on many of the rivers and meres of the plain. Abbott[10] noted birds in summer at Capesthorne, the copper mines at Alderley Edge, Redesmere, the Dean at Adlington Hall, Radnor Mere, Boothsmere, the Bollin at Mottram, Sutton Reservoir and Bosley in various years from 1914 to 1926. Bell[7] still considered that one or more pairs bred regularly around many of the meres in 1962, but later[9] wrote that there were no longer any breeding records from the meres, though breeding still occurred at the Sandbach Flashes. Otherwise he quoted Coward & Oldham to the effect that the species was common on hill-streams and reservoirs.

In a survey of the Peak District between 1977 and 1980, Holland et al. (1982b) considered the Cheshire segment of the population to number only ten pairs; sites were the upper Dane at Danebower (2), Lamaload Reservoir (2), Macclesfield Forest Reservoirs (3), Bosley Reservoir (1), the Dane at Back Forest (1) and Todd Brook (1). Subsequent fieldwork for this atlas has produced records from other sites on the Dane and at Wildboarclough, but has not revealed any further confirmed breeding.

Away from the hills a scatter of records from sand-quarries and flashes, around Chelford, Sandbach and Wybunbury in the south, near Warrington in the north and near Sandiway and Winsford in mid-Cheshire, may indicate another two or three pairs, but only the Chelford sand-quarries are a regular site. Most of the 'possible' breeding records are likely to be birds which have moved away from the hills following early breeding failures or to be late migrants travelling further north. Few of them are likely to be genuine territory holders although birds holding territory in lowland sand-quarries may be overlooked. It is likely that the county population in any one year is under 15 pairs. Results from the BTO Waterways Bird Survey show that the population is very stable from year to year.

This seems far removed from Coward & Oldham's "abounds in summer on all the streams and reservoirs in the hill country...and on all the meres and many of the rivers of the plain it is common", and the decline, though not quantified, is surely genuine. It is, moreover, a decline which is matched in the lowlands of neighbouring counties (Holland et al. 1982b). Various authors have blamed the increased level of recreational disturbance, particularly from

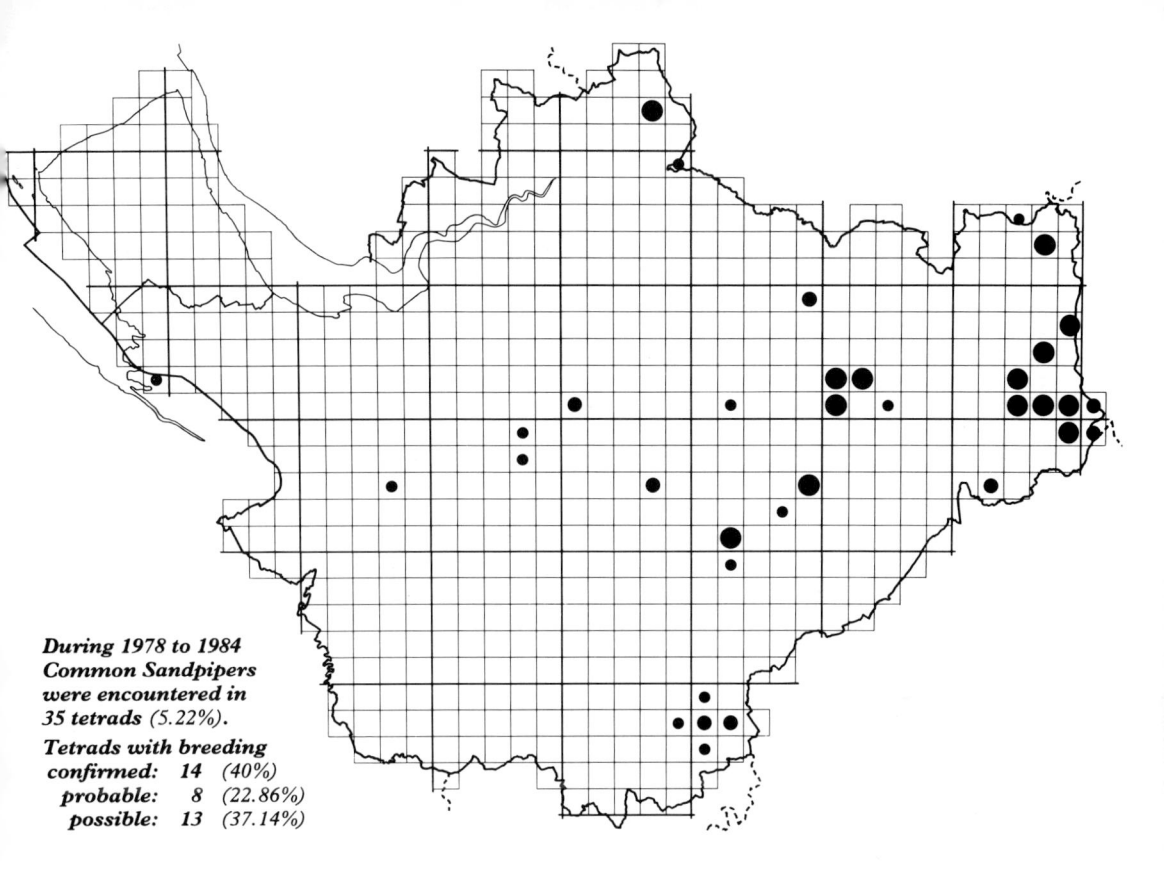

During 1978 to 1984 Common Sandpipers were encountered in 35 tetrads (5.22%).

Tetrads with breeding
confirmed: 14 (40%)
probable: 8 (22.86%)
possible: 13 (37.14%)

anglers and yachtsmen ([7,47]; Holland et al. 1982b; Yalden 1984) but birds have also deserted Rostherne Mere and several of Abbott's sites which are free from disturbance.

In the Peak District study, adult survival (81% per annum) and hatching success (90%) were both good. Survival of the chicks, however, seemed very poor. Interruption of their feeding or exposure to marauding corvids at this stage is critical (Holland et al. 1982a, 1982b). Certainly there is evidence that recruitment is barely able to balance adult mortality (DWY). Feeding takes place on shingle banks and wet flushes rather than in the water, with beetles, flies and ants as regular prey. Anglers standing in the riparian zone for long periods are thus a potent threat, as are picnickers. Even quite small refuges would help this species. DWY

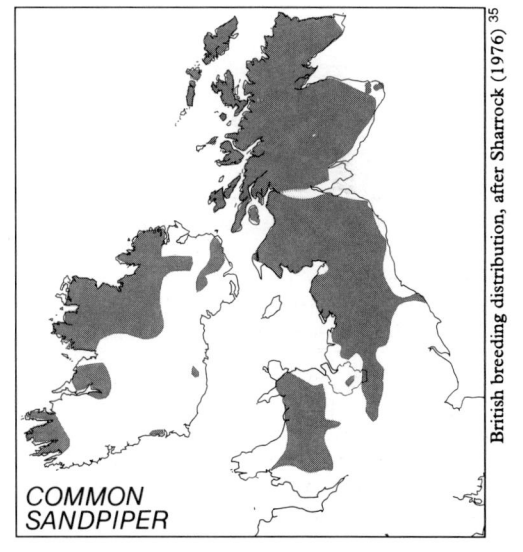

COMMON SANDPIPER

British breeding distribution, after Sharrock (1976) [35]

REFERENCES

Holland, P. K, Robson, J. E. & Yalden, D. W. (1982a): The Breeding Biology of the Common Sandpiper (*Actitis hypoleucos*) in the Peak District. *Bird Study* **29**: 99-110.

Holland, P. K, Robson, J. E. & Yalden, D. W. (1982b): The Status and Distribution of the Common Sandpiper (*Actitis hypoleucos*) in the Peak District. *Naturalist* **107**: 77-86.

Yalden, D. W. (1984): Common Sandpiper numbers and recreational pressures in the Derwent Valley. *Magpie 3*.

Little Gull
Larus minutus

THE Little Gull is an east European and Asiatic species. Bell[7] could trace only 30 records from 1869 to 1960. Since the 1960s the species has increased considerably in Liverpool Bay, particularly on spring passage, so that by the start of this survey counts in excess of 100 birds were being reported from the coast. In recent years several hundred birds have been caught and ringed by the South West Lancashire Ringing Group, marked birds indicating that this population comes from Finland where there has been a commensurate increase in the number of breeding birds.

In parallel with this increase an inland passage has developed. This has occurred most markedly in spring, with birds moving along the Mersey valley at a time when Black-headed Gulls are nesting. As a consequence birds have been seen at the Woolston gullery in every summer since 1980. Many of these birds are immatures, but a few adults have also been recorded, reminiscent of the time when two adults were identified by the 'keeper on 7th June 1941 at the Oakmere gullery[4]. There must be a possibility that this species may yet breed in the county, in view of nesting attempts by Little Gulls in other British gull colonies. The birds at Woolston reacted to human disturbance by hovering above the intruder uttering alarm calls in much the same way as described for Little Gulls in the Netherlands in the early 1970s, where they subsequently bred (Veen 1980).

REFERENCE

Veen, J. (1980): Breeding Behaviour and Breeding Success of a Colony of Little Gulls (*Larus minutus*) in the Netherlands. *Limosa* 53: 73-78

Black-headed Gull
Larus ridibundus

THE Black-headed Gull has always been very local as a breeding species in Cheshire because of its rather exacting habitat requirements at this season. Large numbers – up to several hundred pairs – will settle in colonies where conditions are suitable however. The birds almost invariably nest on the ground, where their pale coloration makes them conspicuous to foxes and other predators. Consequently the favoured nest-site is on a tussock of rushes or other emergent vegetation in a marsh, or on an island.

Birds arrive at the colonies in March or early April, most leaving by the end of July. It is not unusual to find immature birds anywhere in the county during April and May, and less often in June. Adult birds showing full breeding plumage should be watched closely at this time for signs of nesting intent. Odd pairs may take up territories in spring around any suitable stretch of water, and sporadic breeding is not exceptional. It is possible that an occasional pair of prospecting birds was ignored during this survey as not being in suitable habitat. Otherwise the map for this present survey is probably very accurate, with most of the confirmed breeding records coming from two distinct areas: the Dee marshes and the inner Mersey valley.

However, the most traditional breeding area in Cheshire is around the meres and flashes of Delamere. The first recorded mention of this gullery comes from William Webb in King's "Vale-Royall" of 1617 where he speaks of "Puits or Sea Mawes in the flashes". A steady population decline during the nineteenth century brought the Black-headed Gull close to extinction in Britain[35] but birds were breeding in Delamere in the 1860s and a colony established on flashes in the area in 1887 was believed to have moved from a neighbouring site[2]. At least 15 pairs were present in 1899, 29 nests in 1901 and some 300 pairs by 1907, the increase being attributed in part to the Wild Birds Protection Acts of 1880 and 1894 and subsequent

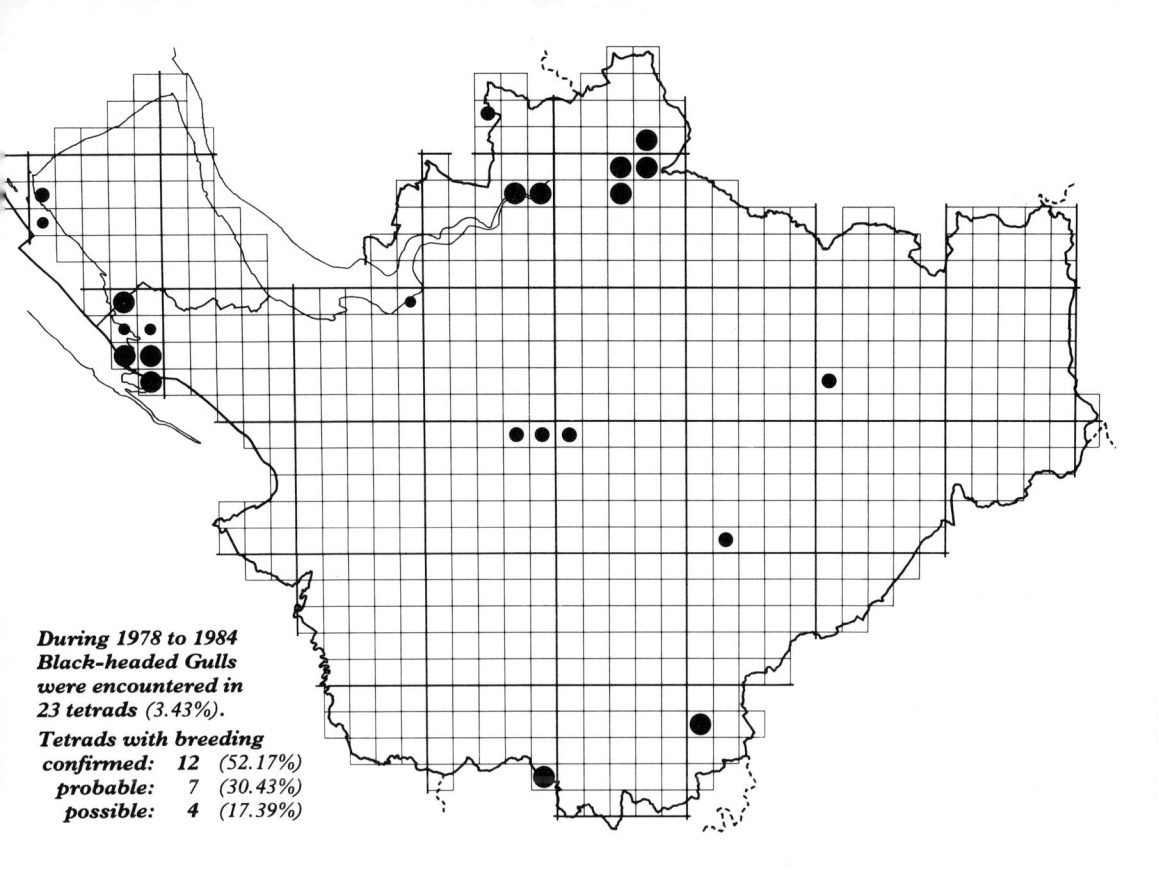

During 1978 to 1984 Black-headed Gulls were encountered in 23 tetrads (3.43%).

Tetrads with breeding
confirmed: 12 (52.17%)
probable: 7 (30.43%)
possible: 4 (17.39%)

Orders of 1895 and 1898. At that time this was the only documented colony in the county.

The location of the Delamere colony is known to have changed from one year to another according to local conditions such as water-levels and reed-cutting, with the major sites being the Gull Pool on Newchurch Common, Oakmere, Fish Pool (now drained) to the south-west of Oakmere and, sometimes, Crabtree Green, a "submerged forest" to the north-west of the Gull Pool. Occasionally there were smaller, satellite colonies on other flashes and meres in the area. In 1915 Abbott[10] saw several hundred pairs at Newchurch Common. Hardy[4] and Hollom (1940) noted that since 1918 numbers varied between 10 and 300-400 pairs at Oakmere. The best year seems to have been 1931 with 500 pairs at Newchurch Common. Some natural disasters befell the birds, as well as sporadic human persecution, with the colony once deserted because of a forest fire. In 1934 foxes entered the colony following a drought and the only birds rearing young were two pairs which nested in low fir trees rather than on the ground[5], a rather unusual occurrence. Counts of 302 in 1938 (Hollom 1940) and up to 300 pairs in 1958

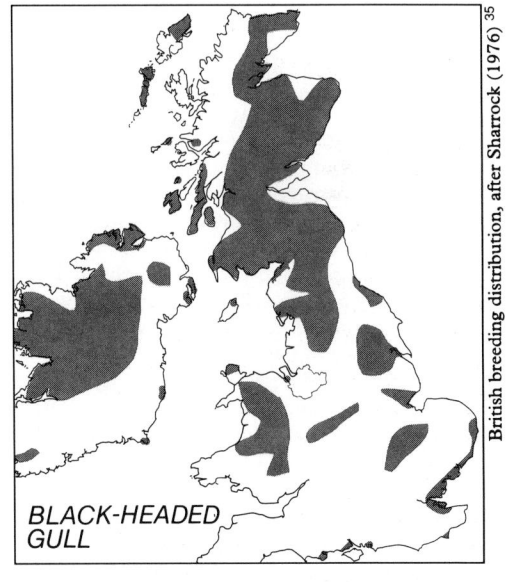

BLACK-HEADED GULL

British breeding distribution, after Sharrock (1976)[35]

(Gribble 1962) were entered in the national censuses of breeding Black-headed Gulls.

By 1965 the mid-Cheshire colonies were deserted[9] and there are no later reports of nesting in Delamere. In late May 1981 a few pairs took up territory and were seen displaying at flooded sand-quarries, where

125

provision of suitable nesting islands or rafts might encourage them to nest and revive the association of this species with the Delamere area.

Coward & Oldham[2] wrote that formerly one or two pairs had nested on Burton Marsh, although saltmarsh had begun to develop only towards the end of the nineteenth century following reclamation work. The *L&CFC Report* for 1949 records ten nests here in 1944, and there has been sporadic breeding in various years since: about 50 pairs in 1949, up to 20 in 1958, 24 in 1968, 100 or more in 1970, but only a few in 1980[8].

At the turn of the century Coward & Oldham remarked upon the great increase of Black-headed Gulls at all seasons in the Mersey valley, thinking this due in part to the wide waterway of the Manchester Ship Canal. Indeed the construction of the canal has had a marked bearing on the bird's breeding status in the county. In 1928 a few pairs nested on a canal dredging sludge-pool at Thelwall Eye, and by 1945 some 200-300 pairs were breeding in June[3,5]. Unknown numbers bred at Frodsham sludge-pools in the 1950s.

Recently the Woolston/Thelwall Eyes have held the largest colony in the county; in 1978 some 100-150 adults summered and 10-12 nests were recorded; 170 pairs bred in 1979; 250-300 pairs in 1980; 110 pairs attempted to breed in 1981 although, in common with many other Black-headed gulleries elsewhere in Britain, the site was inexplicably deserted in late May with only 60 pairs eventually breeding. In 1982, 350-400 pairs bred and, in 1983, an estimated 450 pairs, although a freak hail-storm destroyed many nests. In 1984, up to 400 pairs bred.

Some 30 pairs settled at Risley Moss in 1981 as an offshoot of the main colony at Woolston and in response to raised water-levels, but this site was also inexplicably deserted in late May and it was believed that birds which subsequently took up territory in Delamere that year had been displaced from Woolston and Risley[8]. In 1983, 13 birds appeared at Risley in late April but left in late May without nesting, and four pairs which built nests at Gatewarth Sewage Farm deserted.

In the south of the county between 50 and 100 pairs bred at Baddiley between 1943 and 1946 with 14 pairs there in 1958, although there were none in 1949[3]. At Combermere the gullery was formed in 1949, presumably replacing that at Baddiley, when 200-300 birds were present. This subsequently increased although by 1958 there were only 50 pairs at the most. A small colony existed in 1980 after what was thought to be a period of absence. At Doddington Pool a single pair raised young during the early years of this survey. At Farmwood Pool (Chelford) a pair was present in May 1982 and at Watch Lane Flash, Sandbach a pair was on an islet during April and early May 1978. A pair was holding territory at Batemill sand-quarry in 1983 but offered no evidence of breeding success.

REFERENCES

Gribble, F. C. (1962): Census of Black-headed Gull colonies in England and Wales, 1958. *Bird Study* **9**: 56-71.
Gribble, F. C. (1976): A Census of Black-headed Gull Colonies in England and Wales in 1973. *Bird Study* **23**: 135-145.
Hollom, P. A. D. (1940): Report on the 1938 survey of Black-headed Gull colonies. *British Birds* **33**: 202-221, 230-244.

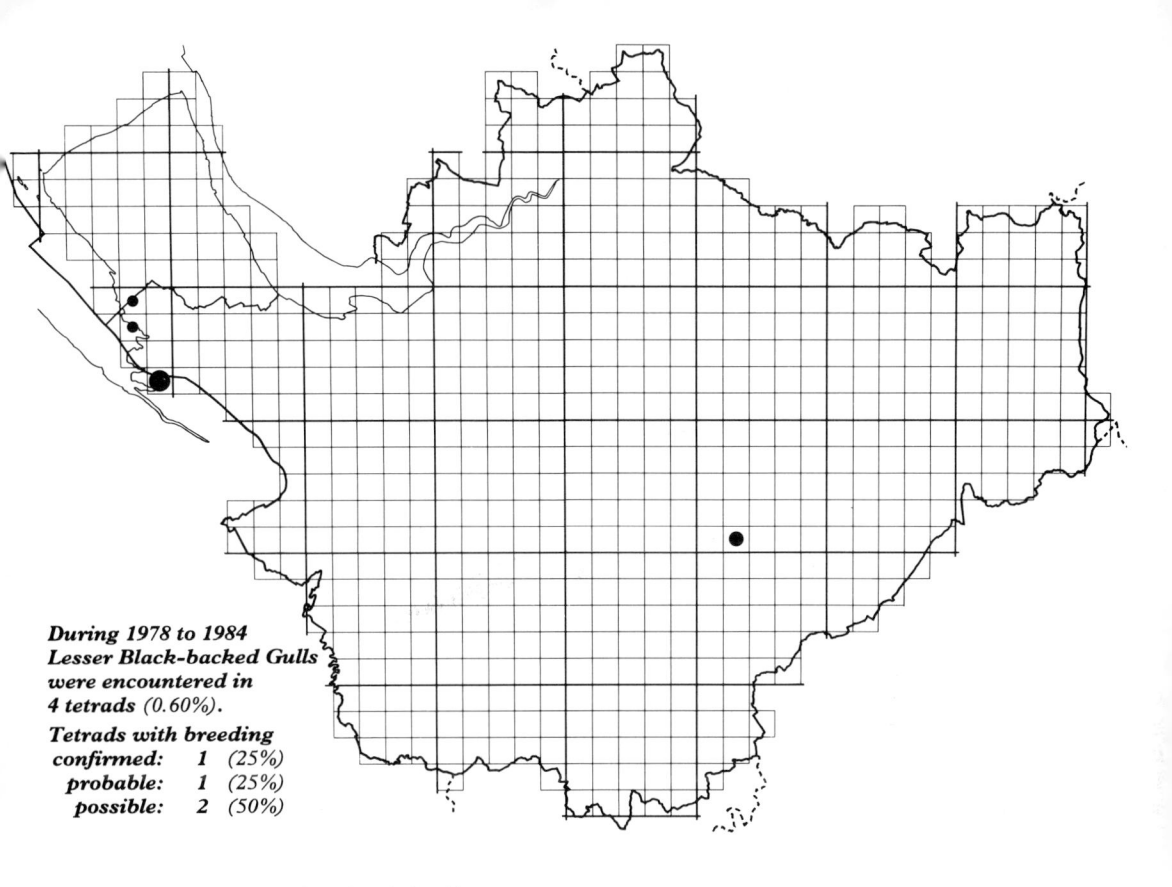

During 1978 to 1984 Lesser Black-backed Gulls were encountered in 4 tetrads (0.60%).

Tetrads with breeding
  confirmed: 1 (25%)
  probable: 1 (25%)
  possible: 2 (50%)

Lesser Black-backed Gull
Larus fuscus

COWARD & Oldham[2] stated that Lesser Black-backed Gulls occurred all year round at the coast, and that adult birds were seen in the Mersey estuary in May when most are at nesting colonies. However, at that time the nearest such colony was on Puffin Island, Anglesey. Inland, they occurred more frequently than Herring Gulls, and adult birds were seen at Marbury (Northwich) in each of the months March, April, May and June. These birds, when disturbed, sometimes wheeled high overhead uttering the scolding "ha, ha, ha" call usually only heard at nesting colonies. In 1907 birds were at Marbury from early March to mid-August.

The numbers of Lesser Black-backed Gulls in Cheshire increased considerably from 1928, and this was more marked than for any other gull species[6]. Bell[7] quotes an unsubstantiated reference from the MNA's report to breeding on Burton Marsh prior to 1939. However, the first documented record of nesting was in 1943 when a pair attempted to breed on an islet at Witton Flashes (Northwich). Eggs were laid but failed to hatch, although the birds continued to sit for 53 days, from 26th May until 8th July, twice the normal 26-day period[6]. MNA reports for 1956 and 1957 refer to three pairs nesting on Burton Marsh in 1955 and 1956. Subsequently, the species seems to have reverted to the status of a non-breeding visitor, with many hundreds of birds passing through the county and lesser numbers staying in winter. Only very small numbers of adults are seen in early summer, and all such birds should be watched for signs of breeding intent.

During the present survey a single pair bred, again on Burton Marsh, in 1979, and there have been records of pairs holding territory and/or displaying at Witton Flash, Top Flash (Winsford), Watch Lane Flash (Sandbach), Farmwood Pool (Chelford) and Whirley sand-quarry. Migrant birds not infrequently circle over meres and other waters, uttering their laughing breeding calls, and it may be that only the lack of islets for secure nest-sites is preventing the recurrence of inland nesting.

Herring Gull
Larus argentatus

ON 9th June 1945 a nest with two eggs, presumed to be of this species, was found on Little Hilbre. The *L&CFC Report* for 1943-49 states that "...the bird had not been seen actually on the nest, but as Herring Gulls were constantly present and Lesser Black-backed very occasionally, there is no good reason to doubt that it was a Herring Gull's nest...". A pair bred, successfully rearing one chick, on a roof-top in Hoylake in 1971. The following year the birds returned and attempted to breed but were discouraged by the house owners[8]. This record occurred during the period of the BTO Atlas and was shown to be an isolated case between the coastal towns of north Wales and the Ribble[35]. During a national census of large gulls nesting on buildings in 1976 there were records of a pair at West Kirby, "nesting" at Heswall and a nil return for Cheshire (Monaghan & Coulson 1977).

REFERENCE
Monaghan, P. & Coulson, J. C. (1977): Status of Large Gulls Nesting on Buildings. *Bird Study* **24**: 89-104.

Common Tern
Sterna hirundo

"BROCKHOLES was of the opinion that the Common Tern formerly bred in Wirral; we have failed to obtain confirmation of this supposition, and it certainly does not nest in Cheshire at the present day...". So wrote Coward & Oldham in 1910.

Subsequently, nesting was first reported from the rapidly growing saltmarsh at Burton in 1918, and by 1934 some 50 pairs were present[12]. Numbers have fluctuated ever since: 24 pairs in 1940, 80 in 1941, 40 in 1944, 60/70 in 1953, 30 in 1957, 40 in 1958, 24 in 1959 (although there were only about 10 nests), 30 pairs in 1960 and 60/70 pairs in 1966. Breeding success was often poor, the highest tides destroying many nests in some years, and, as the area was readily accessible by foot, it seems likely that human interference also had a restricting influence on the annual success[7,9]. 1934 and 1966 were welcome exceptions. In 1934, Farrar[12] noted that birds were sitting hard during the first week of July and that, due to the unusually low tides, more young were hatched than ever before or since. Similarly, in 1966, many free-flying young were present in mid-July and there were many nests containing chicks and eggs[8].

CBRs refer to 46 pairs at Burton Marsh on 28th June 1970; 20 pairs with 20+ young in August 1972 and some 25 pairs in late June 1974. However, because later *CBRs* have adhered more strictly to the county boundaries no breeding records have qualified, and there is considerable doubt as to what proportion of the Burton Marsh records was actually within the Cheshire county boundary.

Since 1970 the Merseyside Ringing Group has provided nesting platforms in cooling-

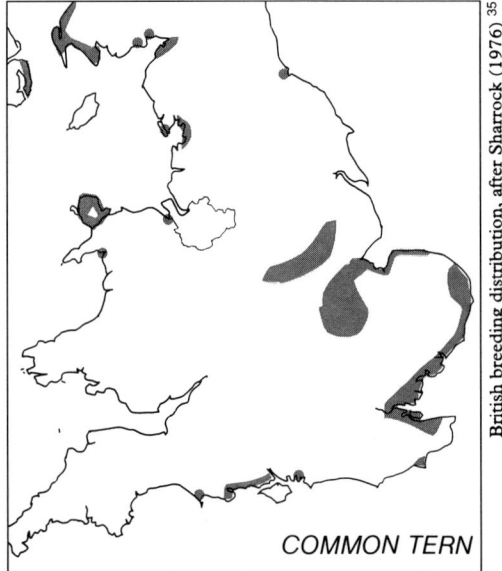

British breeding distribution, after Sharrock (1976)[35]

COMMON TERN

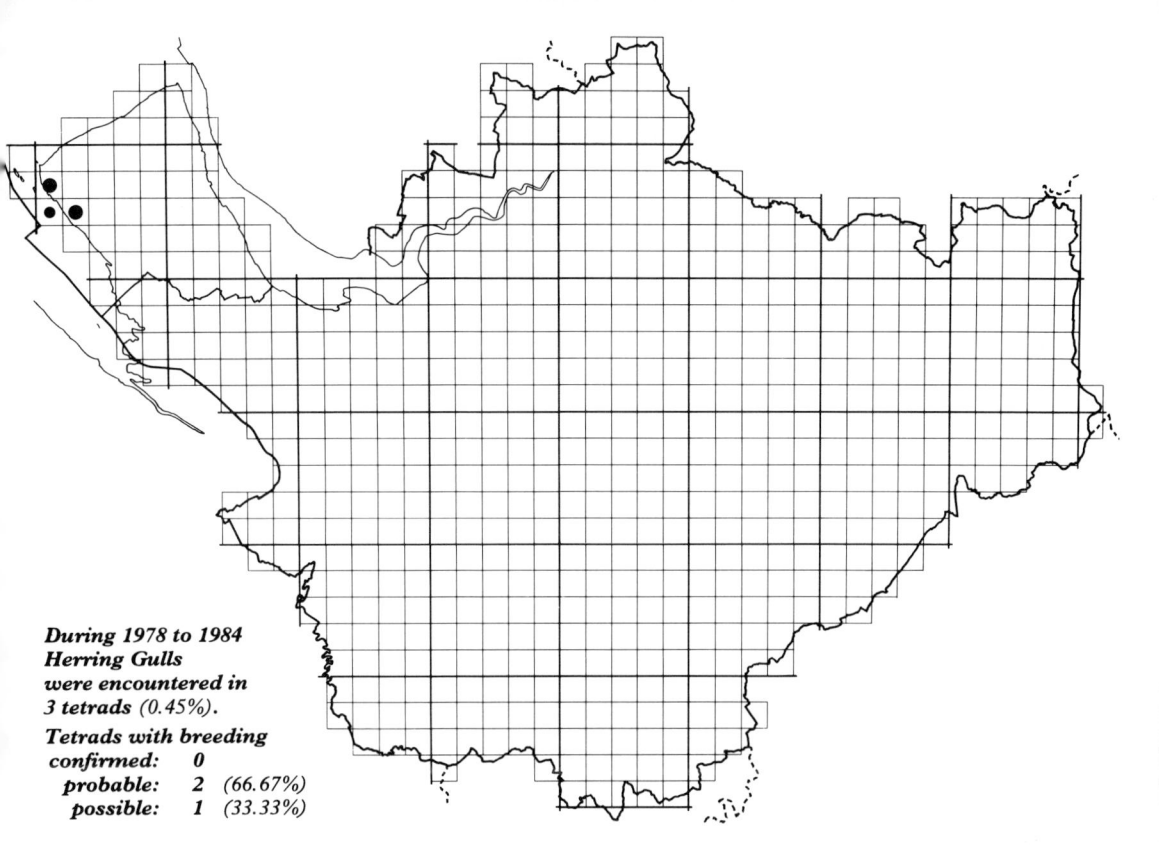

*During 1978 to 1984
Herring Gulls
were encountered in
5 tetrads (0.45%).*

Tetrads with breeding
  confirmed: 0
  probable: 2 *(66.67%)*
  possible: 1 *(33.33%)*

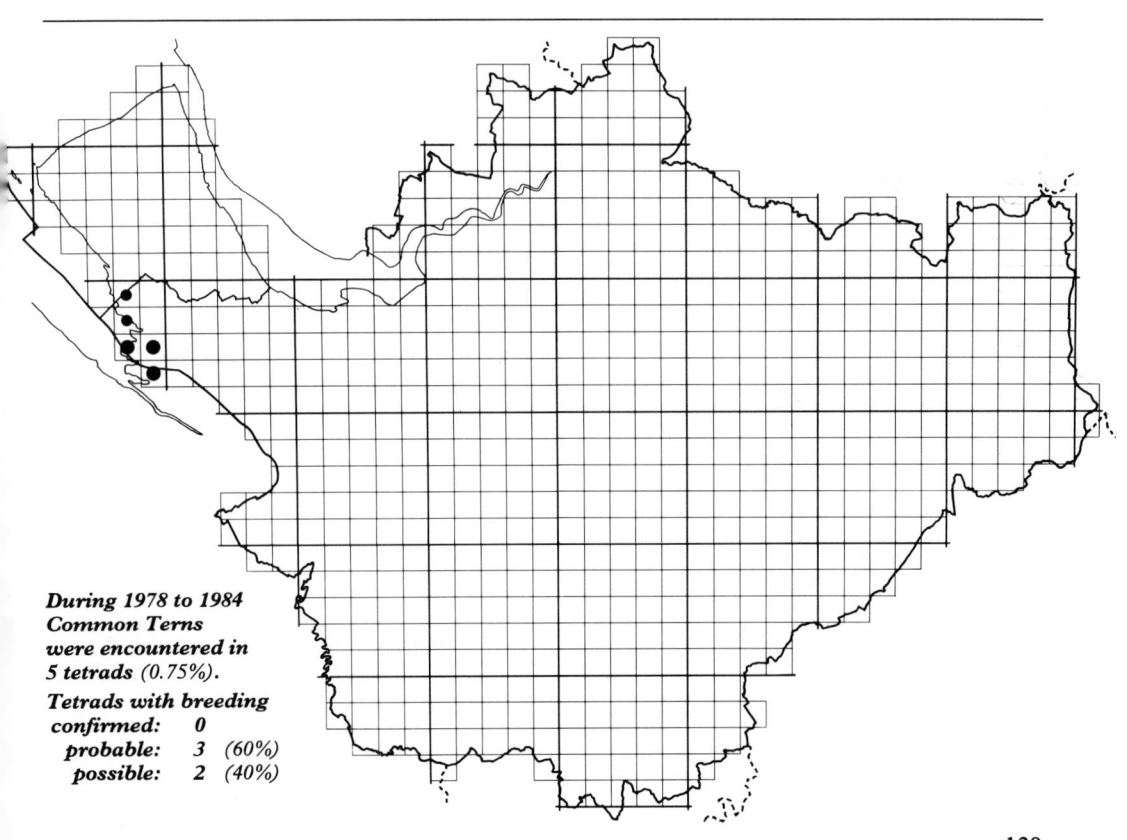

*During 1978 to 1984
Common Terns
were encountered in
5 tetrads (0.75%).*

Tetrads with breeding
  confirmed: 0
  probable: 3 *(60%)*
  possible: 2 *(40%)*

water lagoons of the steelworks at Shotton, Clwyd, just outside the Cheshire county boundary. It appears that all the Burton Marsh birds have gradually moved to this safe breeding site, with the colony now between 150-200 pairs strong, producing well over 300 young each year (Birch 1984, Norman 1987). In 1971 the MRG's efforts were rewarded by the presentation of the Prince of Wales Award.

Inland nesting has occurred in Derbyshire since 1956[45], in the West Midlands since 1952[47] and in Greater Manchester since 1979[46]. Birds are seen inland in Cheshire each year, but no suitable nest-sites are available. The provision of islets or nesting rafts in sand-quarries or meres could possibly encourage birds to breed again within Cheshire.

REFERENCES

Birch, R. R. (1984): The Shotton Tern Colony. Merseyside Ringing Group report for 1984, pp23-30.
Norman, D. (1987): Are Common Terns successful at a man-made nesting site? *Ringing & Migration*, **8**: 7-10

Little Tern
Sterna albifrons

COWARD & Oldham[2] had no evidence that the Little Tern had ever nested in Cheshire, although it did so at several sites in north Wales. However, Farrar[12] stated that two or three pairs nested annually on some sandy knobs, sparsely covered with marsh grass, on the very edge of the Dee saltings. Although eggs were laid, none of the terns succeeded in rearing a family, the nests being washed out by the highest tides. The birds apparently nested late in the season, scarcely ever sitting until the middle of June. Hardy[4] echoed Farrar's statement. For how long this nesting continued and whether it took place in the Cheshire part of the marshes, is unclear. In more recent times there has been no evidence whatsoever of nesting intent, nor does there appear to be any suitably undisturbed habitat within the county for this declining species.

Feral Pigeon
Columba livia

ALL Feral Pigeons are ultimately descended from the Rock Dove despite their varying mongrel plumages. Perhaps because of these plumages, the bird's predominantly urban existence, its dependence on man and its pest status, or perhaps just out of tradition, the Feral Pigeon is ignored by almost all bird-watchers. Yet to many people, particularly those living in the most densely populated parts of the county, this is perhaps the most familiar bird. Coward & Oldham[2], discussing "the alleged occurrence of the Rock Dove...in Cheshire", discounted the record of birds nesting in a rabbit hole on Middle Hilbre and those mentioned by Brockholes which "frequented the high portions of the river bank between Eastham Ferry and Hooton" as "...undoubtedly erroneous", suggesting they may have referred to Stock Doves rather than this species because the latter were often called "Rock Doves". However, they added the rider "...unless they were feral dovecot Pigeons". Subsequently, Hardy[4] saw birds nesting on the cliffs of Hilbre in 1934 and 1938. Otherwise almost no mention is made in the literature until Harrison & Rogers[24] wrote of "small parties of town pigeons, presumably from the Manchester conurbation" which occasionally fed on ploughed land around Rostherne.

The county bird reports for 1981 and 1982 listed a few counts of flocks and this appears to have stimulated some interest, for in 1983 a considerable number of records came to light, particularly from Wirral. The

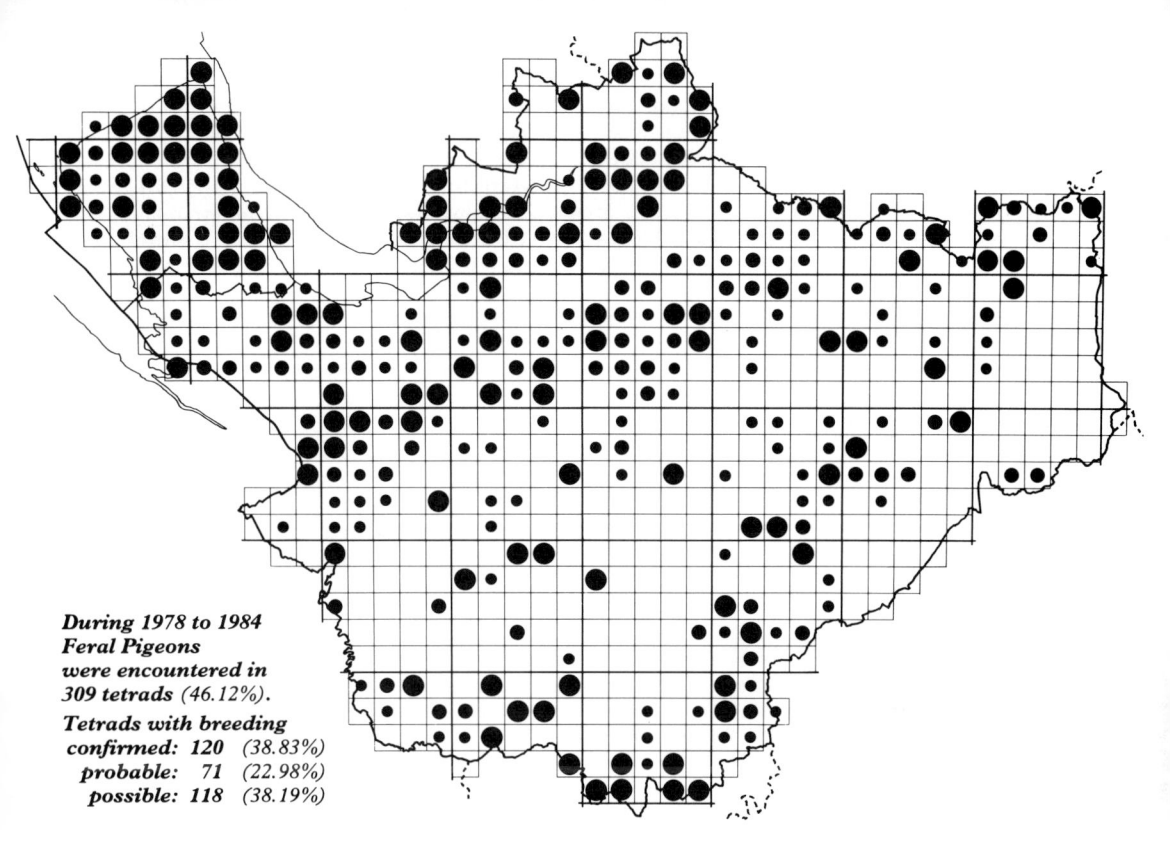

During 1978 to 1984 Feral Pigeons were encountered in 309 tetrads (46.12%).

Tetrads with breeding
confirmed: 120 (38.83%)
probable: 71 (22.98%)
possible: 118 (38.19%)

largest concentrations appear to be around the grain terminals in Birkenhead and Wallasey docks: six flocks here in March 1983 totalled 1500 birds. Flocks from north Wirral and from the Manchester conurbation visit the outlying areas from time to time to feed in stubble fields and elsewhere. Thus 112 were counted near Alderley Edge in December 1981 and 700 at Ness Gardens in October 1983. 120 at Woolston, also in October 1983, presumably flew out from Warrington, and 400 at Wybunbury sand-quarry in January 1983 were probably from Crewe or Nantwich.

Most country towns also hold smaller populations. Recent winter counts refer to 61 in Macclesfield, 46 in Winsford and 70 in Sandbach. Flocks of a similar size are also known to be present in Knutsford, Middlewich and Northwich, and doubtless elsewhere.

Ledges or cavities in buildings, particularly in the older and more derelict parts of towns, provide nest-sites. The ancestral Rock Dove is a bird of rocky coastlines, nesting especially in sea caves, a habit recalled by those pairs of Feral Pigeons that gain access to the roof space of buildings via a loose or missing slate or tile. Coward (1903) referred to an undated, but old account of birds being shot at Wilmslow Church: "...in the church accounts...Two pence was allowed out of parish funds for powder and shot for the churchwardens to execute the pigeons which had become a nuisance in the church". Recently several churches in the county are known to have been fitted with wire-netting to prevent birds from nesting in the bell tower.

Although primarily an urban bird, there are also many pairs of Feral Pigeons which nest in outbuildings even in the most rural areas and feed on grain put out for farmyard fowl. These birds are neither more nor less dependent on human benevolence or tolerance than their urban relatives, but unfortunately were often ignored by observers during this survey. A study by Murton & Westwood (1966) showed that Rock Doves from Flamborough Head (Yorkshire) and Feral Pigeons from Leeds had the same food requirements as Stock Doves, so that competition could be expected between the two species in areas where their ranges overlap. For this reason Feral Pigeons deserve attention in future years so as to clarify their status. It seems that it is only in London that they have been studied in detail, their colours, diet and breeding being summarised by Simms (1975).

The map, although undoubtedly incomplete, shows concentrations of breeding records around most of the urban areas in the county. Holland et al.[46] estimated the Greater Manchester population at 5400 pairs assuming 100 pairs per urban tetrad and 12 in suburban squares. They are much less numerous in Cheshire, however, and only in the Birkenhead area might densities of 100 pairs per tetrad be reached. An average figure of 20 to 30 breeding pairs per tetrad in towns, six to ten in suburbs and two or three in rural tetrads would give a county total of 1700 to 2500 pairs. Birds apparently pair for life and some may breed in any month of the year, so their reproductive potential is high and their numbers probably limited only by food supplies, provided either deliberately or inadvertently by man.

REFERENCES

Coward, T. A. (1903): *Picturesque Cheshire*. Sherratt & Hughes, London and Manchester.
Murton, R. K. & Westwood, N. J. (1966): The Foods of the Rock Dove and Feral Pigeon. *Bird Study* **13**: 130-146.
Simms, E. (1975): *Birds of Town and Suburb*. Collins.

Stock Dove
Columba oenas

A bird of parkland or wooded farmland, the Stock Dove breeds widely across lowland Cheshire, numbers thinning out in the hills towards the eastern boundary. Most nests are in cavities in trees, often high above the ground. In undulating country, trees on the tops of hummocks with a good view are frequently chosen. Occasionally ledges on little-used or abandoned farm buildings are occupied, such sites being particularly frequent in the eastern hills where suitable trees are scarce. Ledges on quarry-faces or holes amongst boulders on rocky slopes may also be used. Hardy[4] states that thatched roofs were used as nesting sites and quotes a record of Boyd's from a church roof at Birkenhead. Nest-boxes and the tangled epicormic shoots of lime trees may be occupied. The latter resemble the disused nests of Magpies which were not infrequently adopted in Boyd's day[5] although only one such site was reported during this survey, at Siddington in June 1983. There is undoubtedly some competition for nest-sites with other species such as the Kestrel, Jackdaw and Tawny Owl, and also with grey squirrels.

Many bird-watchers have a blind spot for pigeons and there is little doubt that the Stock Dove is more widespread in Cheshire than the map suggests. Birds are easily located from February onwards by song and by the downward-spiralling display flight. In 1983 such display was noted at Rostherne as early as 6th January. Pairs return to their nesting trees during March and April and may be seen inspecting future nest-sites. A pair with two eggs at Eaton Hall on 23rd March 1943 was unusually early[3] although most birds (80%) attain full breeding condition in March (Lofts *et al.* 1966). If a nest is approached once eggs have been laid, the sitting bird will often throw itself off the nest as soon as the observer's back is turned, and circle at a discreet distance. The nest is often inaccessible, resulting in a high proportion of 'ON' records, but adults with swollen crops

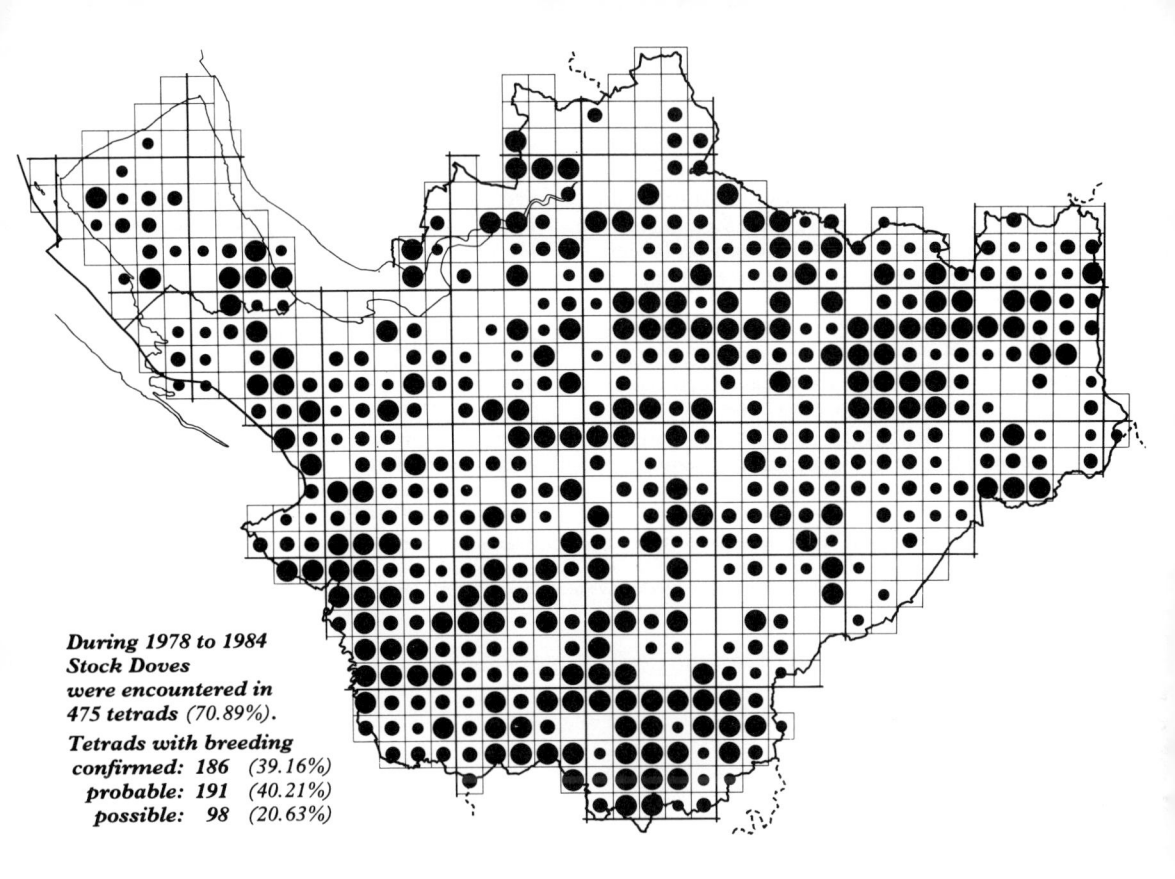

During 1978 to 1984 Stock Doves were encountered in 475 tetrads (70.89%).

Tetrads with breeding
confirmed: 186 (39.16%)
probable: 191 (40.21%)
possible: 98 (20.63%)

may be seen flying back to nest-sites. The breeding season is prolonged with up to three or four broods, and young may be found in the nest at any time from May until September. Boyd[6] found newly-hatched young in a Magpie's old nest as late as 18th September and a pair was tending a chick in a barn along the upper Dane at 1350 feet above sea level in late September 1984.

Coward & Oldham[2] noted that at one time the Stock Dove was a common resident on the Wirral coast, where it nested in rabbit burrows in the sand-dunes; at Meols, Stock Doves were called "Sand Pigeons". However, with the growth of West Kirby, Hoylake and New Brighton and the conversion of the sand-dunes into golf-links, many nesting sites had been destroyed. However, at that time it still nested in rabbit burrows, holes in the marl cliffs and in the sandstone rocks at Burton Point. It occasionally nested in these rocks during the next thirty years[4]. Brockholes stated that, on Caldy Hill (where Stock Doves were known as "Hill Pigeons"), birds sometimes nested beneath gorse bushes as well as in rabbit burrows. In Coward & Oldham's time Stock Doves were thinly distributed, nesting in trees in other parts of Wirral and throughout the plain, although they were locally plentiful in parkland.

Bell[7] noted that a number of localities on the eastern plain had been deserted during the previous 15 years, and birds continued to breed in small numbers in Wirral[18]. In 1965, the editor of the *CBR* drew attention to the scarcity of reports, and subsequent information suggested it was then largely restricted to the west and south of the county. This contraction of range coincided with a national decline recorded between 1950 and 1961 when the population was estimated to have fallen by 90%. This was largely attributed to egg-shell thinning due to the use of organo-chlorine pesticides in the 1950s and 1960s, particularly in intensively farmed habitats (O'Connor & Mead 1984). However, numbers have increased steadily in recent years, estimated nationally to be by a factor of five by 1980. Birds returned to the Wilmslow area in the mid-1970s and subsequently spread back into many hill areas. *CBRs* from 1976 indicate a general increase in east and south-east Cheshire also. National CBC figures show that between 1964 and 1984 the national population increased in every year except 1982.

Generally the Stock Dove is a sedentary species. Of birds ringed by Boyd during the

1930s, 22 were recovered or retrapped, some of them several times. Of these, 19 were caught in the same traps (some in more than one year). The other three were shot or found dead within two miles of his home[6]. Nationally only 27% of full-grown birds were found more than ten kilometres from the place of ringing whilst 38% of ringed nestlings were subsequently found more than ten kilometres from their natal site (O'Connor & Mead 1984). Following restrictions on organo-chlorine pesticides this degree of juvenile dispersal will have aided recolonisation by an expanding population.

Even though the species now breeds throughout the county, the map shows the south and west still to be the most favoured areas. In 1981 JPG considered that Stock Doves outnumbered Woodpigeons around Harthill. Scarcity of nest-sites in the mossland belt north of Warrington probably explains absence in that area, and on the eastern side of Wirral and in urban areas elsewhere competition for food with feral and dovecote pigeons may be a limiting factor. Allowing two pairs for each tetrad with confirmed or 'probable' breeding the county population is estimated at about 750 pairs.

REFERENCES

Lofts, B., Murton, R. K. & Westwood, N. J. (1966): Gonad Cycles and the Evolution of Breeding Seasons in British Columbids. *J. Zool.* **150**: 249-272.

O'Connor, R. J. & Mead, C. J. (1984): The Stock Dove in Britain, 1930-80. *British Birds* **77**: 181-201.

Woodpigeon
Columba palumbus

WOODPIGEONS increased very greatly nationally during the nineteenth century, largely due to the expanding agricultural practice of growing crops such as turnips and clover which provided an ample supply of winter food. The exact status of the bird in Cheshire at that time is not known, though Coward & Oldham[2] noted an increase in numbers over the 50 years up to 1910. According to the Greater Manchester Atlas[46] the Woodpigeon was described in 1860 as rare in that region and it did not expand away from woodland habitats until the first quarter of this century. In 1910 Coward & Oldham stated that it was a well-known resident throughout Cheshire, being most abundant in the woodlands of Wirral and the plain, but with large numbers also in the plantations and conifer woods of the hills.

Boyd[6] considered it exactly suited to the agricultural countryside around Great Budworth, nesting not just in coverts, but hedgerows also, and feeding on the farmers' crops. This is essentially the pattern to this day, though birds also breed in large suburban gardens, a habit which had started by 1941, when Hardy stated that they nested in Birkenhead Park and in Shrewsbury Road, Birkenhead, and became widespread in the 1960s. The species now nests throughout the county, with the possible exception of urban areas such as Warrington and Widnes and the dockland of Birkenhead.

Woodpigeons have a very prolonged breeding season. The display flight, in which the bird claps its wings in a noisy ascent followed by a stall and downward glide, may be seen in any month of the year provided the weather is mild. In rural areas nesting may begin during the second half of March whilst in urban areas it may be as early as mid-February[36]. Eggs are laid from April, sometimes March, until September or October. There is in fact a tendency for many birds to nest in late summer and autumn

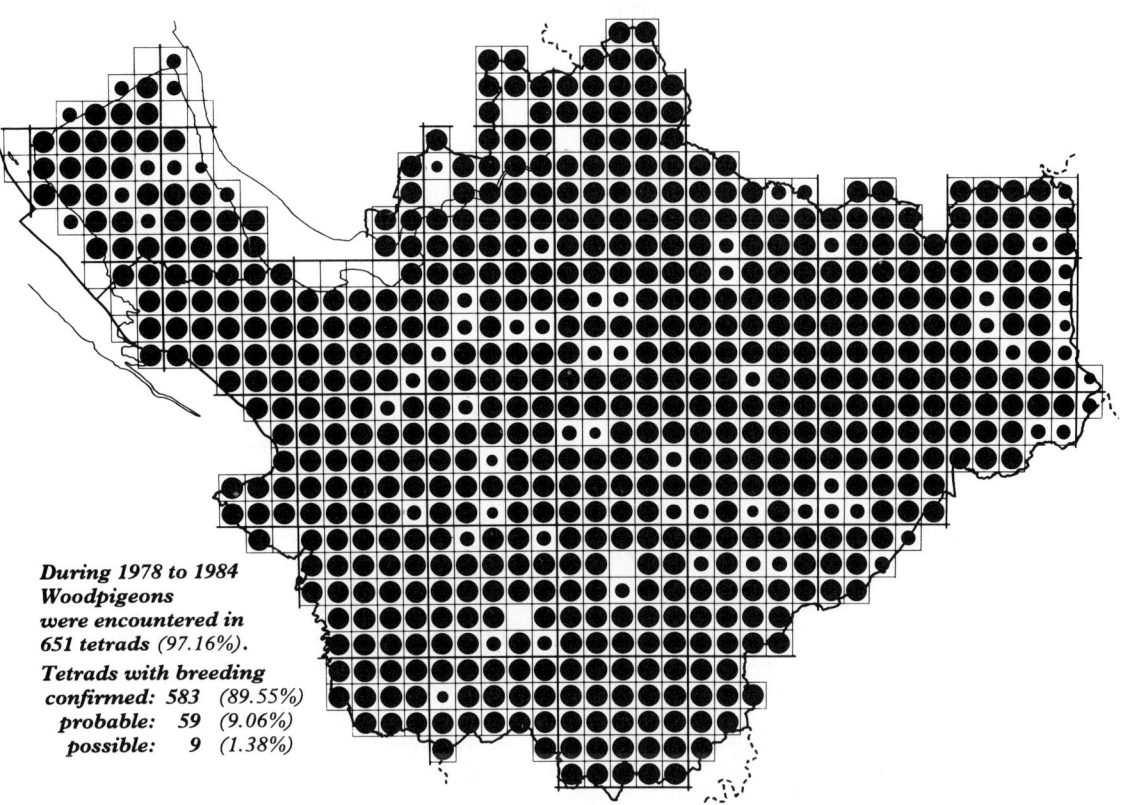

During 1978 to 1984 Woodpigeons were encountered in 651 tetrads (97.16%).

Tetrads with breeding
confirmed: 583 (89.55%)
probable: 59 (9.06%)
possible: 9 (1.38%)

when grain and other foods are abundant, which gives a greater chance of nestlings being successfully reared to the flying stage (Murton 1966). It is not unusual to encounter fat squabs still sitting on the flimsy nest in late October; Boyd[5] records seeing a Woodpigeon sitting on the nest in mid-October and two young, almost ready to fly, on a nest on 22nd October 1936. On the other hand large flocks may be seen feeding together after winter flocks of other resident species have split up for the summer: 200 were recorded at Pickmere on 6th May 1981.

Woodpigeons are easy to confirm as breeding birds and over 90% of the mapped records refer to nests. The majority of nests in Alderley Park appear to be robbed by Jays, who also evidently experience little difficulty in finding them. Many nests, which are often little more than flimsy platforms, are built in tall scrub or thickly branched trees at a height of between 10 and 20 feet. Boyd[5,6] found that conifers and other evergreens were favoured in spring and early summer, with many more deciduous species being used towards autumn. Hawthorns contained 25 out of 110 nests of which he kept note, and overgrown hedges are still of great importance, though even neatly trimmed hedges sometimes provide sites – during the present survey a Woodpigeon was flushed off a nest some three feet off the ground in a hedge bordering a cereal field. Good numbers of Woodpigeons inhabit Macclesfield Forest, where they nest in the denser conifer stands, particularly the Sitka spruce which is their favoured conifer in the British Isles.

Sharrock[35] quotes densities in excess of 50 pairs per square kilometre in arable lowland England, and an average of 1000 pairs per occupied 10-km square over the whole country. No counts are available for Cheshire, but given that the species was very scarce in some agricultural tetrads in central and south Cheshire, such high average levels (40 - 200 pairs per tetrad) appear improbable. Numbers are undoubtedly higher in better wooded areas in the east and west of the county and, generally speaking, occupied tetrads may hold between five and 50 pairs. Taking an average of 20 - 30 pairs per tetrad, the county population would lie between 13,000 and 20,000 pairs.

REFERENCE

Murton, R. K. (1966): Natural Selection and the Breeding Seasons of the Stock Dove and Wood Pigeon. *Bird Study* **13**: 311-327.

Collared Dove
Streptopelia decaocto

THE spread of the Collared Dove from the Balkans since about 1930 to its first breeding in Britain in 1955 is one of the outstanding ornithological phenomena of the twentieth century. In our area there was an unsubstantiated report of a bird at Clatterbridge, Wirral, in June 1959, with documented records following in 1960. These were singles at Runcorn in March, Caldy in May and Ness in June, with a pair at Hatherton, Nantwich, from July to October.

In 1961 a pair which took up residence at Whitby, Wirral went missing in July, only to reappear with a young bird in August – the first circumstantial evidence of breeding in the county. In the same year the species began to consolidate elsewhere in Wirral: two pairs settled at Wallasey, and a pair at Ness, and by the end of that year at least ten were roosting in trees along Wallasey Promenade and four frequented gardens around Rose Mount, Birkenhead[18]. In 1962, five pairs were located in Wallasey; in 1963, seven pairs; and in 1964, 24 pairs, as well as four pairs on Bidston Hill and four pairs in Birkenhead Park.

Given the north-westward thrust of the Collared Dove's expansion and its apparent preference for coastal areas it is understandable that Wirral should have featured prominently in the early years. Both Derbyshire[45] and Greater Manchester[46] recorded their first breeding attempts in 1963. The latter, at Sale, was the start of a nucleus of breeding birds which later spread into north and east Cheshire.

Highlights in the subsequent colonisation of the county were as follows:

1962-63 Nesting at Stockton Heath, Warrington.
1964 Present at Ellesmere Port, Stanlow Point and Upton-by-Chester.
1965 Widespread in larger gardens of Oxton/Claughton area of Birkenhead (W. T. C. Rankin; [3]).
1966 Widespread in Wirral where the wild stock may have been joined by birds liberated at Chester Zoo in 1965. Pair bred Hough Green (Chester). Regular at Lymm and Frodsham.
1967 Pair bred at Runcorn Hill.
1969 Pair building at Leftwich (Northwich) in April: the first evidence of breeding in mid-Cheshire. Birds settled at Sandbach.
1970 Pair bred at Gorstage: first confirmed breeding in mid-Cheshire.
1971 Pair resident at Hough, Wilmslow.
1972 First record from Rostherne Mere NNR.
1973 First records from Sandbach Flashes and Teggs Nose. Bred at Knutsford.
1975 At least three pairs bred at Wilmslow.
1976 More widespread around Knutsford and bred at Mobberley and Dean Row, Wilmslow.

Subsequently there has been continued spread from initial points of colonisation. For example, breeding began around Sandbach Flashes only in the late 1970s, and numbers in Disley increased considerably during the present survey. Once an area has been colonised, birds, particularly breeding adults, become more sedentary.

The arrival of the Collared Dove aroused considerable interest in rural areas because of its unfamiliarity and confiding nature. The birds have been variously named "Grey Doves", "French (i.e. foreign) Doves" and "Ring Doves", this last being a new application of an old name for Woodpigeons.

Collared Doves are strongly associated in the breeding season with mature gardens and suburban parks with ornamental conifers, hollies and other evergreens, and these, plus ivy-covered tree trunks, form typical nest-sites. Numbers of breeding birds near the centres of towns are inversely correlated with the size of the built-up area, partly due to competition from other doves and pigeons[36]. In rural areas, isolated farmsteads, particularly where grain is put down for livestock feed, are preferred to more open country where birds are generally absent. The song, a triple "coo" with the accent on the middle syllable, and sometimes mistaken for a Cuckoo by the lay person, soon betrays the bird's presence, and the male has a conspicuous display flight, towering upwards from some prominent perch then sailing down with tail fanned. Overhead wires and supporting poles or

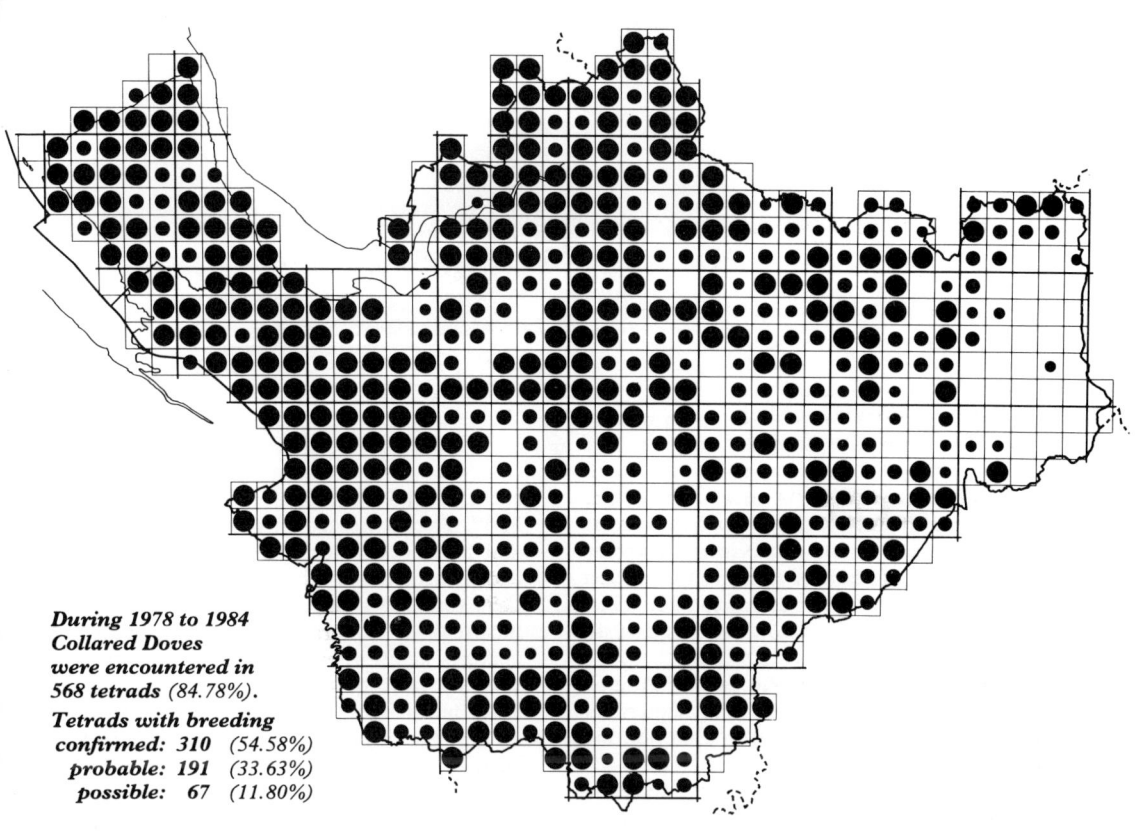

During 1978 to 1984 Collared Doves were encountered in 568 tetrads (84.78%).

Tetrads with breeding
confirmed: 310 (54.58%)
probable: 191 (33.63%)
possible: 67 (11.80%)

pylons provide suitable perches and cantilever street lights are often favoured and used by the pair for mating.

The Collared Dove has a prolonged nesting season. In 1967, a bird on Bidston Hill was "brooding continuously from 18th January until 3rd February"[8]. The nest was then deserted until 23rd March when a bird began to sit again. In 1976 one in a Noctorum garden was sitting on eggs at Christmas[8]. Young are invariably capable of breeding the following spring, sometimes within months of hatching[36]. No doubt this aided the species' spread although its dramatic expansion westwards is still not fully understood.

Unlike the related Turtle Dove which feeds largely on wild seeds, the present species depends heavily on man for food, albeit indirectly; farmyards, chicken runs, flour mills, zoos and wildfowl collections are all favoured, especially in winter, though many birds take food put out in gardens. In a few cases contractions of range have taken place where food supplies have dried up. For example, a small flock at Ashley disappeared in 1971 when the free-range pigs, whose food they shared, were taken away.

Collared Doves may be remarkably tolerant of human activity near the nest. The first mid-Cheshire nest at Gorstage was in a large roadside holly continually brushed by passing lorries, and a nest in a shed at Wrenbury in 1983 was repeatedly blasted by a tractor exhaust during silage operations – the bird sat on unperturbed.

In recent winters some very large flocks have been reported: 130 at Sandbach in January 1978; 320 at Cotton Edmunds in October 1981; 180 at Calveley in December 1981; 280 at Neston in November 1982 and 250 at Frodsham in 1982-83. Whether these are all locally bred birds is unclear, passage being a regular feature at the coast in April and May, with a total of 670 at Hilbre, for example, in spring 1984. The general trend of movement is northwards, with Cheshire-ringed birds found in Cumbria, County Down (Northern Ireland), Glasgow, Dumbarton and Aberdeen (Scotland), although one moving to West Germany is puzzling.

Given the tendency of the species to nest in loose groups it seems likely that many of those tetrads with breeding confirmed will contain between five and ten pairs, with fewer pairs in lower category tetrads. The county population may thus be in the region of 2000-3500 pairs.

Turtle Dove
Streptopelia turtur

THE Turtle Dove is one of the last summer visitors to arrive, usually during May, although birds occasionally arrive during the second half of April[6,7], the earliest date on record being 15th April 1980. The average arrival date in the Great Budworth area over a period of 24 years was 9th May[6]. Most birds leave before the end of August although some may stay until late September and are occasionally reported in October, the latest Cheshire date on record being in the third week of October 1961. It may be confused with its close relative, the Collared Dove, if plumage details are not seen well, but is easily distinguished as soon as it starts its purring song which may be heard throughout most of its stay in this country.

Turtle Doves were not recorded in Cheshire until 1851, with the first recorded breeding at Alderley Edge in 1870. They did not colonise the Great Budworth area until the 1890s[6]. Sharrock[35] attributed the increase and spread in nineteenth-century England to new methods of arable cultivation developing through the country. By the turn of the century, birds were breeding throughout Wirral and lowland Cheshire, even plentifully in places, though remaining scarce in the east of the county[2]. Abbott[10] noted the Turtle Dove infrequently around Wilmslow between 1915 and 1926, and plotted no territories at all on his detailed maps, which covered a total of 22 square kilometres, for 1919, 1921 and 1922.

Boyd[5] talks of birds coming to his hen pens, apparently in good numbers. A bird ringed at his home in June 1932 was trapped there again in May 1937 and no fewer than 36 of the birds ringed by him during the period 1931-39 were recaught in a subsequent year, some of them several times. Hardy[4] considered numbers to be increasing north of the Mersey, with nesting first reported at Houghton Green, Warrington in 1937. A pair nested in Birkenhead in 1934 and birds occasionally nested in thickets on Frodsham Marsh. The population appears to have remained more or less stable until the outbreak of war in 1939 when agricultural changes again affected the species' fortunes. With the need to increase home production of food, tall hedges were laid to reduce shading around the edges of fields. This practice was particularly widespread in the winter of 1941-42 and in the following summer Boyd noted fewer birds around Great Budworth than in the previous 20 years. Bell & Samuels[17] recorded a simultaneous decrease in the Wilmslow area to which birds have not since returned.

The Turtle Dove declined further throughout the county during the next 20 years and by 1967 was absent as a breeding species from eastern Cheshire[7,9], its stronghold then being the area in the west, south of Chester. The *L&CFC Report* for 1961-62 described the Turtle Dove as common in Wirral, where some slight expansion of range was also suggested in the late 1960s[8]. In the early 1970s a few pairs bred regularly in the Chelford and Marton (SJ86/87) areas from which they had largely retreated prior to this survey. The map shows no records in the Great Budworth area during this survey, in sharp contrast to Boyd's experience.

At the time of this survey the Turtle Dove was a rather scarce visitor to the county. Tall scrub or overgrown thorn hedges are the typical nesting site, and weed seeds of arable or disturbed ground form the bulk of the bird's diet. As the map shows, breeding birds are widely if thinly scattered wherever this combination of food supply and nesting cover prevails. The much quoted correlation with the distribution of the plant fumitory does not

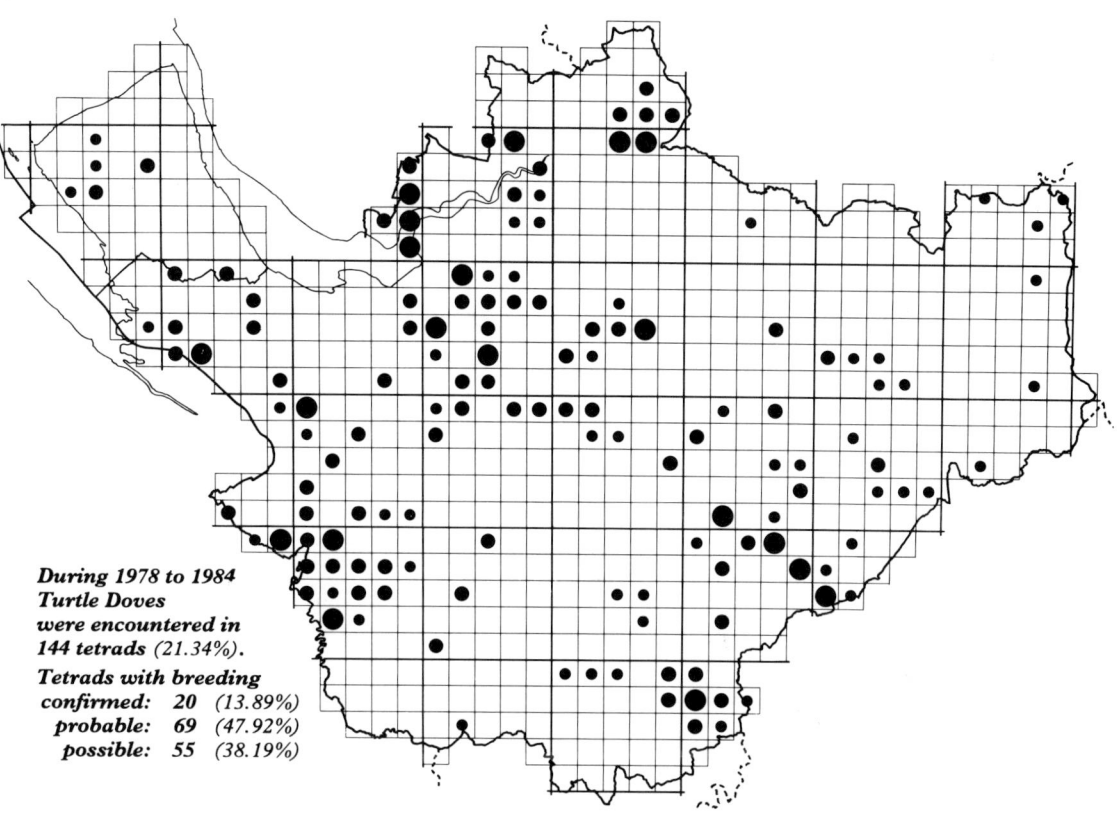

During 1978 to 1984 Turtle Doves were encountered in 144 tetrads (21.34%).

Tetrads with breeding
confirmed: 20 (13.89%)
probable: 69 (47.92%)
possible: 55 (38.19%)

hold good in Cheshire, the plant seldom being abundant enough to provide the steady supply of food which it does in parts of southern England where it may form 30-50% of the bird's diet (Murton et al. 1964). Indeed, the use of herbicidal sprays has reduced the availability of weed seeds in arable crops and may be responsible for the recent decline in numbers. Overgrown sand-quarries and sludge-pools are favoured haunts at the present time because of the abundance of trefoils, goosefoots and other food-plants.

The map undoubtedly overstates the distribution in any one year. A few occupied tetrads are known to hold several pairs of birds but in many cases only single pairs or even unmated birds were suspected, and the county population in any one year may be only 80-100 pairs. Many of the 'possible' breeding records will refer to over-shooting migrants and few, if any, will have been breeding birds. Cheshire is close to the north-western limit of breeding for the Turtle Dove, and therefore will be noticeably affected by any contraction of range, which may be linked to long-term climatic changes[36]. The underlying trend appears to be downwards as agriculture becomes more intensive and the species' food supplies diminish. Furthermore Turtle Doves face hazards on their migrations across the spreading Sahara and from shooters around the Mediterranean.

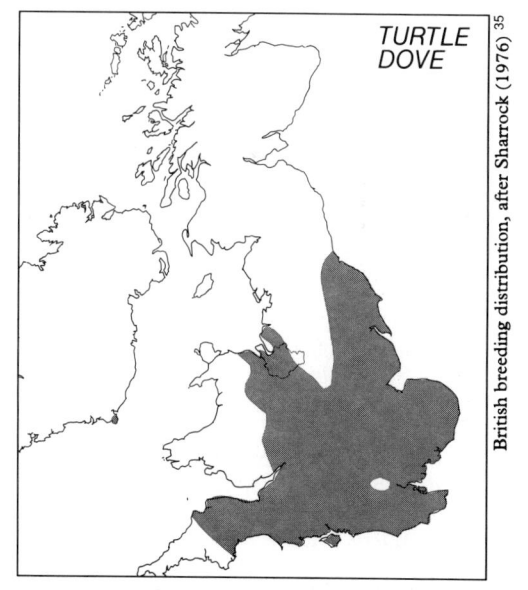

TURTLE DOVE

British breeding distribution, after Sharrock (1976) [35]

REFERENCE

Murton, R. K., Westwood, N. J. & Isaacson, A. J. (1964): The Feeding Habits of the Woodpigeon *Columba palumbus,* Stock Dove *C. oenas,* and Turtle Dove *Streptopelia turtur. Ibis* **106**: 174-188.

Ring-necked Parakeet
Psittacula krameri

DESPITE the presence of breeding Ring-necked Parakeets in a semi-feral state in adjacent parts of Greater Manchester during the years of this survey, there have been few sightings of this species in Cheshire to date. In 1979 one or two were seen along the Ladybrook valley on the Greater Manchester boundary from January into April and again in December. Single birds were seen at the Sandbach Flashes in 1980 and 1981, and at Woolston in 1983. The only record at all suggestive of breeding was of two birds which "appeared to be interested in a hole in a tree" at Hale in July 1983.

A number of other parakeet species are kept in captivity, and a proportion of free-flying birds will not be Ring-necked. Identification is not therefore quite so simple as many bird-watchers appear to believe. Nevertheless there remains a possibility that Ring-necked Parakeets may become a feature of suburban areas within Cheshire or Wirral in years to come.

Cuckoo
Cuculus canorus

UNLIKE any other British species the Cuckoo is parasitic, using the nests of other insectivorous species in which to lay a single egg. A bird reared by a particular host species usually becomes host-specific preferring only to parasitise nests of that species. A female will normally lay up to 25 eggs in a season within its "home range" which may spread over more than one tetrad. Males are not normally territorial like other species but "patrol" a home range, although they may become more territorial when the density of males is higher. The mating system is complex and not well understood but, based on the wide overlap of song and egg-laying "ranges", is thought to be mainly promiscuous, although the possibility of monogamy, polygamy or polyandry cannot be ruled out. Several males and females may breed in the same area, each male vying with its rivals to court every female which, in turn, may copulate with several males in any one laying period [36].

Cuckoos occur throughout the county in all habitat types. However, there were relatively few instances of confirmed breeding during the present survey and out of a total of 51 such records 28 were recorded as 'FL' and some of these may relate to young hatched in adjacent tetrads. Whilst Cuckoos have undoubtedly decreased in recent times the lack of confirmed records largely reflects the reluctance of modern bird-watchers to disturb nests, it being necessary to examine the nests of likely host species to confirm breeding. Thus, many of the 'probable' breeding records can be taken to indicate breeding Cuckoos.

Coward & Oldham[2] had an extensive list of host species: "...the Cuckoo usually foists its eggs upon the Tree or Meadow Pipit. Less commonly the Robin, Hedge Sparrow, Pied Wagtail, Yellow Wagtail, Yellow Bunting, Sedge Warbler or Whitethroat is imposed upon...". They also recorded that a young Cuckoo was said to have been reared by House Sparrows at Northwich in 1861, that a young Cuckoo was reared by a pair of Swallows, which also reared one of their own young, at Malpas in 1892, and that a Cuckoo's egg was found in a Wheatear's nest in a rabbit's burrow on Hilbre Island in 1901. Over 100 different host species have been recorded throughout Europe [36].

Hardy[4] found the Cuckoo to be particularly common on the dunes of Wirral where it parasitised Meadow Pipits and, sometimes, Whinchats. Boyd[6] knew the Dunnock as the favourite foster parent and found Cuckoos in their nests on twelve or more occasions, four times in Pied Wagtails' and only once in the nests of Reed Warbler and Reed Bunting. He had also heard of Robins, Yellow Wagtails and Song Thrushes

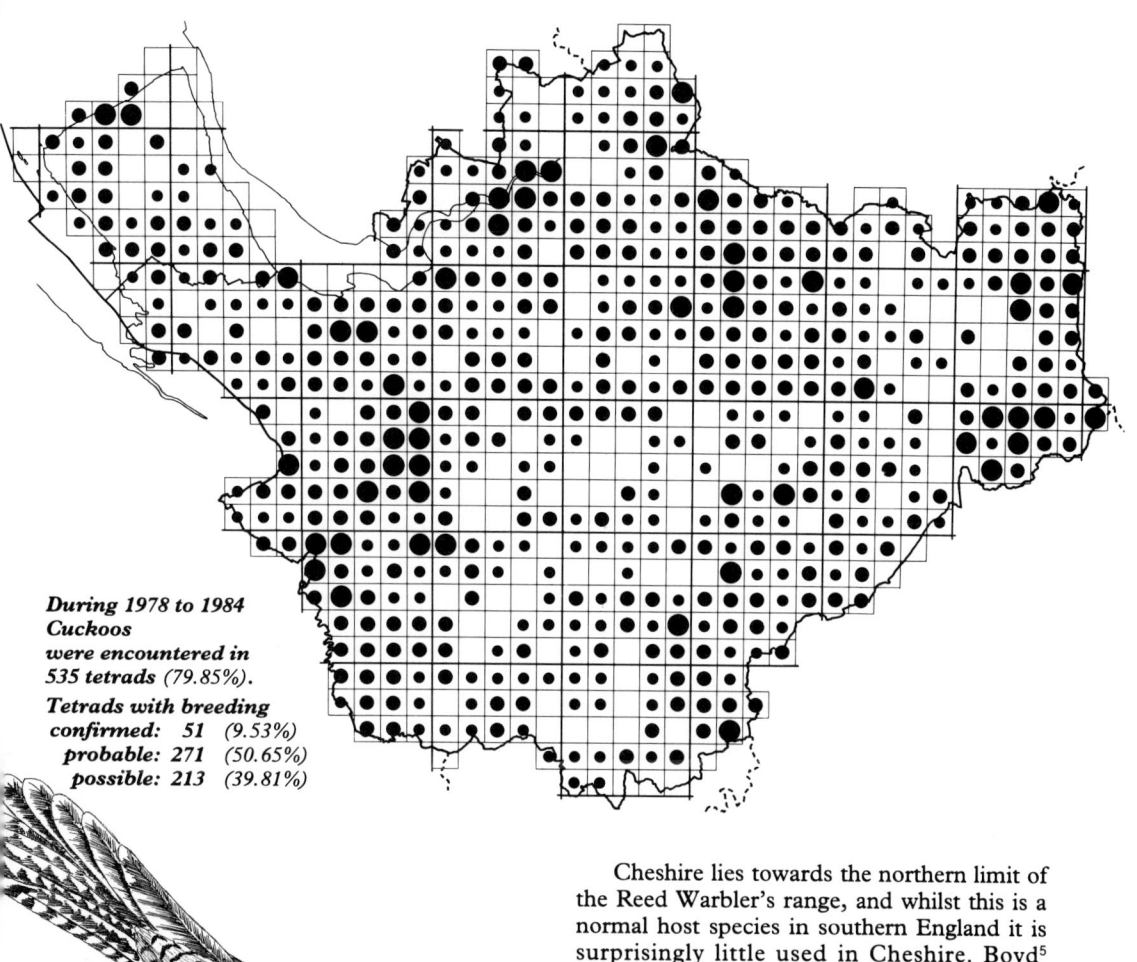

During 1978 to 1984 Cuckoos were encountered in 535 tetrads (79.85%).

Tetrads with breeding
- confirmed: 51 (9.53%)
- probable: 271 (50.65%)
- possible: 213 (39.81%)

being parasitised, but Meadow and Tree Pipits were by then scarce in the Great Budworth area.

Recent records point to just two species being commonly parasitised, the Dunnock and Meadow Pipit. During this survey the host species was recorded only for nine of the 51 confirmed breeding records: six parasitising Dunnocks in the lowlands and three parasitising Meadow Pipits, one at Frodsham and two in the hills. Many of Coward & Oldham's listed host species are now much less numerous, but it seems at least probable that Pied Wagtails and Robins are still used to some extent.

Cheshire lies towards the northern limit of the Reed Warbler's range, and whilst this is a normal host species in southern England it is surprisingly little used in Cheshire. Boyd[5] examined several hundred nests but only once found a Cuckoo's egg, at Marbury Mere in 1934. The intensively studied colony of Reed Warblers at Rostherne Mere NNR had hosted only one Cuckoo's egg (in 1977) in over 500 nests in 15 years (Calvert 1983). Cuckoos at Woolston, Kingsley and elsewhere have often been seen in reed-beds and are suspected of parasitising Reed Warblers, but no proof had been forthcoming by the end of the survey.

The decline of the Tree Pipit and several other former host species may be expected to have affected Cuckoo numbers. In 1921 Abbott saw several young birds around the Davenport Green sewage farm – no such concentrations have been reported in recent years. *L&CFC Reports* indicate that the species was "abnormally scarce throughout the county" in 1960, with further declines in the early 1960s. Some recovery was suspected in Wirral by 1966, but it remained very scarce in north Cheshire.

Cuckoos are now perhaps most easily seen in the eastern hills where they perch prominently on rocky outcrops, dry-stone walls, trees or wires. Frequently they are mobbed by Meadow Pipits, and at Saltersford in 1983 one was watched removing a young pipit from a roadside nest – possibly to encourage the adult to lay a replacement clutch which the Cuckoo could then parasitise. Similar observations were made on the outskirts of Knutsford in 1955 when the eggs of a Dunnock were apparently eaten by a Cuckoo (Wright 1955).

The typical "cu-coo" call of the male is familiar to most people but far fewer would recognise the female's liquid bubbling call, often given after laying an egg or in response to the male's advertising call. Cuckoos are often most vociferous in the early morning and only for a short period of their brief stay in this country. Thus, the apparent absence from parts of the southern plain may be artificial because visiting observers were seldom active in this area until later in the day and visits were not necessarily timed to fit in with the species' main song period. However, because of the open nature of much of this area, and the distinct lack of birds in general thereabouts, the map is likely to be reasonably accurate. Conversely, Cuckoos will penetrate suburban areas, where small passerines are more numerous – in 1933 a youngster was reared in a Birkenhead park[4].

It is difficult to assess just how many birds breed in the county. Because of their wide-ranging "territories" it is quite likely that individuals may have been recorded in more than one tetrad and some records may have been of passage birds. It is also likely that the map overstates the true distribution in any one year. With these factors in mind it is estimated that the Cheshire population may lie in the range 250-350 breeding birds.

REFERENCES

Calvert, M. (1983): Reed Warbler nests and instances of re-use for further breeding attempts. *North Western Naturalist* pp8-10.
Wright, R. C. H. (1955): Cuckoo eating whole clutch of Dunnock's eggs. *British Birds* 48: 456-457.

Barn Owl
Tyto alba

"THE artificial environment created by man in the British countryside during the last two thousand years or so would appear to suit the Barn Owl very well indeed. It can never have been a woodland bird, since its method of hunting requires the existence of open spaces with a thick ground layer of vegetation to provide a suitable habitat for its small mammal prey." (Bunn *et al.* 1982)

In Cheshire, however, as in many parts of the country, the Barn Owl is in decline. Its white, ghostly form is now seldom seen hunting during the late afternoon or crossing the car headlights at night. Coward & Oldham[2] knew it as a common but "curiously local" resident, being rare in the hill country and in certain lowland districts, for example around Knutsford. They also described it as being most abundant in the vicinity of houses, breeding in the suburbs of south Manchester, in Chester, where a pair had nested annually in the cathedral tower prior to 1893, and in the populous areas of Bebington and Birkenhead Park. Brockholes had described it as "not so common as formerly" in Wirral. Nest-sites included church towers, barns, dovecotes, the roofs of houses and frequently hollow trees, which were also used as day-time roosts, in parkland and elsewhere.

Boyd[5] made several references to Barn Owls and their continuing decline, emphasising their value to farmers in destroying rats and mice. Analysis of 143 pellets carried out by Coward, including 37 from Great Budworth, showed that a variety of shrews, mice and voles was taken, in particular common shrews, wood mice and field voles, but no rats, although Boyd had seen birds with rats in their talons. Analysis of pellets collected in 1949 again showed the preponderance of the smaller mammals but also several rats. These analyses also showed that Barn Owls took birds such as Starlings, Blue Tits, Greenfinches, Blackbirds and, in

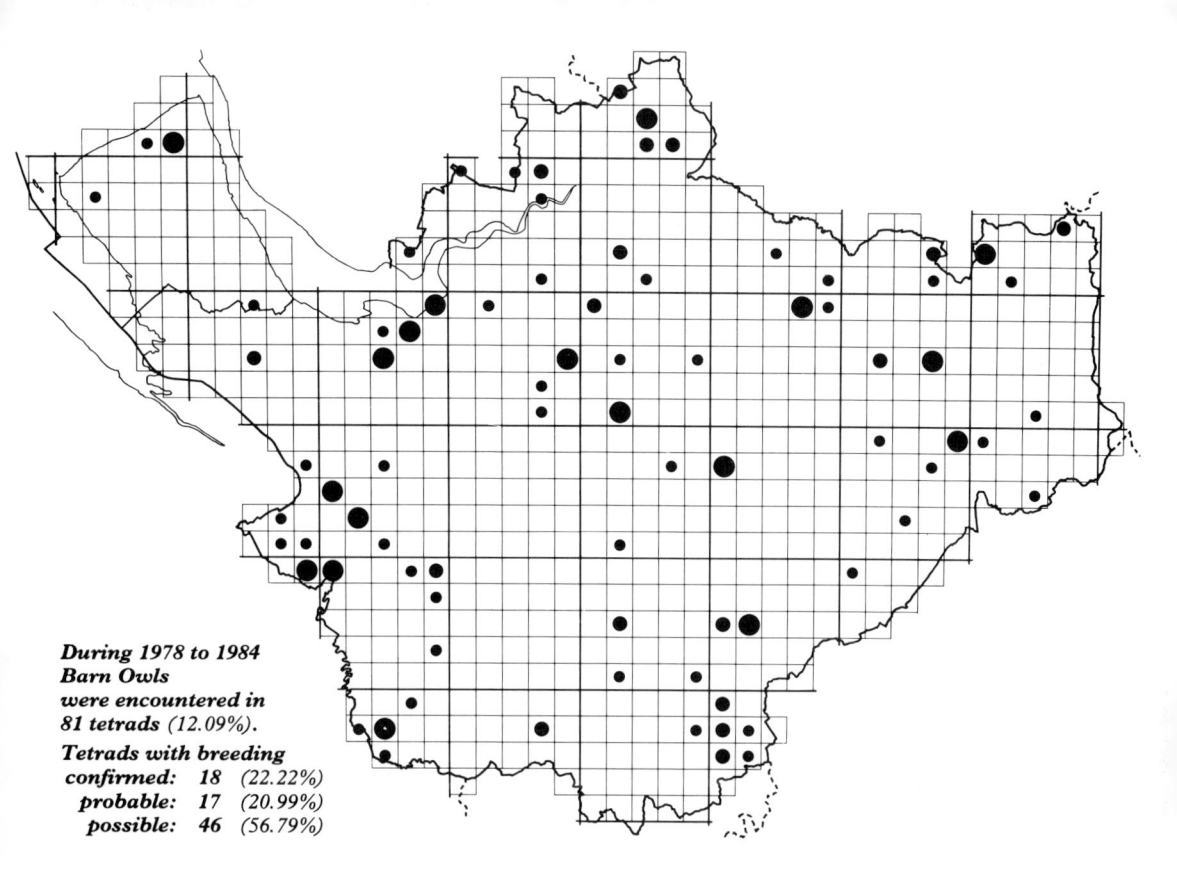

During 1978 to 1984 Barn Owls were encountered in 81 tetrads (12.09%).

Tetrads with breeding
confirmed: 18 (22.22%)
probable: 17 (20.99%)
possible: 46 (56.79%)

the Great Budworth district, particularly sparrows and finches. On one occasion Boyd was given a pellet containing the bones of a bat. When corn was stacked prior to threshing it attracted many mice and rats into farmyards at the season when owlets were in the nest.

Boyd[5] noted in his country diary in 1933 and 1934 that Tawny Owls had moved into sites, including buildings, formerly occupied by Barn Owls, and deplored the fact that many birds were still being shot despite legal protection. In 1935 he knew of three pairs in four and a half square miles around Frandley, each of them nesting in a pigeon-cote. At one time each farm had a pigeon-cote, often in the gable end of a barn, but as these fell into disrepair suitable Barn Owl nesting sites were lost. Shortage of nest-sites appears to have been a significant factor in the decline; a nest-box which Boyd installed in his own loft was soon occupied and held birds from 1935 until at least 1951. At that time he knew of ten or more inhabited pigeon lofts around Great Budworth. Some of these had been in almost constant use for many years, but nevertheless he considered the owls to be probably less plentiful than formerly[6].

Boyd[6] ringed 30 broods over 25 years. Most clutches were of five eggs, occasionally three or four and once six. His earliest record was of four eggs on 13th May, but often eggs were not laid until June. In 1948 a nest held five eggs on 29th June; three young were ringed on 28th August and their "snoring" continued nightly until early November. It was exceptional for all the young to be reared: eggs hatch at intervals and the smaller owlets tend to lose out when food is scarce and may starve to death or be trampled on only to be eaten by their older siblings. Of the broods ringed by Boyd, two had five young, three had four, eleven had three and ten had two.

Griffiths & Wilson[13] found Barn Owls to be sparingly if widely distributed in north Wirral into the 1940s. By the 1960s, Bell[7] considered them to be thinly spread throughout the county but rare in the hills, and suspected that they were probably not so common as fifty years previously. In his supplement, however, Bell[9] reported that single birds had been recorded in many localities in all regions of the county. However, breeding records were scarce and he listed only four instances in 1964, two or three from within the present county boundary in 1965, and two or three in 1966.

Of 102 south-western Cheshire farms visited in 1967, only two held breeding Barn

Owls. 19 which had held birds at some time during the previous five years were deserted[8]. This pointed to a recent, rapid decline, coincidental with those of the Sparrowhawk and Stock Dove, implying that pesticides may have been responsible. A number of pigeon-cotes and other sites had recently been destroyed, but other suitable sites had also lost birds. Only three nesting pairs were reported in 1967 in the whole of Cheshire, with sight records from 17 other localities.

Over the five years from 1967, with interest in the species aroused, more records were forthcoming[8]. In 1968, young were reared at nine known sites with seven further localities providing records during the breeding season, and in 1971 "fourteen breeding sites..." were notified, though to what level of evidence this refers is unclear. Subsequent *CBRs* contain no detail until 1978, the start of this survey.

In 1969 Wilson (Bunn *et al.* 1982) surveyed 48 farms in Cheshire. Of these four held successful breeding birds; 13 had birds present which did not attempt to nest; and sixteen, no longer occupied, had held birds in the past with breeding formerly at four of these. At ten of the fifteen farms which had never had birds the buildings appeared suitable. However, Bunn *et al.* (1982), studying Barn Owls in Lancashire and Cumbria, suggested that loss of habitat and nest-sites were the major factors.

The seven years of fieldwork for this survey furnished a cumulative total of only 18 tetrads with confirmed breeding, although birds were seen in a further 62 tetrads at some stage. It is probable that a number of pairs was missed, but undoubtedly the map reflects the scarcity of breeding birds compared with Boyd's day. The Cheshire breeding total at the end of this survey is likely to be under ten pairs. CADOS[28] commented that the distribution map for their area in the west of the county "...grossly exaggerates today's picture" and that records of Barn Owls in a fifth of their tetrads was "...totally out of the question.". They considered that the majority of their records referred to unattached birds roaming over a large area and being reported in several years. Their final estimate was a maximum of four pairs and there was some doubt as to whether birds still bred on a regular basis.

Recent nests have been reported from modern warehouses and factories, and in 1984 a pair is said to have nested in a scrap car, hinting at some considerable adaptability, but, in general, destruction of nest-sites is a continuing problem. A derelict factory at Ettiley Heath, Sandbach, which held a breeding pair from 1973 or earlier until 1977 has since been demolished, and a known nesting site at Mobberley was stopped up in 1979 because the "snoring" of the owlets was disturbing children sleeping in the room beneath. A bell-tower was used at Ravenscroft Hall, Middlewich until 1983 when the building (a listed one) was restored. Removal of hedgerows and tidying of rough grassland have caused further losses, for example in the Dean valley at Woodford where a former "beat" was converted into a golf-course. Severe winters, the noticeable lowering of average temperatures in April and May since the mid-1950s and wetter summers may also have contributed to the demise of what is essentially a southern species, and the habit of hunting over roadside verges has led to the deaths of many owls, killed by traffic.

Attempts are currently being made to re-establish birds in parts of the county which still appear suitable, through the Barn Owl Breeding and Release Scheme (S. Jeacock *in litt.*) and by the provision of nest-boxes where suitable sites are lacking. Such a scheme is possible, as has been shown at Rostherne Mere NNR, and there is considerable evidence that fledglings will disperse in all directions to take up suitable nest-sites. Five of Boyd's ringed nestlings were recovered more than ten miles from the nest site: one was found dead at Holmes Chapel ($11\frac{1}{2}$ miles SE) two months after fledging, and another was recovered there when it was seven months old; a third nestling was shot at Lytham, Lancashire (42 miles NNW) eight months after in late winter; a fourth was recovered at Adlington ($16\frac{1}{2}$ miles E) two years after ringing and the fifth at Churton (16 miles SW) four and a half years after ringing. He considered the first bird was "...proof that they are apt to wander off as soon as they are fully fledged", a supposition which has been borne out throughout the country (Bunn *et al.* 1982).

While attempts, such as the release scheme, to revitalise Cheshire's Barn Owl population are to be welcomed, the long-term success of the species may depend more on a return to sympathetic agricultural practices.

REFERENCE

Bunn, D. S., Warburton, A. B. & Wilson R. D. S. (1982): *The Barn Owl*. T. & A. D. Poyser, Calton.

During 1978 to 1984 Little Owls were encountered in 375 tetrads (55.97%).
Tetrads with breeding
  confirmed: 146 (38.93%)
  probable: 113 (30.13%)
  possible: 116 (30.93%)

Little Owl
Athene noctua

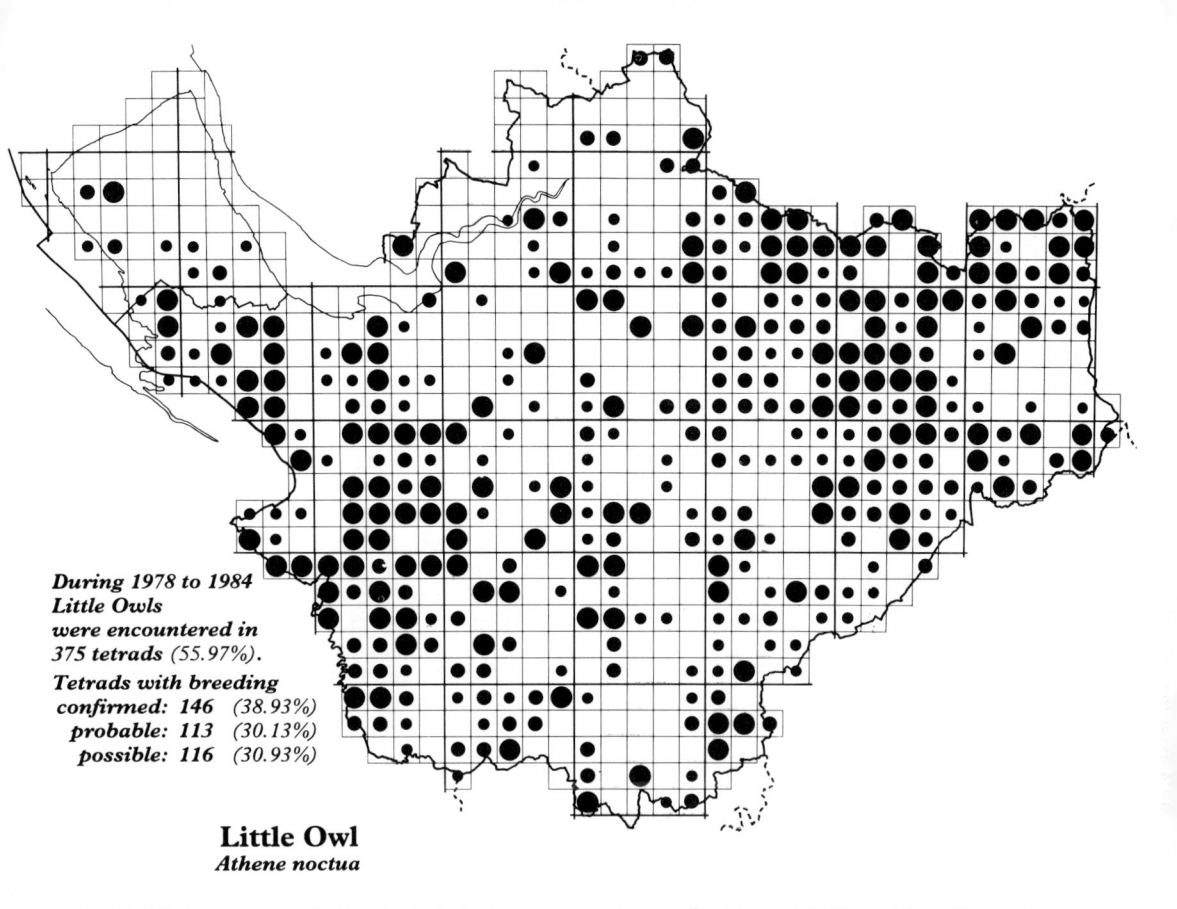

THE Little Owl population in Britain has its origin in introduced birds. The first Cheshire birds were said to have bred at Eaton Park in the late 1880s and one was seen at Tabley some time prior to 1887, in which year one was shot at Arley[2]. It is difficult to relate these records to known introductions to Yorkshire in 1842 (failed), to Kent during 1874-80, in Northamptonshire during 1888-90, and in Yorkshire, Hampshire and Hertfordshire in the 1890s ([37], Witherby & Ticehurst 1908).

Few records are then available until 1921 when three young were found at Alderley Edge, the first definite instance of breeding in the county[7]. Thereafter, the species colonised rapidly, with breeding at Lach Dennis and Newton-by-Daresbury in 1924. Abbott[10] found birds at Mobberley and Warford in 1922, North Rode in 1926 and Redesmere in 1927. At Arley, where the 'keeper first saw a Little Owl in 1926, between 30 and 40 were shot in 1930-31 and 30 on 3000 acres in 1933.

Hardy[4] found the species to be widely distributed and increasing, though not so numerous in Wirral as in the Chester area or eastern Cheshire. He listed 13 nesting localities. *L&CFC Reports* detail seven or eight nests within half a mile of Eaton Hall in 1949 and an increase in the Bidston area of Wirral in 1948. Boyd[6] described the species as common since the 1930s in the Northwich area, and this probably also applied to much of the county. By the 1960s the species was thinly distributed throughout Cheshire[7].

Lightly wooded farmland or parkland with

scattered trees is the preferred habitat, and typical of much of the county. Exceptions are the built-up north side of Wirral, the open mossland with coverts north of the Mersey (this forming better habitat for the Tawny Owl which may be a controlling predator there – Mikkola 1976), and parts of the southern plain where hedgerow trees are either few in number or too young to provide nest-holes. Typical nest-sites are in hollow trees, rabbit burrows, under tree roots, and in disused buildings. In the eastern hills, rocky screes, stone walls and cavities in buildings are used.

Little Owls are not infrequently encountered by day, perching prominently on fences, trees, telegraph poles or wires, but can easily be missed during a visit to a tetrad unless dusk visits are made to listen for calls. Diurnal duets are not uncommon, and are particularly frequent during early spring when the passing of a shower is often marked by much calling. At Capesthorne, in March 1982, birds yelped "keeow" from five territories simultaneously, and the Curlew-like hoot of the males was also uttered. Such hooting may be heard regularly from the middle of February to the end of May, less often in early June, and infrequently in November and December (JPG). To hear several males echoing each other's song at night is an eerie and unforgettable experience.

Birds often perch near the nest-hole which is then fairly easily found. The young may be fed by day, particularly after leaving the nest, and breeding is therefore not difficult to prove, although the map shows many records of 'probable' breeding only. Little Owls feed on earthworms, beetles and other invertebrates, small ground-feeding birds up to the size of thrushes, and small mammals. Birds are only important during the nesting season (Mikkola 1983). Little Owls occupied some of Boyd's nest-boxes and he recorded remains of unusual prey items such as a frog, a Snipe and "...a little break-back mouse trap, which presumably had held a mouse when it was taken" [5]. They took a lot of Tree Sparrows, thus damaging a nest-box colony of which he was particularly proud.

Although perhaps genuinely less common in central Cheshire than elsewhere, the scarcity there, suggested by the map, is almost certainly accentuated by sparse coverage for this survey, particularly in early spring. Fieldwork for the BTO Winter Atlas survey revealed birds in several 'unoccupied' tetrads around Tarporley and Winsford between 1981 and 1984, and since they are largely sedentary[41] there seems little doubt that Little Owls were overlooked in this area during the breeding season. However, the former prevalence around Arley can surely no longer apply. At Moore, where seven birds were heard answering each other in the spring of 1958, there are now only one or two, so it seems that a general decrease may have occurred in the north of the county. The effects of pesticides on the populations of small mammals and seed-eating birds in the 1950s/60s undoubtedly had some effect on Little Owl numbers[37].

Sharrock[35] estimated a British population of 7000-14,000 pairs during the BTO Atlas survey (1968-72), based on a "conservative 5-10 pairs per occupied 10-km square". Densities in Cheshire are frequently much higher with clusters of pairs living in loose colonies – for example seven territories along one mile of the Dean valley in 1977 and three territories along half a mile of ridge at Allgreave in 1984. It is also possible that the milder climate in Cheshire compared with eastern or south-eastern counties, where severe winters are more likely, is beneficial to the Little Owl, especially as Britain lies at the northern edge of the species' range.

Early CBC results showed the Little Owl to be more widespread in west central England than elsewhere (Williamson 1967) and the map in the Winter Atlas[41] revealed a similar distribution. National CBC figures for 1984 suggest a breeding total of almost twice that recorded during the BTO Atlas and this, combined with the western bias in its numbers, may indicate 30-40 pairs per occupied 10-km square in Cheshire, giving a county total of some 700-1000 pairs.

REFERENCES

Mikkola, H. (1976): Owls killing and killed by other owls and raptors in Europe. *British Birds* **69**: 144-154

Mikkola, H. (1983): *Owls of Europe*. T. & A. D. Poyser, Calton.

Williamson, K. (1967): The bird community of farmland. *Bird Study* **14**: 210-226

Witherby, H. F. & Ticehurst, N. F. (1908): The spread of the Little Owl from the chief centres of its introduction. *British Birds* **1**: 335-342.

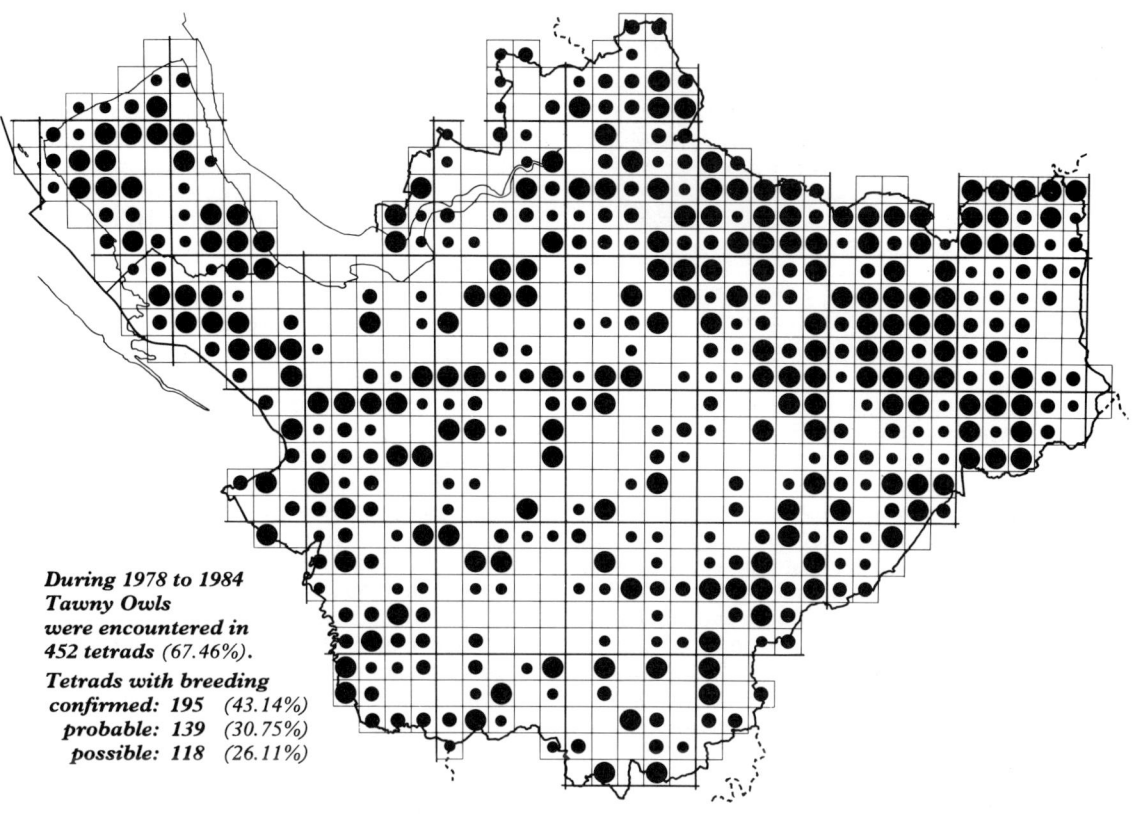

During 1978 to 1984 Tawny Owls were encountered in 452 tetrads (67.46%).

Tetrads with breeding
  confirmed: 195 (43.14%)
  probable: 139 (30.75%)
  possible: 118 (26.11%)

Tawny Owl
Strix aluco

THE quavering hoot and sharp "kewick" of a Tawny Owl on a moonlit night in late winter are our most familiar nocturnal bird calls. Tawnies are very much the owls of mature woodlands, whether broad-leaved, mixed or coniferous. Wooded gardens, churchyards, parkland and copses on farmland are also frequented.

In Greater Manchester this species was very rare a century ago, perhaps due to persecution in this densely populated area[46]. No comparable scarcity is recorded for Derbyshire[45] or the West Midlands[47]. Brockholes[2] did not include it in his Wirral list but Coward & Oldham[2] knew it as a common resident in the parks and woods of the lowlands, including such sites in south Wirral as Burton, Upton, Chorlton, Stanney and Stoke. In the hills it was "...perhaps not so plentiful as the Long-eared Owl..." but bred near Disley and in the Dane valley near Wincle.

Griffiths & Wilson[13] found Tawny Owls fairly commonly all over north Wirral by 1945, and subsequent authors have all described it as the commonest owl species. When it achieved its current prevalence in the eastern hills is not known, but the stands of spruce and pine in Macclesfield Forest, dating from the 1920s onwards, now support a thriving population, and birds may be found wherever there are stands of trees in the hill areas, perching on fence-posts or stone walls to hunt in relatively open country. The Georgian parkland belt of the north-east stands out clearly on the map. As stands of timber trees in these parks and elsewhere reached maturity this century, Tawny Owls must have found them ideal habitats.

The Tawny Owl is the earliest of all the owls to lay, with a mean first egg date, based on BTO nest record data, of 25th March[35]. They are single brooded, laying three to four eggs at daily intervals, but in years of food scarcity will not breed. Perhaps the most frequent nest-site is inside a hollow tree trunk, especially those shattered trees where the cavity is like a chimney, permitting entry from the top. Nest-boxes are readily occupied. Old nests of Crows, Magpies and Sparrowhawks, and even squirrel dreys are occasionally used. This habit presumably enables the species to occupy conifer plantations where cavities are scarce. From the 1930s onwards Boyd referred to Tawnies nesting in buildings formerly occupied by Barn Owls, and this practice is now frequent. Tawny Owls use similar sites to those of Jackdaws, Stock Doves and Kestrels, though whether any of these species could win an argument over a hole with the highly territorial Tawny Owl is doubtful – a female Kestrel was found dead in a Tawny Owl's nest at Moore and remains of another Kestrel were found in a pellet from Danes Moss in 1979. Such finds are not uncommon (Mikkola 1976). A Little Owl was attacked by a Tawny at Petty Pool in 1980. A dead Magpie was found in a nest at Kelsall in 1944[3] and a freshly dead adult Jay at Burton in 1965.

Initial location of this species by call is easy provided visits are made at the right time. The hooting song is usually heard from December to June and again in September and October. In early spring, hooting increases in intensity, with phrases terminating in a wavering crescendo. Coward & Oldham[2] noted that birds can occasionally be heard in August but their latest date was 18th July. In 1983 a bird was hooting at Dean Row, Wilmslow on 28th and 29th July. Diurnal hooting is not unusual, though generally half-hearted. Some day-time observations were made by finding an adult – usually the male – being noisily mobbed by small birds which had found its off-nest roosting site. Nevertheless the species will have been under-recorded since many tetrads received inadequate coverage at dusk for birds to be detected.

Dusk is also the best time to hear the hunger calls of the owlets – a hoarse, whispered version of the adult "kewick" –, which provides easy confirmation of nesting from May to August. Records of such hunger calls in Cheshire span the period 23rd April to 9th August. Birds with young to feed are sometimes seen hunting by day, and on 13th May 1981 an adult carrying a vole flew, within feet of JPG, to a nest in a hollow tree at Little Budworth. The following day similar behaviour was noted at Wettenhall[8].

Small mammals, particularly mice and voles are the main prey of the Tawny Owl. However, in some urban areas a high proportion of birds, particularly House Sparrows and Starlings, is taken (up to 89% in Manchester according to Yalden & Jones (1970). A nest at Grappenhall contained the remains of a Blue Tit and many Starlings (DN). Boyd[5] recorded a young Jackdaw, a Blackbird, a Starling, a Yellowhammer, a Song Thrush, House Sparrows, a Robin, frogs, rabbits and beetles as well as rats, mice, shrews and voles. Subsequently, he detailed the analysis of the contents from 96 pellets collected between 30th March 1947 and April 1948[6]. They contained 22 house mice, six wood mice, four field voles, four bank voles, three shrews, four rabbits, eight rats, 21 frogs and 55 beetles.

The map clearly shows the species' preference for wooded areas, though it may have been under-recorded in south central Cheshire as there seems to be no reason why fox-coverts in this area should not contain Tawny Owls. Local persecution may explain absences in some areas. Tawny Owls are generally sedentary although movements up to ten kilometres are not uncommon and it is likely that most potential sites would be filled by juvenile dispersal.

On the night of 20th June 1974 some 14 territories, including three where young were heard calling, were located between Marton and Wilmslow – a distance of 17 kilometres – and in 1976 five birds were calling at Mottram Hall Woods on 2nd March and six were heard at Westparkgate a week later. In April 1982, ten birds were calling in four tetrads near Malpas, and three pairs held territory within earshot of Appleton Reservoir[8]. Five pairs has been suggested as a normal population at Rostherne[24]. An average of just two pairs in each tetrad with confirmed breeding and one pair in all other occupied tetrads would give a county total of some 550 pairs.

REFERENCES

Mikkola, H. (1976): Owls killing and killed by other owls and raptors in Europe. *British Birds* **69**: 144-154.
Yalden, D. W. & Jones, R. (1970): The food of suburban Tawny Owls. *Naturalist* **914**: 87-89.

Long-eared Owl
Asio otus

THE Long-eared Owl is widely distributed in Britain throughout all kinds of arboreal habitats but has declined very considerably this century. Its decline has mirrored the increase in the numbers of Tawny Owls and has been ascribed to interspecific competition[35]. The species now has the slenderest of footholds in the county. Unfortunately, this scarcity causes the birds to be much sought after by "birders". More than one Cheshire nest has been deserted following publicity, so no map is published for this species for reasons of conservation.

Coward & Oldham[2] described its distribution in Cheshire in detail: "...though nowhere abundant, [it] occurs in all parts of the county, frequenting fir woods and even isolated clumps of Scotch firs. In Wirral, Brockholes described it as resident at Bidston, Prenton Mount, Storeton, Ness, Burton and Ledsham... In the Grosvenor Museum, Chester, there are young birds in down from Saughall and Burton. In the woods at Burton and Irby it is common. It breeds in Stanney Wood near Thornton-le-Moors, and on the Eaton Estate, as well as at Delamere, where in some woods it is fairly plentiful. Butts Clough in the Bollin valley, the Moss Covert at Plumbley, Rudheath, Gawsworth, Alderley Edge...may also be cited as localities in the lowlands where it nests. In...the wooded portions of the hill-country it is the commonest Owl, and used frequently to be captured in pole-traps when these were set on the moors."

Griffiths & Wilson[13], writing in 1946, recorded Long-eared Owls from those parts of north Wirral where there were conifer woods, and mentioned one nest on a ledge behind a gorse bush in a quarry at Newton, and others at Frankby and Thurstaston. Boyd[5] had only once seen the species in the Great Budworth area however – a probable pair at Stretton Moss in 1920 – so the decline in inland districts would seem to have taken place early this century. Bell[7] could quote only one recent breeding record, at Bidston in 1949. However, Raines[18] stated that a few pairs bred in the conifer woods and on the heaths of Wirral – a very similar situation to that recorded by Griffiths & Wilson fifteen years earlier. A pair on Bidston Hill failed to breed successfully between 1957 and 1959, but birds continued to be recorded here until 1964 when one was found shot[9].

A Forestry Commission survey of Delamere Forest made in 1962 stated that birds were almost certainly present in stands of pine but were very difficult to locate – there are no recent records. Subsequently *CBRs* include the following breeding season records: one at Thornton Manor estate in June 1964, one or two pairs present and breeding in Macclesfield Forest from 1967 (and probably for several years before this) until 1972 (Vince 1968) and one, presumably a migrant, in a Wallasey garden in May 1973. The 1975 *CBR* stated that the Long-eared Owl, "may now be extinct as a breeding bird in the county". However, fieldwork for this survey revealed birds present in a total of nine widely scattered tetrads, with old pellets found at a further two sites.

Scrubby woodlands with younger or more spindly trees seem to be favoured, the lack of old timber with suitable nest-sites for Tawny Owls perhaps discouraging this competitor species. The maturation of conifer plantations and development of a closed canopy in other woodlands this century must have helped the Tawny and hindered this species (cf. Blackcap, Chiffchaff or Nuthatch). Long-eared Owls often breed in the old nests of other species, particularly those of Magpies, Carrion Crows and Woodpigeons, and occasionally on the ground. The other major requirement would appear to be extensive areas of rough open ground over which to hunt. They are longer-winged than Tawny Owls, and hunt more in the open, although taking similar types of food (voles, mice and birds).

Long-eared Owls are more strictly nocturnal than other owls, although JPG has heard them hooting by day on the continent. Nocturnal visits to suitable localities in all parts of the county might have been rewarded by hearing the wing-clapping display, the low hoot or later the squeaky hunger calls of the owlets, said to sound like an unoiled hinge.

There appears to be ample opportunity for the species to occupy other suitable areas, since British-ringed nestlings and juveniles have been shown to wander some tens of kilometres: a nestling ringed in Cheshire on 1st June 1980, for instance, was found dead at Sleaford, Lincolnshire (150 km E) on 9th March 1983. Although confirmation of breeding came from a total of only four tetrads during the seven years of this survey, the species is undoubtedly overlooked, and the county population may stand at between five and ten pairs.

REFERENCE
Vince, R. (1968): Observations on a Pair of Nesting Long-eared Owls in Macclesfield Forest. *CBR* pp26-27.

Short-eared Owl
Asio flammeus

SHORT-EARED Owls nested regularly on Carrington Moss (now GMC) until 1893 by which time all the mossland had been reclaimed, and other mosses in the far north of the county must have been eminently suitable. Boyd[6] considered that it may have nested on the unreclaimed Whitley Reed some 100 years previously. By 1910, however, Coward & Oldham doubted whether the bird bred anywhere in lowland Cheshire, although it had been seen in summer on the Eaton estate. To the grouse moors in the east it was a familiar winter visitor and there was a suggestion that an occasional pair remained to nest, since birds bred on the neighbouring Derbyshire moorlands. In fact Frost[45] could trace only eleven years since 1905 in which Short-eared Owls bred on the Peakland moors in Derbyshire, so this suggestion may have been optimistic. Nevertheless in 1908 a bird hunting at Knight's Low, Lyme Park, was suspected of having a nest nearby.

The only possible evidence of breeding quoted by Bell[7] was in Macclesfield Forest, then young plantations, in 1936, and on the moors south-west of Buxton, Derbyshire in 1937. However, Hardy[4], writing in 1941, stated that young had been seen in a nest on Holcroft Moss, east of Warrington.

The Short-eared Owl is now a winter visitor in some numbers, with at least 20 regularly present in the Mersey valley and Dee areas. In 1964 one or two birds were seen regularly on Frodsham Marsh from 27th July and there was some evidence that breeding had occurred. Subsequently there has been an increase in the number of summer sightings, many of wintering birds staying around the estuaries into May, and others of birds in the eastern hills wandering from nest-sites outside the county boundary.

In 1973 one or perhaps two pairs remained and displayed on Burton Marsh until late April. Two years later a nest with one egg and a small chick was found in a marshland area by the Mersey. In 1976 a pair remained on Frodsham Marsh throughout the summer and breeding was considered to have been attempted.

During the present survey there were no confirmed breeding records, but recent plantations in the eastern hills show promise for the future. Coastal nesting from Norfolk to Kent has developed during this century[35] and there seems no reason why the undisturbed parts of the Mersey and Dee marshes should not be occupied regularly, especially as a pair bred successfully on the north side of the Mersey in 1982. Regrettably both the adults of this pair were shot and the young were hacked back to the wild at the Wildfowl Trust centre at Martin Mere, Lancashire (A. Duckels pers. comm.).

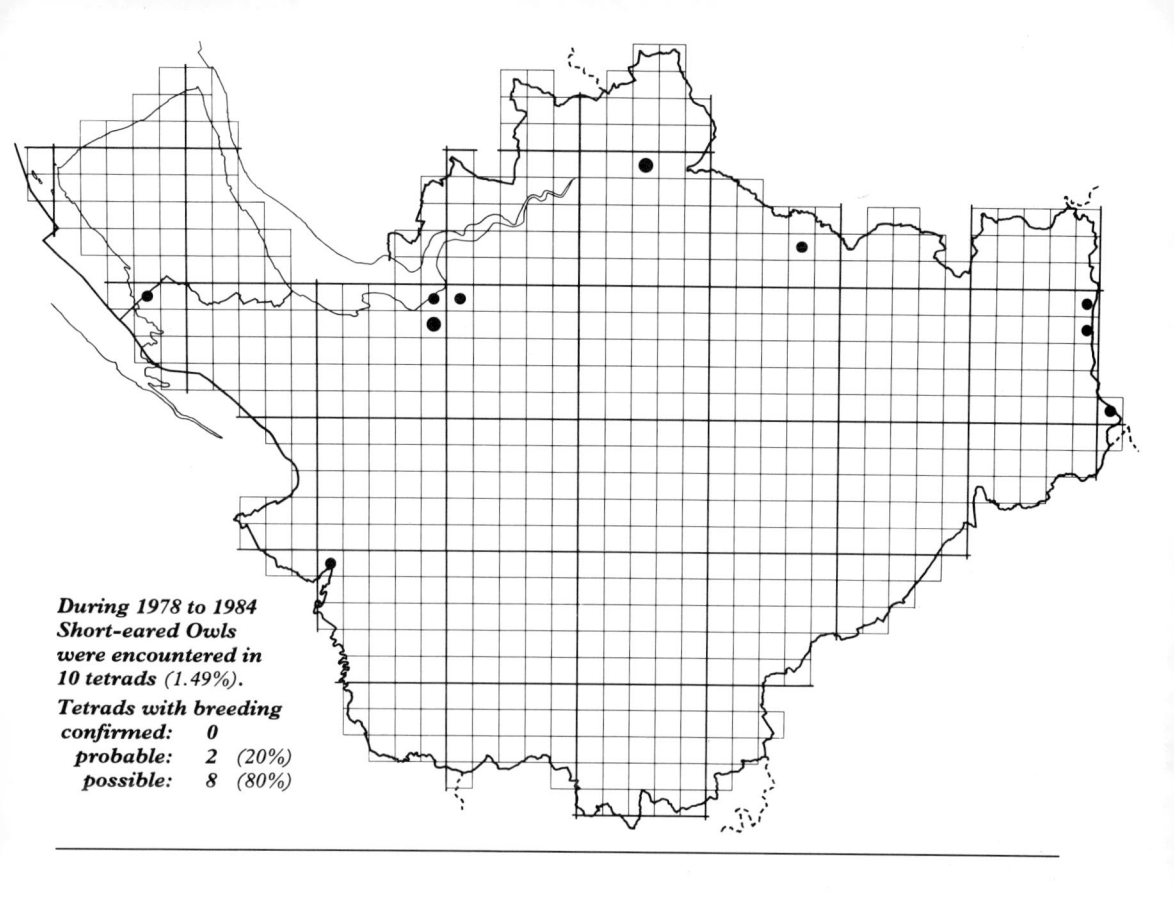

During 1978 to 1984 Short-eared Owls were encountered in 10 tetrads (1.49%).

Tetrads with breeding
confirmed: 0
probable: 2 (20%)
possible: 8 (80%)

Nightjar
Caprimulgus europaeus

THE eerie sight of this large crepuscular bird catching insects must formerly have been quite familiar to country people, as shown by the number of old dialect names for the Nightjar: Goatsucker, Fern Owl, Bracken Owl, Moth Owl, Evening Jar, Jenny Spinner, Nighthawk and Lich Fowl. However, of these names "Lich-fowl" (pronounced "Leitch"), once peculiar to the Peckforton Hills, is now the most appropriate for its status in Cheshire – it shares its origin with the word "lych-gate", and means "corpse-bird"! All the indications are that the Nightjar is now extinct as a Cheshire breeding species, with only a handful of records during this survey.

Coward & Oldham[2] knew it as a local but widely distributed summer resident. It bred regularly in Wirral in the conifer woods of Bidston, Storeton, Ness and Burton, and on low heather-clad hills as at Bidston, Caldy and Thurstaston. It occurred in the Peckforton and Bickerton Hills and was very common in Delamere Forest, frequenting thickets as well as open glades. Half-a-dozen birds might have been heard along a couple of miles of forest road. On Little Budworth Common and Abbots Moss it was even more abundant, and it also occurred in some numbers on Frodsham and Helsby hills. The 1917 *L&CFC Report* described it as very abundant on Risley Moss and near Rixton six or seven years previously when as many as six were recorded on the railway fence between Padgate and Glazebrook. Abbott plotted three territories on Alderley Edge in 1919, one at the Row of Trees, Alderley in 1921 and five pairs on Lindow Moss in 1922. He heard a bird or two with some regularity on Lindow Common and also at Styperson. Birds frequented patches of rough ground and open woods in many other parts of the plain: Soss Moss, Marton, Mere Moss, Moss Covert at Plumley, Butts Clough near Ashley, Rudheath (Allostock), Knutsford, Somerford, Tabley and Petty Pool. It occurred throughout the hill country from the Goyt to the Dane.

Sharrock[35] refers to a widespread decline in Britain which became general about 1930 and very pronounced from 1950. The absence of records in *L&CFC Reports* from 1937 to

1949 suggests a similar trend in Cheshire although detailed information is scarce. The main reason for this decline, which parallels those of the Red-backed Shrike and Wryneck, is thought to be a change in climate: there has been a noticeable lowering of average temperatures for April and May since the mid-1950s and summers have tended to become wetter (Gribble 1983). However, many heathland and mossland sites, formerly occupied, have been planted with trees or become overgrown with birch scrub. Others have been dug out for sand or opened up for leisure pursuits. Hughes (1983) saw up to six adult birds annually in Delamere Forest from 1951 to 1957, after which date they disappeared. Clegg had seen birds here annually until he left the district in 1952. Bell[9] refers to presence at Abbots Moss from 1962 onwards, at Newchurch Common in 1964 and another site near Northwich (probably Rudheath) in 1964 and 1965. Hughes also found birds at Abbots Moss in 1967, 1968 and 1969. A single bird was reported from Delamere in June 1974.

Most recent records have been of males uttering their peculiar, far-carrying, "two-toned", churring call, formerly likened to the sound of a spinning-wheel (Jenny Spinner) or more recently to a distant motor-cycle, and bearing some similarity to stridulations of a cricket. This is generally delivered from a "song post" where the bird will perch, in elongated fashion, parallel with the branch. In 1967 and 1968 a churring bird was reported from the sandstone hill at Bickerton, and again, but for one night only, in May 1971. A male was churring on Frodsham Hill on 13th May 1974 and another was recorded near Nantwich on 27th July 1976. However, birds will churr when migrating and these records may refer to birds on passage.

In Wirral, Griffiths & Wilson[13] found nests at Caldy, Bidston and Thurstaston Hills, as well as at Storeton, but 15 years later the local population was said to have fallen to only two or three pairs[18]. Thereafter, only two single birds were reported in 1961; single birds in 1963 at Heswall Dale and Caldy Hill (where they had been absent for five years); single churring birds, or at best a pair, were reported on Thurstaston Hill between 1966 and 1969; and the last record was of two churring birds at Frankby on 4th June 1973. A bird recorded at Leasowe on 28th August 1977 was undoubtedly on passage.

Elsewhere, old Wilmslow residents reported churring birds on the Lindow mosslands sometime prior to the 1939-45 war, and birds were present on Danes Moss until 1950[26]. Bell[7] mentioned some evidence of presence at Brereton Heath in the preceding four years, this site having been occupied annually in the 1930s. The last record from here was of a pair in 1964. In 1963, birds were heard at Marton Heath, Cocksmoss and in the vicinity of Thorneycroft Hall. Only one of these was occupied in the following year and none in 1965 or 1966 (Bell pers. comm.). In 1972, one was churring near Langley in July, and isolated records in 1974 from Eaton quarry, Eaton Hall and Moreton Hall (all SJ86) undoubtedly all refer to the same migrant individual seen at a site near Congleton. A bird was reported from the moorland above Lamaload Reservoir in the early 1970s (A. Booth pers. comm.).

During this survey the only evidence of breeding came from Risley Moss where a male was present in 1978, a pair in 1979, and a pair plus an unmated male in 1980. There were isolated records from Crewe Memorial Hospital and Weaverham in June 1979. None whatsoever was located in Cheshire during the national BTO Nightjar census in 1981, although this may have reflected the reluctance of observers to make night-time visits to areas where they felt there was little chance of success. In June 1982 an unmated male churred in a cleared area beneath mature larches in Macclesfield Forest and in mid-July 1984 a single bird was churring on Congleton Cloud.

Typical Delamere territories ranged from about 1.6 to 16 hectares and were usually newly planted with young conifers and bounded by mature trees (Hughes 1983). Ground-cover consisted of heather, gorse, birch scrub, bilberry and bracken. Many such areas remain in the forest, and given the species' presence at Chat Moss (GMC) and continued prominence on Cannock Chase (Staffordshire) its absence from Delamere seems a little surprising.

REFERENCES

Gribble, F. C. (1983): Nightjars in Britain and Ireland in 1981. *Bird Study* 30: 165-176.
Hughes, W. (1983): Nightjars in Delamere Forest, 1951-58. *CBR* p67.

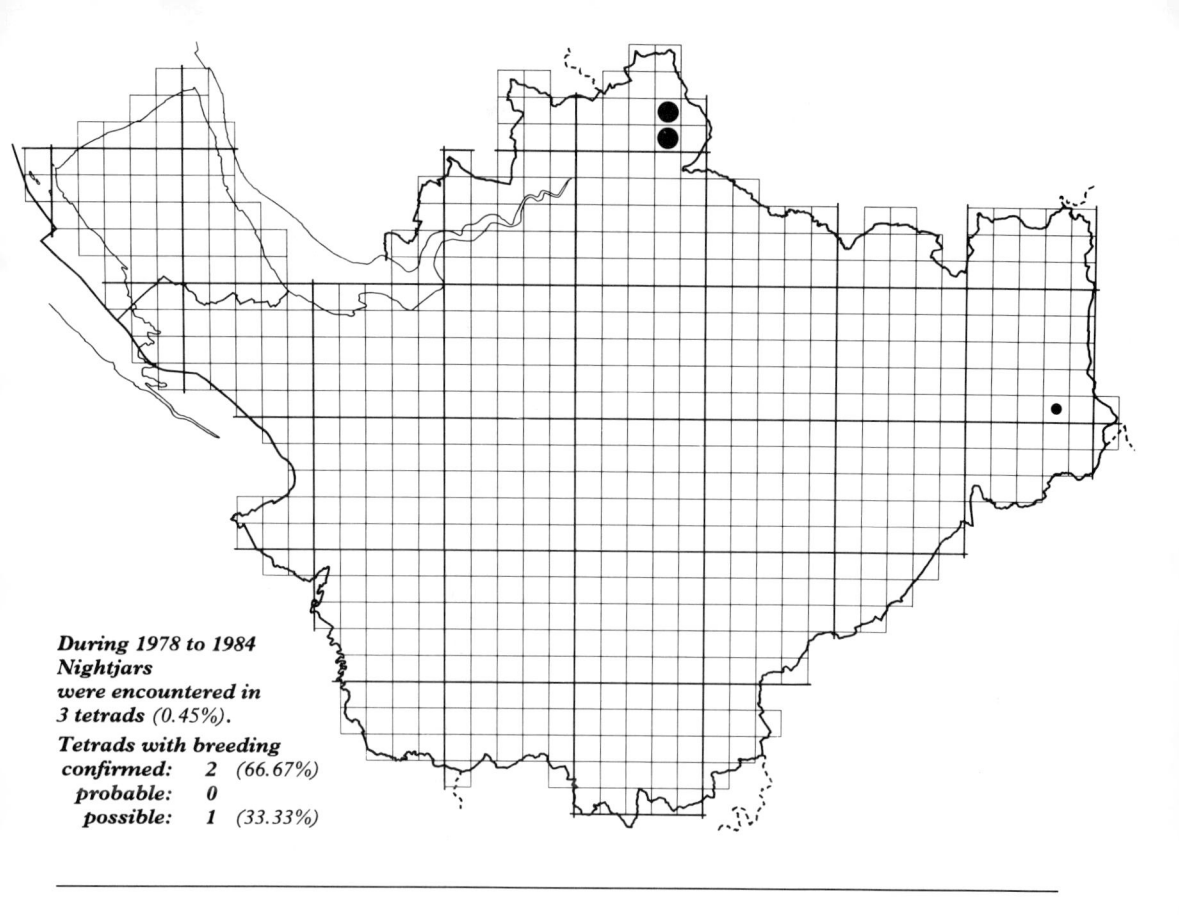

During 1978 to 1984 Nightjars were encountered in 3 tetrads (0.45%).

Tetrads with breeding
confirmed:	2	(66.67%)
probable:	0	
possible:	1	(33.33%)

Swift
Apus apus

SWIFTS are present in the county for only a brief period in summer – shorter than any other bird – traditionally arriving on or about May Day and leaving in early August. In fact the first birds generally arrive in late April, with numbers increasing rapidly in May when parties of displaying birds scream around the buildings where they nest.

Such display continues throughout the summer, chiefly in the early morning and evening, and provides the first hint that birds may be nesting. The nest is often in a cavity in the walls or roof-space of a building, entered generally through a small gap under the eaves. Since the birds are almost exclusively aerial, only wind-blown material such as dry grass or feathers can be used in nest construction, but as nests may be reoccupied in successive years, large amounts of material may accumulate.

The bulk of Cheshire's Swift population breeds in suburban areas, mostly in older housing or other buildings with suitable cavities. Modern estates, more suited to House Martins, are largely avoided, although birds may be seen prospecting. Boyd[6] stressed the paucity of nesting sites in a rural district. He knew of some 15 pairs nesting in Great Budworth school, but only two or three farms in the parish of Antrobus held pairs, with the odd pair elsewhere. In nearby Northwich and Hartford however both nest-sites and Swifts were plentiful.

Eggs are laid from the middle of May to the end of June, and hatch in about $19^1/_2$ days. Lack (1956) showed that from hatching to fledging takes from five to eight weeks depending on weather conditions and consequent availability of food. In poor weather the growing chicks can go torpid to reduce their energy needs and retard their rate of development. In the cold, wet summer of 1978 the breeding season was abnormally prolonged. Whereas birds are normally scarce in the county after the end of August, a pair was still bringing food to a Grappenhall nest on 20th September. Others were around nesting sites at Nantwich in mid-September and at Ashley on 3rd October. Many hundreds of birds were seen in the county during September, with stragglers into November. It appears likely that in this exceptional year eggs laid in early July did not produce fledged young until late September, if at all. A further late breeding record in 1980 referred to a young bird which fell from a nest in Sandbach on 3rd September.

The map shows the main areas for breeding Swifts to be in and near the major towns, including Crewe, Chester, Warrington, Runcorn, Wilmslow, the east Wirral conurbation from Ellesmere Port to Wallasey and the west Wirral towns from West Kirby to Heswall. There are confirmed breeding records from most smaller towns such as Winsford, Middlewich, Holmes Chapel, Sandbach, Congleton and Nantwich, as well as a sprinkling of records in some villages throughout the county, but most noticeably around Chester and south-west Cheshire.

During the summer, feeding Swifts wander far from the nest and can be seen feeding over the eastern moors and the coast alike, according to where airborne insects are concentrated by weather conditions (Lack 1956, Elkins 1983). Vast feeding concentrations have frequently been found at Frodsham or Woolston, with up to 2000 birds present at a time. Ringers have caught large numbers of birds as they swoop low over the banks (e.g. 180 on 11th July 1981). Retrap analyses of many catches indicate that the total number of individual Swifts visiting such sites is well in excess of 10,000, a figure several times greater than the entire Cheshire breeding population. This gives some indication of the enormous catchment area of these good feeding sites: breeding birds may, exceptionally, forage up to 100 miles from their nest in bad weather (Bromhall 1980), such a journey taking several hours' flying time, and their chicks may then be fed only once or twice a day. The Weaver Bend (Frodsham) area lost its attraction for Swifts towards the end of this survey period, possibly an indirect result of excessive use of the river by powerboats whose powerful wake eroded ten metres or more off the vegetated banks, greatly reducing the area of land supporting insects. The Woolston feeding flocks have been observed flying west at dusk and returning eastward at first light, probably spending the night on the wing over the sea.

Reduction in smoke pollution since the 1960s is thought to have had a beneficial effect on insect populations and consequently on Swifts in the Greater Manchester conurbation, and whilst most Cheshire towns are too small for any significant change to have been noticed, it is conceivable that increases may have taken place in the Birkenhead and Warrington areas. However, there must have been some local decreases as, in contrast to Boyd's experience, there are now few nests in Northwich and no breeding records from Hartford or Great Budworth. Renovation of older properties in some areas may have reduced the number of available nest-sites and caused local declines.

Swifts nest more or less socially, with loose colonies in favoured streets or groups of buildings. B. Martin has made detailed counts around Warrington, finding around 100 pairs in four urban/suburban tetrads, including ten pairs in a church steeple - a classic "Swifts in a Tower" site! All of these nests are in buildings at least 30 years old, notably in the pre-war estate at Westy. Breeding birds seem very faithful to particular sites and many streets remain unoccupied even though the houses appear identical in every respect. The species will use nest-boxes (Douglas-Home 1977), so their provision could enable Swifts to colonise newer buildings.

The closely watched population near

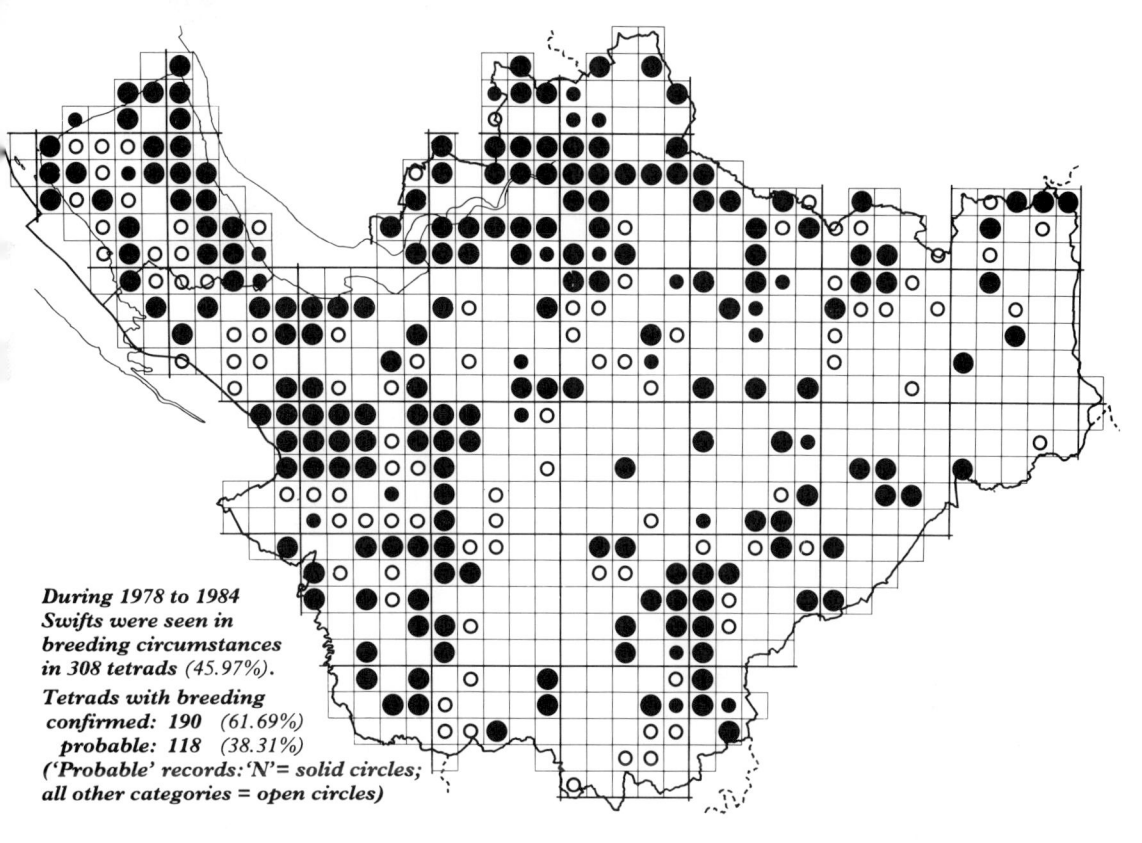

During 1978 to 1984 Swifts were seen in breeding circumstances in 308 tetrads (45.97%).
Tetrads with breeding
confirmed: 190 (61.69%)
probable: 118 (38.31%)
('Probable' records: 'N' = solid circles; all other categories = open circles)

Warrington is remarkably stable, with no observable fluctuations from one year to the next. Swifts are long-lived, with one locally-ringed bird retrapped at 17 years of age, and they usually do not breed until their fourth year. Some of the large population of non-breeders stay in Africa, but others return here and may investigate future nest-sites. This "pool" of non-breeders may ensure a stable population level, if shortage of nest-sites is a dominant factor.

This species presents several problems in recording breeding status. The adults do not feed their young once they have left the nest, and fledged young fly great distances, reportedly leaving the country within a day or two (Lack 1956). White-faced juveniles may occasionally be seen in the first few days of August, but this provides no proof of breeding in the tetrad, or even in the county. Also, the adults flash past so quickly that it is nearly impossible to see their throats bulging with insects collected for their brood. Thus, almost all confirmed breeding records came from sightings of birds at nest-sites. Fortunately, it is not difficult to see Swifts entering likely nest-sites (but it may be hard to discriminate between 'ON' and 'N'), and sites are usually occupied year after year. Additionally, non-breeding birds may prospect for new nest-sites. 'Probable' breeding records other than 'N' are unlikely to refer to birds actually nesting, although they may indicate areas which could be occupied if there were suitable nest-holes. First column records are not mapped for this highly mobile species.

The only detailed count away from Warrington is of over 50 pairs at Chester City Hospital in May 1981 (DN). Such high counts are unlikely to apply to the county as a whole. Sharrock[35] estimated 40 pairs per 10-km square with proven breeding. This would indicate a Cheshire population of 1000-1200 pairs, averaging five or six pairs per tetrad with confirmed breeding or visiting probable nest-sites ('N'). A good census of the county's Swift population would be valuable.

REFERENCES

Bromhall, D. (1980): *Devil Birds*. Hutchinson, London.
Douglas-Home, H. (1977): *The Birdman*. Collins, London.
Elkins, N. (1983): *Weather and Bird Behaviour*. T. & A. D. Poyser, Calton.
Lack, D. (1956): *Swifts in a Tower*. Methuen, London.

Kingfisher
Alcedo atthis

RINGING "chee" or "chi-kee" calls often alert the riverside bird-watcher to the approach of a Kingfisher and a moment later the bird dashes past showing either a brilliant blue-green back or a bright orange belly. With luck, excited chases may be watched in spring as two birds shoot rapidly along the stream, wheeling round over an adjacent meadow and returning back and forth along a favoured stretch of water. Later in the season a fish may be visible in the bill during the all-too-brief flypast – an indication that the bird may be feeding young nearby.

The Kingfisher does have a song, "a sweet trilling whistle, almost a warble" [2] which is heard mostly in March, before fieldwork for this survey started in earnest. Most records, in fact, either referred simply to birds seen in suitable habitat, or to confirmed breeding. Other than birds carrying fish for the young, this latter category referred mostly to occupied nests in river or canal banks, or to family parties seen soon after fledging. To see up to half-a-dozen Kingfishers fishing together is one of the highlights of any lucky bird-watcher's year.

In Victorian times Kingfishers were much sought after to fill the cases of stuffed birds which were then fashionable, and "the fatal brilliancy of plumage" made them a prime target[2]. Despite this hunting and persecution by fishing interests, the species held its own in the county: Coward & Oldham[2] described it as a widely distributed resident, though nowhere abundant. It then nested sparingly throughout Wirral and the plain and along the streams of the eastern hills.

It seems unlikely that the bird's status has changed much, if at all, this century. Boyd[5] wrote that "in a land of meres and streams like Cheshire it is plentiful", and referred to Kingfishers as plentiful throughout the Great Budworth district[6], which remains a stronghold, with birds particularly conspicuous at Marbury Mere. Griffiths & Wilson[13] mentioned several pairs in the Thornton Hough district, but the Kingfisher is not often seen in north Wirral.

The population crashes periodically following hard winters: severe weather in 1947 and 1963 seriously reduced numbers[3], and persistent frosts during the present survey in early 1979 and December 1981 will have resulted in somewhat depressed numbers. They may however be able to raise two or three big broods in a good year and thus recover quickly from a low level. Although usually associated with running water, the species often hunts fish or invertebrates in meres, reservoirs or marl-pits, hence its vulnerability in frosty weather, although some birds survive by moving to the coast. Birds often feed well away from the nest. Those that nest in the banks of the River Bollin, for example, appear to feed mainly at marl-pits, and birds nesting in the banks of the Mersey and the Manchester Ship Canal feed in streams, ditches and ponds nearby. Coward & Oldham[2] recorded a nest at Rainow over a quarter of a mile from the nearest water.

River pollution is undoubtedly a key factor restricting presence along the Mersey for example, and some of our other rivers are not as clean as they might be, the lower reaches of the Weaver being an obvious example. In recent years the River Dean has deteriorated because of aviation spirit running off the airfield at Woodford, and, by the end of this survey, held fewer Kingfishers than a decade earlier. Other of our rivers, such as the Gowy, have low banks and consequently lack suitable nest-sites.

Kingfishers feed by perching on a branch,

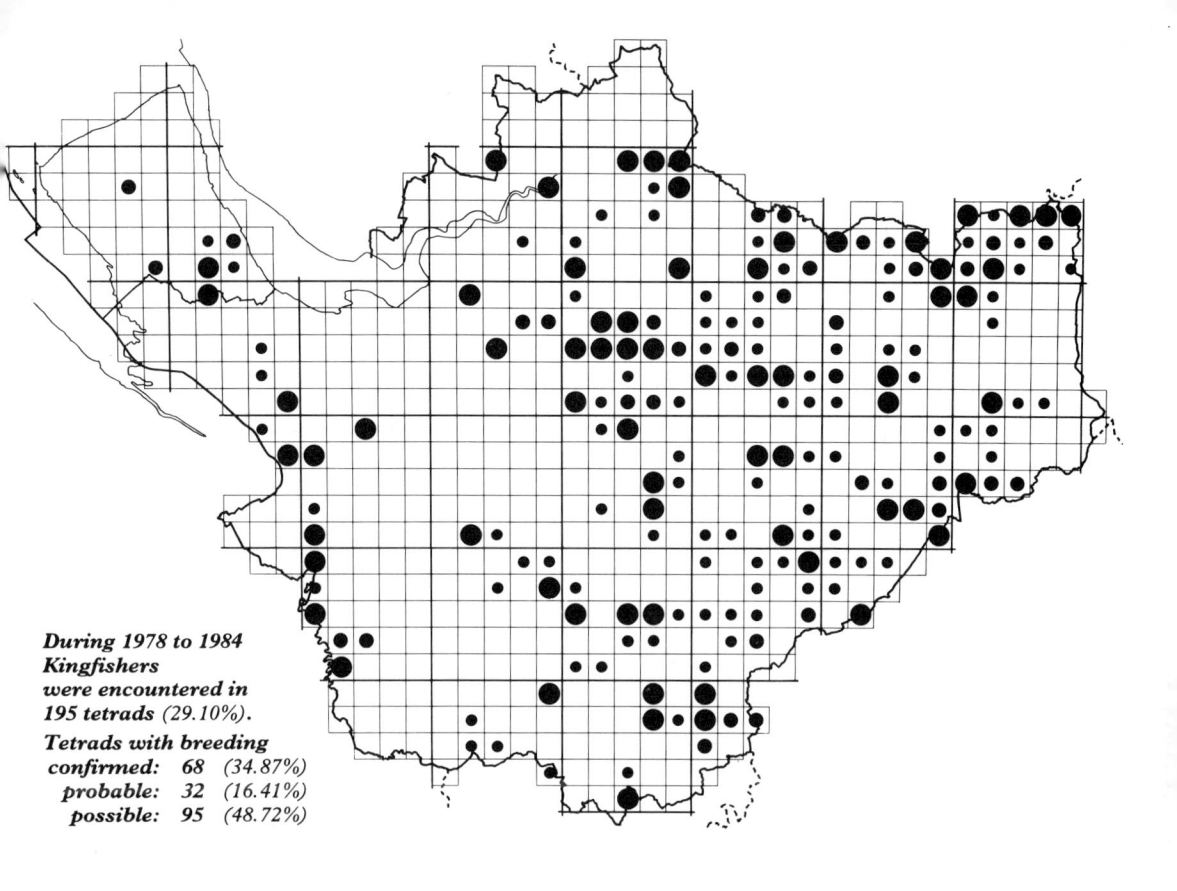

During 1978 to 1984 Kingfishers were encountered in 195 tetrads (29.10%).

Tetrads with breeding
confirmed:	68	(34.87%)
probable:	32	(16.41%)
possible:	95	(48.72%)

post, or even the end of a fishing rod, and wait for a suitable fish to come into view. They may hover occasionally also. Birds probably abandoned their nesting attempt at Woolston in 1983 when the trees alongside their feeding stream had their overhanging branches lopped, leaving no suitable fishing perches.

Kingfishers fly low and appear vulnerable to traffic where roads cross frequented brooks. Thus two birds were killed at Arclid in 1983. Windows are another frequent cause of death, particularly to young birds during post-juvenile dispersal. A further hazard is vandalism at the nest-site, a proportion of which are dug out. Any behaviour which might draw attention to the nest should be discouraged: the species is specially protected by Schedule 1 of the Wildlife and Countryside Act (1981).

The map may not depict the distribution entirely accurately, for a combination of reasons. The survey, running as it did over seven seasons, could exaggerate the picture slightly in that some birds may have bred in adjacent tetrads in different years. On the other hand the severe winters experienced during this period reduced numbers somewhat. It seems unlikely that many tetrads with confirmed breeding will more hold more than one pair of birds, although many of those with less satisfactory evidence will hold pairs: more systematic searching of rivers would probably have revealed more birds, as many stretches of riverbank are not accessible by public footpath. There were probably 100 to 150 pairs of Kingfishers in Cheshire during this survey, their population kept in check by winter mortality and the limited number of nest-sites adjacent to good fishing areas.

Hoopoe
Upupa epops

THIS is a passage migrant which has never been proved to breed in the county although there was an unsubstantiated record during this survey. Breeding is attempted in Britain less than annually[43].

Bell[7] stated that Hoopoes had been recorded about 18 times up to 1960, eight of the total occurring between 1951 and 1960. There were a further 35 records up until 1984 although it is possible that a few of these may have related to the same bird being seen at two localities in the same year. These records show an increased tendency towards occurrence in spring and early summer, probably relating to over-shooting individuals. One record is undated (early summer), but the rest fall as follows: March (1), April (9), May (11), June (5), July (3), August (2), September (3).

One at Willaston, Wirral in 1958 had been present for about ten days before its presence was reported on 26th April, and remained a further two days[18]. One again frequented Willaston during May 1968. One remained at Henbury Hall for about a week in late April and early May 1971.

During this survey there was a report of a pair nesting near Spital in Wirral (R. J. Raines pers. comm.). Initially a bird was seen feeding on a compost heap. Subsequently, two birds were reported and, eventually, a family party which frequented the edge of a strip of mature woodland. Dr Raines, who visited the site on two occasions in late June/early July, was shown a Jay on the first visit but on the second saw a Hoopoe which he considered to be a fully-fledged bird.

Wryneck
Jynx torquilla

IN the mid-nineteenth century the Wryneck bred throughout most of England and Wales, including Cheshire where it was considered to have been rare (Monk 1963). A steady decline, comparable to that of the Red-backed Shrike and Woodlark, then set in, and regular breeding in this country ceased in 1968, with only sporadic occurrences since (Taylor *et al.* 1981).

The only known breeding records for Cheshire are: Byerley[2] stated that it had bred at Saughall Massie; about 1884 a bird and seven eggs were taken from a nest in the trunk of an old poplar at Oakmere; in May 1925 a nest was found with four eggs at Upton, Macclesfield. Hardy[4] gives nests with young at: Thornton Hough in 1934, which was also listed by Monk (1963) and at Blakeway Nurseries, Bromborough on 16th June 1939. The 1971 *CBR* gives a second-hand, unconfirmed report of a pair nesting at Chelford.

Wrynecks are now known only as scarce passage migrants through Cheshire and Wirral, usually in autumn and probably en route from Scandinavian breeding grounds. However, a very few pairs, undoubtedly of Scandinavian origin, have bred in Scotland since 1969.

REFERENCES

Monk, J. F. (1963): The past and present status of the Wryneck in the British Isles. *Bird Study* 10: 112-132.

Taylor, D. W., Davenport, D.L. & Flegg, J. J. M. (1981): *The Birds of Kent*. Kent Ornithological Society, Meopham.

During 1978 to 1984 Green Woodpeckers were encountered in 235 tetrads (35.07%).

Tetrads with breeding
confirmed: 54 (22.98%)
probable: 69 (29.36%)
possible: 112 (47.66%)

Green Woodpecker
Picus viridis

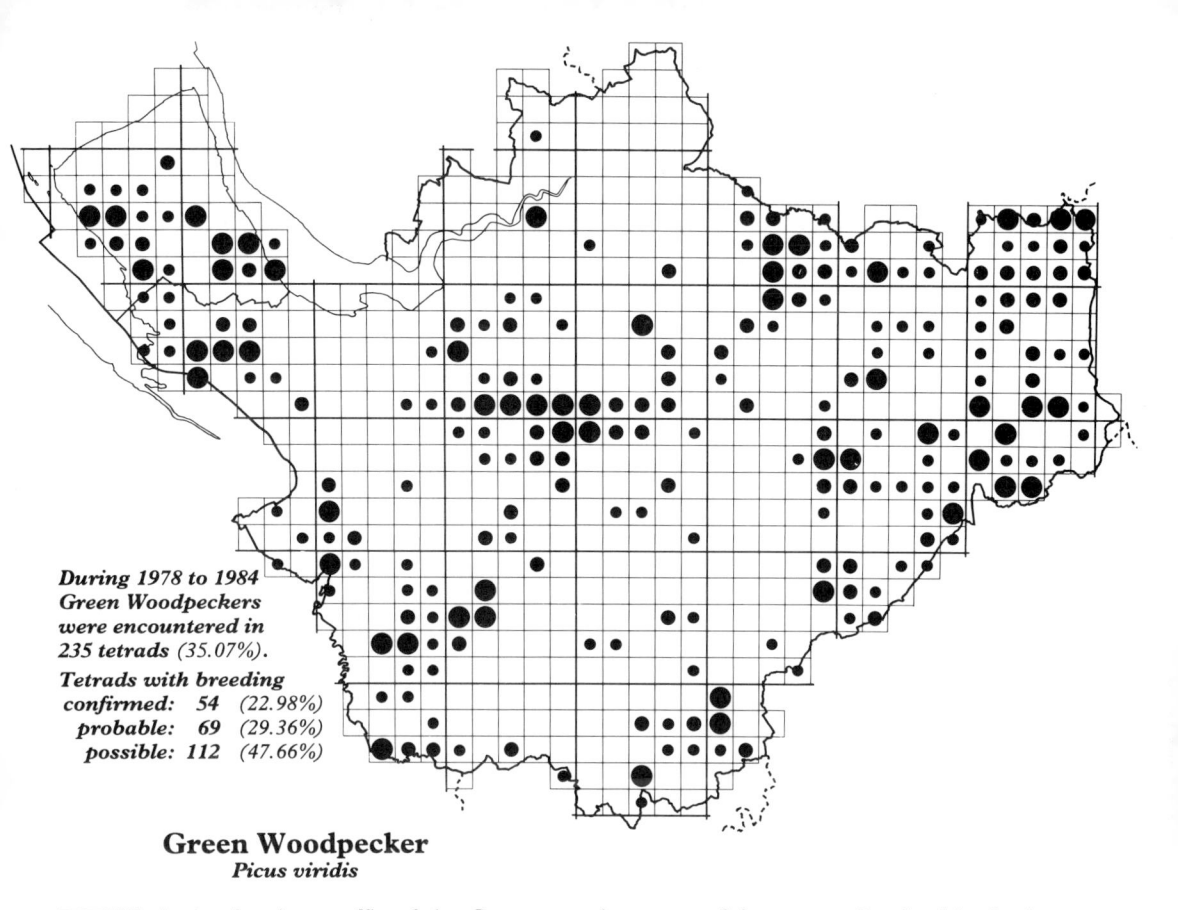

THE ringing laugh or yaffle of the Green Woodpecker, emanating from a distant stand of trees, is often the first indication of the species' presence unless a bird is surprised on the ground and watched bounding away, revealing a conspicuous yellow rump. Green Woodpeckers feed very largely on the ground, especially on ants, and, as these insects prefer areas with sparse vegetation or short turf, the birds' distribution follows closely that of permanent pasture, as in the eastern hills, or heathland as in Delamere and west Wirral. Parkland is important for the same reason, with several east Wirral sites in this habitat, and Tatton Park has long been regarded as a stronghold. The species becomes scarce towards the Mersey valley, and is very sparsely distributed in the counties of Greater Manchester[46] and Merseyside[35].

Birds have been recorded in autumn and winter at places such as Woolston and Frodsham Marsh, and regularly turn up in woodland sites as at Alderley Edge or Styal during the winter months when ants are less active and other invertebrates may be more important. Such sites may be occupied into the early spring, perhaps giving rise to 'possible' breeding records, but it appears that the nature of the surrounding land is of prime importance in the nesting season. Nests are always in trees, of course, but some territories, particularly in the eastern hills, are only sparsely wooded and even isolated trees are used on occasions.

In heathland or mossland areas the nest is usually excavated in an alder or birch tree; elsewhere the partially rotted wood of old hardwood trees such as oak is favoured. In Delamere, nests have been found in dead pines, although generally speaking such trees are cleared away by conscientious foresters.

Proof of breeding is difficult to obtain until the young hatch and call from the nest, or later when the scaly-plumaged youngsters can be seen nearby accompanying their parents. The Green Woodpecker does not commonly drum as do our other woodpeckers, though a bird in Delamere on 17th April 1982, which had been yaffling from the top of an alder in which it had its nest, gave a rapid series of sharp taps as the observer approached.

Its overall distribution appears to have changed little this century. Coward & Oldham[2] knew the Green Woodpecker as a local resident, plentiful in a few localities where old timber was abundant and scarce in Wirral although it had occasionally nested there. It was plentiful around Delamere, and many of the other sites they listed are still occupied from time to time. A few pairs then nested annually at Alderley.

Despite this apparent long-term stability, there have been marked fluctuations in the species' fortunes. Such fluctuations were often brought about by severe winters which interfere with feeding, and Coward & Oldham[2] considered the Green Woodpecker had "... undoubtedly suffered from the increase of the Starling". Abbott recorded his first sighting at Alderley Edge in 1923. In the Great Budworth area birds "...disappeared from many old haunts in the 1920s and all to be seen up to about 1936 were stragglers...In 1936 they returned to Tabley and have gradually found their way back to woods at Arley, Marbury and Holford Moss..." [6]. Hardy[4] noted that the Starling "often usurps occupied nests of Green Woodpecker at Oakmere". The *L&CFC Report* for 1949 stated that there had been "a quite remarkable extension of range..." with birds abundant in Delamere and the south of the county and cited records from a long list of sites including several in south Wirral. Bell[7] recorded a noticeable increase in Wirral in the 15-20 years up to 1960, and a similar extension eastwards into the wooded hills east of Macclesfield but the severe frosts of early 1963 caused a sudden crash in numbers. The species was still recovering at the start of the present survey (1978), when birds were again colonising the woods around Marbury (Great Budworth) and 21 sites were found to be occupied in the east of the county, predominantly in the foothills from Disley southwards to Wincle. The severe winter which followed set back the recovery, and in 1979 only eleven sites were tenanted in the same area[8]. It seems likely that the frosts of December 1981 were similarly catastrophic, although no comparative figures are available.

This survey probably gives a somewhat distorted picture of the bird's distribution. In 1978 the county's population was at its highest level for 20 years or more, and many of the mapped eastern records came from this year. Central and southern districts were generally not covered until 1979 or later, after the population had fallen in the intervening hard winter. However, on the basis of 125 tetrads holding birds probably or definitely breeding, with few squares holding more than one pair, the Green Woodpecker population of Cheshire and Wirral may have been around 150 pairs in 1978, falling to perhaps half this number by 1982.

Great Spotted Woodpecker
Dendrocopos major

THE "rattle" of a Great Spotted Woodpecker from a nearby wood is a familiar sound in most parts of the county where there is mature timber, but it was not always so. During the seventeenth and eighteenth centuries the bird bred as far north as the Scottish Highlands, but by the start of the nineteenth it had apparently disappeared from much of England north of Cheshire and Yorkshire[35]. Extensive coppicing and felling, and competition with the newly flourishing Starling were amongst reasons given. In Cheshire, the private estates with their plentiful supply of mature timber doubtless helped the species to survive in the county.

A resurgence occurred nationally in the late nineteenth century and although Coward & Oldham[2] devoted much attention to the woodpecker's struggles with Starlings, they could nonetheless describe it as a widely distributed if rather scarce resident, confined in the breeding season to those woods where it could find suitable nesting places.

In Wirral, breeding was almost unknown: a pair nested at Bromborough in 1860; four young were obtained at Hooton in 1865; and a pair was shot at New Brighton in 1887 – a further reason for the species' scarcity may be deduced here! Birds occurred throughout the year at Eaton Park and Delamere Forest, and breeding may have occurred at Edge (Malpas) in 1893. The fir woods at Alderley Edge were frequented, as were the parks at Alderley and elsewhere on the plain. The only reference to birds in the eastern hills were at Goyt's Bridge and Taxal (both now in Derbyshire).

Griffiths & Wilson[13] were aware of the scant nature of old records but added that the species had increased considerably in north Wirral over the ten years up to 1945 and the *L&CFC Report* for 1949 recorded that it had greatly increased in Wirral and was by then to be found in almost every woodland. Hardy[4] considered that this increase was due to the spread of birchwoods. Around Great Budworth, Boyd[6] found it by far the commonest woodpecker, occurring in almost every wood throughout the district, and he made no mention of a long-term increase, implying that the bird had long been common here. On moving from Delamere in 1952, G. H. Clegg found that the Great Spotted was much commoner than the Green around Nantwich, rather suggesting that their status in Delamere was the reverse. By 1962 the Great Spotted was the commonest woodpecker "for a greater part of the county...including the wooded valleys in the eastern hills"[7].

Subsequently it has remained our commonest woodpecker as this survey clearly shows. Though chiefly associated in summer with larger stands of broad-leaved or mixed woodland, birds will nest in dead conifer stumps and other trees around Delamere Forest, where they now outnumber the Green Woodpecker. In Macclesfield Forest the species is scarce, and is chiefly seen in winter carrying off spruce cones. The agricultural areas of the southern plain lack timber, although one nest was found in a small free-standing tree in a field hedgerow near Church Minshull and odd pairs may have been missed in the larger fox-coverts. There is a similar lack of woodland in the highest hills and to the north of the Mersey which would explain absences in these areas. However, Great Spotted Woodpeckers, unlike some other arboreal species, achieve a widespread distribution in the counties immediately to the north[35].

The distinctive loud "tchick" call is often the first indication of presence (but beware a similar softer call of breeding Lesser Spotted Woodpeckers). Birds may also be heard tapping on trees, to obtain insects or, more rarely, sap (although this has yet to be reported in Cheshire). In addition, drumming, a short series of rapid strikes on a branch or tree-trunk, is heard regularly from mid-February to the middle of May, occasionally in January and June, and exceptionally as early as November – two such records, from Handforth and Tatton in November 1976, were on foggy days (JPG). This early drumming may be associated with pair formation. Birds often travel in pairs during the winter months and excited chases may be seen at this season. Display is more frequent in spring however: on 19th April 1920, Abbott[10] recorded three birds in Windmill Wood, Alderley Edge "...making a great noise and chasing each other – probably courting." and on 20th April 1983 two pairs were disputing territory at Redesmere, the males chasing each other continuously for three-quarters of an hour or more. In June the far-carrying calls of hungry youngsters draw attention to many nests. Young have been recorded in the nest from 21st May to 14th July and fledging has taken place from 8th June. The young can fly before they are fully grown, and records of three-quarter-sized woodpeckers with red crowns occasionally cause identification problems!

Nests are often in birch, alder or pine, or in hardwoods such as oak or beech where the heartwood has been softened by fungal attack. A nearby bracket fungus of the Razor-strop on birch or Sulphur Polypore on oak sometimes forms a sheltering porch above the hole. Worn bark below the entrance hole is a useful sign that a nest has been recently occupied. Most holes are excavated at between ten and twenty feet from the ground, but one in a willow on the Eaton estate in 1967 was estimated to be 40 feet up. A nest at Alsager in 1982 was only one foot higher up a sycamore stump than an active nest of a Lesser Spotted – on several occasions a Great Spotted was seen to feed the Lesser Spotteds' young.

Outside the breeding season, when their diet includes more seeds and plant material, Great Spotted Woodpeckers will often move far from woodland, frequenting scrub, hedges and gardens. A juvenile ringed at Woolston on 29th July 1981 was found dead after cold weather on 27th December 1981 in Blackburn (Lancashire) town centre shopping precinct, 40 kilometres north. Late wandering birds in spring and dispersing juveniles in early autumn may account for many of the 'possible' breeding records. However, many occupied tetrads will hold three or four pairs. Six to eight pairs are usual along the Overleigh Drive (Eaton estate) and six pairs were found in Tatton Park on 28th April 1978. The Bidston Hill CBC plot (18.2 ha.) held from three to five pairs in every year of this survey. National CBC figures suggest that this species is about twice as numerous now (1984) as during the BTO Atlas survey period (1968-72). Allowing just three pairs for each tetrad with confirmed breeding and one for those with 'probable' breeding would give an estimated county population of some 900 pairs.

Lesser Spotted Woodpecker
Dendrocopos minor

THE Lesser Spotted Woodpecker is typically associated with river- or lake-side alder carrs where it excavates its nest holes in the soft timber. Mature woodland with suitable rotting trees is also frequented. Birds usually feed higher in the trees than the other woodpeckers, finding insects even amongst the thin twigs at the ends of branches. Coniferous trees appear unsuited to this feeding habit and it may be for this reason that such plantations are avoided.

The map shows this species to be widespread throughout much of rural Cheshire. It shuns the higher ground; in the

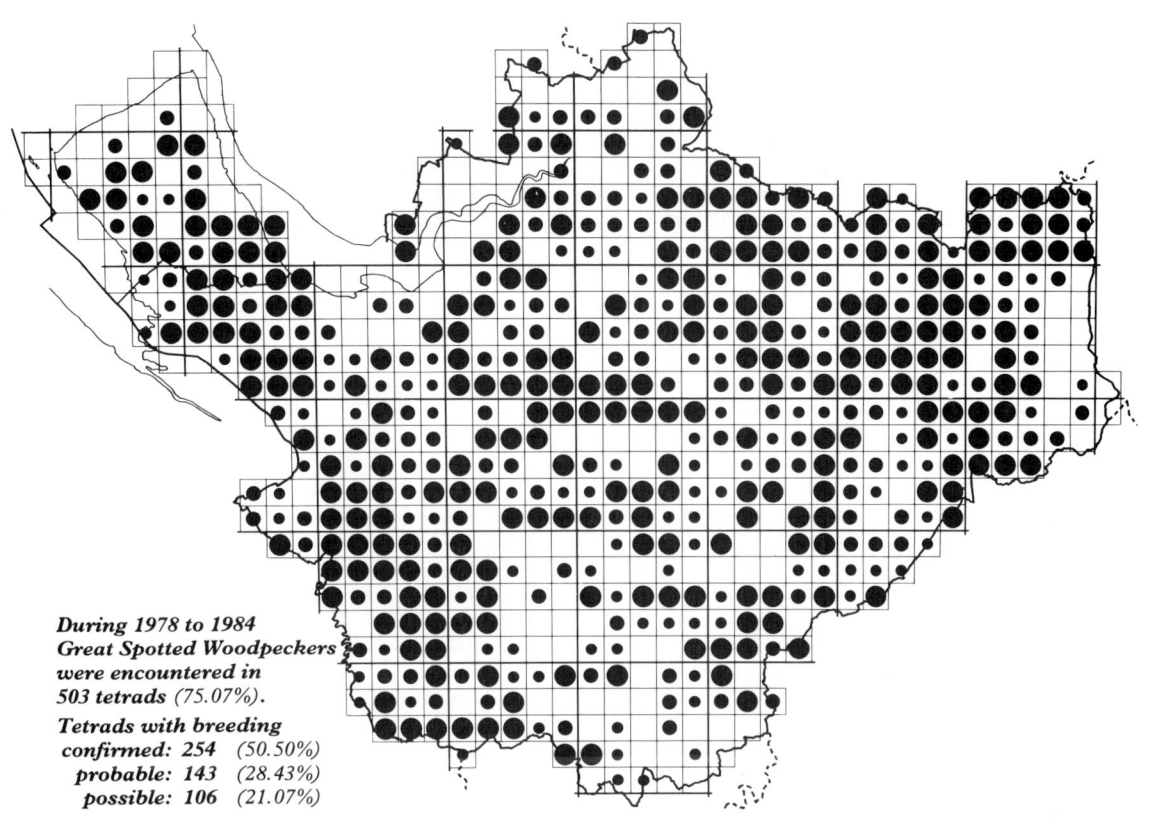

*During 1978 to 1984
Great Spotted Woodpeckers
were encountered in
503 tetrads (75.07%).*

Tetrads with breeding
  confirmed: 254 *(50.50%)*
   probable: 143 *(28.43%)*
   possible: 106 *(21.07%)*

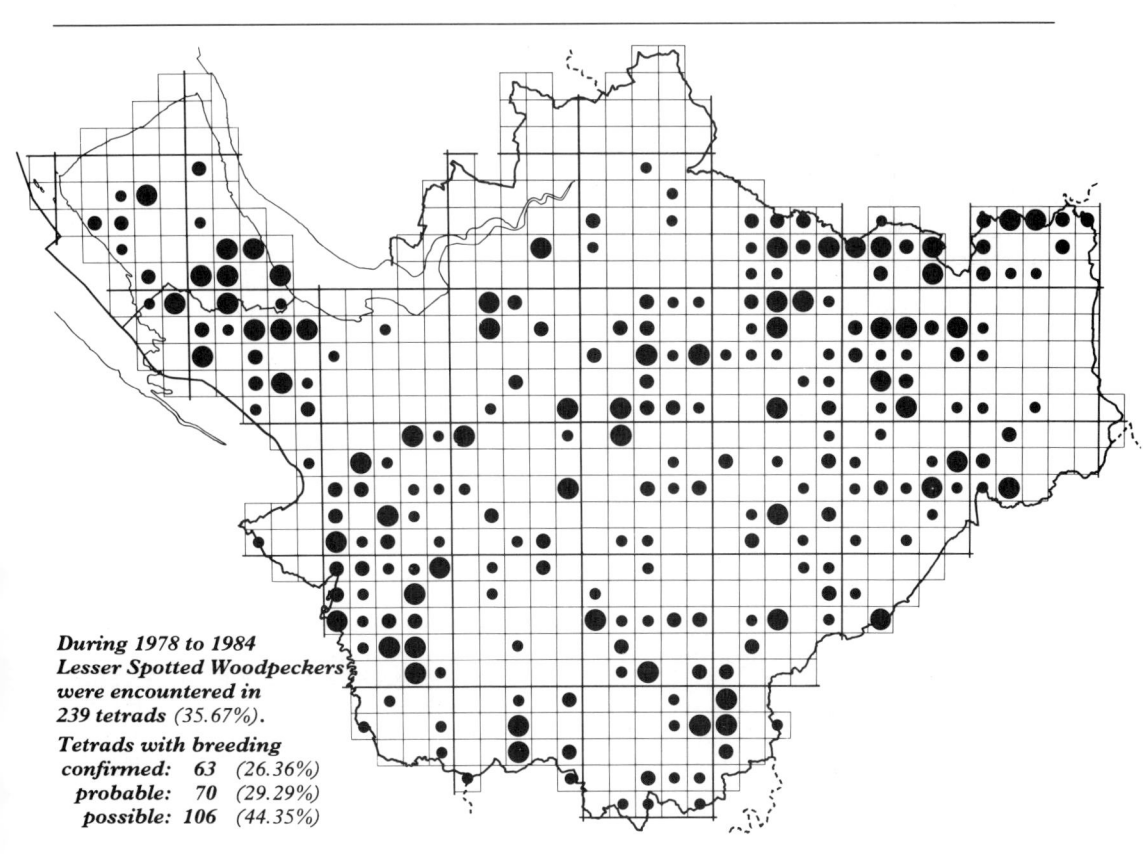

*During 1978 to 1984
Lesser Spotted Woodpeckers
were encountered in
239 tetrads (35.67%).*

Tetrads with breeding
  confirmed: 63 *(26.36%)*
   probable: 70 *(29.29%)*
   possible: 106 *(44.35%)*

eastern hills and the sandstone ridge of Peckforton and Delamere it is largely absent. Lesser Spotted Woodpeckers also seem to be absent from much of the lowest-lying land, including the north Wirral coast, the Gowy/Helsby/Frodsham marshes and the northern part of the county around the Mersey valley. They are missing from a large area from the Mersey north to the Ribble[35]. All the gaps in Cheshire are presumably attributable to a lack of suitable trees.

Because of their small size and their habit of feeding in the tree-tops, the birds are often overlooked unless calling, and the majority of records come from the period February to April when woodpeckers are displaying. Most records of drumming fall between mid-March and mid-May, although other reported dates include 27th December 1943 at Vale Royal[3], 28th November 1974 and 29th January 1983 when two were drumming near Tattenhall. Interestingly the Rev. Wolley-Dod, who knew the species to be plentiful at Edge, near Malpas, in 1896, stated that "it begins to rattle at the end of January or early February."[2]. It was not until the early 1930s that it was finally accepted that woodpecker drumming is not vocal, but a rapid series of tapping notes on trees: each burst of drumming, lasting for $1^1/_2$-2 seconds, may sound quite different, depending on the resonance of the boughs used. The normal call, a high-pitched ringing "pee-pee-pee-pee..." of seven or eight notes is heard especially in March and April, less often in February and May, with some resurgence in June and July as the young, which will also call, leave the nest. Autumnal calling is not too infrequent in late September and October. A remarkable butterfly-like display flight with birds flying up from the tree-tops and swooping back down is sometimes seen in early spring: on 25th April 1975 two males and a female were seen displaying in this manner in the Bollin valley at Mottram St Andrew.

The typical nest-site is in an alder or birch trunk between six and 15 feet off the ground, but nests have been found at greater heights in rotten branches of oak, sycamore and other trees. The calls of the young draw attention to many nests, and nesting adults have an anxious "chip" call, softer than the normal call of the Great Spotted Woodpecker. One frequent form of evidence during this survey came from recently excavated holes of characteristically small size (only likely to be confused with those of the Willow Tit) noted after leaf-fall in the autumn. Many nests are taken over by Starlings however, often within days of excavation, so discovery of a recent nest hole is not conclusive evidence of breeding. Indeed one January JPG watched a bird making a hole for a roost in a rotten ash stump.

Coward & Oldham[2] regarded this woodpecker as a scarce resident although recognising that it was much overlooked. Hardy[4] thought it to be widely distributed, but often overlooked, and listed seven sites in Wirral where it had nested in addition to Eaton Park and the Moore/Norton area. Griffiths & Wilson[13] knew of no breeding records in north Wirral until 1934 at Bidston and their only other recorded nest was at Storeton in 1939 although birds had been seen in several other places. Boyd[6] believed it to have increased a little during the 15 years up to 1951, but it was still by no means common around Great Budworth. This echoes a statement in the *L&CFC Report* for 1939-42 that it was increasing and more widespread than in the past, but Bell[7], writing in 1962, could find no evidence of continued increase "during the last 20 years". In the early 1970s, *CBRs* carried several statements of increasing abundance to the point where in some areas (as for example the Bollin valley at Mottram) it was almost as frequently encountered as the Great Spotted. Numbers seem to have declined somewhat since, but birds can still be surprisingly obtrusive in spring – JPG encountered eight birds in as many tetrads in ten days in April 1981! There were occasional records during the survey of two or three birds drumming or calling apparently in competition with each other, but most occupied tetrads will hold only one or two pairs and and the county population may lie between 150 and 250 pairs.

Woodlark
Lullula arborea

PARSLOW[34] outlined the history of the Woodlark in Britain as having "fluctuated over the last 100 years or more as markedly as perhaps any other British breeding bird". Sharrock[35] commented that it was said to breed in most counties of England and Wales during the nineteenth century but had ceased to do so in north-western England by the middle of that century.

Byerley, writing in 1854, described it as plentiful twenty years earlier on the authority of Mather, a Liverpool taxidermist, but "now never seen". Coward & Oldham[2] took this, and evidence of earlier presence in south Lancashire, to reinforce scant nineteenth-century records from Cheshire. In April 1859, Brockholes[2] saw a "wild unsettled bird" at Claughton, Birkenhead and in May 1861 there was a pair at Burton but he failed to find the nest. Lord de Tabley, writing in the mid-1860s, referred to the species as all but extinct, mentioning Tabley and Lower Peover as localities. In 1878, Sainter wrote that Woodlarks had recently been seen at Gawsworth where they used to breed 25 years earlier. Other evidence exists from Petty Pool, Alderley, Poynton, and Manley.

In 1934 one was singing at Spital, in Wirral, but subsequently there have been only seven fleeting records of passage migrants, the last in 1979. Nationally the Woodlark has declined such that from 1984 it has been included in the list of species for which the Rare Breeding Birds Panel collects data.

Skylark
Alauda arvensis

THE exuberant song of the Skylark as it climbs steeply in stages skyward has always been regarded as an inseparable part of the rural summer scene, whether over the coastal marshes, the hayfields of the plain, or the heather moors of the east.

Song is heard from mid- or late January until early July. Birds may sing again after the autumn moult from mid-September and through October whenever the weather is mild. At this season, southward-bound passage birds have been heard to sing as they go, casting doubts on the usual territorial explanation of song and suggesting sheer exuberance, although practice song by young birds may be a more likely explanation. Hardy[4] recorded song for the BTO's Bird Song Survey (1937-38) on 222 days, and in the following year's survey on 247 days. In severe winters song may not start until March, although one bird was seen hovering in song-flight at Sandbach Flashes in the teeth of an easterly wind with the ground frozen hard. Song at these flashes often includes phrases learnt from the local Little Ringed Plovers; Frodsham Marsh birds can imitate Redshanks, Lapwings and Meadow Pipits perfectly; and in the Bollin valley at Mottram St Andrew, Green Sandpipers have been mimicked.

Skylarks are one of the first birds to sing in the early morning, sometimes well before first light, although song may be heard at all times of day. Not surprisingly 'song' was the most frequent 'possible' breeding record, easily upgraded to 'territory' on subsequent visits to tetrads. Less frequently, display is observed on the ground or when the cock bird perches on a rock or low wall with his tail erect and crest raised. The nest is situated on the ground in a hay or cereal field, or in a tussock on rough ground such as moorland or coastal marsh. In agricultural areas, many records of confirmed breeding related to adults seen dropping into the middle of a field with a beakful of insect food, or rising from the nest with a faecal sac held in the bill. The game-bird-like cheeping of large young was a less frequent form of confirmation.

Although typically a bird of the countryside, Skylarks will also nest on urban waste-ground. Boyd noted a single pair on a stretch of rough ground by Seacombe Ferry in 1940, and in the mid-1970s birds were heard to sing regularly over a derelict patch by Handforth railway station. At Woolston up to

15 singing males have been counted around the banks of the sludge-pools. Hardy[4] noted that they were especially abundant on the coastal golf-courses and sand-dunes of Wirral, and up to four pairs nest annually on Hilbre[25]. In agricultural areas rotation pastures are preferred, with tall crops in which to nest and short turf on which to forage for invertebrates. The adults also take vegetable matter – Skylarks are a pest in parts of eastern England where they nibble the young seedlings of sugar-beet. Birds were recorded nipping off the buds from sprouting wheat at Little Leigh in 1942[5].

The BTO Atlas[35] showed this to be the most widely distributed bird species in Britain, and suggested an average of 500 to 1000 pairs in each 10-km square in the country, equivalent to 20 to 40 pairs per tetrad. Although some tetrads, as on Frodsham Marsh, hold more birds than this, such a figure on average seems excessive for Cheshire. Some idea of population density can be gained from the ease with which birds were recorded in some areas. During the present survey it proved difficult to locate any Skylarks at all in some tetrads in the south of the plain, and only a few pairs were present in many others. The well covered areas around Chester give further substance to this supposition of relative scarcity, and even in SJ87 and SJ98, two exhaustively worked 10-km squares, by no means all tetrads furnished confirmation of breeding.

These low numbers may in part be due to the effects of severe winters during the survey. Three CBC plots in south-eastern Cheshire held a total of ten territories in 1981 but only one in 1982 after the frosts of December 1981. The Risley Moss CBC held twelve pairs in 1978 but fell to seven pairs in 1979 and just three in 1980 as encroaching birch scrub rendered the habitat less suitable. The spread of house building has also reduced available habitat in some areas, notably in Wirral where the Birkenhead School Natural History Society recorded a decline from 15 pairs in 1963 to just six pairs in 1973 because of this. A factor of much greater potential significance however must be the change from hay to silage making in recent years. With only six weeks between cuts of grass in good growing seasons, and allowing half this time for sufficient cover to develop, there is little chance for larks to nest successfully, particularly in the south of the plain where arable crops are few.

This can be a difficult species to confirm breeding and most 'probable' breeding records will relate to nesting birds. The more suitable tetrads, including the eastern moors, may hold double figures of pairs, but a figure of five to ten may be a more realistic average for the county as a whole. This would give a total of 3000 to 6000 pairs.

Sand Martin
Riparia riparia

COWARD & Oldham knew the Sand Martin as "an abundant summer resident...met with in all parts of the county, frequenting the meres and rivers in large numbers". They referred to springtime roosts of up to a thousand and "incalculable thousands" in early autumn. Such figures are unthinkable nowadays since the Sand Martin has declined to the status of one of our scarcer summer visitors.

It is difficult to tell when the decline started. Boyd (1946) mentioned only spring arrivals and autumn flocks, with no hint that the species' abundance had changed. Bell (1962) suggested that availability of nest-sites governed distribution, but published no

During 1978 to 1984 Skylarks were encountered in 631 tetrads (94.18%).

Tetrads with breeding
confirmed: 321 (50.87%)
 probable: 269 (42.63%)
 possible: 41   (6.50%)

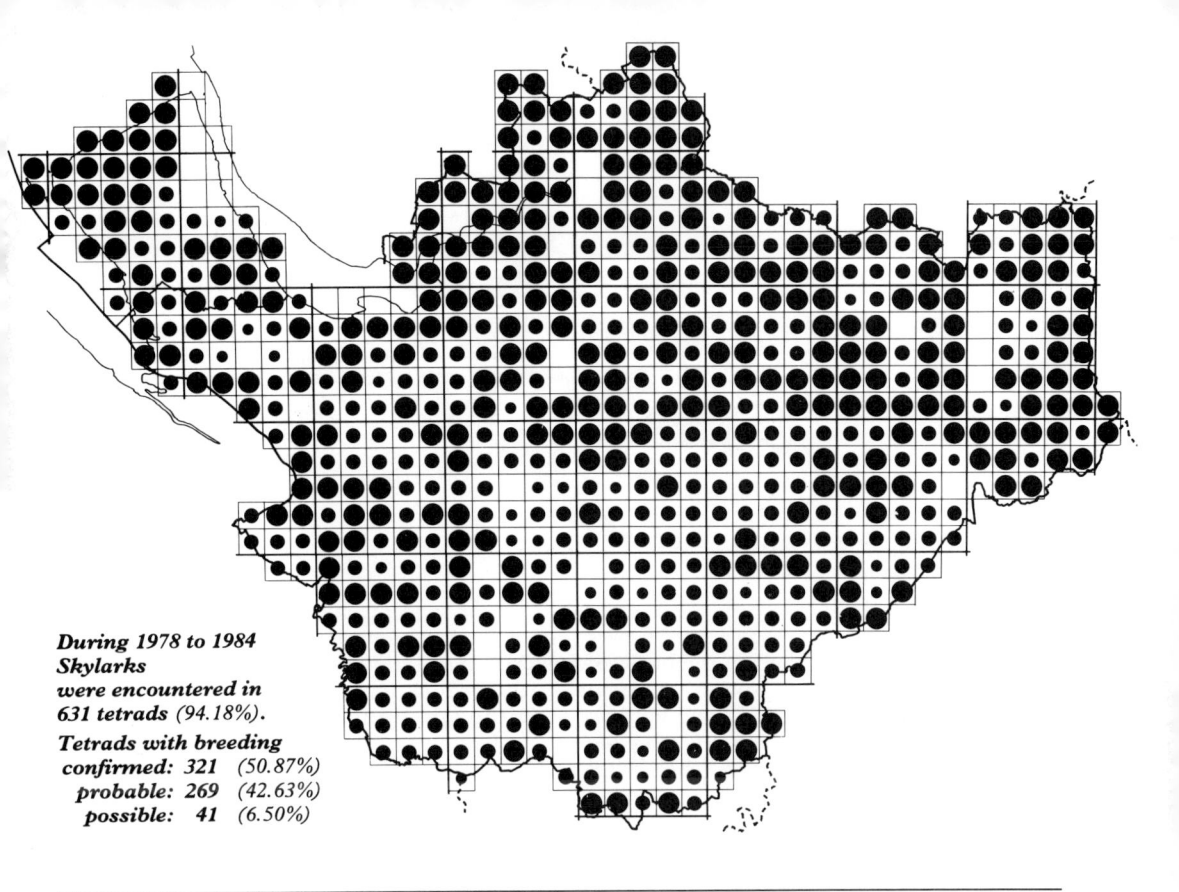

quantitative data. As the Sahelian drought intensified in the late 1960s the numbers of several trans-Saharan migrants fell and the spring of 1969 brought a marked reduction of Sand Martin numbers in Britain, which has been well documented nationally (Winstanley et al. 1974). Unfortunately the species' fortunes were ignored by Cheshire and Wirral recorders, with no mention of breeding colonies until the 1975 *CBR* when 97 holes at Shakerley were thought to constitute the largest colony in the county. No more counts of major colonies were published until 1978 when this survey stimulated some interest. By 1982 almost all breeding concentrations were being censused, with a county population that year of 800 – 1000 pairs. In 1983 they had another good breeding season, with several ringers studying the species and one (DN) handling over 1300 birds including 419 breeding adults. However, recurrent drought in the southern Sahara led to another collapse in winter 1983/84, with the 1984 national breeding population estimated at only 30% of the previous year's figure. By then the British total stood at less than one tenth of that 20 years earlier, and at the end of the survey period the Cheshire breeding population was as little as 300 pairs, the lowest ever recorded.

Sand Martins dig their own burrows in near-vertical banks. They are quite selective, being particularly sensitive to humidity, usually avoiding soil so damp that nests might be flooded and also staying clear of faces so dry that they might collapse. As well as the more orthodox sand or gravel faces, birds also use the softer strata of red sandstone in quarries and railway cuttings. In 1892 some 60 pairs tunnelled into the sandstone of the Manchester Ship Canal cutting at Ince, and at the turn of the century a considerable number of pairs nested in the rock wall of the canal at Lachford whilst at Lymm several pairs burrowed into crevices between the sandstone blocks of the dam wall[2].

Relatively few birds nowadays nest in river banks in the county: a colony in the banks of the Bollin at Mottram fell from 85 nests in 1970 to 40 in 1971, 19 in 1972 and only six by 1974, this reduction perhaps reflecting the Sahelian drought although the bank repeatedly collapsed during spates following spring storms[21]. In 1966 a pair bred in a disused Kingfisher hole at Poynton and extruding field drains are occasionally occupied. Sand Martins find almost all of their food over water, so the many flooded sand-quarries, particularly those still worked by pumping slurried sand, are ideal. The vast majority of our birds nest in the vertical faces of such quarries, freshly exposed faces being preferred with holes often following the lines of particularly favoured strata. In general colonies are protected by quarry personnel, with sand extraction being timed to avoid disturbance in the breeding season, but there have been a few reports of colonies being bulldozed. The rapid exploitation of huge, modern quarries contrasts with the old sand-pit at Gib Hill, Anderton whose owner remembered the site being occupied by martins throughout his life, the face being dug away each autumn[6].

Despite the occasional hazards from quarrying, birds almost always prefer the actively-worked sites. Worked-out quarries are often graded and grassed over, usually to comply with planning regulations, but even if left alone they soon become unsuitable for Sand Martins as faces crumble or are colonised by vegetation. However with the present commercial demand for sand there is no shortage of nest-sites. The main hazards facing the birds in this country come from the weather and predators. Sand Martins are one of the first summer migrants to return in late March or early April, and cold springs will see them flocking at the few insect-rich sites although doubtless some succumb to starvation at this time. Conversely, hot, dry summers can cause havoc when nesting faces become liable to collapse; however most birds have raised the first brood – often flying by early June – before temperatures reach their peak.

The major natural predation appears to be from foxes, who plunder nests by climbing all but the sheerest faces, or dig into the ground above a colony to reach nests built too near the top of the bank. Sparrowhawks are important predators of adult birds whilst Little Owls, Kestrels, Magpies and Crows have all been recorded trying to snatch fledglings from the entrance to their burrows. A much worse problem in some areas is human vandals, who have destroyed some colonies and caused substantial damage at others.

Ringing at Cheshire colonies shows the well-known tendency of adults to return to previous years' breeding sites, with first-year birds often settling at more distant colonies. Juveniles wander widely in their first autumn, even moving north by as much as 100 kilometres in a day and recalling Boyd's (1946) observation of thousands going north or north-west on 15th August 1939. These gregarious birds may keep together on their migrations, like the five juveniles caught roosting at Llangorse Lake, Powys on 22nd July 1983, one of which had been ringed at Fourways sand-quarry on 3rd July, three at Beeston sand-quarry on 4th July and the fifth at Nether Alderley on 7th July.

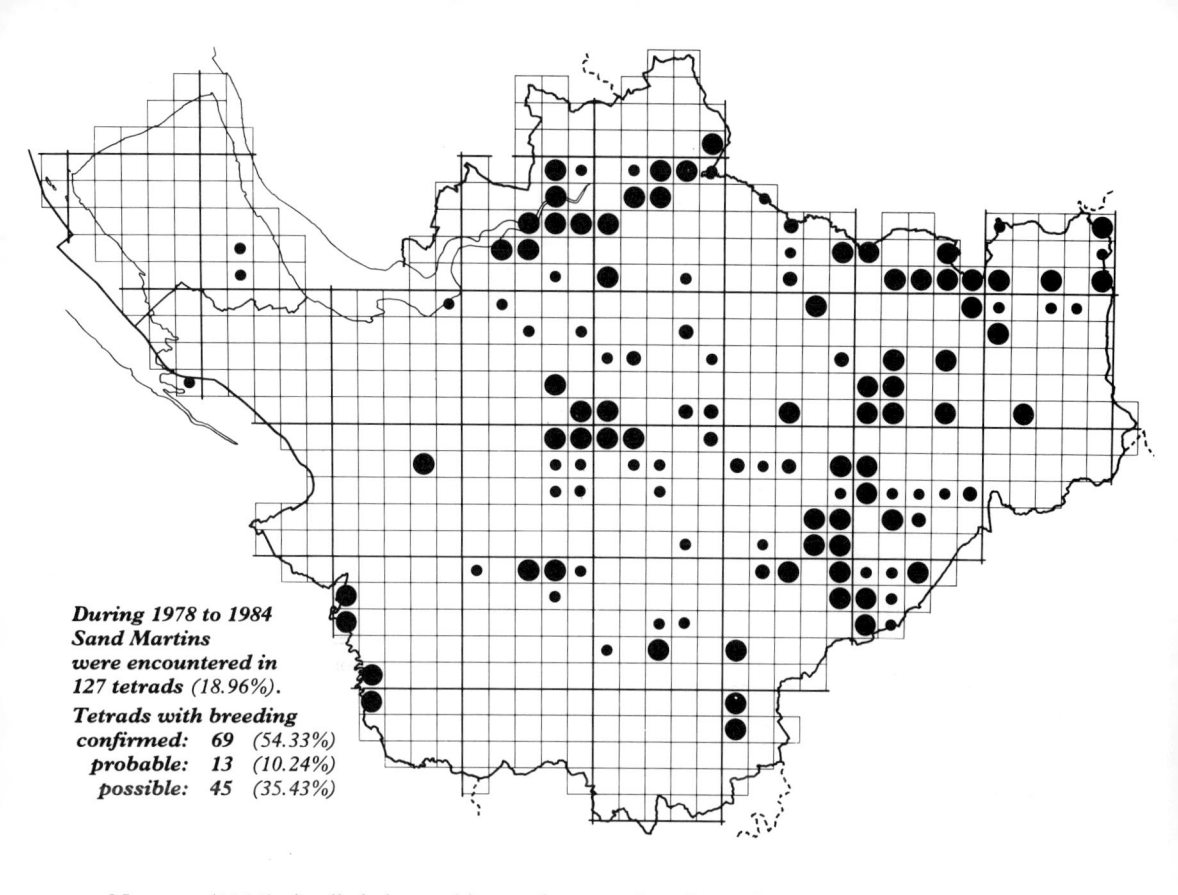

During 1978 to 1984 Sand Martins were encountered in 127 tetrads (18.96%).

Tetrads with breeding
confirmed:	69	(54.33%)
probable:	13	(10.24%)
possible:	45	(35.43%)

Norman (1982) detailed the problems of estimating Sand Martin populations. Old holes from previous years may be used again; many trial borings may be made; one entrance may lead to more than one nest-chamber or vice-versa; birds are highly gregarious and will visit several nearby colonies, up to ten kilometres from the nest-site, and may well all congregate at one sand-quarry and appear to desert the others for periods during the day. The breeding season is long and staggered and varying proportions of females and juveniles will be seen at different times of the season.

However, although it may be difficult to quantify their population, it is easy to confirm that Sand Martins are breeding, with 'B' and 'N' records regarded as confirmed breeding for this survey. The map shows clearly the correlation of colonies with sand-quarries, as well as lines of mostly smaller colonies in the banks of the Mersey and the Manchester Ship Canal, and the rivers Dee, Bollin and Dane. Not all rivers are readily accessible, and a few small colonies in the banks of the Weaver, for example, may have been missed. It is not known whether birds have ever been common in the Wirral peninsula. Most of the sandstone there is so hard that burrowing is well-nigh impossible, although there are a few softer patches. The mapped 'possible' breeding records are unlikely to be breeding individuals and may well relate to birds on passage – some adults are known to wander widely even at the height of the breeding season – or prospecting juveniles. The map represents seven years' data and severely overstates the situation at the end of the survey, when only about one-third of the mapped tetrads were known to be occupied.

DN

REFERENCES

Norman, D. (1982): Some observations at Sand Martin colonies; censusing difficulties highlighted. *CBR*, pp77-78.
Winstanley, D., Spencer, R. and Williamson, K. (1974): Where have all the Whitethroats gone? *Bird Study* 21: 1-14.

Swallow
Hirundo rustica

ALTHOUGH the first isolated Swallows may arrive in Cheshire around the end of March, it is normally the second or third week in April before they are here in any numbers. Boyd[6] noted that birds often paid a brief visit to nesting sites on first arriving, before rejoining the flocks feeding over the meres. However, behaviour at this season depends largely on the weather. In 1981, for example, records became widespread from 8th April as birds returned straight to breeding-sites during warm, sunny weather. In late April the weather turned cold with snow showers and birds were forced to leave their intended nest-sites and congregate in areas where food was less scarce. It seems likely that many birds starved.

Such incidents are not without precedent, the aerial feeding habits of the species rendering it particularly vulnerable to adverse weather conditions. Coward & Oldham[2] noted that in late springs Swallows often take refuge from severe weather in buildings and outhouses where many die from cold and hunger, and that in May 1886 hundreds perished in Cheshire. Prolonged cold, wet weather in May or even June will cause birds to desert their nests for a time. On 3rd June 1936, in bitterly cold weather with almost continuous rain, Boyd[5] found an early brood cold and almost dying, their parents having flown to some mere to seek flies to keep themselves alive. On 28th May 1983, during cold drizzly weather, one of a nesting pair at Lamaload was watched hovering to pick insects off low vegetation (JPG).

Perhaps because of this vulnerability to vagaries of the weather, Swallows are not early nesters. Boyd[6], who ringed several thousand nestlings, knew of no eggs laid before 7th or 8th May and found that many birds do not start to lay until the end of that month or even June. Hardy[4] refers to the work of W. Ritson around Warrington during the national Swallow survey of 1934 and 1935. In both years Ritson had the "earliest nesting date" in the country, 28th and 29th April respectively, with young hatching on 11th and 12th May. It is quite normal for two broods to be reared, and three is by no means exceptional. Young from third broods may hatch in August, but others are still in the nest in October. In 1978 a pair was still feeding young in a nest in a pig hut by Elton Hall Flash on 5th October. On 1st October they resisted any temptation to join the steady southward passage which took several hundred Swallows over their nest-site. Bell[7] gave an exceptional record of two young still in a nest at Poynton on 27th October 1937.

The mud-built nest is typically saucer-shaped, and supported on a ledge or beam, or at least some slight projection, such as a nail, on a vertical wall. Occasionally nests are built without any support whatsoever, resembling those of House Martins but with no attachment to the roof above. Boyd referred to nests on a plate suspended by wires in a shippon, occupied in three successive years, and another in a tun-dish hung in a shed. His extensive notes showed that nearly half of all second broods are reared in the same nest as the first, although one pair reared three broods in three different nests. Old nests are often renovated and may be used for several years, with perhaps an alternative nest for the second brood. Inevitably a few nests do crack and fall, especially in hot summers: five young, whose nest in a house porch collapsed, were placed in a plant-pot on a plank, where their parents successfully reared them.

Of 193 nests noted by Boyd[6] in 1934 and 1935, 143 (74%) were in buildings housing animals, a further 27 (14%) in fowl houses, and 23 (12%) in unoccupied buildings or dwelling houses. Dairy or pig farms are particularly favoured and the main preference is where cattle are milked, calves are penned or bulls are billeted. There is usually an

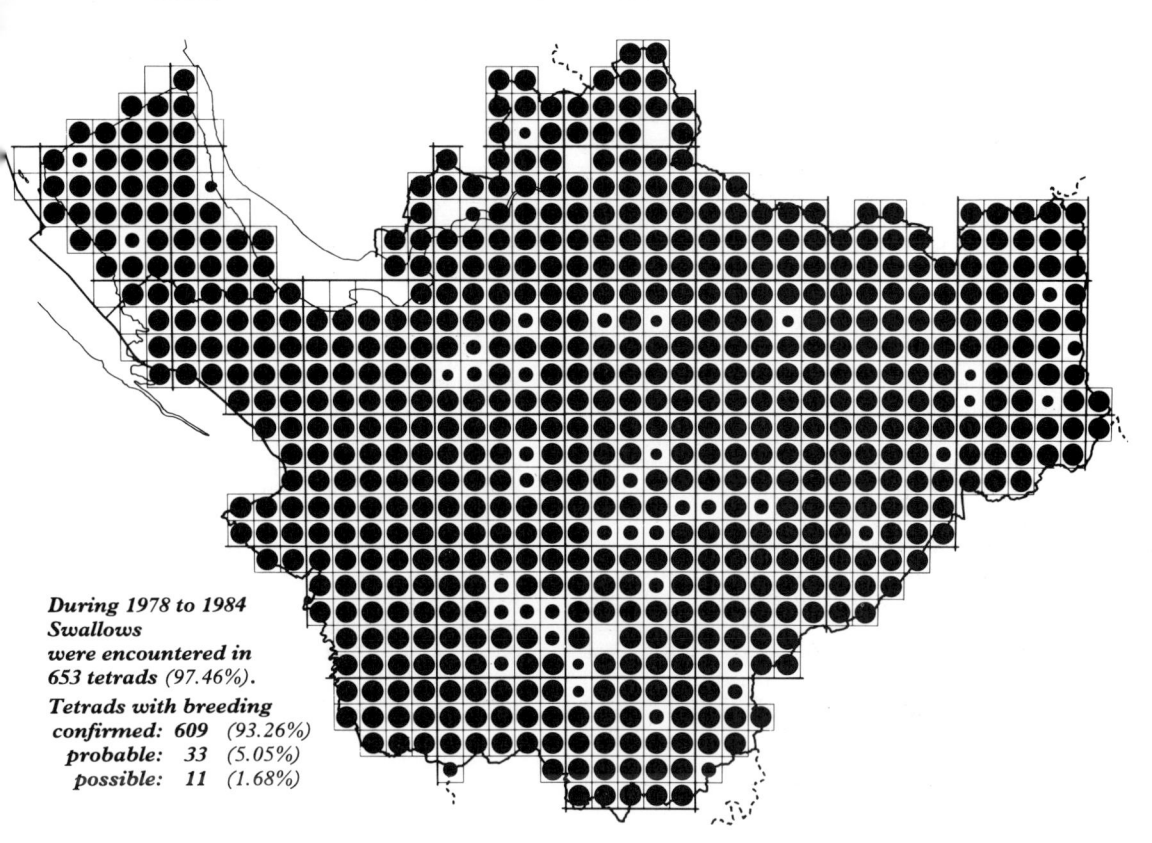

During 1978 to 1984 Swallows were encountered in 653 tetrads (97.46%).

Tetrads with breeding
 confirmed: 609 *(93.26%)*
 probable: 33 *(5.05%)*
 possible: 11 *(1.68%)*

abundance of insects associated with dairy farms as well as mud for nest building. Conversely, stables are largely ignored or avoided. Nests in garden sheds, garages and house porches are not unusual, and church porches are used in more rural areas. In July 1967, a pair was feeding young in an unusual site under the eaves of a house at Ettiley Heath, Sandbach[8]. Since the last century there have been occasional reports of Swallows building in culverts or pipes at little above water-level and in caves. The species nests annually on girders beneath bridges at Sandbach, Frodsham Marsh and no doubt elsewhere.

Boyd[6] suggested that the population depends on the number of available nesting sites. Certainly many sites have been occupied for many decades, with actual nests often outliving one generation and being taken over by the next. At Rostherne a pair was known to have nested in the boathouse since the early years of this century[24], and continued to do so until 1981, although none appeared in 1982 or 1983.

There is much information on clutch and brood sizes from the surveys of 1934 and 1935. Ritson's 62 broods around Warrington averaged 4.1 young[4], and 201 broods around Antrobus averaged 3.98 young[6]. Boyd regarded five eggs as a normal clutch, with six not unusual, first broods being the largest. Clutches of nine and seven eggs were considered the product of two hens. Such large sets of eggs often resulted in no more than a normal brood.

The faithfulness of Swallows to one nest-site in successive years is well known, but really applies only to adult birds, although year-old birds often return to within a few miles of their birthplace. Thus a nestling ringed in August 1932 nested half a mile away in 1934 and in the same shed again in 1935. A nestling ringed at Rostherne in July 1975 was controlled in 1976 as a nesting female at Pott Shrigley, Macclesfield, 20 kilometres away from its place of ringing. The oldest bird ringed by Boyd survived for at least four years.

White Swallows are hatched from time to time. In 1867, an albino was reared in a nest at Gayton with three normally plumaged birds[2]. More recently a young albino was seen at Ashley in August 1971[21] and in 1977 a pure albino was reared by normally plumaged parents at Long Lane Farm, Peover[22]. In July of the same year a white bird with some grey on the forehead and rump was present at Crabmill Farm, Sandbach[21], similar birds having been noted in Wirral in 1966[8].

The Swallow was one of the easiest species to locate and to confirm as breeding during the present survey. The twittering song draws attention to birds overhead at any time from their arrival in spring right up to autumn when the birds gather in communal roosts. Often birds will sing from wires outside the building which houses the nest, song even starting well before dawn, although few field-workers were out at this hour. Once young are in the nest, the adult birds may be watched swooping back through a doorway or broken window pane to feed their offspring. Once these have fledged they may be seen waiting to be fed, perched on wires or the roof of some outbuilding, or even flying up to take food from their parents in flight.

The distribution of the Swallow in Cheshire has changed little since Coward & Oldham[2] found it to be universally distributed, "hawking for insects on the bleak hilltops as commonly as in the cultivated plain". Hardy[6] regarded the Swallow as a common nester in suburban areas and stated that it nested occasionally on Hilbre Island. Craggs[22] mentions no recent records of its doing so however. There has also been some withdrawal from the urban fringe in recent decades. As early as 1945 Griffiths & Wilson thought numbers had decreased in north Wirral. The main gaps on the survey map come from the east Wirral conurbation and the centres of Widnes and Warrington.

Boyd's census of the Swallow population in the townships of Antrobus and Sevenoaks (2717 acres) revealed 88-90 pairs in 1934, i.e. 33 pairs per tetrad, and 84-85 pairs in 1935 (30 pairs per tetrad). In 1978 CBC work in south-east Cheshire found some 37 pairs in 393 hectares. Some farms support many pairs: in 1977, 120 young were ringed at Boothbed Farm, Goostrey, and a total of 1400+ within the two 10-km squares SJ77 and SJ78. The best farms may hold nine or more pairs although three to four is more usual. In all, the mapped area probably holds some 15,000 – 20,000 pairs.

House Martin
Delichon urbica

HOUSE Martins arrive back from their winter quarters during April or May, and, while birds often pay a fleeting visit to their old nest-site as soon as they return, they do not usually start nesting until two or three weeks later. Breeding may then continue until September and it is not unusual for young birds still to be in the nest in October, though these may be abandoned by their parents if the food supply dwindles with the onset of cold weather. In 1974, young were still in a nest in Macclesfield on 14th October, the same date as two broods in Chester in 1980, and, in 1979, well-grown young were still being fed at Norley on 15th. Griffiths & Wilson[13] had a similar record for 16th October.

House Martins feed exclusively on flying insects, often at greater altitudes than the other hirundines. Consequently, their distribution is less closely influenced by land use: they are just as likely to occur over suburban areas as over open country. Clean air regulations have had the effect of spreading pollutants more thinly across town and country alike and, with extensive tree planting in some towns, insect populations there must have increased. In 1980 a single pair of House Martins bred in Warrington town centre for the first time in many years and by 1982 ten pairs were present at the same site.

The mud-built nest is typically constructed under the eaves of buildings, although any overhanging ledge may be used. Boyd noted over 100 nests on the Dutton railway viaduct (SJ57) in 1943 and, in 1971, 33 nests were built under a new bridge over the M56 motorway at Runcorn. House Martins in many areas seem to prefer to nest on newer houses, provided that they have suitable eaves. All recorded nests in Cheshire have been in man-made sites although in the early 1970s a pair nested at the top of a quarry face at Pott Shrigley under an overhang of turf, and in 1980 birds were seen prospecting a similar overhang in a sand-quarry at Nether Alderley (JPG). Birds are fairly often found visiting Sand Martin colonies, where some may be seen clinging to

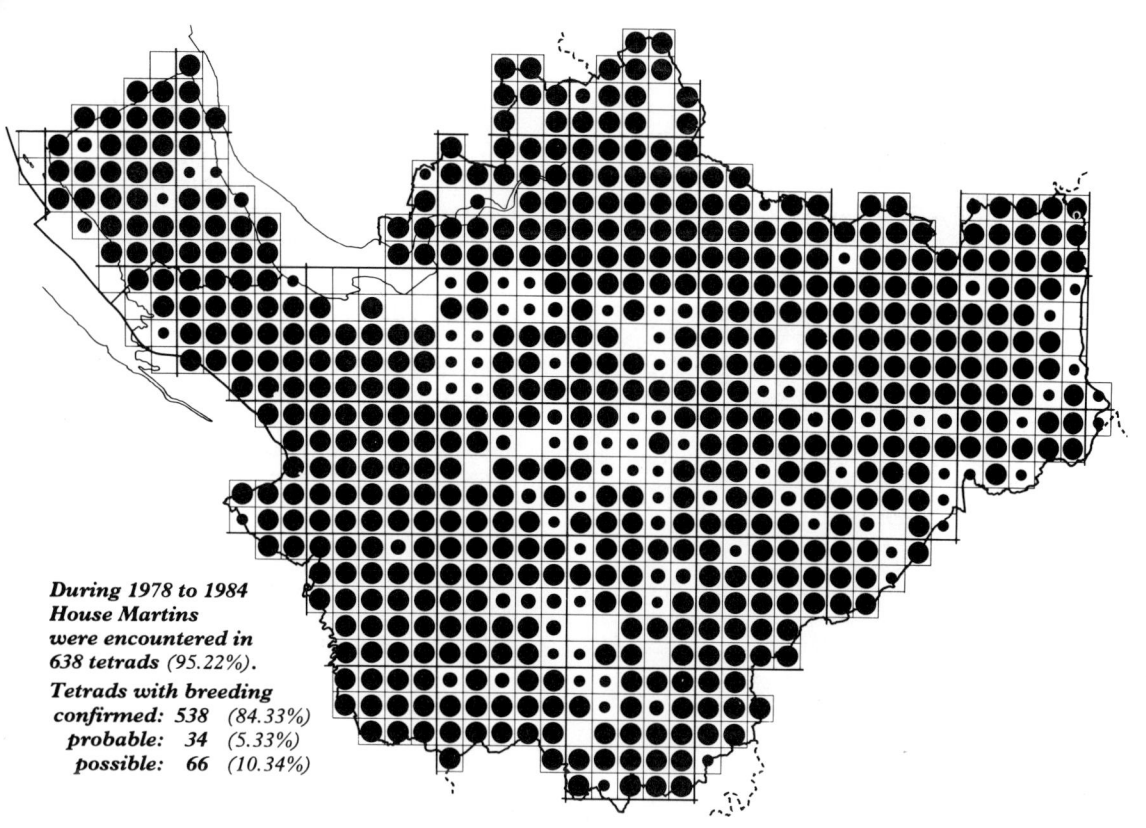

During 1978 to 1984 House Martins were encountered in 638 tetrads (95.22%).

Tetrads with breeding
confirmed: 538 *(84.33%)*
probable: 34 *(5.33%)*
possible: 66 *(10.34%)*

the quarry face, possibly assessing potential nest-sites of the above type. Boyd often found nests inside dutch barns where the heat from the metal roof was apt to cause mud nests to crack and fall. No such nests have been reported since, perhaps because such barns have become obsolete. In the early part of this survey DE found many nests inside open-fronted farm buildings in the Knutsford/Goostrey area. In some instances there were up to 50 or more nests although many were apparently not in use. One of the Warrington nests in 1982 was extraordinary in being built inside the roof-space of an office building where it appears that two broods were raised (Martin 1982).

The availability of mud for constructing nests is seldom a restricting factor given the Cheshire climate. Nevertheless artificial nests of cement and sawdust are regularly occupied and at Rostherne birds have also been encouraged to breed in nests made from coconut shells. In view of the fact that nests from previous years are often renovated, this readiness to adopt artificial substitutes is perhaps not surprising. Boyd had a note of one nest built on top of an old Swallow nest, and another on a wall below the eaves which had evidently been started too low down – the nest was completely rounded with a bottle-shaped entrance pointing upwards. Nests are occasionally taken over by House Sparrows, but a greater hazard is the all-too-frequent tendency of "tidy"-minded householders to break down nests built on their walls.

Birds will feed at considerable distances from the nest, particularly during wet or windy conditions when flocks may gather in the lee of trees to snap up insects dislodged by the wind, and several hundreds may often be found at suitable sites such as Woolston or the Weaver Bend. Thus a proportion of 'possible' breeding records during the survey will have been of wandering birds, perhaps especially including first-brood juveniles. The chick from Rostherne (7th August 1981) whose ring was found in a Sparrowhawk pellet at Royden Park, Wirral (50 km W) on 30th June 1982 may give some indication of the extent of the movements within the county. It was generally easy to find some confirmation of breeding however, either by finding the conspicuous nests or by watching birds collecting mud, or gathering grasses and feathers to line the nest. Some birds may fly a mile or more to collect suitable mud, and some 'B' records may refer to adjacent tetrads.

This survey showed the House Martin to be our third most widespread summer visitor (after the Swallow and Willow Warbler). Martins are missing from the lowest-lying marsh areas and the highest eastern hills, which are devoid of suitable nest-sites. However, it is not easy to explain the absence of breeding House Martins from other tetrads such as Birkenhead, the west Delamere hills and a scattering of squares in south-east Cheshire. Perhaps the most intriguing, and inexplicable, absences are those in a large number of tetrads closely following the River Weaver through the centre of the county. A detailed study of this area, to attempt to discover why some areas are occupied whilst adjacent ones are avoided, could be very instructive.

The species is often colonial, and the number of pairs nesting in any tetrad may vary from year to year. At Alderley Park where counts of nests have been made since 1978, the following totals are available:

	1978	*1979*	*1980*	*1981*	*1982*	*1983*	*1984*
No. nests	55	59	61	81	81	85	51

Some rural tetrads held only single pairs but the county average is probably ten to 15 pairs per tetrad with confirmed breeding, giving a Cheshire total in the range 5000-8000 pairs. This species is not covered by any of the standard BTO breeding surveys, and there is no information on any short- or long-term population changes.

REFERENCE
  Martin, B. (1982): Unusual nesting site of House Martins. *CBR* p78.

Tree Pipit
Anthus trivialis

IN the eastern hills the Tree Pipit still nests in good numbers along lightly wooded slopes or the edges of woodland. At Alderley Woods, an outlier of the Pennines, a few pairs breed annually in glades kept open by the tramping of human visitors. There are small populations on relict mosslands in the plain as at Lindow and Risley, but otherwise most birds are restricted to the sandstone ridge from the Delamere Forest, where many pairs nest in the young plantations, down through Peckforton to Bickerton Hill.

Coward & Oldham[2] described it as an abundant summer resident, generally distributed throughout Wirral and the plain. It was, then as now, absent from the bare moorlands but occurred freely on the hillsides wherever there were plantations. It was particularly plentiful in many of the parks, and in those parts of Delamere Forest where oaks then predominated the Tree Pipit and Wood Warbler outnumbered any other species.

The diaries of Norman Abbott, and in particular his maps of singing summer visitors in the Wilmslow area, compiled in 1919, 1921 and 1922, show just how abundant the species then was. His study area of 22 square kilometres contained 58 or 59 singing males, many of these on farmland. Boyd remembered regular occurrences around Frandley until about 1930, birds having sung from hedgerow trees, and he remarked on a general decrease since then which he suspected was due to agricultural changes. His list of breeding haunts in the Northwich

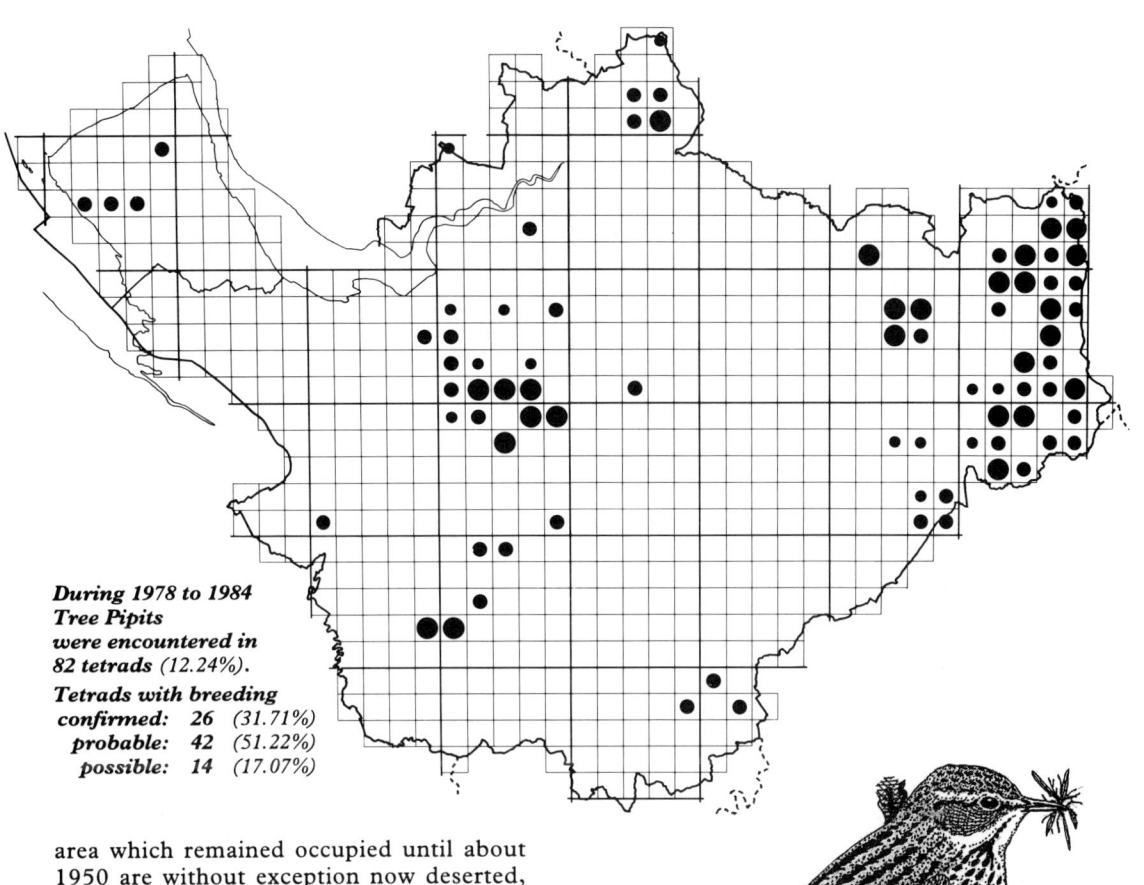

During 1978 to 1984 Tree Pipits were encountered in 82 tetrads (12.24%).

Tetrads with breeding
confirmed: 26 (31.71%)
probable: 42 (51.22%)
possible: 14 (17.07%)

area which remained occupied until about 1950 are without exception now deserted, showing that the decline has continued since: Whitley Reed, Stretton Moss, Arley Park Moss and nearby parkland, rough wasteground at Holford, a railway embankment at Hartford and a few pairs in the Dane Valley at Northwich and at Belmont Park. Bell[7] remarked on the species' disappearance from the Wilmslow area during the 1950s. Coward & Oldham[2] regarded the Tree Pipit as the normal host for the Cuckoo in lowland areas but Boyd knew of no such records.

Griffiths & Wilson[13] regarded the species as scarce in Wirral by the 1940s and by 1960 it was reported to have ceased to breed there[18]. However the *CBR* for 1968 mentions three pairs at one unspecified site, probably Haddon Wood, and a pair or two were found at Thurstaston Hill and Caldy Hill until 1972. The four Wirral records plotted on the map refer to only one or two birds each year, and the Tree Pipit is still a very scarce breeder on the peninsula.

Whilst agricultural changes during the last fifty years have been extensive, other suitable habitats remain superficially unaltered and other contributory causes of the decline in Cheshire should be sought. The synchrony of the decline with that of other trans-Saharan migrants throws suspicion on the state of the winter quarters. On the other hand, although there has been a general decrease in southern and eastern Britain this century, Tree Pipits have colonised much of Scotland in the last hundred years to become one of the commonest birds in the highlands[35]. The national CBC figures, which are biased towards the south-east, indicate a gradual drop in numbers over the last twenty years.

Tree Pipits are most easily located by the song-flight of the male, given from arrival in mid-April until the middle of July, or the hoarse "zeez" call-notes. When young birds are in the nest, the parents will often perch conspicuously with food in their bill uttering the alarm-call "plip . . . plip". The nest, always well-hidden on the ground, is traditionally one of the more difficult to find.

In 1981, 14 territories were located in the young plantations of the Delamere area and a fuller census might have revealed up to 40-50 pairs. Most of the birds here are south of the main "switchback" road, on the edges of the cleared areas: stands of trees below twelve feet tall were the most favoured. Five were singing at Brown Knowl (SJ45/55) in 1982 and a maximum of ten pairs may be estimated for the Peckforton—Bickerton ridge. Three pairs were on Risley Moss in 1980 but the reserve report for 1982 implies breeding here to be irregular. In 1982 some twelve pairs were located around Lyme Park and the adjacent Bollinhurst Brook, and the eastern hills generally probably held around 80 pairs. In all the county population may stand at 125-150 pairs, but a tiny fraction of that of 70 years ago.

Meadow Pipit
Anthus pratensis

A CHARACTERISTIC bird of marginal, "untidy" land, nesting in large numbers amongst heather on the eastern moors and rough grass on hill-pastures. Sharrock[35] correlated many of the areas of thin distribution in Britain with heavy clay soils which the species seems to avoid, perhaps explaining the absence of breeding Meadow Pipits from much of lowland Cheshire. It is noteworthy that the species nests widely along the Mersey valley with its alluvial, peat and river terrace soils, and more especially along the sandstone ridge to the south. In Wirral, breeding is concentrated along stabilised sand-dunes, heaths and saltmarsh edges. Hardy[4], writing in 1941, mentioned breeding on Hilbre Island, where from seven to ten pairs nested each year between 1963 and 1969[25].

Elsewhere, Meadow Pipits are very localised in summer, although small numbers breed on relict lowland mosses (for example Lindow Moss and Danes Moss), and also on the sloping banks of extensive industrial sludge-pools as at Frodsham, around Northwich, and, until the recent introduction of sheep and resultant close grazing, at Cledford, Middlewich. Boyd[6] referred to birds breeding at Anderton, Northwich, at the site of long-destroyed salt works, saying "there alone in the district the Meadow Pipit breeds".

Breeding Meadow Pipits are easily located by the silvery tinkling song given in display flight and audible over considerable distances. However, such song is often uttered by migrant birds from March onwards into May before they reach the breeding areas. Such passage birds may linger, often in flocks, for weeks at favoured localities and account for most of the records of 'possible' and 'probable' breeding in scattered lowland tetrads (although odd pairs occasionally stay to breed in isolated "waste" patches, as at Watch Lane Flash and the Dean Valley at Adlington in 1975).

In those hill areas where pipits are abundant, it is not unusual for a walker crossing the moors to flush a bird off its nest in a grass tussock, purple moor grass being

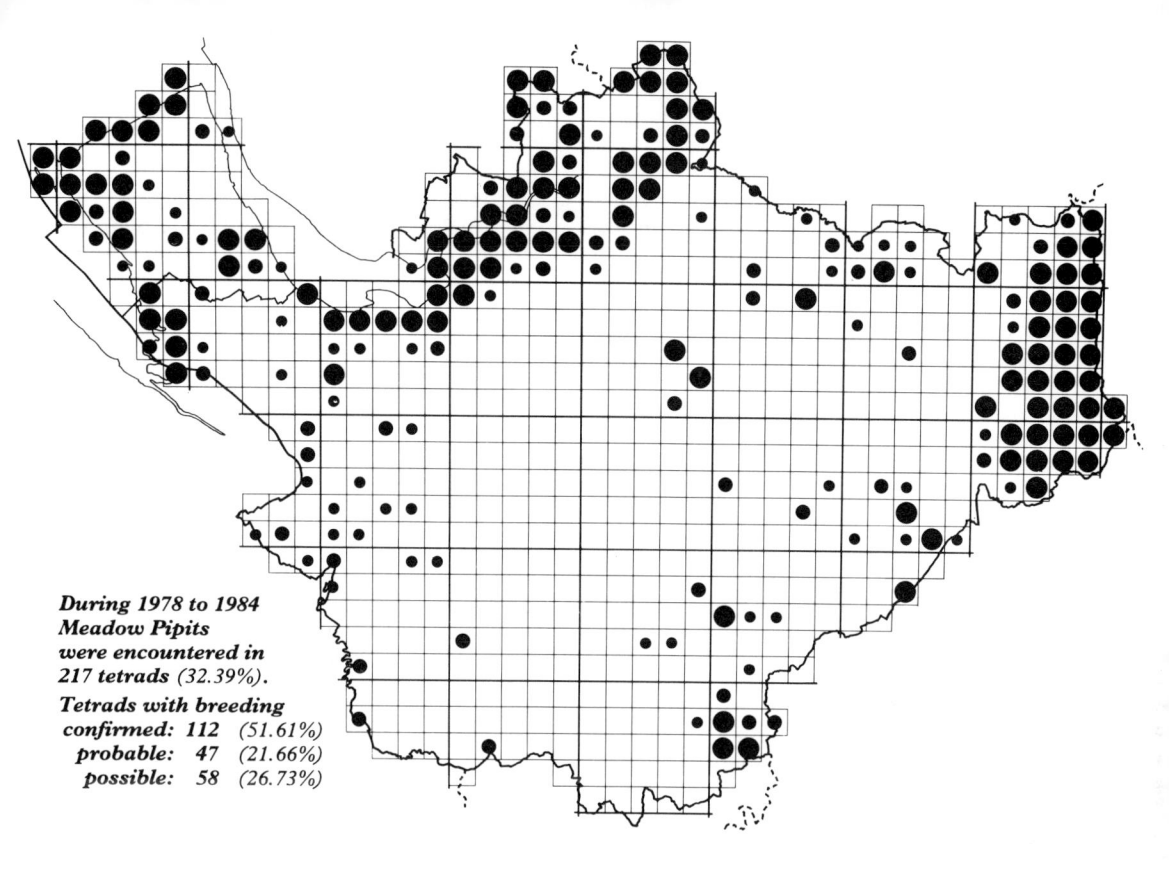

During 1978 to 1984 Meadow Pipits were encountered in 217 tetrads (32.39%).

Tetrads with breeding
confirmed: 112 *(51.61%)*
probable: 47 *(21.66%)*
possible: 58 *(26.73%)*

particularly favoured. Proof of breeding is generally obtained when young are in the nest however. Then the adults perch conspicuously with food held in the bill, and call anxiously, "pit....pit....". Meadow Pipits will sometimes fly more than half a mile to the nest, carrying food from a feeding area. The moorland breeders and birds at Frodsham Marsh form important (and perhaps locally the major) hosts for young Cuckoos, and pipits may often be seen noisily pursuing Cuckoos.

In recent years several Cheshire ringers have invested much time in studying breeding Meadow Pipits. On Frodsham Marsh most broods are of four or five chicks with an average of 4.3 (DN), while in the eastern hills JSAH marked 84 broods averaging 3.98 young. Birds fledge up to a week later in the hills, showing the usual variation of timing and clutch-size with altitude. The mean hatching date was 17th May at Frodsham and 20th May in the east of the county. JSAH noted a few true second broods, but most later broods followed failed first attempts.

Birds from the Cheshire moorlands have been found wintering as far south as Africa, with a chick ringed at Higher Disley on 26th May 1980 dead in Morocco on 3rd January 1982, and another from Bakestonedale Moor on 20th May 1978 caged in Morocco on 15th January 1979.

The distribution in the county has perhaps not changed markedly this century. Certainly, a few pairs bred around Northwich in Boyd's day as now on "tumbled land". Although Sharrock[35] records confirmed breeding in five additional 10-km squares (SJ45, 56, 64, 78, 87) compared to the present survey, these doubtless referred to isolated pairs. There is some indication of a decline in the eastern hills in recent years however, probably linked to liming of moorland and general upgrading of pastures. Cold springs also have an adverse effect on breeding success. Following persistent easterly winds and drizzle in late spring 1978 many young were found chilled.

On 31st May 1983 JSAH counted 38 pairs in SJ98 between Higher Disley and Pott Shrigley, allowing an estimation of the total population of that 10-km square to be made at 50 pairs, almost all of these in seven tetrads. In the eastern hills there may be 300 to 500 pairs, probably nearer the lower figure, with a further 300 along the Mersey valley and around Wirral and 20 or so pairs scattered elsewhere. This would give a county population of around 700 pairs.

Rock Pipit
Anthus petrosus

MOST of the few known nesting records came from Hilbre Island in the last century, and are listed by Coward & Oldham[2]. Brockholes[1] had seen eggs from there, and three nests were found by a Mr Walker on 24th May 1858. In May 1894 Coward and Oldham saw "several pairs of nesting birds on the edge of the low, sandstone cliffs". Griffiths & Wilson[13] stated that Rock Pipits nested on Hilbre occasionally, but gave no specific detail. Hardy[4] found a bird feeding young in May 1940 and Bell[7] refers to one or two pairs nesting there. Craggs[25] refers only to "old but not well-documented records of breeding on the Hilbre islands".

In the Mersey estuary Hardy watched three pairs carrying food to holes in the stone embankment of the Manchester Ship Canal above Eastham in May 1936 and there is a record of a pair at Mount Manisty on 7th May 1939. The following year birds were seen at Bromborough Dock on 20th March and 16th April[7]. In 1943 Miss Henderson saw a bird carrying food at this latter locality, the only real evidence of breeding on the mainland[13].

No breeding season records were received during this survey.

Yellow Wagtail
Motacilla flava

AT the turn of the century the Yellow Wagtail was an abundant summer visitor, chiefly frequenting water meadows and other low-lying situations on the plain. It occurred in smaller numbers on upland pastures in the east[2]. Smith (1950) studied the species in detail along the River Mersey at Gatley, near Stockport, and in his monograph described Cheshire as "*the* county for the Yellow Wagtail" with the position in south and mid-Lancashire similar. He went on to describe the species' status in Cheshire as "abundant practically everywhere except in forested areas such as Delamere, and in the moorland areas of the eastern hill country", and that it appeared to be increasing. It was listed as "common in the Crewe district; at Macclesfield; over most of the Wirral; at Eaton Hall near Chester; and at Styal". The Mersey valley contained a large breeding population. In addition, in a paragraph on south Lancashire, Smith said that Yellow Wagtails were especially common in the Gowy valley and at Puddington marsh – sites actually in Cheshire. Boyd watched at least half a dozen pairs in Birkenhead dockland in 1940.

Yellow Wagtails typically breed in loose colonies, and the county contains a large number of separate breeding areas. Thus in the eastern hills there are small groups associated with hill-pastures, others frequent the river valleys where particular fields are traditionally favoured, and generally smaller numbers are found in mixed farmland elsewhere. The association with potatoes implied by the dialect names of "potato-dropper" and "tater-setter" is due to the habit of spring migrants dropping into the tilled fields as the crop is being set in April. The species not infrequently nests in potato or other arable fields, usually with pasture nearby. In fact, as the rotation of fields used for growing potatoes progresses from year to year so the wagtails will follow. The nest is often situated under a large leaf of dock, in a tussock of grass in a damp meadow, or some similar site, and the large leaves of potato are ideal for this purpose. In 1982 two pairs nested on saltmarsh at Gayton Sands.

Wagtails with blue heads resembling continental races of this species occasionally breed with Yellow Wagtails. Such records tend to recur within particular local populations, suggesting they refer to genetic aberrations rather than vagrants. The earliest recorded instance was at Great Meols in 1954[7]. Craggs[25], writing in 1982, refers to a pair of blue-headed wagtails breeding in market gardens at Meols "in recent years". In both 1955 and 1957 a blue-headed male paired with a normal female in the Frodsham

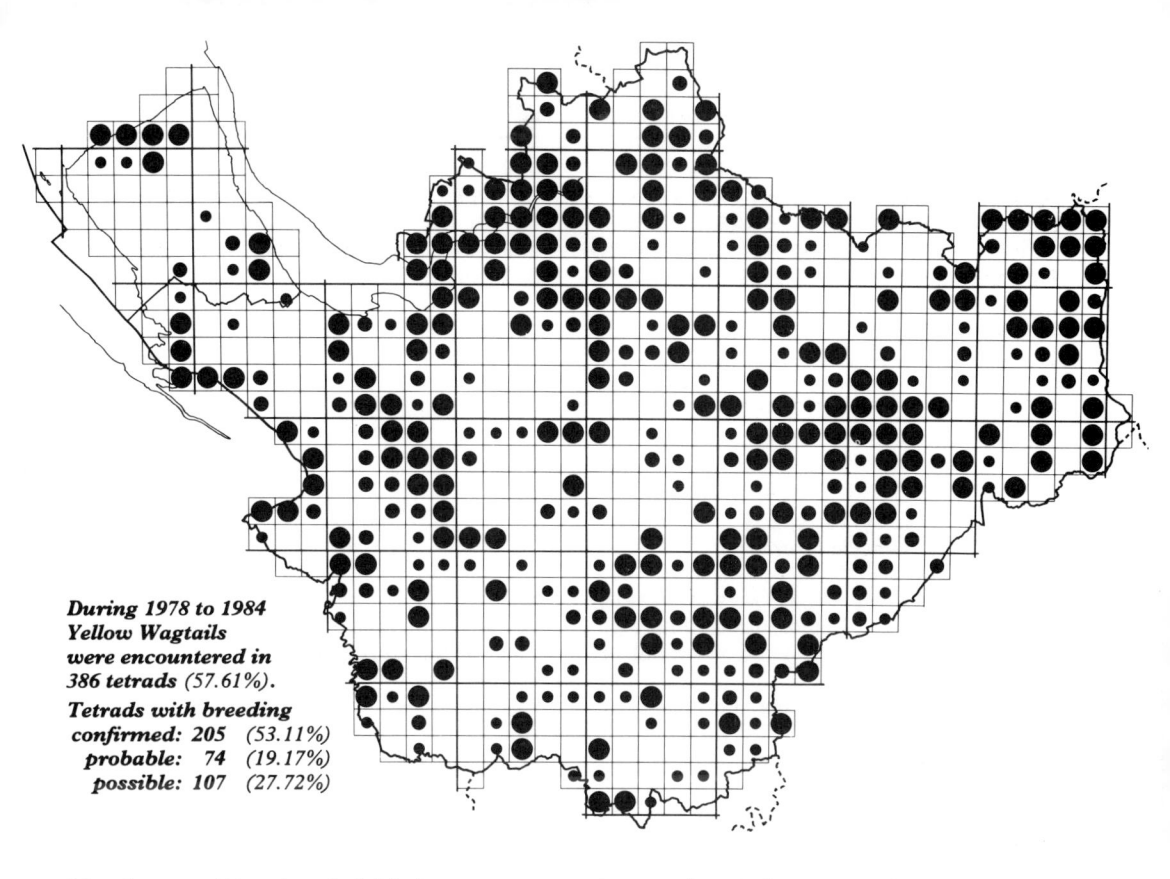

During 1978 to 1984 Yellow Wagtails were encountered in 386 tetrads (57.61%).

Tetrads with breeding
confirmed: 205 (53.11%)
probable: 74 (19.17%)
possible: 107 (27.72%)

Marsh area. Blue-headed birds are seen feeding young every few years both there and around Sandbach Flashes. In 1981, a male with a pale blue head helped rear young at Higher Disley where, in June 1948, an "aberrant" pair was feeding young 750 feet above sea-level[3].

The first birds, almost invariably bright yellow adult males, arrive back in Cheshire from their West African wintering grounds in the middle of April, with the main bulk of birds, including most of the females, turning up in early May. The species is very conspicuous in early June, and breeding may easily be proved by seeing adults collecting insects for their young: they may fly 400 metres or more from the nest, and a small number of 'FY' records may have referred to birds actually nesting in an adjacent tetrad. In July the moulting adults and their fledged young are joined by other wandering or migrating birds and form roosts in reed-beds, mustering up to a few hundred birds on occasions, although such passage roosts have dwindled of late. A proportion of the 'possible' breeding records may derive from single birds on spring passage or autumn migration. True second broods are very uncommon, Smith (1950) recording only one, but Yellow Wagtails will often try again after losing a first brood, prolonging the breeding season, and every year some birds are still found in Cheshire into the month of October.

That some contraction of range has taken place over the last 30 years or so is clear from the map. The bird's strongholds are along the valleys of the Dee, Gowy and Mersey, and on the heavier soils of the undulating country towards the south-eastern border. Sandier soils and intensive grass-growing areas are largely avoided. Large areas of Wirral are now deserted, perhaps due to building development, and the species is increasingly scarce both in intensively farmed areas of the plain and in hill-pastures where farming methods remain relatively unchanged. As with other trans-Saharan migrants, the causes of the decline may lie largely in Africa.

Some eight to ten pairs per tetrad breed on Frodsham Marsh, but the density is generally lower. Assuming an average density of two or three pairs per occupied tetrad, the county may hold between 650 and 1000 breeding pairs which seems not unreasonable in view of Smith's assessment of the county's importance and the estimate[35] of 25,000 pairs nationally.

REFERENCE

Smith, S. (1950): *The Yellow Wagtail*. New Naturalist, Collins, London.

Grey Wagtail
Motacilla cinerea

THE thin song of a Grey Wagtail, delivered from a branch, protruding pipe or similar perch above a weir, waterfall or fast-flowing brook, will often draw the observer's attention in spring. Perhaps little will be seen other than the bird bobbing low over the water to land, with tail wagging, on a pebble in midstream. The song can be surprisingly difficult to pinpoint against the background of rushing water. It is uttered chiefly from March to May, less often during the summer months, but also not infrequently by birds wintering around sewage works and elsewhere.

Nest-building begins in late March. On 23rd March 1975, when the male had still not developed his black chin, a pair collected grass-roots from mole-hills at Adlington. Similar behaviour was noted at Rostherne on 23rd and 24th March 1978, and this is probably a normal source of nest material (JPG). Nesting birds often utter an anxious drawn-out "tsweep" when the nest is approached by a potential predator. JPG has heard this call as early as 12th March by a nesting stream in Tatton Park, and as late as 10th July when fledged young were still accompanied by their parents. Confirmation of breeding is most easily obtained when the adults are feeding young. A pair at Aldford had young in the nest by 17th April, but usually it is mid-May before newly fledged birds are seen, with more in June and July – Grey Wagtails are sometimes double-brooded. From July onwards dispersing juveniles begin to appear at the Sandbach Flashes and elsewhere away from breeding-sites.

In the early years of this century Grey Wagtails bred in considerable numbers on the hill-streams in the east of the county, and sparingly in the lowland[2]. It was nowhere more numerous than on the upper Dane above Bosley where it was the commonest wagtail, though it also nested along the lower reaches where suitable sites were available, as at Buglawton and Cranage. A few pairs bred elsewhere in the lowlands, chiefly in the eastern part of the plain, with favoured sites occupied year after year. Alderley Edge, Siddington, Mouldsworth Station, Winsford Flashes and possibly Marbury near Northwich were listed. Birds had twice been seen in June along the Dee above Chester, although the species was generally uncommon in western Cheshire and Wirral. A pair had nested some years previously at Bache Grounds, Chester, but, in Wirral, Brockholes had seen the species only occasionally. Abbott saw birds in the breeding seasons of 1919 and 1920 at Quarry Bank Mill, Styal, and in 1921 in Wilmslow Park.

Bell[7] reported that during the next fifty years it was found nesting in many localities on the plain, mainly in the east. However, Boyd[6] could give only one breeding record around Great Budworth: in June 1938 a pair nested by a tiny waterfall at Comberbach, formed by damming of a stream. Bell also quoted breeding records from Wirral in 1941 and 1949, where by 1962 it was regarded as a resident in small numbers. This population may have been established earlier however, for Miss Henderson knew of a few pairs breeding annually in the Raby and Spital districts in the 1940s[13].

In lowland areas breeding is limited by the shortage of fast-flowing water, almost all known sites being by weirs, sluices or mill-races. Grey Wagtails often nest within the

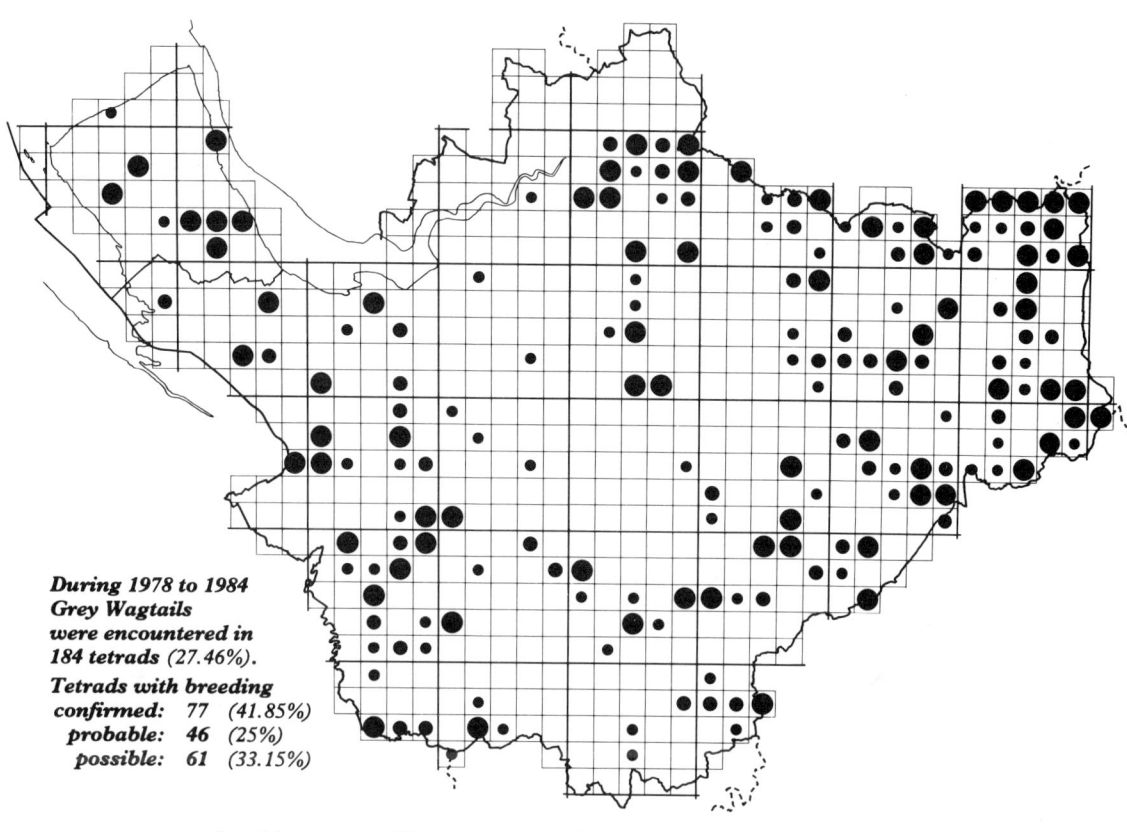

During 1978 to 1984 Grey Wagtails were encountered in 184 tetrads (27.46%).

Tetrads with breeding
confirmed: 77 (41.85%)
probable: 46 (25%)
possible: 61 (33.15%)

constant roar of rushing water. However, in some cases territories are centred on narrow tributary streams to rivers, as at Styal Woods, and records from Hockenhull and Aldford may fall into this category. The degree of cover along the banks seems to matter little so long as mud or shingle banks are not densely overhung – several known sites are within woodland. Some birds in the Warrington area bred alongside the Manchester Ship Canal. Despite the limitations on habitat in the lowlands, there is a suspicion that pairs may have been overlooked in these areas, there being only limited access here compared to the hill-streams. Nevertheless the number of pairs located during this survey came as a surprise to many observers.

It is thought that some of our Grey Wagtails move south for the winter, being replaced by others from farther north, such as the bird caught in Arrowe Park, Wirral in January 1981 which had been ringed as a chick near Aberdeen in 1978. Nevertheless the species often suffers badly in severe winters. The *L&CFC Report* for 1963 stated that the population in the Stockport/Macclesfield/Goyt Valley area (including parts of neighbouring counties by modern definition) had fallen from 24 occupied sites in 1962 to just four in 1963. Just 16 birds were ringed in the latter year as against 98 in 1962 "with no less effort or interest". Similar effects were noted during this survey after severe weather early in 1979 and again late in 1981. The numbers of territories reported annually to *CBRs* as occupied in the eastern hills during this survey reflect this with 20 or more in 1978 but only three or four in 1979, and similarly seven in 1981 falling to just one in 1982 when the South Manchester Ringing Group ringed no nestlings for the first time in its history. Territories along the rivers Bollin and Dean to the east of Wilmslow were also devoid of birds. In the same season however probably three pairs bred in Styal Woods, and several other lowland sites were occupied. By 1984 some twenty sites were again occupied in the hills.

The upper reaches of the River Dane remain the stronghold of Cheshire's Grey Wagtails. R. M. Blindell has estimated the population of the 10-km squares SJ86 and SJ96 at 31 pairs, most of these along the Dane and its tributaries. Territories are not all occupied in any given year, especially following severe winters, yet with birds recorded in 180 tetrads, 119 of these with at least 'probable' breeding, there was probably a peak population of between 120 and 150 pairs in the county during this survey. Their population is probably limited by winter mortality and, particularly in the lowlands, by the availability of suitable nest-sites. A number of lowland brooks were found to be heavily contaminated with silage effluent, raw sewage and other organic waste during this survey. Since the range of aquatic life in such streams is limited, food supplies may then be inadequate to allow Grey Wagtails to breed.

Pied Wagtail
Motacilla alba

THIS most versatile of our wagtails is often associated with water but just as often breeds around buildings in agricultural areas. Birds may be seen flycatching from rocks in streams, but more usually quarter open ground such as shortly grazed turf, the shores of lakes or road surfaces and even mossy roof-tops searching for insects. The nest is sited in a natural cavity in a bank, or more usually in holes in walls and amongst ivy on or cavities in outbuildings. Boyd noted a pair at Frandley in 1928 that built and lined six nests in a series of nine ventilation holes in a shippon wall before finally using one.

Given this diversity of feeding areas and nest-sites, it is little wonder that Pied Wagtails breed throughout the county as the map shows. Indeed, given time, odd pairs might have been found in many of the apparently empty squares, although the species is very scarce in summer in some urban parts of north Wirral and the Mersey valley, where it is, however, common in winter. Despite the apparent absence from parts of south and central Cheshire, JPG's impression is that the species is, if anything, more numerous in this area than in the east. Dairy farmyards with their associated insects are the most typical breeding habitat over much of the county. In late May and June breeding can be confirmed while driving through successive tetrads by seeing wagtails, beaks full of insects, flying up off the quiet country roads where they often hunt. A female on the River Dean repeatedly caught over a dozen midges by flycatching from stones below a weir, dipping each fly in the water before storing it in the back of her bill – the first midge caught might be washed fifteen or more times!

Both Coward & Oldham[2] and Boyd[6] commented that few Pied Wagtails remained for the winter and the latter author noted that spring roosts of immigrants in reed-beds were normal during late March and April. Large flocks, including Scottish birds (Davies 1986), are now spending the winter on sewage works in the county, dispersing from early March, about which time pairs begin to appear around breeding-sites, even at altitude in the eastern hills. Despite the presence of birds throughout winter, and the few reports of spring roosts in reed-beds nowadays, the start of the nesting season does not seem to have moved forward, with the first young still hatching around the middle of May.

Observations at sewage works show that song may be heard at any time during the winter provided the weather remains mild, but its frequency increases towards the spring

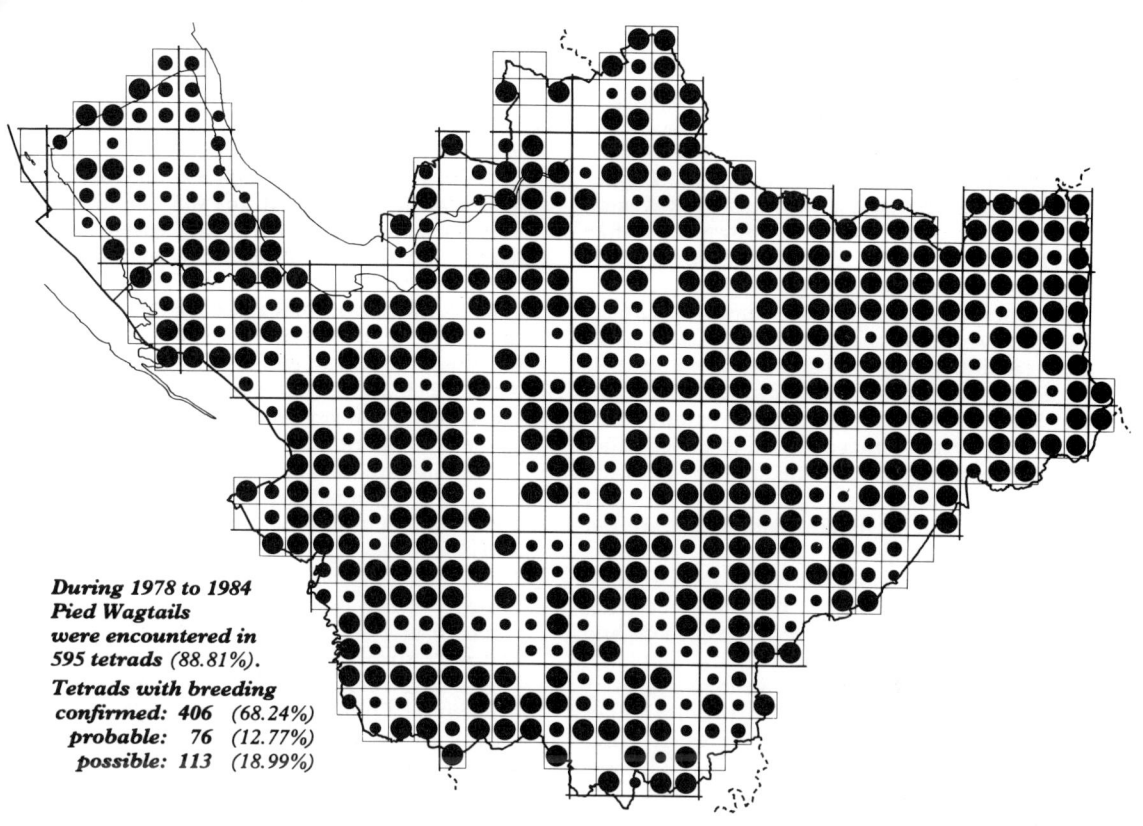

During 1978 to 1984 Pied Wagtails were encountered in 595 tetrads (88.81%).

Tetrads with breeding
confirmed: 406 (68.24%)
probable: 76 (12.77%)
possible: 113 (18.99%)

and reaches a peak during April when the birds are displaying vigorously (JPG). At this time territorial birds can be remarkably aggressive. Boyd[6] had a record of a cock wagtail, annoyed by other birds whilst collecting nest material, which dived on a sparrow from behind and killed it, and on 27th April 1982 at Alderley Park one drove off a stoat by "dive-bombing" it! This species is more frequently seen than any other displaying at its own reflection, perhaps in a car wheel hub-cap, wing mirror or window.

Considerable numbers of birds of the race *M. a. alba,* known as the White Wagtail, migrate through Cheshire en route to their Icelandic or perhaps continental breeding grounds, and very occasionally mixed pairs of White and Pied Wagtails have been reported nesting in the county. On 25th May 1980 a cock White Wagtail was accompanying a hen Pied Wagtail to a nest at Frankby Hall, Wirral, and in the mid-1970s a similar pair was present at Moston near Sandbach.

Sharrock[35] suggested an average of 150 pairs breeding per occupied 10-km square in Britain and Ireland, and national CBC densities range from one to four pairs per square kilometre[40]. While Sharrock's figure (six pairs per tetrad) may be met in some parts of the county, there are also many tetrads where only one or two pairs could be located. The laboratory at Daresbury provides ideal habitat, with extensive close-cropped lawns alongside the canal for feeding, and numerous nesting niches available in the large buildings: five to six pairs in about 15 hectares is usual here (DN). In the absence of other accurate counts it would appear that a more conservative estimate, perhaps 75-100 pairs per 10-km square, is more appropriate for Cheshire, giving a total population of between 1900 and 2500 pairs.

The population drops after hard winters, the national CBC and WBS indices falling by 25% from 1981 to 1982 with smaller decreases from 1978 to 1979. Breeding numbers recover quickly, and there is no evidence of any long-term changes. Both Coward & Oldham and Boyd referred to Pied Wagtails fostering young Cuckoos, but there have been no such records in recent years, perhaps largely because the emphasis on nest-finding has been removed from bird-watching.

REFERENCE

Davies, N. B. (1986): in *The Atlas of Wintering Birds in Great Britain and Ireland* (ed. P. Lack). T. & A. D. Poyser, Calton.

Dipper
Cinclus cinclus

AS a bird of the hill streams, the Dipper shares its habitat with the Grey Wagtail and Common Sandpiper. However, unlike these species which feed from above the water and at the water's edge respectively, the Dipper takes its food from below the surface. The main prey items are aquatic insect larvae – caddisfly, mayfly and stonefly – and their adults, and these are abundant in hill streams (Shaw 1979). To hunt them it requires fast-flowing water with a mixture of pools, riffles and boulder perches. The still waters of the hill reservoirs, often frequented by Grey Wagtails and Common Sandpipers, are not suitable for Dippers.

Given its habitat restriction, the Dipper is an easy bird to census. The adults carrying food to their young are very conspicuous and an easy way to confirm breeding. Similarly, the young are quite vocal in their relatively secure nests, built in culverts, holes under bridges, or in the roots of trees or other vegetation alongside the streams. The low proportion of 'possible' compared with 'probable' and confirmed breeding records reflects this, and it is likely that these are early dispersing young or errant birds from nearby territories rather than overlooked breeding attempts. Dippers start breeding early with over 50% of clutches completed by mid-April (Shaw 1978). They are generally resident in their territories throughout the winter, although they may have to evacuate the higher reaches of streams. Indeed the colder water of winter and early spring carries a larger prey population than the warm water of summer when the larger larvae have emerged as adults. This early food supply is presumably responsible for the early breeding season, and may allow a second brood before the summer drought causes dispersal.

Coward & Oldham[2] regarded the species as practically confined to the hill country of eastern Cheshire, which is still essentially true. They mention one breeding locality on the plain, at Cranage Mill near Holmes Chapel. Boyd[5] describes a nest "with eggs just hatching" in June 1938 at a site in lowland Cheshire, thought to be on the Peover Eye at Astle, Chelford. He had seen birds "for some years" along this stream and concluded "that there are more than one pair...it is an interesting extension of range". More recently birds were recorded here between 1954 and 1967 with adults feeding young in 1960, 1963 and 1966 (R. Harrison pers. comm.). In 1967, a bird was observed gathering leaves at this site on 2nd April and breeding was again recorded in 1969[8]. Dippers returned in 1983 and were suspected of having bred again in 1984. Elsewhere on the plain, a pair was present at Cuddington in mid-Cheshire during May and June in 1974 and nested unsuccessfully in 1975 ([8], Todd 1975).

Coward & Oldham[2] knew of no breeding records in the west of the county. The record mapped near Threapwood on the Dee during this survey is therefore notable, and may be an outlier of the widespread Welsh population. Otherwise the current map looks much as they and Boyd would have expected. The records from Norbury/Lady Brook in the north are especially interesting in the light of their references to the species' breeding on the banks of "a polluted brook at Middlewood".

The current headquarters of the species is obviously the River Dane and its tributaries. A specific survey on 21st May 1983 located 21 territories between Danebower and Hugbridge, a distance of 15 kilometres involving ten tetrads, and the species occurs further downstream to Congleton. This

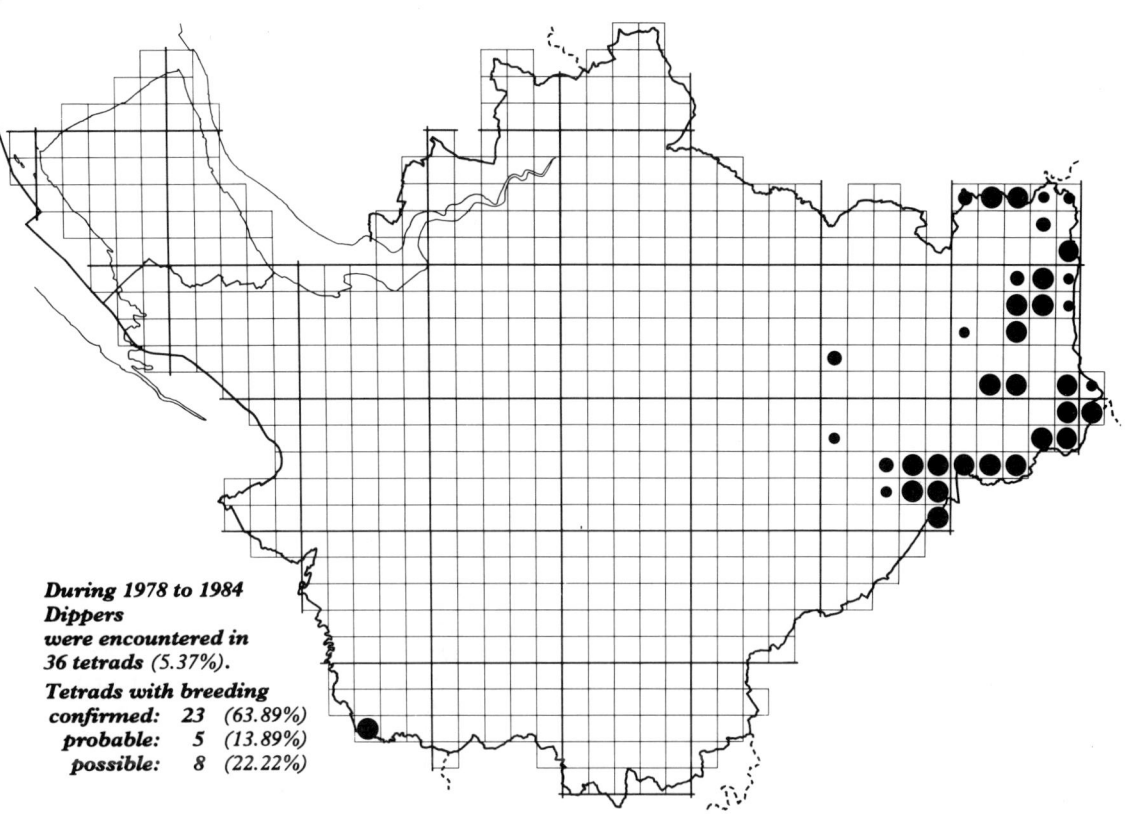

During 1978 to 1984 Dippers were encountered in 36 tetrads (5.37%).

Tetrads with breeding
confirmed:	23	(63.89%)
probable:	5	(13.89%)
possible:	8	(22.22%)

density, 1.4 pairs per kilometre, is rather higher than comparable figures from elsewhere. In neighbouring Derbyshire, Shooter (1970) thought that there were between 97 and 112 pairs in different years – about one pair per mile – while for Staffordshire, Harrison *et al.*[47] report that surveys of the River Dove consistently give one pair per 1.5 kilometres. In analysing results from the BTO's Waterways Bird Survey, Marchant & Hyde (1980) give the highest average density at seven pairs per 10 kilometres of river. At a reasonable estimate then, of two pairs per tetrad in Cheshire (essentially one pair per kilometre of river), the county population would be 70 pairs. This species appears not to be as susceptible to hard winters as might be expected for a resident. The population of Dippers on WBS plots fluctuates very little from year to year, so this is likely to be the population level in most years. It is conceivable that Coward & Oldham might have found the same sort of numbers, for the bird's habitat has altered little. However, Dipper numbers are proving to be affected by increased acidity in Welsh rivers (Ormerod *et al.* 1985), and it is to be hoped that Cheshire birds will not be similarly affected.

DWY

REFERENCES

Marchant, J. H. & Hyde, P. A. (1980): Population changes for waterways birds, 1978-79. *Bird Study* **27**: 179-182.
Ormerod, S. J., Tyler, S. J. & Lewis, J. M. S. (1985): Is the breeding distribution of Dippers influenced by stream acidity? *Bird Study* **32**: 33
Shaw, G. (1978): The Breeding Biology of The Dipper. *Bird Study* **25**: 149-160.
Shaw, G. (1979): Prey Selection by Breeding Dippers. *Bird Study* **26**: 66-67.
Shooter, P. (1970): The Dipper Population of Derbyshire, 1958-1968. *British Birds* **63**: 158-163.
Todd, D. (1975): Dippers Nesting in Mid-Cheshire. *Cheshire Bird Report* p35.

Wren
Troglodytes troglodytes

THE Wren keeps closely to cover at all times of year. For such a diminutive bird however, suitable cover includes not only woodland, scrub and hedgerows, but also stands of tall herbage such as nettles or willowherb, bracken-beds and the dense layer of buckler ferns beneath mature stands of pines in Delamere. Consequently its distribution extends right across the county from the coast up to the edge of the moors. Hardy[4] reported that the species nested on Hilbre, but Craggs[25] quoted only one unlined nest found in May 1965.

As the herb layer in woods dies down for the winter, many Wrens move into reed- or nettle-beds. In severe winters, insect food becomes very difficult to obtain and birds fare badly. After the severe winter of 1962-63, P. H. Oswald, the warden, reported that only one pair survived at Rostherne[24], and only one pair was seen around the Sandbach Flashes[19]. The 1964 *CBR* spoke of "complete recovery" however.

Two extended cold spells occurred during the present survey. Prolonged frosts early in 1979 caused heavy mortalities. On four CBC plots in south-east Cheshire, a 45% reduction in numbers was recorded in the following summer. A 59% decrease was noted on the Alderley Park plot, where birds continued to filter into the woodland and take up territories into mid-April. By 1980 recovery was largely complete. At Risley Moss 64 pairs in 1978 dropped to 40 in 1979 but recovered to 63 in 1980, and at Alderley Park in 1980 numbers were double those of 1979 and almost back to the 1978 level. Census work on three farmland plots in the south-east showed only a partial recovery however, although on average the species had resumed its 1978 status as second most numerous species. This is the classic pattern expected for a species whose preferred habitat is woodland, which overflows into other areas when population levels are high: after a decline, woodland numbers recover more rapidly.

December 1981 brought snowfalls followed by prolonged frosts. Frozen snow-drifts against hedgerows across much of the county will have caused many birds to starve. At Prestbury sewage farm, Wrens were watched tunnelling down along emergent nettle stalks to feed under the snow. However in many habitats, and especially in farmland areas, feeding was impossible. At Dean Row, Wilmslow the population appeared to have been more than halved after the first ten days of frosts, and, by January and February of 1982, JSAH found only one or two birds on each visit to the eastern hills compared with five to eight per visit in the previous November and December. At Risley Moss on 4th February nine Wrens were found dead in a roosting crevice.

In the summer of 1982 breeding numbers at Risley were down again by 30%. Three south-eastern CBC plots held 40 territories as against 49 in 1980 and 56 in 1981 – a 29% decrease after the winter. Numbers at Risley had not recovered by 1983. Ringing has shown that most Wrens are sedentary but a small proportion of the population may attempt to avoid the effects of winter by moving south: a bird ringed at Red Rocks, Hoylake on 1st August 1978 was found at Brixham, Devon (332 kms S) on 22nd January 1980, illustrating the species' capacity for long-distance movement.

Wrens are easily located by their disproportionately loud song which is uttered at any time of the year, but most frequently in spring and summer. The male starts to build early in March, generally completing several "cock's nests". The female then chooses one of these and lines it before laying. Coward & Oldham[2] and Boyd[6] gave extensive lists of nest-sites in sheds, ivy or creepers; hedges and banks; amongst the twigs growing from the trunk of an oak or alder; in the exposed roots of a fallen tree; or in an adapted Swallow's nest. Sites of all these types are still used, but the thatched roofs of sheds and haystacks are no longer generally available. In 1890, a nest containing six eggs was found inside the old nest of a House Martin at Prestbury. In July 1898 a nest, supported between the drooping wing and body of a dead Sparrowhawk, was taken from a keeper's gibbet at Carden Park, and in 1899 a nest at Capesthorne was built in the head of a Brussels sprout. A typical nest is domed or egg-shaped, but Coward twice found nests in crevices in the bark of an old poplar tree with an arch of nesting material around the entrance hole, their roofs formed by the bark of the tree. During the current survey Wrens were reported building over nests of a Robin at Mottram Hall and of a Swallow in the boathouse at Rostherne. At Alderley Edge in May 1971 a bird was seen entering a nest in a crack in a rock-face.

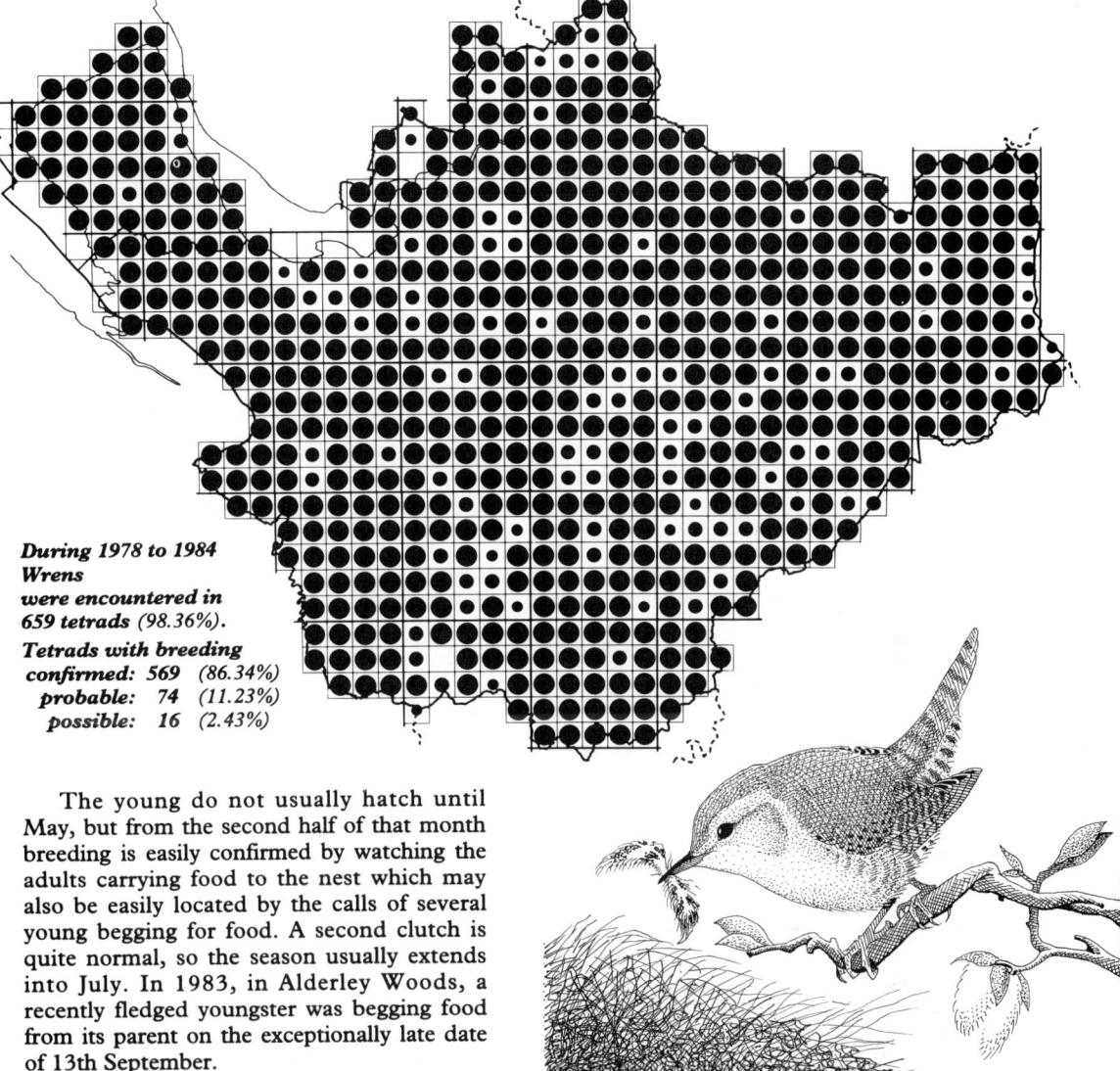

During 1978 to 1984 Wrens were encountered in 659 tetrads (98.36%).

Tetrads with breeding
confirmed: 569 (86.34%)
probable: 74 (11.23%)
possible: 16 (2.43%)

The young do not usually hatch until May, but from the second half of that month breeding is easily confirmed by watching the adults carrying food to the nest which may also be easily located by the calls of several young begging for food. A second clutch is quite normal, so the season usually extends into July. In 1983, in Alderley Woods, a recently fledged youngster was begging food from its parent on the exceptionally late date of 13th September.

Wrens were found in almost every tetrad during this survey, although breeding was not proved in some. They are sparsely distributed in the eastern hills, and there are other noticeably thin areas on the map: in the Stanlow area, north of Warrington, and in large parts of south-east Cheshire. The species probably is rather less common in these places than in the rest of the county.

At Rostherne Mere NNR from 50 to 71 pairs were found between 1976 and 1978 in 22.8 hectares of mixed woodland. The 13-hectare plot at Alderley Park held from 11 to 27 pairs between 1978 and 1980, with larger territories among the conifers than in rhododendrons. 18.2 hectares of deciduous woodland, heath and scrub on Bidston Hill held between 27 and 39 territories from 1975 to 1984. In 1978 the six south-eastern CBC plots on wooded farmland held a total of 173 territories in 393 hectares. At Risley Moss, 73 hectares, including 25 hectares of woodland, have held between 40 and 64 pairs.

Such figures suggest peak densities per square kilometre of between 200 and 300 pairs for woodland habitats, up to 90 pairs for wooded mossland, and 40 pairs on farmland with copses. Many wooded tetrads are likely to hold 200 pairs or more, with perhaps a third of this number or less in intensively farmed areas. This would give a Cheshire and Wirral population of around 70,000 pairs, not very far short of Sharrock's[35] estimated national average of 3000 pairs per 10-km square. At its peak population, the Wren is one of our commonest birds.

Dunnock
Prunella modularis

WHILST being one of our most numerous breeding species, the Dunnock or Hedge Sparrow is also one of the most neglected by bird-watchers. *CBRs* from 1964-1977 contain no breeding information whatsoever. In fact the breeding behaviour of "Britain's most boring bird" is bizarre and probably the most unusual of all British passerines (Davies 1987). Males and females separately set up "territorial" ranges in spring and, depending on the cock's ability to monopolise one or more hens, the resulting mating systems range from polygyny (a male with two or three females) through single pairs to polyandry (a female with two or three males) or even several males associated with several females. The commonest units are pairs and "trios" – one female with two males. In such trios, one (dominant) male guards the female and the other (subordinate) male may or may not succeed in mating: only if he has been successful does he help in feeding the young.

Coward & Oldham[2] knew it as one of the commonest residents, nesting everywhere in hedgerows and gardens, and even in gorse bushes at a considerable altitude in the hills. Only the windswept moorlands were avoided. Their description of its distribution is equally true today. Dunnocks are absent only from the highest hills and Frodsham Marsh, with other gaps on the map coming from under-recorded tetrads in Widnes and central southern Cheshire, and the bare saltmarshes of the Mersey and Dee.

Dunnocks are essentially scrubland birds, occurring wherever woodland is regenerating, be it scrub on railway embankments, the edges of disused sand-pits, felled woodland springing back from the stumps, or young plantations of conifers, and where the succession to woodland is held in check – for example, ornamental shrubberies in parks and gardens, trimmed hedgerows on farmland or in suburbia, or stands of gorse, controlled by burning, in the hills or on heathland. Mature woodland is avoided where the canopy is closed and undergrowth is sparse. Thus in Delamere, and at Alderley Park[8], conifer plantations are deserted when the trees reach perhaps 20 feet tall and the lower branches are lopped. Removal of rhododendrons from woods at Rostherne may account for reduced numbers of Dunnocks there in recent years: six to eight territories in 1979-83 as against 10-14 in 1976-78. At Alderley Park it was also noticed that mature woodlands were deserted in winter, whereas birds remained in the evergreen cover of young conifer plantations.

Boyd[6] noted nests in thorn hedges, holly, yew, gorse, etc., and for several successive years a nest was built in a roll of wire leaning against a wall in a garden. A pair built a complete cup inside a disused Song Thrush nest and reared a brood. Records during this survey showed that sites in lower vegetation are also frequent, for example in nettle-beds, gooseberry bushes, and even amongst garden cabbages.

During winter, small flocks of Dunnocks may feed together in favourable sites. Thirty were found in a commercial gooseberry patch at Daresbury in December 1982 and a party of eight or more fed on nettle seeds in a weedy market garden at Sandbach in January 1975. By mid-February birds revert to their various territories[21]. Boyd[6] recorded triangular contests in February and March, with birds chasing each other and sometimes resulting in flight. In March, rapid and excited chases in which only two birds engaged took them through the tree-tops; and in April he saw the female stand with wings shivering and tail erect while a male pecked beneath her tail. However it is only in recent years that the details of the species' complex breeding biology have fully been sorted out.

Boyd had nearly 100 recoveries of birds within a few hundred yards of the place of ringing, and only two found some eight miles away, showing just how sedentary the species is. It is most unusual to see a bird in sustained flight, though Dunnocks turn up regularly on Hilbre with a small spring movement and marked autumn dispersal, there being nearly 300 October records over the twenty years to 1977[25]. One of the Hilbre birds, ringed on 9th April 1974, was found in Warwickshire on 9th March 1978.

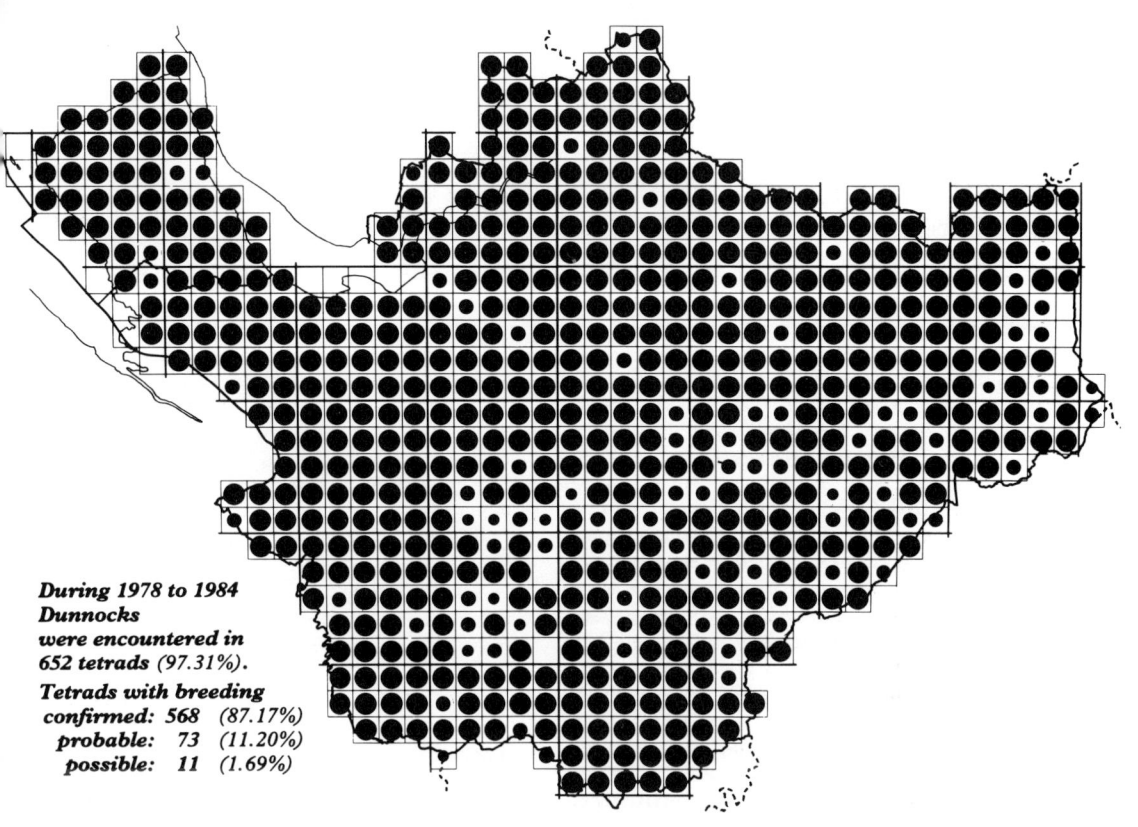

During 1978 to 1984 Dunnocks were encountered in 652 tetrads (97.31%).

Tetrads with breeding
confirmed: 568 (87.17%)
probable: 73 (11.20%)
possible: 11 (1.69%)

Song is not usually delivered from a high perch. More often the flat top of a clipped hedge is used. Birds have occasionally been recorded singing by night, as at Hoole, Chester in May and June 1965, but usually this takes the form of a short burst only, presumably when a roosting bird is disturbed.

Despite the species' ubiquity, confirmation of breeding was not always easy to obtain. Food is carried to the young in the bill, but often the adults seem to sense the observer's interest and swallow the food themselves to conceal their intentions. Dunnocks are quite frequently used as foster parents by Cuckoos, although Cuckoos have not yet evolved a blue egg to mimic their hosts. This is, therefore, presumed to be quite a recent development (on an evolutionary timescale), but it could be that no Cuckoos actually specialise in using this species, cuckolding them only if they cannot find nests of other, more favoured hosts. Most studies of Cuckoos have concentrated on the colonial species, and parasitism of Dunnocks has received rather less attention.

Numbers may be reduced by prolonged frosts. Hardy[4] reported that many were found dead in the hard winter of 1940. Three CBC plots in south-eastern Cheshire showed a 19% reduction from 37 territories in 1981 to 30 in 1982. After that same winter no more than four were heard in song around Lyme Park, and birds were scarce elsewhere in the eastern hills, suggesting some contraction of range from the higher ground. The severe winter of early 1979 however had no significant effect: four south-eastern CBC plots held 56 territories in 1978 and 52 in 1979, whilst no change was noted in a woodland plot at Alderley Park.

Sharrock[35] suggested a figure of 1500 pairs per 10-km square, i.e. 60 pairs per tetrad, as a conservative estimate. In 1978, six CBC plots on wooded farmland totalling 393 hectares held 90 territories[23], closely in line with the regional average figure for farmland CBCs of 24 pairs per square kilometre[42]. At Rostherne some 72 hectares of farmland held from four to seven pairs between 1978 and 1982. At Risley Moss 18 or 19 pairs in 25 hectares of woodland seems usual, as against six to 14 pairs in some 23 hectares at Rostherne. At Alderley Park, G. B. Hill found seven pairs in about thirteen hectares of woodland in 1978 and 1979. Intensively agricultural areas, with few hedgerows, which cover large parts of the centre and south of the county may scarcely reach even the figure of 30 pairs per tetrad, though elsewhere, as in the south-eastern CBC plots, Sharrock's average may be far exceeded. Most census methods count singing males, which are usually somewhat more numerous than the breeding female Dunnocks. The number of "pairs" may not therefore be easy to assess, and the county population probably lies between 20,000 and 30,000 nesting females.

REFERENCE

Davies, N. B. (1987): Studies of West Palearctic birds: 188. Dunnock. *British Birds* **80**: 604.

Robin
Erithacus rubecula

CONTRARY to popular belief, rather few Robins nest in old kettles although a pair did so at Macclesfield in 1984. Perhaps the most typical nest-site is in a hedge-bank or the side of a ditch, concealed amongst grasses or ferns. Many others are built in ivy or other creepers on walls. Coward & Oldham[2] often found nests in the loosely built walls beneath hedges in Cheshire lanes, or beneath a tussock of grass on the ground in woodland. They and Boyd[6] also listed nests on shelves and ledges in sheds and outbuildings; supported upon the handle of a garden fork leaning against the wall of a shed; on a roll of wire; in an open nest-box used by Spotted Flycatchers in the previous year; four feet up in a cypress; flat on the ground; and once in a Blackbird's old nest.

More freak nest-sites are recorded for Robins than for any other bird. Nests in sheds have – for example – been placed in the pocket of an old duffel-coat (JPG!) and in the saddle-bag of a bicycle. In 1978 a bird was incubating on a nest inside a busy factory at Langley, in a room full of steam, only two feet away from a working machine, and at Crewe in 1983 a pair built inside a hanging-basket within three days of it being put up.

Robins sing throughout the year, pausing only during the moult which, for most birds, falls in July and August. Song may however diminish considerably towards the end of May. Autumn song has a melancholy quality absent from the song of courting birds in spring. Birds often sing beneath street lights and the platform lights on railway stations, this habit having spread to many parts of the county particularly since the 1960s when councils made the decision to leave street lights on all night. On 26th January 1969 a bird was singing at Wilmslow at 5.45 a.m., and at Dean Row in the late 1970s song was heard regularly until after midnight and again from well before dawn, for example at 4 a.m. on 3rd March. Some of these nocturnal songsters are erroneously reported as Nightingales, having replaced the declining Sedge Warbler in this respect.

Both male and female Robins have a territorial song during the autumn months, and for the earlier part of the winter will drive away any intruding Robin from their patch. Such squabbles are familiar to all who put out food for the birds. From around Christmas until February pairing takes place and from then on the two birds may be seen together around the bird-table. Courtship feeding is often seen in which the cock passes a grub to the hen.

Nest-building begins in late February or early March, but nests may be ready well before eggs are laid[6]. Thus, nests complete on 26th February and 1st March did not contain any eggs until 26th and 24th March respectively. Unseasonal nests are sometimes reported in mild winters. Coward & Oldham[2] mentioned a nest with six eggs at Hale Barns on 5th January 1901, and one with four eggs at Sale at Christmas 1908 – both localities are now in Greater Manchester. Hardy[4] had a record of a nest with five eggs at Prenton, Birkenhead on 24th January 1932, and in 1975 a Robin was reported on eggs at West Kirby on 1st February.

The typical clutch varies in size from five

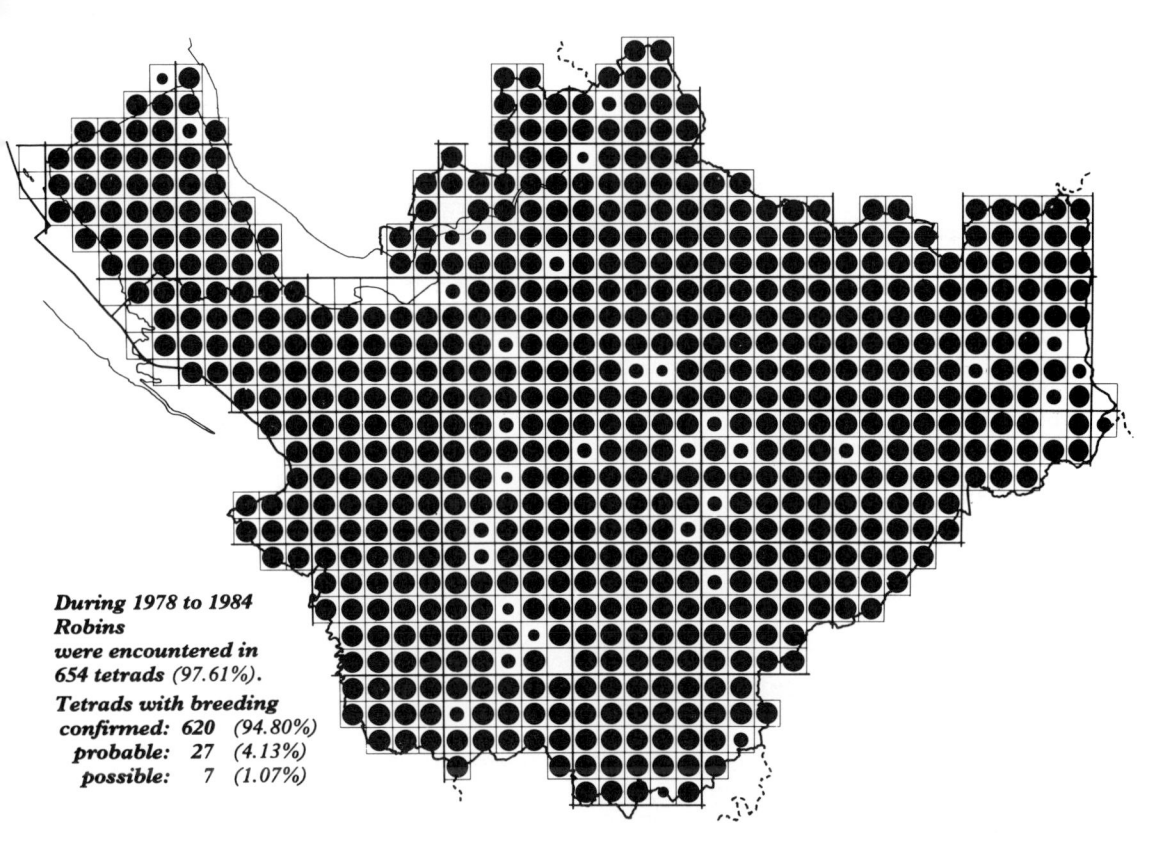

During 1978 to 1984 Robins were encountered in 654 tetrads (97.61%).
Tetrads with breeding
confirmed: 620 (94.80%)
probable: 27 (4.13%)
possible: 7 (1.07%)

to seven eggs. Boyd ringed a brood of seven at Marbury in May 1924, and in 1983 a nest in the Overleigh Cemetery at Chester held eight young on 14th May. Two broods are often raised, generally in different nests, although Boyd[6] recorded at least one instance of the second brood being reared in the same nest as the first. At Rostherne in 1977 two broods were reared from the same nest in a garden shed. In 1943 a pair at Plover's Moss is said to have had four broods, three of them successful, the fourth being taken by a Jay[3].

Robins vigorously defend their breeding areas, birds at the boundaries of their territories puffing out their breasts to show as much red as possible. An extraordinary display of violence was recorded at Higher Bebington, Wirral in 1977 when a Robin whose young had just hatched proceeded to attack and kill ten-day-old Song Thrush chicks in a nest 15 metres away (Coffey & Boyd 1978).

Coward & Oldham[2] stated that, "except upon the bare hilltops in the east of the county, the Redbreast is at all seasons one of our best known birds; it is common in the woods and game coverts in the open country, as well as in the gardens of houses in the towns....". There has clearly been little change in status since, and the Robin remains one of our commonest birds in farmland, woodland and suburb.

Most survey workers experienced little difficulty in confirming breeding. Anxious Robins give a high-pitched wheeze which is often an indication that a nest is nearby. Later the adults carry grubs back to the nest, and finally the spotty-plumaged young, out of the nest, beg noisily to be fed.

The Robin is one of our most widespread birds. Most of the tetrads without proof of breeding were urban (Wallasey, Runcorn, Widnes, Warrington, Northwich, Crewe, Middlewich, Macclesfield) – where they definitely breed but it may be difficult or embarrassing for survey workers to search for birds. Other gaps are seen in SJ55/56, where some tetrads were poorly covered for this survey, although the species probably is scarcer there than elsewhere. The only parts of Cheshire truly lacking breeding Robins are the barren hills, as was the case a century ago.

Most British Robins are sedentary, sometimes spending all their adult life within an area of less than half a hectare, as exemplified by a bird inhabiting a ringer's garden at Bidston, being trapped on over 70 occasions between 1977 and 1983. On the other hand, another Robin ringed there on 30th August 1981 was found dead in the Isle of Wight on 10th January 1982, perhaps a response to the exceptionally hard weather in December 1981.

In 1978 six wooded farmland plots in south-east Cheshire totalling 393 hectares held 132 pairs. At Risley Moss, between 36 and 44 pairs were found each year from 1978 to 1980 on 73 hectares (including 25 hectares of woodland). 18.2 hectares at Bidston Hill held from 28 to 38 pairs between 1975 and 1984. 22.8 hectares of broad-leaved woodland at Rostherne held from 34 to 47 pairs between 1976 and 1978. At Alderley Park 13 hectares of conifers and woodland with rhododendrons held 15 pairs in 1978 and 16 in 1979.

The above figures suggest densities of between 90 and 190 pairs per square kilometre in woodland habitats and 33 pairs per square kilometre on wooded farmland, similar to the national CBC results. Open farmland of a type that covers large areas of the county may hold fewer birds. An estimate of 35,000-40,000 pairs in Cheshire and Wirral compares closely with Sharrock's[35] suggested national average of 1500 pairs per 10-km square.

REFERENCE

Coffey, P. & Boyd, A. (1978): Robin Killing Nestling Song Thrushes. *British Birds* 71: 463.

Nightingale
Luscinia megarhynchos

CHESHIRE has always lain just outside the normal range of the Nightingale: birds have traditionally bred further north on the eastern side of Britain, in south Yorkshire and Lincolnshire, but stopped at Shropshire on the west side. The reason may be climatic, with cooler, damper summers than those experienced in the south-east. However, the dense undergrowth produced in coppiced woods forms the species' favoured habitat, and coppicing has not been widely practised in Cheshire, at least in recent times. Board of Agriculture returns for 1905 show a smaller area of woodland regarded as coppice in Cheshire than in any other English county (Rackham 1976).

Nightingales have nonetheless overshot into Cheshire from time to time, taking up residence in scrubby thickets of varying description. Coward & Oldham[2] listed eleven occurrences from within the present recording area from 1862 until 1908. No nests were found, but they considered it undoubted that the species had bred on some occasions. Perhaps the best evidence, in 1896, came from Romiley (now GMC) where a bird was seen carrying food. One at Lymm in 1865 so captured the public interest that a special Nightingale train was started from Manchester to hear it! Boyd[5] mentioned the "popular delusion that any bird singing by night must be a Nightingale", but he had heard only two in Cheshire – at Warmingham, Crewe, in 1912 and at Barnton, Northwich, in 1926.

Bell[7] lists a further twelve records up to 1961, though he considered one of these doubtful, and a record of a nest in a garden at Hatherton, where the bird was never heard, also seems suspect. A record in May and June 1954 referred to a pair at Pulford where a bird was again present in 1956. All these records came from the south, west or central parts of the county, as did a further three records up to 1966[9].

Subsequent *CBRs* refer to two singing near Crewe in May and June 1968, a locality where breeding was said to have occurred in the past, and once again at Crewe Hall in 1974. A passage bird was seen on Frodsham Marsh in August 1973.

During the present survey a bird was ringed at Bidston Hill on 24th April 1978 and remained until 28th April; one was reported at Risley Moss from 27th to 31st May 1978; a migrant was on Hilbre Island on 15th May 1979 and in 1980 a bird sang infrequently from willows and thorn scrub at Fodens Flash, Sandbach, between 14th and 17th May, and on 16th May one was singing at Caldy, Wirral.

REFERENCE

Rackham, O. (1976): *Trees and Woodland in the British Landscape*. Dent.

Black Redstart
Phoenicurus ochruros

COWARD & Oldham[2] could cite only three records. One of these was a report of a pair in old trees in Eaton Park on 7th May 1888 although they considered that a very late date and it was more likely to have referred to [Common] Redstarts. Bell[7] gives five records between 1910 and 1939, one of them in the breeding season, and mentions 15 further records up to 1960 "mostly from Wirral for January, March, April, May, July, October and November".

In the 24 years up to 1984 there were approximately one hundred further records, predominantly between February and May, and in October–November. Along with this general increase in sightings came the first breeding records.

1973 A male was singing on territory at Chester in early May with a female seen occasionally. A nest with four eggs was found deserted and the female was subsequently found trapped in a building. Following her release on 13th June a second attempt to nest was thought to have been made, for the pair was still present on 25th July. In late July an old nest, probably of this species, was found nearby.

1974 A nest found with five eggs on Hilbre on 23rd June was vandalised the same day. Only the female was ever seen. A pair at Prestbury from 15th June to 7th August left when vandals entered the derelict farm building in which they were thought to be nesting.

1976 A male was singing from the top of an office block at Wilmslow on one day in June.

1977 A female was watched feeding newly fledged young in the Birkenhead Docks in July and a female was seen in the same area on 16th August.

There have been no further breeding attempts reported in Cheshire or Wirral.

In Greater Manchester two young were reared in 1977 and five territories were located in 1981[46]. Black Redstarts bred in Liverpool in 1979. The species is also established in the West Midlands with nine pairs in 1977 and six in 1978. The preferred habitat of "large, sometimes derelict, industrial installations near to water, especially canals"[47] is amply represented in Cheshire, so future colonisation remains a possibility. A female ringed in the Netherlands on 28th March 1982 was found only five days later, trapped in a stairwell at Upton High School, Upton-by-Chester, having travelled an average of over 100 kilometres a day.

Redstart
Phoenicurus phoenicurus

THE Redstart is primarily a bird of the upland parts of the county, being most typical of the mixed oakwoods of the Pennine foothills where it also occurs amongst hawthorn scrub with scattered trees, extending westwards along the Dane valley to Timbersbrook and with an outlier on the sandstone escarpment at Alderley Edge. Small distinct populations inhabit the Bickerton and Peckforton Hills, and Delamere Forest where birds frequent the ageing broad-leaved trees which border the conifer plantations. Elsewhere, a sprinkling of pairs inhabits the Dee valley south of Aldford, and in 1981 a pair reared two young in a small stand of oaks bordering a brook through the plain near Wettenhall.

In the eastern hills, many nests are situated in cavities in drystone walls or in natural holes in trees – rowan being especially favoured. In 1919 a pair built their nest in a heap of dry stones in Windmill Wood, Alderley Edge. In Delamere however, nests on the ground amongst dead bracken are more normal (for example seven nests out of eight located in 1964 and all those found – at least thirteen – in 1965). By way of contrast, C. G. Bennett (pers. comm.) found a nest, from which the young flew, 18 metres up an elm tree in Bramhall Park (now GMC) in

193

about 1960. Coward & Oldham[2] knew of nests built on beams beneath the eaves of buildings, Boyd[5] heard of a pair that built inside a lamp outside a friend's garage in 1938 and a pair raised a brood of six from a clutch of seven eggs inside a stable lamp at Plover's Moss in 1948[3] – such sites are indicative of how familiar the species must once have been.

Coward & Oldham[2] stated that numbers were increasing throughout the centre and east of the county, and in Delamere and certain well-wooded parks it was abundant. Abbott plotted six singing birds in Alderley Woods in 1919. It was decidedly common in the eastern hills and the *L&CFC Report* for 1949 suggested that it was commoner than recent notes had implied, and had possibly become more numerous there in the preceding years. In Wirral, the species was and still is scarce, Hardy[4] mentioned Oxton, Bebington and Eastham as localities where they were found during the last century. However, odd pairs bred in at least five years between 1955 and 1965 and probably also 1968, 1970 and 1971. Around Chester, where the bird had formerly been common, it had become rare, although a few pairs still bred in Eaton Park early this century. By 1943/44 it was considered to be a notable absentee near Eaton Hall although there were four pairs within 100 yards of a house at Kelsall in 1943[3]. A bird seen between Church Minshull and Worleston, where Redstarts are rarely found, in July 1945 was possibly on passage.

Bell[7, 9] makes no reference to subsequent nesting in the Chester area, although four or five pairs were recorded around Aldford in 1965; the 1971 *CBR* stated that it "remains a common species" along the Dee south of Aldford and in the Peckforton area, and in 1972 several pairs were breeding in an area with overgrown hedgerows and scattered trees in the Dee valley south-west of Chester[8]. The species was, and still is, common in the sessile oakwoods of Clwyd just a few kilometres west of the Dee.

In common with other trans-Saharan migrants, numbers have been falling for fifty years or more. Boyd[4] commented in his country diary for 21st April 1934 on the Redstart's scarcity around Great Budworth in "recent years". However[6], he considered the species was plentiful in Delamere Forest until considerably later. Twelve pairs were nesting alongside the main "switchback" road through Delamere in 1964 and thirteen in 1965, but in 1969 following severe drought in the Sahel region of the southern Sahara, only one or two pairs appeared. Numbers had recovered to some five pairs by 1973, but further drought in the Sahel was thought to have brought about a corresponding reduction in 1974 – although six males turned up in spring, only two found mates. By 1982, the population here had recovered to five pairs only. National CBC figures for 1984 indicated that Redstarts were continuing to recover and Marchant (*BTO News* Sept-Oct 1985) suggested that this species was perhaps less dependent on rainfall in West Africa than had previously been thought. Pine-clad heath is said to constitute the ancestral habitat of the Redstart, and the loss of wooded heathland this century may have played a part in the species' decline.

Records submitted to the *CBR* in recent years suggest a county population in the region of 105 pairs, 90 of these in the eastern hill areas, where, for example, at least 19 territories were occupied (eleven of which were located in Lyme Park) in 1982 and at least 30 pairs were breeding in SJ96 in 1983. The decline in numbers is not yet so noticeable in this stronghold.

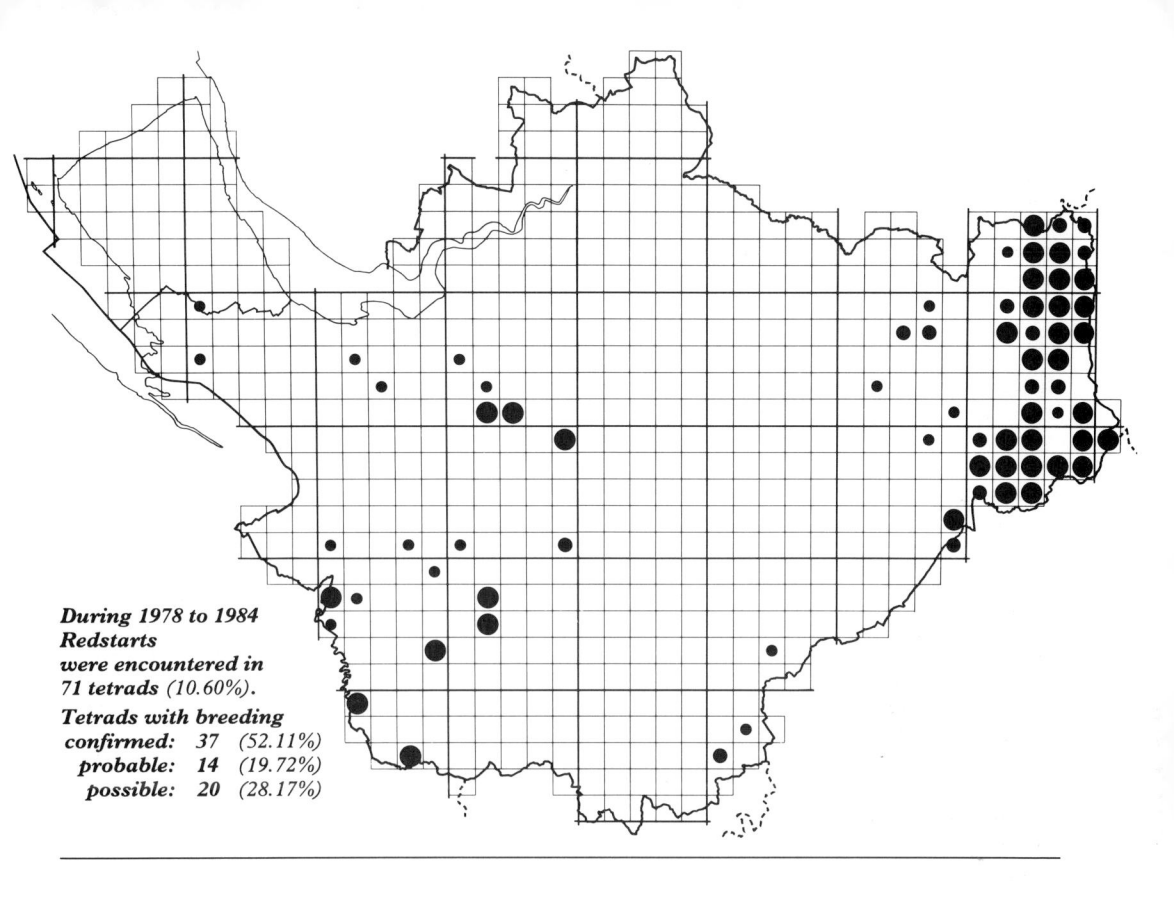

During 1978 to 1984 Redstarts were encountered in 71 tetrads (10.60%).

Tetrads with breeding
confirmed: 37 (52.11%)
probable: 14 (19.72%)
possible: 20 (28.17%)

Whinchat
Saxicola rubetra

AT the turn of the century, the Whinchat was a widely distributed breeding bird throughout Wirral and lowland Cheshire generally[2]. It was especially plentiful in low-lying meadows along the Mersey between Stockport (now GMC) and Warrington, and in the marshy fields of the Gowy valley. A few pairs bred in the hills east of Macclesfield, but it was not so numerous here as in low-lying areas, although birds did breed at 1260 feet on the summit of Bosley Minn. Abbott's maps show 36 territorial males in 22 square kilometres around Wilmslow in 1919-22: five along a railway embankment, 15 on mossland, one on heathland, 12 on farmland and three at a sewage farm. All the farmland sites were on the periphery of Lindow Moss, an area still characterised by small, damp fields. Abbott also saw two or three males at Bagmere on 19th June 1926 – a site occupied into the 1970s – and adults with young in Danebower Hollow on 2nd August 1926.

Boyd[6] noted a decrease from around 1930 but could still list five or so sites in the vicinity of Northwich where a few pairs bred in 1951. Hardy[4] also commented that the species was becoming scarcer as a nester, listing fields at Hale, the Gowy valley, Mount Manisty and the Frodsham–Helsby marshes as breeding sites. Griffiths & Wilson[13], noting Coward's assessment of the bird's status in Wirral, commented: "There has evidently been a considerable reduction in the breeding population". The only nesting site of which they were aware was at Leasowe, where there were some six pairs in 1945. At Woolston A. R. Sumerfield considered the Whinchat to be a common breeding bird around 1950, but by 1981 it was known essentially as a passage migrant (*WECG Report* 1981). At Sandbach Whinchats ceased nesting about 1951[19]. The 1963 *L&CFC Report* noted breeding at Millington and presence in the Northwich district, but absence from Prestbury,

Wilmslow, Morley and the Nantwich district.

The decline continued up until and during the present survey. In 1974 pairs bred at Bagmere and Cledford lime-beds, but by 1977 the species was restricted to the hills as a breeding species in eastern Cheshire[21]. The reasons for this decline are obscure, but related in part to more intensive land usage. Boyd blamed cultivation of an area of rough grass and brambles for reducing the numbers at Whitley Reed, and Bell[7] later speaks of "restricted nesting requirements" as regulating the population; wasteland, mosses and rough upland pastures then being its chief haunts. Indeed, mowing of roadside verges and "tidying" of rough ground have been blamed for a decline throughout southern and eastern England in the last 60 or 70 years[35] and remaining populations are largely in upland areas of north and west England, Wales and Scotland. In Cheshire at least, the decline has outstripped the rate of habitat destruction, however, and other factors must be contributing to its demise. On Carrington Moss (now GMC), R. Harrison reports that birds formerly bred in rhubarb fields until the plant was replaced by synthetic chemicals in commercial jam-making, and doubtless the same crop was once favoured around Lindow Moss.

In the eastern hills, the Whinchat is now very scarce, having declined further during the course of the survey. Remaining territories are often associated with stands of bracken or young plantations where cessation of grazing provides temporarily favourable sites. Lowland territories are scattered along the valleys of the Dee, Gowy and Mersey where suitable rough ground persists, the most important colony being on the grassy banks of the Frodsham Marsh sludge-pools. DN took a particular interest in this latter site, noting that during this survey period the population declined dramatically, with nine pairs in 1982, seven pairs in 1983, and only two pairs in 1984, followed by just two unmated males in 1985, and no birds at all in 1986. An article in a cage-bird magazine described a rough and ready method for finding nests at Frodsham, but such illegal behaviour is not thought to have brought about the decline there.

At Risley Moss eight pairs were present in 1978, with six pairs in the following two summers. With probably fewer than 50 pairs left in Cheshire and still declining, the future looks bleak. However, the population in the neighbouring Clwyd hills seems to be buoyant, with fifty or more nests found annually in an area of only a few square kilometres, the numbers found being limited only by the energy and spare time of the ringers concerned! It is far from clear why the Cheshire population should be declining so rapidly, but the withdrawal from the east and centre of the county recalls that of other trans-Saharan migrants such as the Sedge Warbler and Whitethroat.

Stonechat
Saxicola torquata

THE cheerful song of a Stonechat from its perch on a flowery gorse bush is all too scarce a sound in Cheshire at present. This is a consequence of severe winter weather in 1979 and 1981 leading to starvation in this essentially insectivorous bird. Few Stonechats migrate, most preferring to stay and battle through the British winter. In fact their distribution follows the milder west coast – a pattern reflected by the coastal bias in Cheshire records. Coastal heathland forms the typical habitat, the sandstone hills and sand-dune areas of Wirral being ideal, though areas of rough grassland with some scrub of

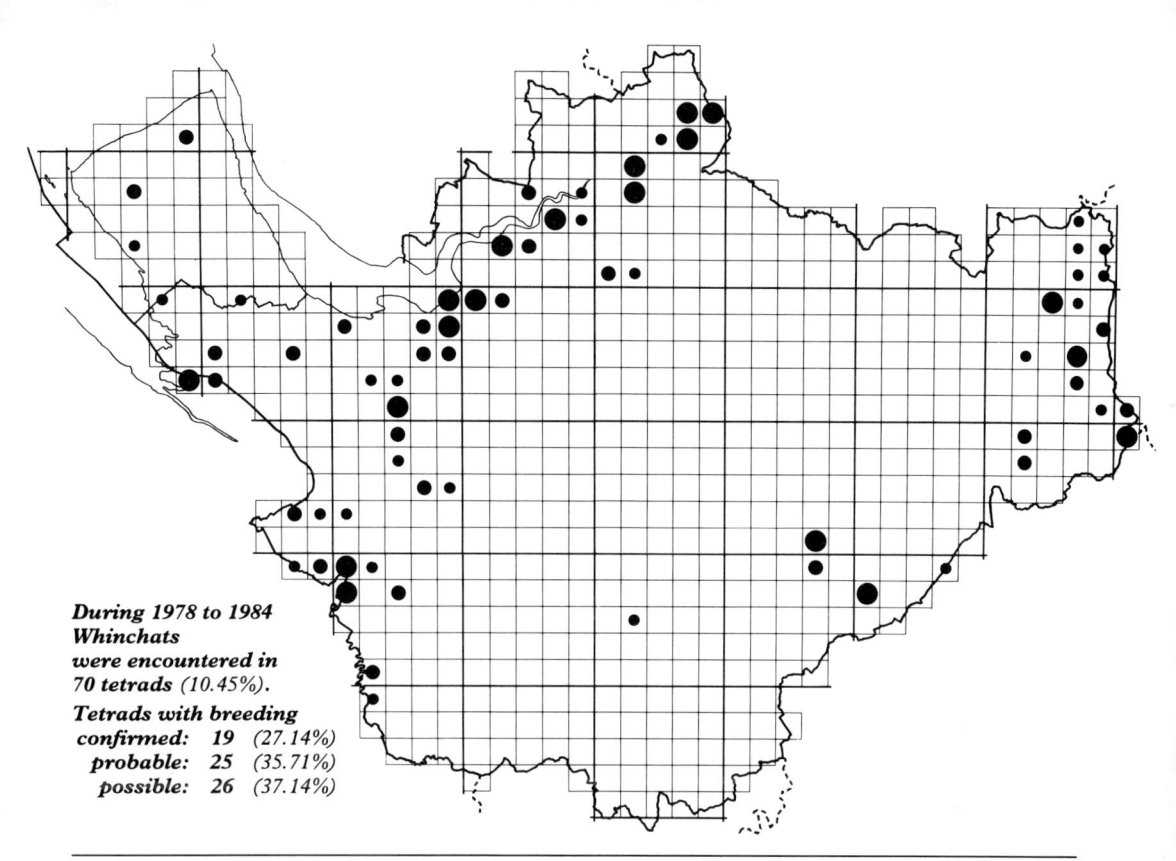

During 1978 to 1984
Whinchats
were encountered in
70 tetrads (10.45%).

Tetrads with breeding
confirmed: 19 (27.14%)
probable: 25 (35.71%)
possible: 26 (37.14%)

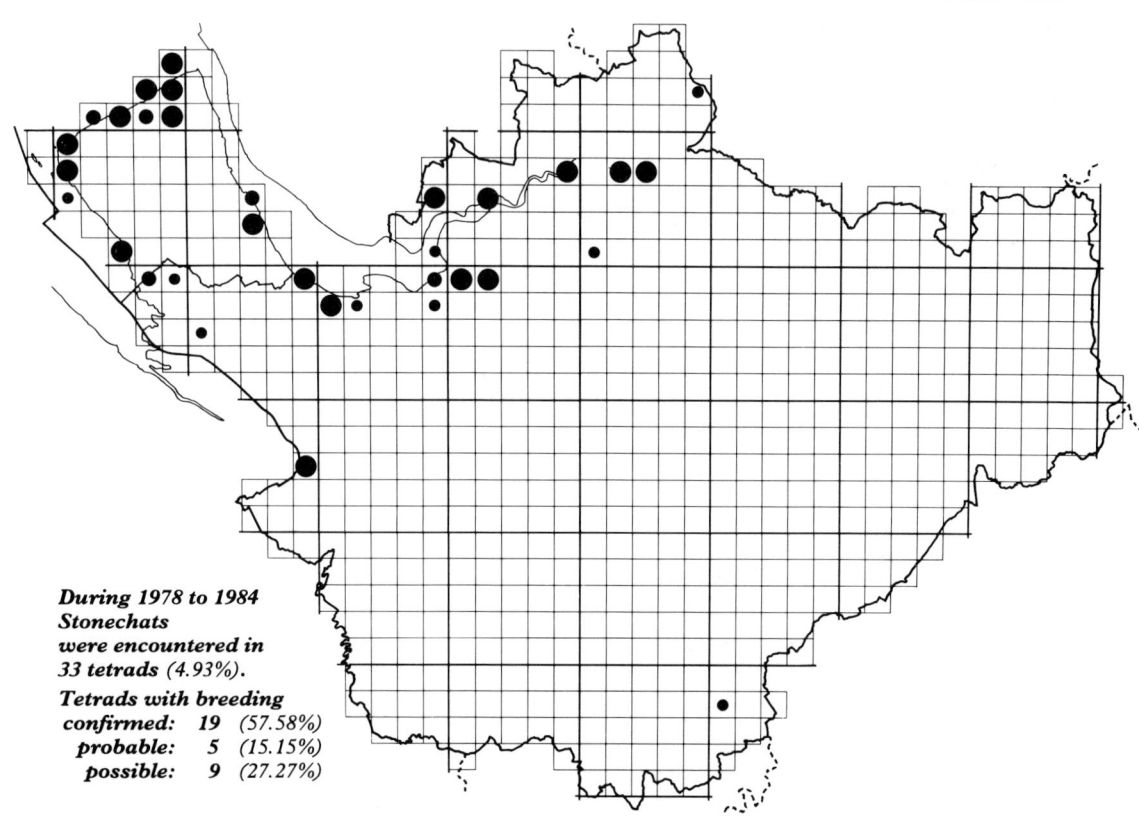

During 1978 to 1984
Stonechats
were encountered in
33 tetrads (4.93%).

Tetrads with breeding
confirmed: 19 (57.58%)
probable: 5 (15.15%)
possible: 9 (27.27%)

gorse or bramble are also occupied, as on the banks of settling-tanks at Frodsham, Rocksavage and Woolston. Numbers have always fluctuated according to the severity of winter weather, but the destruction of heathland habitats for golf-courses, forestry, agriculture or building accounts for an underlying downward trend.

In Wirral, Brockholes[2] described Stonechats as abundant in summer in suitable localities and Coward & Oldham[2] found them not uncommon (though not so abundant as Whinchats!) nesting on heathy hills and warrens near the coast. In his diary for 1935 Boyd referred to their breeding freely at the coast, and this situation pertained, with nesting in the hills at Caldy, Bidston, Heswall and Thurstaston, and on some of the golf-links, until the severe winters of 1939-40 and 1940-41. In 1940 only one pair was found on Caldy Hill, and none at all in north Wirral in 1941. In 1942, a pair resumed nesting at Wallasey[13]. Over the next 15 years breeding was only intermittent, and in 1960 only one breeding-site was found[18].

There then followed a period of five years in which no breeding was reported, ending in 1966 when a pair reared two young on Bidston Moss. Subsequently, *CBRs* list up to four pairs or more each year up to the start of this survey in north Wirral. In 1968 a census of the coastline on 24th March by D. L. Clugston and J. R. Mullins revealed thirteen pairs, though only three remained to breed[8].

Inland, there is less of a tradition of breeding. Coward & Oldham[2] mentioned a few pairs nesting annually on the Peckforton Hills and in June 1905 Oldham saw a fledged youngster on heathy ground by Oakmere. Birds nested in a few places in the hills east of Macclesfield, as at Shutlingslow, but some measure of scarcity can be gleaned from their correspondent, N. Neave, who had never seen the species at Rainow. In the late eighteenth century (1789) however, Stonechats were described as common in the High Peak area of Derbyshire – just over the county boundary[45].

In 1927, Stonechats were again reported to be nesting regularly at Oakmere, and Boyd[5] found nests more than once on the heaths of Delamere Forest around this time, perhaps at this same site. Birds were present by the Manchester Ship Canal at Lymm in several years between 1913 and 1943 with breeding twice confirmed here. Hardy[4], in addition to a list of Wirral sites, mentioned presence at Burtonwood until the frosts of 1940. Over the next twenty years or so inland breeding was noted only from the area of the Weaver estuary[7], and presumably only infrequently at this site. A pair attempted to nest in Runcorn Docks in 1961.

The next inland breeding followed in 1971 when pairs bred at Moss Side (Moore) and probably also in the Frodsham area where young birds were seen in July. In 1972, several pairs nested around the sludge-pools at the latter site; in 1973 two pairs were present with a third pair on Frodsham Score; and at least single pairs were present in each season up to the start of this survey. Moss Side was also occupied in 1975 and 1976, and in 1977 pairs were recorded from Stanlow and New Ferry, as well as Parkgate and Denhall, well inside the Dee estuary.

Thus the population at the start of the survey was higher than at any time since the 1930s, but severe weather in 1979 brought a marked decline. Whereas seven or more territories were reported in the 1978 *CBR*, only one or two were reported in 1979. By 1981 numbers had largely recovered and seven or eight pairs were notified, but a prolonged cold spell up to Christmas of that year seems to have wiped out the species as a Cheshire breeding bird for the time being. No territorial birds were reported up to 1984.

In 1980, a pair nested in a garden centre at Meols, Wirral, and successfully reared five chicks. In 1981, a pair was present in late spring at a site south of Crewe. There have been no recent summer records from the hills, although birds were seen in the mid-1970s on the adjacent gritstone moors of Derbyshire and north Staffordshire, and odd pairs bred in the Clwyd hills even after the 1981/82 winter.

Not all territorial records during the survey found their way into *CBRs*, and indeed the map, showing 19 tetrads with confirmed breeding and a further thirteen with minor evidence, suggests that in the best years (1978 and 1981) there may have been 15 to 20 pairs in the county. Records from inland sites along the Mersey valley date from these years.

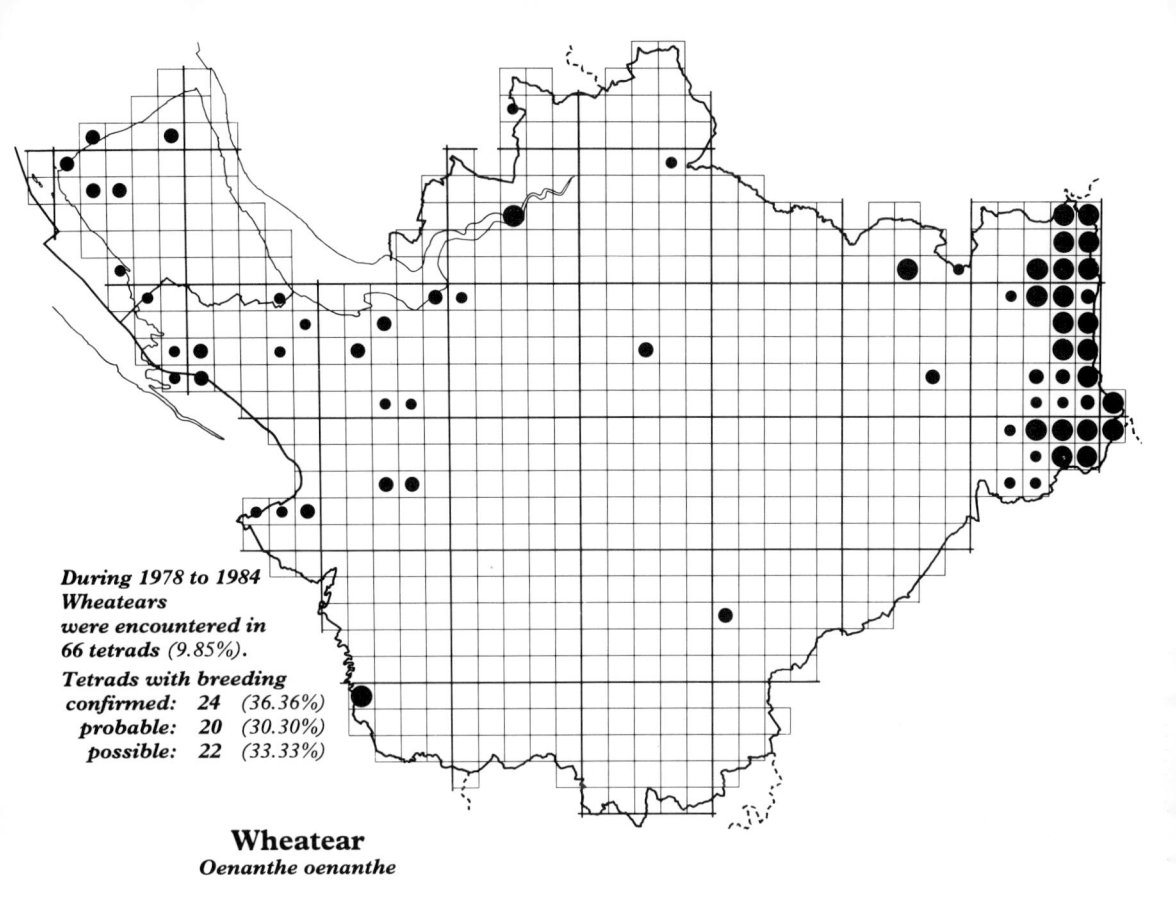

During 1978 to 1984 Wheatears were encountered in 66 tetrads (9.85%).

Tetrads with breeding
confirmed: 24 (36.36%)
probable: 20 (30.30%)
possible: 22 (33.33%)

Wheatear
Oenanthe oenanthe

THOUGH nowadays largely restricted to the hill country of Cheshire, the Wheatear is a bird associated with short, tightly grazed pasture wherever that may be found. In southern England it is found on chalk downland when these are grazed by sheep and rabbits, and the sandy heaths of the Breckland are also a stronghold. Brooke (1971, 1981) has described the manner in which Wheatears use their territory for hunting, running forward across the short turf and scanning it for ants, beetles, larval insects and other prey. Occasionally they vary this technique by standing on a boulder or fence-post to watch for flying insects and hunt in the manner of a flycatcher.

Nests are built on or below ground – in rabbit burrows, under boulders, or within tumbled gritstone walls. When near to fledging, the young gather at the entrance to the nest in a noisy party begging for food, and recently fledged young, accompanied by their parents, are also very conspicuous. These habits make it easy to confirm breeding in this species, and the regularity with which second broods are reared means that there often two chances of doing so each season.

Coward & Oldham[2] regarded the species as plentiful in the eastern hills and well known as a breeding species in Wirral, including Hilbre Island, and along the coast to the Frodsham marshes. Specified nesting sites elsewhere included Heswall Hill, Woodley (Hyde – now GMC) and Lyme Park. Coward also mentioned the Dee Marsh as a favourite nesting haunt, but he failed to find any there on 30th June 1917[3]. Hardy[4] described the bird as a decreasing nester in Wirral, with records from Storeton, Mount Manisty, Bromborough, irregularly at Hilbre, and elsewhere. In 1948 a pair bred at Bidston

Moss, but it was by then a scarce breeder on the peninsula[3]. Bell[7] agreed that the species had declined in Wirral but added that it was still present, and *CBR* for 1968 stated that two pairs bred that year in the sea-wall at Parkgate, and that a pair was thought to have bred on Bidston Moss. Craggs[25] however does not mention it as a breeding bird on Hilbre Island, and it must have died out there some time ago. On the other hand the scatter of mapped records in Wirral, particularly of 'probable' breeding recalls Coward & Oldham's account of its distribution. This would match the decline in lowland England documented by Sharrock[35], attributed to the reduction of sheep farming in downland areas, and disappearance of rabbits following myxomatosis in 1953-54, both changes resulting in the loss of its favoured short-grassed habitat.

The results of the present survey make it clear that the hill country is now the stronghold of the species in Cheshire: all but two of the tetrads with confirmed breeding are there. It is difficult to evaluate the scatter of records elsewhere in the county. Boyd[5] noted that they often turned up at sites where they did not breed, as on 21st April 1934 when he "saw them in two places which are visited annually on passage, although the birds never breed there". There would appear to be plenty of time for some birds to wander around, since they may arrive from mid-March, with the bulk of birds seen in April, yet young are seldom hatched before June. Coward & Oldham remarked that they often saw Wheatears in the plain, with no evidence of breeding activity, and assumed that these were wanderers or passage birds: young from first broods could account for some of these records. However Wheatears often linger at industrial sludge-beds in spring, and have been suspected of breeding intent at Sandbach, Witton and elsewhere. In 1968 a pair bred in the Frodsham area and breeding was suspected at Wallerscote. The BTO Atlas included a record of confirmed breeding in SJ67.

From fieldwork in the hill country, it is unlikely that each tetrad contains more than two pairs, and many will hold only one. In 1982, JSAH found ten pairs in SJ98 and R. Blindell has estimated ten pairs in SJ96. There may be another twenty pairs in the remainder of the Cheshire hills, plus a sprinkling of pairs elsewhere. Combining the 'probable' and confirmed breeding records it seems that the Cheshire population is no more than 45-50 pairs.   DWY

REFERENCES

Brooke, M. de L. (1979): Differences in The Quality of Territories Held by Wheatears (*Oenanthe oenanthe*). *Journal of Animal Ecology* **48**: 21-32.
Brooke, M. de L. (1981): How an Adult Wheatear (*Oenanthe oenanthe*) Uses its Territory When Feeding Nestlings. *Journal of Animal Ecology* **50**: 683-696.

Ring Ouzel
Turdus torquatus

THE Ring Ouzel is a frustrating bird for the ornithologist to study. When nesting the bird is vocal enough; its harsh chattering announces clearly its presence. Prior to this however the song of the male, though loud and distinctive, is very inconsistent. Whereas most male song birds sing in competition with each other, enabling the listener to score territories, Ring Ouzels seem instead, nowadays, often to sing solo. They can also be very secretive when approaching or leaving the nest, though with time and patience they are easy enough to "watch back". The choice of nest-sites, often on steep, rocky hillsides which do not make for easy walking, adds to the difficulty of confirming breeding unless the adults are seen carrying worms back from a nearby field.

The species is confined in Britain to mountain and moorland areas, that is to the north and west, but its habitat is essentially the edge of such moorland. However, the species has been in a modest decline in both range and numerical status in this country,

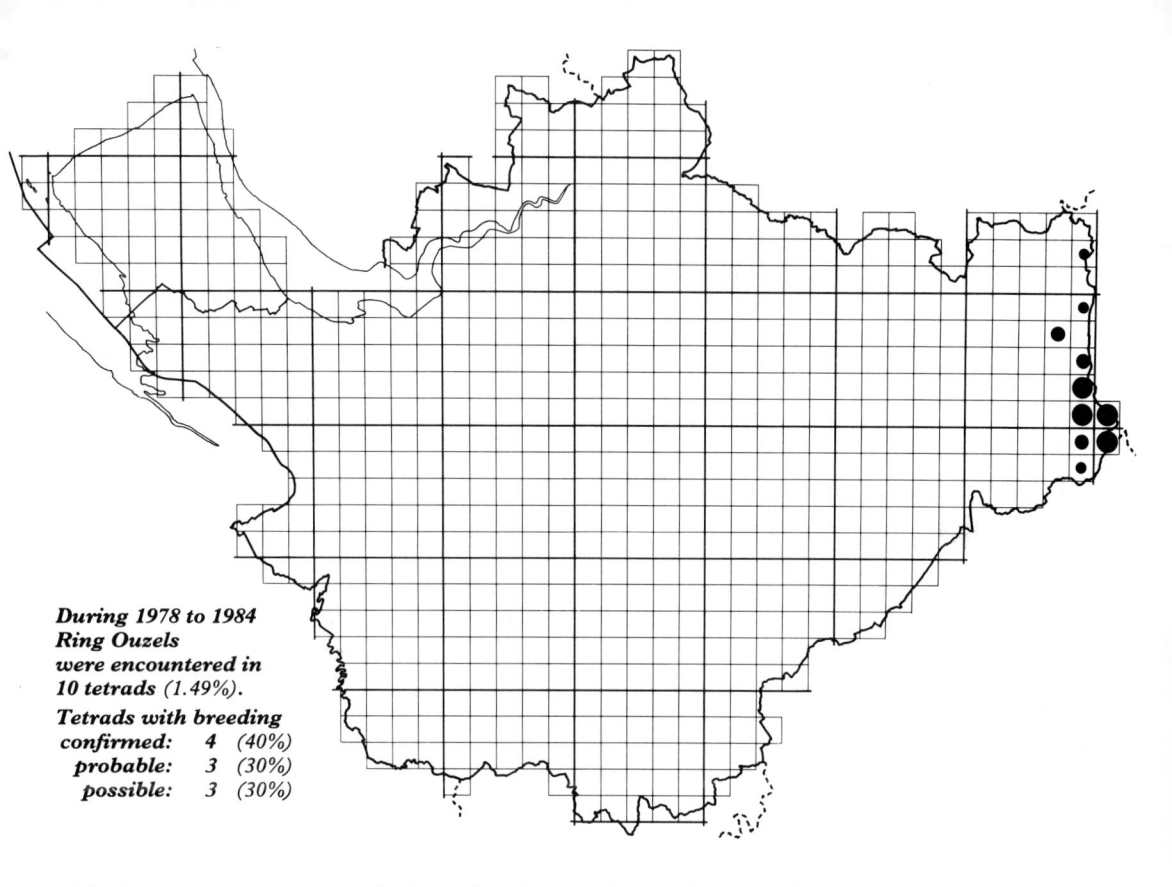

*During 1978 to 1984
Ring Ouzels
were encountered in
10 tetrads (1.49%).*

Tetrads with breeding
confirmed: 4 (40%)
probable: 3 (30%)
possible: 3 (30%)

with the retreat most obvious in Ireland and in Cornwall[35]. There is some evidence that this decline has been the result of climatic change and competition with increasing populations of Blackbirds and thrushes: certainly this change has affected the local Peak District population of which the Cheshire fringe is a part. Coward & Oldham[1] regarded the species as abundant in the hill country from Disley southwards to Bosley and eastwards to the Derbyshire border – half a dozen males might have been heard singing at one time. They also quoted a few records of breeding from the Bickerton Hills, Overton Hills and Helsby Hills in the nineteenth century. Bell[7] clearly felt it was much less common, and confined to the hill country, which is the current status. Indeed it seems that Ring Ouzels may be retreating to even higher ground.

The Cheshire population seems to be limited to between nine and twelve pairs each year, though few of these were confirmed as breeding pairs during this survey. Observers covering the hill tetrads noticed a decline even during this survey, and during the 15-20 years up to 1984, territories at Shutlingsloe, Whaley Moor and Saltersford were deserted. Spread between ten tetrads the density of surviving birds is clearly very low. One must note too that there is some opportunity for mapping pairs twice in one breeding season, for Durman (1977) studying colour-ringed birds in southern Scotland, showed that birds which lose a nest to predators may move 500 metres or more to a new territory for their second breeding attempt.

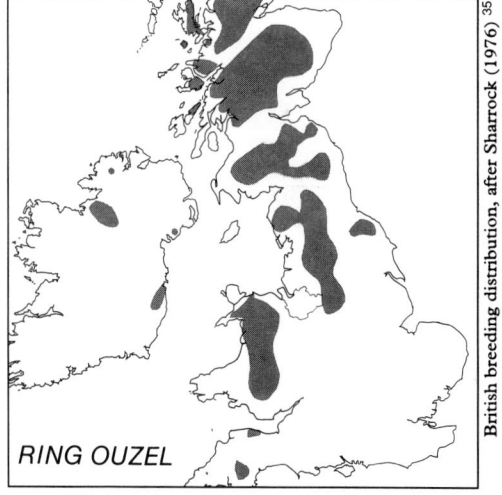

The relatively sparse population in Cheshire and the retreat from the lowlands

RING OUZEL

British breeding distribution, after Sharrock (1976)[35]

matches the pattern in the neighbouring counties. Both Frost[45] and Harrison et al.[47] report that the species formerly bred in limestone areas, but now very rarely does so. In Derbyshire the population may be 200-250 pairs, but in Staffordshire is no more than 20 pairs. There are several reports of flocks feeding on berries of bilberry in July and rowan in August and September, and it may be that feeding up on these is an essential prelude to migration southwards to their winter quarters in southern Europe and northern Africa. In that case another possible cause of the decline is the agricultural change which has taken place, in particular the increase in grazing and the loss of moorland from the limestone areas (Anderson & Yalden 1981). Clearly this is a species whose future needs to be carefully monitored, both within Cheshire and nationally.   DWY

REFERENCES

Anderson, P. A. & Yalden, D. W. (1981): Increased Sheep Numbers and the loss of Heather Moorland in the Peak District. *Biol. Cons.* **20**: 195-213.

Durman, R. F. (1977): Ring Ousels in the Pentlands. *Edinburgh Ringing Group Report* **5**: 24-27.

Blackbird
Turdus merula

LIKE all more recent Cheshire authors Coward & Oldham[2] regarded the Blackbird as "...exceedingly abundant, the fertile plain supplying the conditions it loves". It was and still is plentiful in the wooded valleys of the east, nesting up to the edge of the moorlands, where its breeding ground overlaps that of the Ring Ouzel. Indeed there has been some upward spread by this species in parallel with the Ring Ouzel's withdrawal. Recent conifer plantations in the hills have provided cover in areas which were formerly inhospitable to the bird. At the opposite end of the county two or three pairs nest annually on the Hilbre islands[25].

According to Parslow[34] the Blackbird was still restricted to its original typical woodland habitat until the mid-nineteenth century, being then unknown as a breeding bird in the neighbourhood of houses. At the beginning of this century it was described as suspicious and wary in the Greater Manchester area[46] and Coward & Oldham[2] specifically mentioned that, "occasionally the song is uttered from the ridge-tiles, gable-end, or chimney of a house". Subsequently the species has become

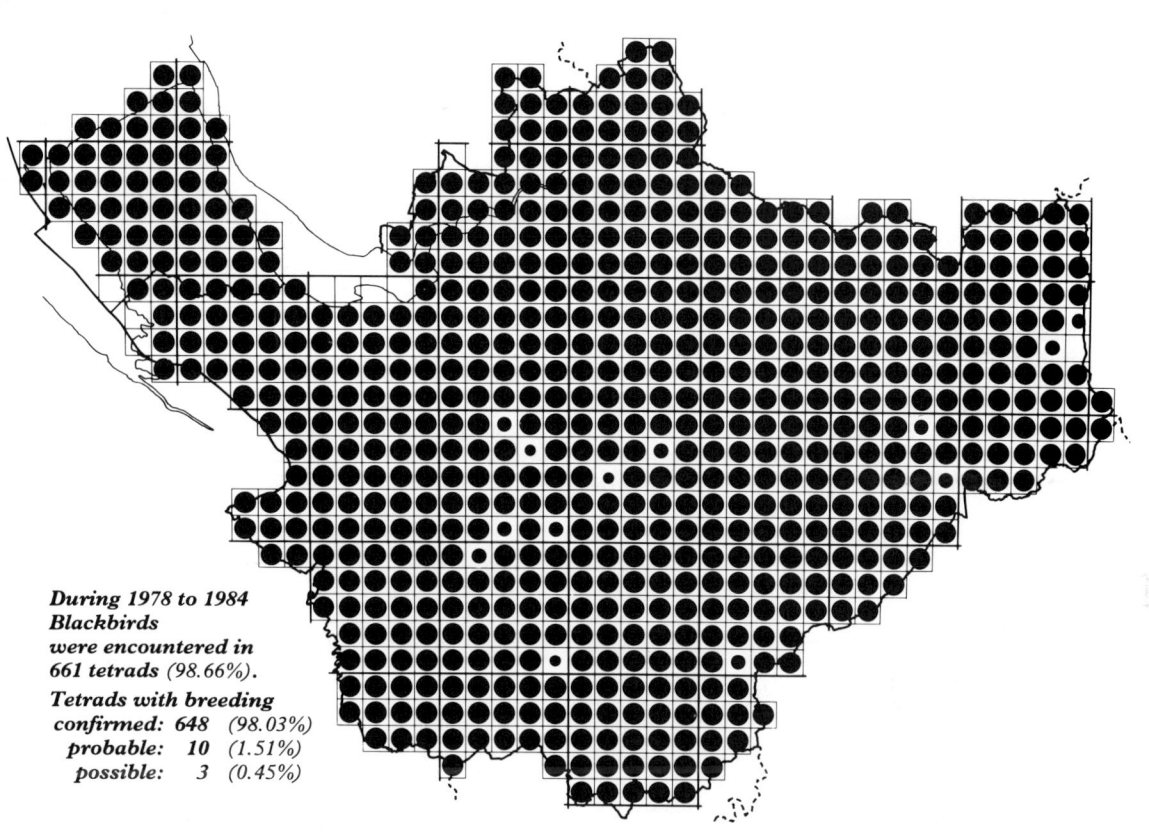

During 1978 to 1984 Blackbirds were encountered in 661 tetrads (98.66%).
Tetrads with breeding
confirmed: 648 (98.03%)
  probable: 10 (1.51%)
  possible: 3 (0.45%)

one of our most familiar garden birds, even in the centre of towns, and to the above list of song-perches, all now in frequent use, may be added television aerials and telegraph poles. The traditional tree-top remains as popular as ever.

Woodland edge seems to have formed the bird's original preferred habitat. At Alderley Park it has been noted that Blackbirds are most numerous in those parts of the woodlands with an understorey of rhododendrons. Many game-coverts and ornamental woods elsewhere offer similar cover, as do shrubberies in parks and gardens. Traditional Cheshire farmland with its network of hedgerows and scattered coverts supports many pairs. Numbers remain high in many parts of the county, but the species is becoming scarce in those southern parts of the plain where hedgerow trees are being felled and mechanical trimming of hedgerows prevents young saplings from taking their place. Blackbirds must have song posts. A hedge without trees does not meet this requirement.

The rich fluty song of the cock Blackbird is heard from February until July. In mild winters, and in the slightly warmer urban areas, occasional snatches may be given in January. Cold springs may hold back song until March. Once fully under way, however, it may even be given during brief flights, on deep Jay-like wingbeats, from one perch to another. In wet summers, when worms are more readily available, the breeding season may be extended into August and a third brood reared, song then being given until the end of that month. Shorter utterances from young, brown-plumaged males in June and July may also be heard.

In spring there may be noisy competition between the males for the females. Coward & Oldham[2] once saw six cocks chasing a single hen. More recently six cocks were fighting at Dean Row on 12th April 1983.

Boyd[6], who ringed 158 broods in the Frandley area in twenty years once saw a bird carrying nest material on 25th January, but, although nests were often built in March, he never saw eggs before the 17th of that month. Twice (in 1935 and 1950) pairs reared three broods in the same nest. Hardy[4] had a record of a nest at Mickle Trafford that was used for two broods in 1929 and three in 1930 – five broods from one nest. In 1944 a two-year-old nest was used at Wilmslow[3]. Adult birds may be seen carrying food for their young from the first week in April, and the first fledged young appear in the second half of that month. Males take over the responsibility of looking after fledged young from the first brood, allowing their mates to get on with a second brood almost immediately.

Blackbirds have been known to nest outside the usual season. In 1967, a nest with one egg was reported from Langley on 31st January. In December 1973, a pair took up residence in a heated, artificially lit factory at Ellesmere Port. The cock sang regularly and nesting might have been attempted had not the heating been switched off over the Christmas holidays. Between 20th and 25th December 1980, eggs were laid in a nest in a multi-storey car park in the centre of Warrington. The eggs hatched in January but the young disappeared when a few days old.

Most nests are built in hedgerows and low shrubs, or among shoots low down on the trunk of a tree. In 1984, a pair built in a clump of hemlock water dropwort at Neston reed-bed. A few nests are built on the ground. Several have been found in such situations at Danes Moss where suitable bushes are in short supply[26]. On Bidston Hill members of the Birkenhead School Natural History Society found Blackbirds breeding at from 25 to 60 feet up in pine trees, evidently in response to human disturbance[9].

Any species which associates so closely with human habitation is likely to use a wide range of artificial nest-sites. Nests are often built amongst the clutter to be found in any garden shed or suburban garage, provided access is available. Occasionally the nest of a Song Thrush is adapted. Boyd[6] recorded two such instances after the young thrushes had flown, but at Wistaston in 1968 a Blackbird was strongly suspected of expelling the thrush's young which were found dead beneath the nest[8]. At Rostherne an old nest was found in an owl nesting box.

Earthworms are a particularly important summer food for Blackbirds although they can adapt to such a varied diet that dry summers apparently cause them little trouble. In suitable areas they feed their young on caterpillars that have fallen from trees, sometimes to such an extent that the chicks' skin turns yellow rather than the usual dull pink (Snow 1958). Fruit is often fed to young in the nest, particularly ivy in April and May, and cherry in June and July. A Blackbird has even been recorded taking a fully-grown frog in a Birkenhead garden (Raines 1955), although this is unlikely to have been fed to chicks.

The Blackbird is noted above all our birds for frequency of albinism. Many cock birds have at least a few white feathers, leading to false records of "Ring Ouzels", although pied hens and complete albinos are comparatively rare. In April 1964 a pure white male with pink legs was photographed at Worleston, Nantwich, where it reared four young after the female disappeared. On 10th June 1983 an almost completely white cock was feeding young in a hedge at Frandley (JPG). Such birds might be expected to attract more than their fair share of attention from Magpies and other predators.

Sharrock[35] suggested an average probably in excess of 2000 pairs per 10-km square, i.e. 80 pairs per tetrad, throughout Britain. All available Cheshire CBC figures exceed this value. In 1978 six farmland plots in the south-east of the county, totalling 393 hectares were found to hold 172 Blackbird territories, although a 22% decrease was evident in 1979 following a severe winter. 73 hectares of Risley Moss, mostly mossland but with 25 hectares of woodland, have held between 14 and 21 pairs. Between 27 and 41 territories have been located in 18.2 hectares of deciduous woodland, heath and scrub on Bidston Hill from 1975 to 1983, though only 19 in 1984[29]. In 13 hectares of woodland at Alderley Park, 16 or 17 territories were found each year from 1978 to 1980. 23 hectares of woodland at Rostherne held between 12 and 37 pairs from 1976 to 1983, the highest figures of 30 and 37 pairs falling before the clearance of rhododendrons, with a maximum of 20 since.

All these plots contain above-average levels of woodland, reflecting an inevitable tendency for census workers to choose plots with at least reasonably varied bird populations. In the Chester area, although the species was found to be breeding in each of 115 tetrads, it was considered "not very numerous" over much of the farmland. A similar scarcity was found in large areas of southern Cheshire and probably a quarter of the county's tetrads may each hold fewer than 20 pairs. Nonetheless the Blackbird remains one of our most numerous birds with a likely county population in the region of 40,000–45,000 pairs.

REFERENCES

Raines, R. J. (1955): Blackbird taking frog. *British Birds* **48**: 185.

Snow, D. W. (1958): *A Study of Blackbirds*. British Museum, London.

Fieldfare
Turdus pilaris

THIS species has been known to breed in Derbyshire in five of the years since 1969[44], and in Staffordshire from 1974-1977[47]. There have been June and July records in Cheshire in 1968 and 1976, and in 1984 one was seen feeding in a small meadow at Cut-thorn Hill between 27th May and 2nd June.

Song Thrush
Turdus philomelos

THE Song Thrush is one of the few birds to sing regularly during the winter months, although the effort put into the performance is proportionate to the mildness of the weather. Prolonged frosts will cause a temporary cessation for a few days or even weeks. Boyd[6] recorded song throughout the period from November to July and once on 8th August 1924. On 17th September 1940 he heard a Song Thrush singing gently in a city park (probably Birkenhead). Each thrush has his own repertoire of phrases, some of them obviously picked up from the local environment. Thus Boyd[6] heard one imitate a Lapwing, and another copied his whistle well enough to deceive his own terrier! A bird at West Parkgate, Lyme Park, in the spring of 1983, often incorporated a Peacock's "mew" into its song – Peacocks being kept at a nearby farm – and on 27th April 1984 a bird at Alderley Park mimicked a Curlew.

Coward & Oldham's[2] summary of the bird's distribution deserves examination. They described it as "abundant everywhere in Cheshire from spring to autumn, from the sea-coast to the hills in the east, where it nests at an altitude of over one thousand feet in the wooded cloughs and even in the stone walls which separate the upland pastures. The large area of land under cultivation, plentifully supplied with woods and coverts, affords the bird ideal conditions of existence, and though no species suffers more at the hands of bird-nesting children, there is no diminution in its numbers, even in the immediate vicinity of towns."

Parslow[34] gives circumstantial evidence indicating a decline in this species, in many parts of Britain, continuing throughout much of this century and accentuated by the cold winters of the 1940s. For example during the years 1931-1939, Blackbirds accounted for just 46% of the combined total of nestlings of the two species ringed in Britain, yet in the years from 1948-1956 the proportions had changed so that Blackbirds made up 60% of the total. Whalley[19], writing in 1963, referred to a general reduction in numbers ringed around the Sandbach Flashes.

Song Thrushes suffer in severe winter weather. In 1917, "many were singing early in January, but most had left or were dead by the end of February" as a result of prolonged cold. Numbers largely recovered by autumn however[3]. Both Coward & Oldham[2] and Boyd[6] implied that the species was to a large extent a summer visitor, the latter author noting that, "many breeders go south and west, returning with some regularity in February, often in the second and third weeks". There is ample evidence of movement from ringed birds, such as the chick ringed at Mobberley on 25th April 1978 and found in Gironde, France, on 4th January 1979, and the adult ringed at Frodsham on 31st July 1981 that was in Northern Ireland the following March.

Boyd also stated that the cold of January and February 1947 almost wiped out his local residents, only one pair nesting in the following spring instead of the usual eight or so pairs. Cold weather in early 1929 had had a similar effect. An old neighbour of Boyd's, born in the 1830s or 1840s, had often noticed periodic reductions during his lifetime. Song Thrushes suffered from the great frost of 1963, but by 1966 they had regained their former numbers[9].

The species seems to be at a very low ebb at present, even allowing for the recent severe winters. It is now relatively scarce on the open hills and in intensively agricultural areas. An association with gardens is strongly marked both in summer and especially in winter. Dense shrubbery in urban and rural settings is also favoured, but the provision of food and the slightly warmer climate near towns probably account for the thrush's preference for suburbia.

Because of the scarcity of lime in Cheshire soils, the county supports relatively few snail species. Consequently the thrush's "anvils", so frequently encountered in the Derbyshire dales for example, are not such a common sight here. However, the garden snail occurs commonly in nettle-beds, perhaps explaining the thrush's relative abundance around sewage works, rubbish tips, market gardens and allotments, and along riverbanks.

Boyd[6] had a record of a nest with two eggs in a holly at Frandley on 2nd February in about 1880. He himself occasionally saw signs of nesting in February, but generally not until March. At the beginning of March 1937 a bird started building under blizzard conditions[5]. The earliest clutch, of four eggs, that he saw was at Wallerscote on 2nd March 1934. Two young flew from this nest on 31st March. A nest with eggs was found at Knutsford on 10th March 1975. By late April many early broods have flown, but two or even three broods are not unusual. Of 242 broods which Boyd ringed the latest in the nest flew on 10th August 1943. Most clutches were of four eggs, with one of six in June 1930. May and June were the months of greatest productivity, with broods of five young then not uncommon. No broods of five were seen in April, and a bird that sat on two eggs in August hatched only one. A recent record concerns a bird seen still feeding young at Sandbach on 13th September 1981.

Typically the nest is placed in a thorn hedge, a bush, or in the lower parts of a tree. Ivy on walls is often chosen, as are evergreens such as holly. Other nests are built on hedge banks, on ledges in sheds, and elsewhere in the clutter around human habitation. Boyd[6] knew of a thrush that built on a Swallow nest in a pig-cote. Occasionally a nest is built on the ground, where it is vulnerable to the attentions of predatory mammals in addition to the Magpies which rob many nests. Coward occasionally found nests on the ground in the willow-beds at Rostherne[24]. Coward & Oldham[2] noted that in early spring the large nest is often absurdly conspicuous, and the increase in Magpies over recent decades may well have played a part in the species' decline. Song Thrushes sometimes seem to get "caught short", and odd eggs may be found deposited on the bare soil of garden borders or at the side of a footpath.

The nest is unique among those of our breeding birds in that it has a lining of mud or similar material moulded inside. Occasionally an unlined nest is used, however. Hardy[4] considered such nests not unusual in a dry spring, mentioning one at Wallasey in 1938. Sitting birds will often allow a very close approach, as though relying on camouflage to escape detection. Boyd had his fingers pecked at and shaken by birds on at least three occasions, and another buffeted him on the head as he approached its nest. One April he watched a thrush sitting unperturbed on a nest in a hawthorn hedge while a Sparrowhawk tried to force its way in! The same nest is often used for successive broods and even in successive years[6].

The Alderley Park 13-hectare CBC plot has held between seven and eleven pairs (1978-1980), although the species is always unobtrusive there in the thick conifer woodland and rhododendrons. Song is infrequent and the birds are rarely seen. At Risley Moss (73 ha.) between seven and eleven pairs have been located. 22.8 hectares of mixed woodland at Rostherne Mere NNR held between five and ten pairs between 1976 and 1980. 18.2 hectares of deciduous woodland, heath and scrub on the edge of suburbia at Bidston Hill held between ten and fourteen pairs annually from 1975 to 1983, then fell to seven pairs in 1984. Fewer figures are available for farmland. The six south-eastern plots, which include a number of small woods, held 47 pairs in 1978 on 393 hectares. Four plots censused again in 1979 (241 ha.) showed a reduction from 21 to 15 pairs after the severe winter.

Sharrock[35] allowed a mean figure of 1000 pairs per occupied 10-km square nationally –

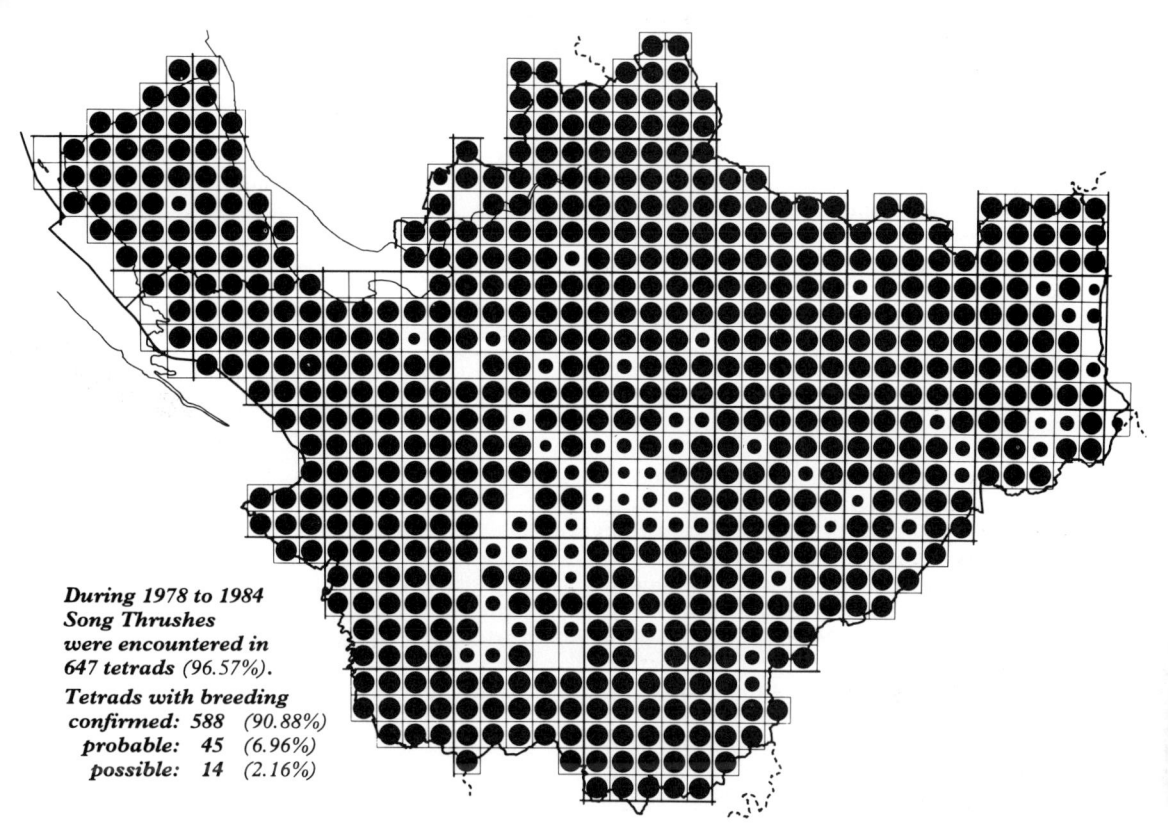

During 1978 to 1984 Song Thrushes were encountered in 647 tetrads (96.57%).
Tetrads with breeding
confirmed: 588 (90.88%)
probable: 45 (6.96%)
possible: 14 (2.16%)

equivalent to 40 pairs per tetrad. A minority of tetrads in suburban areas with large gardens may exceed this figure several times over, and a few well wooded tetrads may hold in the region of 100 pairs. However only the best of the farmland tetrads are likely to hold many more than 40 pairs, and perhaps most will hold far fewer. Between the time of the BTO Atlas and the end of this survey the national CBC index shows the Song Thrush population to have almost halved. The county population is unlikely to exceed 12,000 pairs.

Redwing
Turdus iliacus

UP until the 1960s it was considered exceptional for Redwings to be seen in Cheshire as late as mid-April, the latest date which Bell[7] could quote being 17th April 1960. April records since have been more numerous.

The first May record for the county was of one or two at Styal on 1st May 1976. This was followed during the present survey by records in four years from twelve localities, the latest being at Lindow Moss on 18th May 1978, Alsager on 18th May 1981 and Prestbury on 14th May 1984. One remained at Red Rocks, Hoylake from 11th to 24th May 1980. The Rare Breeding Birds Panel report for 1983 recorded two birds calling at dusk in Mottram Hall Woods on 3rd May. An outstanding record, on 5th May 1981, was of an agitated individual in a derelict coppice at Walley's Green which flicked its wings nervously and called softly, and seemed unwilling to fly far (JPG).

Mistle Thrush
Turdus viscivorus

BOYD[5] suggested that the great variety of vernacular names used for the Mistle Thrush was due to the species having spread into this country after the first English settlers. Coward & Oldham listed ten dialect names and variants in use in the county in 1910, including Stormcock, Sedgecock, Shrillcock and Stone Thrush. Parslow[34] stated that the species was rare in northern England until the end of the eighteenth century. By the start of this century however Mistle Thrushes were permanent residents, generally distributed throughout Cheshire, though much less common than Song Thrushes or Blackbirds. They were more plentiful in the eastern hills than in other parts of the county[2].

Both Coward & Oldham[2] and Boyd[6] mentioned that the Mistle Thrush would often nest close to houses, the former authors specifically mentioning suburban areas. The timing of its spread into the middle of our towns and cities went unrecorded, but Hardy[4] found it to be a common nester in suburban parks, and it is now quite normal for the loud fluty song to be delivered from a high perch on an urban office block or the upper branches of a tree in a city park. The bird sings even in strong winds, hence the name "Stormcock". Unlike a Song Thrush or Blackbird, which seem reluctant to move far from cover unless in a flock, a solitary Mistle Thrush will often feed well out in the open. Indeed the bird seems almost reluctant to feed close to cover. Its habitats thus encompass a wide range of open country with scattered trees, urban playing fields, and gardens with large lawns. The hill country, with its scattered trees, is ideal.

Throughout this century it has been noted that a considerable proportion of our Mistle Thrushes, perhaps young birds, leaves in the autumn. Smaller numbers remain over winter however, and territories may be established in the autumn or, perhaps more usually, early in the new year. Boyd[6] had few records of autumn song: a few notes on 7th August, a brief snatch on 30th September, and three records in mid-November, but then nothing until mid-December. There are more recent records for many dates from 15th September to the end of the year, provided the weather remains mild. Song may then continue throughout the winter, although severe weather will force birds to concentrate on defending their food supply, and song may then be subdued until February. During the winter of 1933-34 Boyd[5] watched a solitary bird which roosted in a holly. On 19th January it was joined by a mate and he heard the first song of the year. Two days later one spent most of the morning singing from a tall pear tree.

The breeding season begins early for the Mistle Thrush, with nest-building often starting in the first two weeks of March. In 1974 a nest was found at Rostherne on 25th February. Boyd[6] saw his first egg on 18th March in a nest from which the brood flew on 19th April. It is not unusual for birds to be feeding young by the first or second week in April. It has been suggested that this early nesting is a strategy to avoid the main season when predators are searching for eggs (Snow 1969). The Mistle Thrush is renowned for its boldness in defence of its nest, its angry chattering when confronting Magpies, Jays and Kestrels being a useful source of evidence during this survey. Birds have been known to attack a man climbing a nesting tree, and to grapple with a Magpie. Nevertheless a proportion of nests is robbed by grey squirrels and avian predators, and many broods from replacement clutches are still being fed in late May, with second broods following in June or early July. Boyd ringed a brood on 19th July from a clutch laid in a nest that had proved unsuccessful in March.

Feathers are often built into the foundation of the untidy nest, and Boyd[5] several times referred to the habit of weaving paper, rags, or bits of string into its structure. More recently lengths of coloured nylon twine and strips of polythene sheet have been used. Nests have been reported at between four and 40 feet off the ground, but usually well above

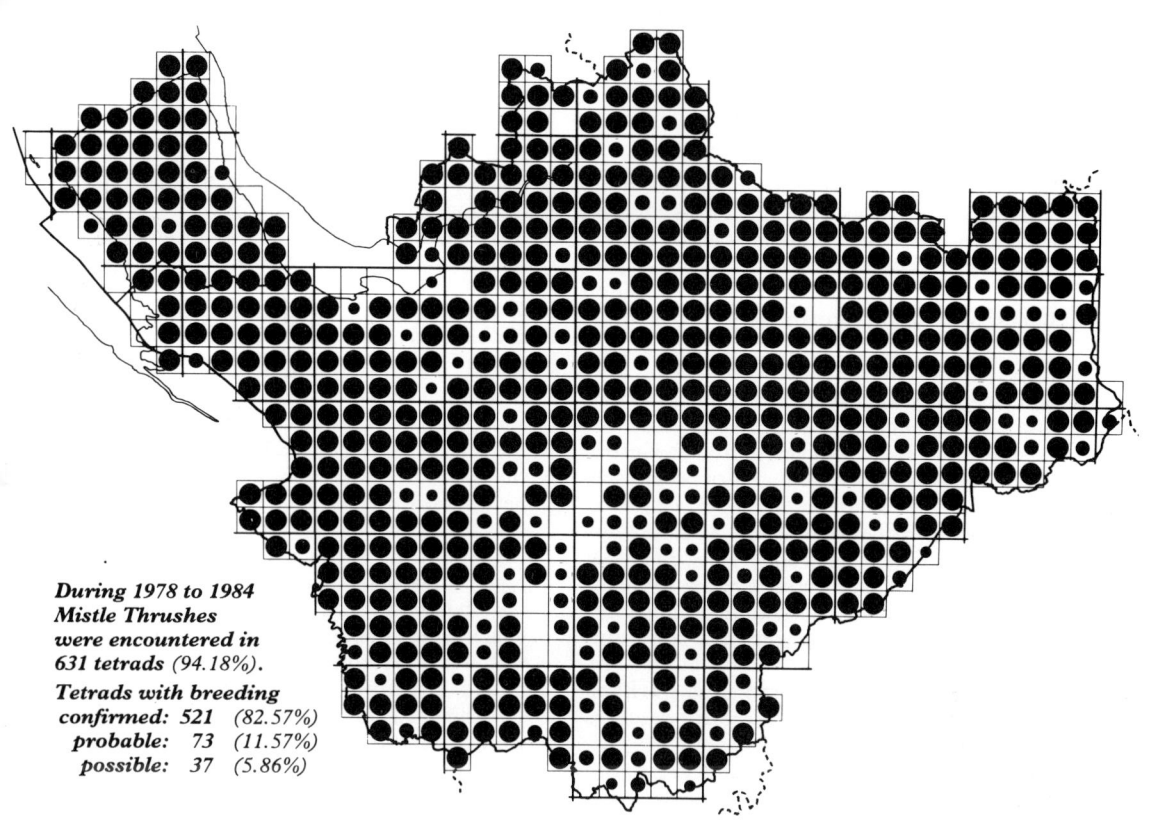

During 1978 to 1984 Mistle Thrushes were encountered in 631 tetrads (94.18%).

Tetrads with breeding
confirmed: 521 (82.57%)
probable: 73 (11.57%)
possible: 37 (5.86%)

head height. Typically they are built in a main fork of a tree. In urban areas poplars are often used, and in several cases nests built in a low fork of a roadside flowering cherry have proved successful. Other trees are used according to availability. Hedgerow oaks and ashes are much used in the lowlands, and sycamores or rowans in the hill areas. A few, exposed nests are blown down by high winds. At Wybunbury vicarage in 1979 an adult was seen to brood three dead chicks whose nest had fallen from a tree[8]. Boyd[6] noted three nests on pergolas. It is not uncommon for them to build on man-made structures, such as inside derelict warehouses at Bromborough and on pylons near Frodsham.

As early as mid-June small flocks of Mistle Thrushes assemble, and by July gatherings of up to 40 or so are not unusual. By August, with the breeding season long finished, larger concentrations may occur around rowan trees. Thus, while this was an easy species to locate during fieldwork for this atlas, visits in the earlier part of the season were more likely to provide evidence of breeding.

Sharrock[35] allowed 100-200 pairs for each 10-km square occupied during the national BTO survey (i.e. four to eight pairs per tetrad). In 1978, six wooded farmland CBC plots in south-east Cheshire, totalling just under a tetrad in area (393 ha.), held 16 pairs, and a search of the Saighton tetrad (SJ45J) during this survey revealed eleven pairs. Between one and four pairs have been located in 73 hectares of mossland and woodland at Risley Moss, and one to three pairs in Mere Covert (10.5 ha.) at Rostherne Mere NNR. Three to five territories were mapped on the 18.2-hectare CBC plot at Bidston Hill in the years 1975-84. However, the usual breeding density of the Mistle Thrush is generally low and the species was difficult to locate in many lowland agricultural tetrads, particularly in the south of the county. Taking an average of perhaps six pairs per occupied tetrad would give a county population of 3500-4000 pairs. Some fluctuation in numbers is noted after severe winters although CBC indices show no long-term change over 20 years. O'Connor & Shrubb[42] demonstrate clearly that Mistle Thrush clutch-size and fledging success both increase in areas with mown grassland, which exposes more invertebrates. The switch to silage cutting may well benefit this species.

REFERENCE

Snow, D. W. (1969): Some vital statistics of British Mistle Thrushes. *Bird Study* 16: 34-44

Cetti's Warbler
Cettia cetti

THIS species reached as far north as Cheshire just in time to be included in this survey. The first county record was of a bird trapped near Frodsham on 20th April 1984, having apparently arrived with a fall of Willow Warblers (Norman 1984). After being measured, ringed and photographed, it was released and not seen or heard again. It is perhaps noteworthy that, in the first years of this species' colonisation of the Channel Islands "On every occasion that a Cetti's Warbler has been trapped....its presence was not suspected beforehand; on release, each one flew immediately to cover and was not seen again. The chances of proving breeding by, say, one or two pairs therefore seem small" (Long 1968). This comment, however, ignores the bird's song, which is loud and distinctive and may be uttered at almost any time of the year, although it is perhaps unfamiliar to many Cheshire bird-watchers and might pass unrecognised.

Cetti's Warblers have been gradually spreading northwards across Europe since the turn of this century, crossing the Channel to England for the first time in 1961. Breeding was first proved in Kent in 1972 and the species has now nested in most of the south and east coastal counties, with the peak British population around 300 pairs. Birds do not migrate, and are insectivorous all the year round, making them susceptible to freezing weather. The winter following the end of this survey period (1984/85) saw a reduction of 75% in the Kent and Suffolk populations[41], so the possible colonisation of Cheshire may have been set back.

This species inhabits low, tangled vegetation in damp areas – a habitat well represented in the county, particularly close to the Mersey valley. This area also has a climate which is less harsh than much of Britain, and it is conceivable that this attractive newcomer could at some stage establish itself in Cheshire. Although only one bird has been found, it was in the breeding season in suitable habitat, and a 'possible' breeding record is justified. DN

REFERENCES

Long, R. (1968): Cetti's Warblers in the Channel Islands. *British Birds* **61**: 174.
Norman, D. (1984): Cetti's Warbler. *Cheshire Bird Report* p87.

Grasshopper Warbler
Locustella naevia

THE high-pitched song of the Grasshopper Warbler resembles the sound of an angler's reel, hence the customary term "reeling". The bird seems largely crepuscular, at least where song is concerned, for this is heard mainly in the early morning and late evening, less frequently in the middle of the day, and not infrequently by night: one was heard reeling at Swanley at 2.40 a.m. on 29th April 1984. Grasshopper Warblers are generally regarded as very skulking birds, but while the reeling bird often remains hidden in the midst of a bramble thicket infuriatingly close to the bird-watcher, at other times it will perch on a low branch of a tree or on top of a low bush in full view. At such times the changes in pitch and

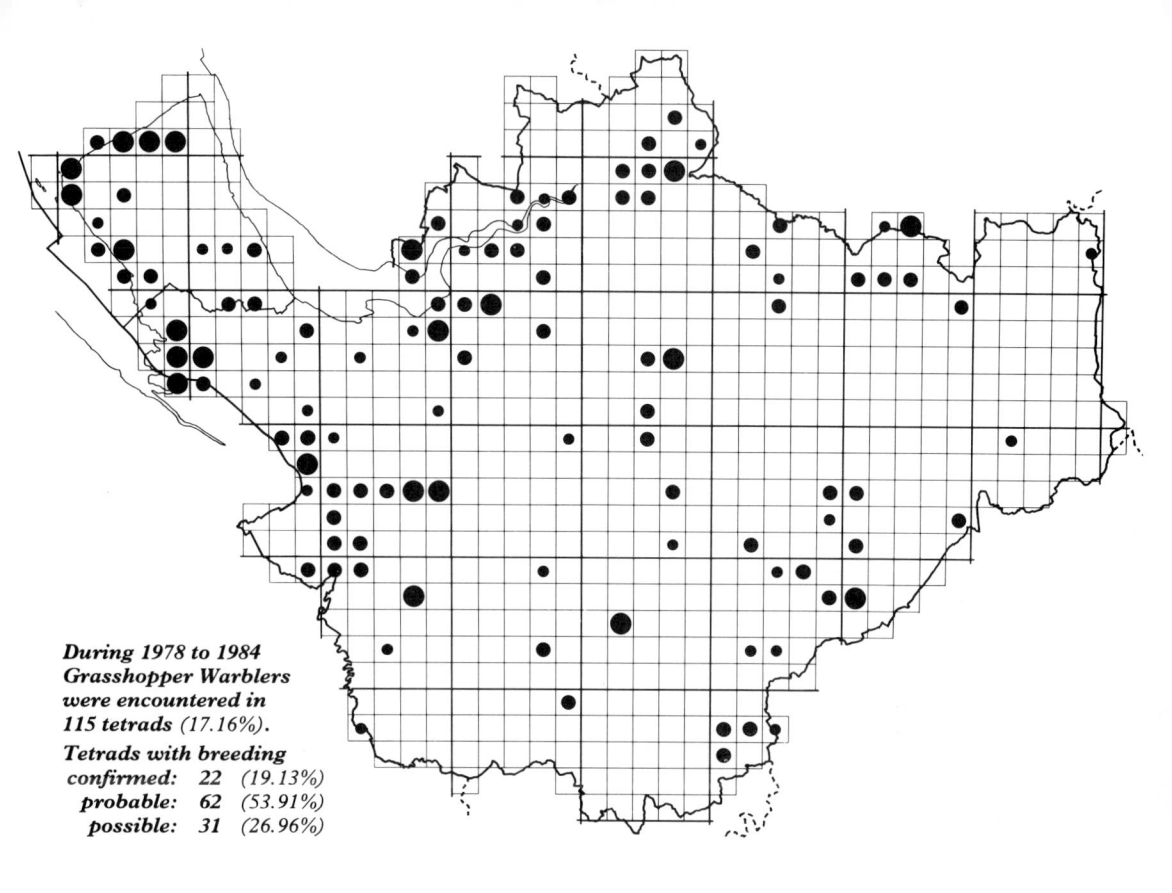

During 1978 to 1984 Grasshopper Warblers were encountered in 115 tetrads (17.16%).

Tetrads with breeding
confirmed: 22 (19.13%)
probable: 62 (53.91%)
possible: 31 (26.96%)

volume of the song can be seen to be due to the bird turning its head back and forth which may also make it sound ventriloquial. Some birds sing in short bursts of a few seconds at a time, while others will reel for a minute or more without a break.

Reeling is heard from the first birds' arrival in April and through to June with something of a resurgence in July and August. A bird at Dean Row in 1972 which held territory around a patch of willowherb reeled every evening from 10th August until 4th September, while a bird at Sandbach sang on 7th September.

The earliest spring record was of a bird at Meols on 11th April 1981, though generally birds do not start to arrive until ten days or a fortnight later than this. In autumn very few are reported after mid-August when most song ceases. In 1979 mist-netting near Frodsham revealed the last adult on 11th August with the last juvenile on 16th September. A bird was noted at Risley Moss on 29th September 1983.

Following initial location and establishing that birds are on territory, confirmation of breeding may be difficult. Watching from a distance may reveal the cock bird carrying insects to feed the female or later both birds carrying food for the young. Several records during the survey however were due to surprise encounters with agitated parents flicking their wings and tail and uttering a sharp "tchick" alarm-note, despite their beakful of flies. Grasshopper Warblers are probably under-recorded, owing to their unobtrusive nature and crepuscular habits. Although the distinctive song can be far-carrying, some males will sing only on overcast, drizzly days when observers are less likely to be present.

Typical habitat consists of low dense herbage with scattered bushes of bramble or gorse for song-posts. Thus rough grassland on sand-dunes or the banks of sludge-beds, nettle-beds as at sewage works and rubbish tips, and scrubby urban waste ground may all be suitable, and conifer plantations in the first ten years of growth provide the right combination of undergrowth and small trees.

Many of these sites are transitional by nature and support birds only for a few years before being cleared. Thus the overgrown orchards to the west of Knutsford Moor held birds from the 1960s or earlier until 1976 when the area was built on. Similarly five territories around Northwich in 1981 could all be lost if derelict industrial land were restored. A site of major importance for this species during the survey, adjacent to the Woolston Eyes SSSI, was due to be used for waste disposal.

The stronghold of the species in recent decades however, is amongst the sand-dunes and rough coastal grassland of north and west Wirral. Such habitat was formerly much more widespread than at present, yet surprisingly Brockholes[2] knew the Grasshopper Warbler only as "a rather scarce summer visitor", and listed only Bidston, Bebington and Puddington as sites where it had occurred, with a nest found near Upton. Coward & Oldham[2] also knew of records from Liscard and Wallasey, and while the suburbs of Birkenhead had then spread over some former haunts, the bird still occurred near Heswall and Little Saughall where it bred in 1904. Subsequently the species appears to have increased despite urbanisation and has been considered fairly common in western Wirral for most of this century[13,18]. In 1941 three pairs bred at Puddington with five pairs at Heswall the following year. By 1968 it was still regarded as fairly common in west Wirral and a similar situation pertained until 1975, which was generally regarded as a poor year throughout the county. In 1976 only one pair was found at Heswall where often several pairs had been present, and this scarcity continued into the present survey.

Away from Wirral, Coward & Oldham[2] described the Grasshopper Warbler as a regular visitor to the Chester district, nesting annually at Ince and Thornton-le-Moors. It was local on the plain generally, but probably overlooked. In Delamere birds occurred regularly at Hatchmere, Little Budworth, Oulton and Newchurch Common. Knutsford Moor and other localities around Knutsford and Chelford held nesting birds. It was evidently rare in the hill country, with neither of the records listed falling within the present county boundary. Boyd[5] only once encountered the species around Great Budworth, and then for one day only.

In recent times there appear to have been more birds inland. At least nine males were singing around Delamere Forest in 1964; two pairs were reported at Antrobus and a nest with five young was found at Great Budworth in 1966; and sites around Northwich now hold birds in most years. At the Sandbach Flashes, where only one bird was recorded prior to 1964[19], birds have occurred in all but three subsequent years[27].

The mapped results of this survey show little difference from the distribution described at the beginning of this century, allowing for some increase inland. The species remains very scarce in the eastern hills where Bell[7] could add only one record, from Macclesfield Forest in 1946. The Wirral and Chester areas remain popular. The Mersey valley forms an additional stronghold with plentiful rough grassland and waste-ground, notably the Woolston Eyes where up to 18 pairs were present in 1982. County-wide, numbers fluctuate considerably from year to year with 1980 being above average perhaps due to prolonged south-east winds in spring. Six territories were then reported on Frodsham Marsh and five males settled in an area of overgrown market gardens near Sandbach, although these were much harassed by bird-watchers and moved on. Ten or more males were reeling between Denhall and Burton in early May.

This species is not covered by the CBC, WBS or other standard surveys, and there is no quantitative information on the population density. Based on counts over 18 years at coastal observatories, Riddiford (1983) found a marked decline in Grasshopper Warbler numbers in 1973 and subsequently, possibly linked to climatic changes in its African wintering grounds, although the timing of this drop differs from that in the well documented Sand Martin and Whitethroat populations. However, the wintering area of Grasshopper Warblers is still not known (Mead & Hudson 1986) and may differ from the wintering areas of those species. No long-term changes in population have been noted in Cheshire. On average, tetrads with confirmed or 'probable' breeding may hold one or two pairs. Many tetrads with 'possible' breeding evidence may be attributed to passage birds only. This would give a county population of around 100 pairs.

REFERENCES

Mead, C. J. & Hudson, R. (1986): Report on bird-ringing for 1985. *Ringing and Migration* 7: 139-188.
Riddiford, N. (1983): Recent declines of Grasshopper Warblers *Locustella naevia* at British bird observatories. *Bird Study* 30: 143-8.

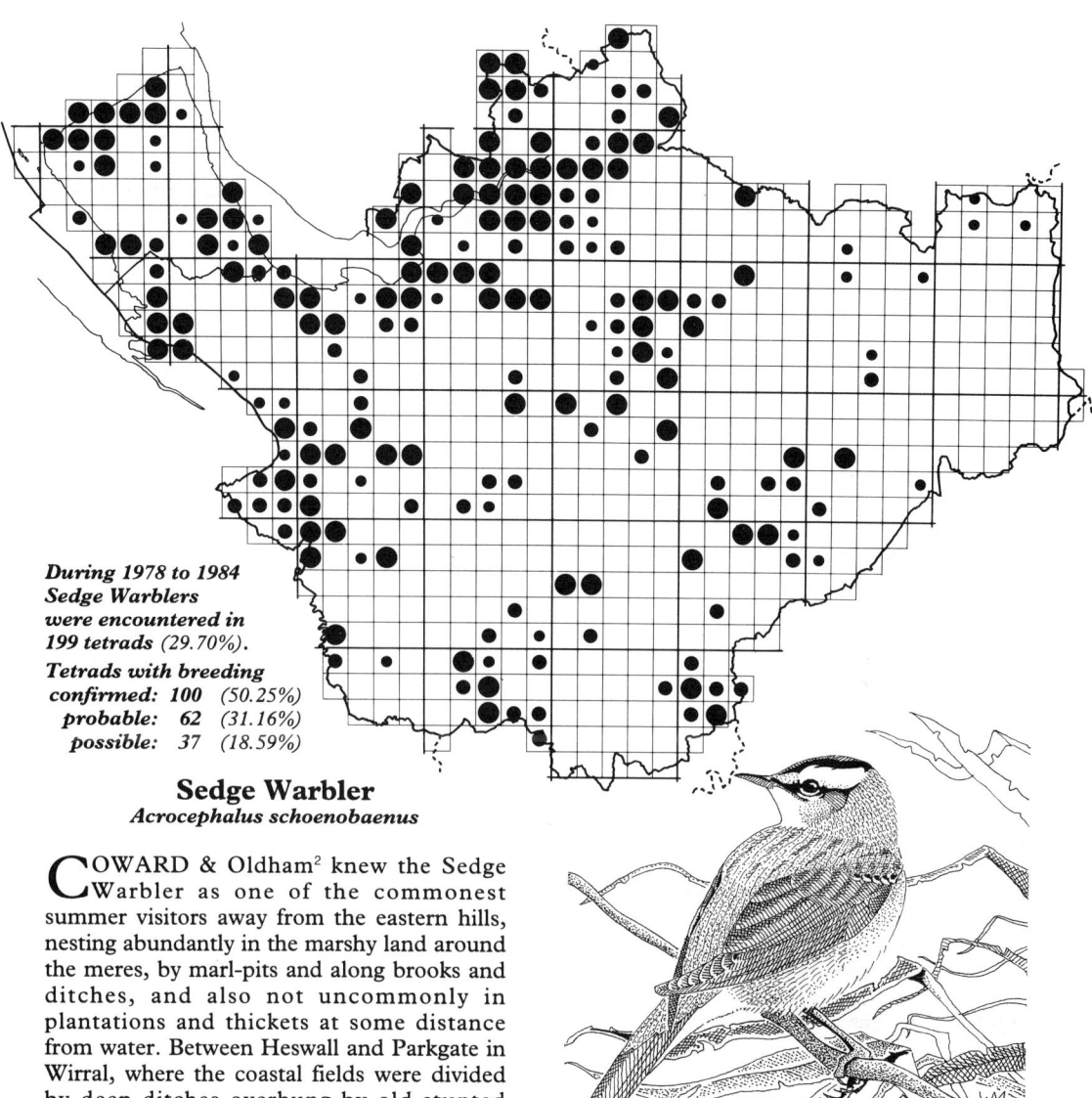

During 1978 to 1984 Sedge Warblers were encountered in 199 tetrads (29.70%).

Tetrads with breeding
confirmed: 100 *(50.25%)*
probable: 62 *(31.16%)*
possible: 37 *(18.59%)*

Sedge Warbler
Acrocephalus schoenobaenus

COWARD & Oldham[2] knew the Sedge Warbler as one of the commonest summer visitors away from the eastern hills, nesting abundantly in the marshy land around the meres, by marl-pits and along brooks and ditches, and also not uncommonly in plantations and thickets at some distance from water. Between Heswall and Parkgate in Wirral, where the coastal fields were divided by deep ditches overhung by old stunted thorn bushes, they stated that "this dense cover affords shelter for innumerable Sedge Warblers, and nowhere in the county is the bird more plentiful". In the eastern hills it bred "sparingly, and then only in sheltered situations in the valleys". They added that a few pairs bred in the Dane valley near Wincle, and that it was fairly plentiful on the marshy ground around Bosley Reservoir.

Abbott[10] mapped only two territories in the 22 square kilometres around Wilmslow and Alderley Edge studied by him during the years 1919, 1921 and 1922; one each at Willow Ground Wood, Styal and at Davenport Green Sewage Farm. From 1924 until 1930 he took a particular interest in this species finding it breeding at Redesmere, Great Warford, Siddington Mill Pond, Astle (by the church), Astle Pool, Lower Peover Mill, Capesthorne Pool, Arley Pool, Goostrey Mill, Brereton Mill, Bagmere, Knutsford Waterworks, Holford Mill and Knutsford Moor. Apart from the last site, where there are still one or two pairs, all of them have been deserted because of improved drainage, dereliction (particularly the mill ponds) modernisation or overgrowth. Griffiths & Wilson[13] found them commonly in the "many ponds or 'pits', fringed with reeds, bramble and thorns" in north Wirral and they remained common around the ponds at Meols Fields in 1956-57 (A. Booth *in litt.*).

In 1934, when Boyd[5] visited Padgate near

Warrington to investigate a report of a Nightingale, he found it to be a night-singing Sedge Warbler and was informed by the farmer that there was a pair by every pit on the farm. The same author suspected a reduction in numbers by 1951, and there has been a marked contraction in range since, although Bell[7] thought it "still to be widely distributed" in those habitats described by Coward & Oldham.

By the time of the present survey it had disappeared from large areas of eastern Cheshire, even where habitat remains suitable, and is seldom found east of Knutsford. In the remainder of the county it is now mainly restricted to wet, marshy strongholds by flashes, canals, meres, in sludge-pools and the like. It has ceased to breed on farmland, but spring migrants are occasionally encountered singing from the scrub around marl-pits, from rhododendrons and other scrubby areas. The drought in the southern Sahara which brought about the crash in Whitethroat numbers in 1969 also affected the Sedge Warbler, but the decline in Cheshire had set in some 20 years previously, possibly influenced by the Sahelian drought of 1940-44.

In recent years the combination of restricted habitat and distinctive chattering song given even at night makes Sedge Warblers easy to locate, and when feeding young they will scold the observer noisily. A few isolated pairs may have been overlooked, but otherwise the map is undoubtedly very accurate. The birds build close to the ground, the nest usually supported by strong weed stems such as those of nettles or willowherbs, and although males normally need song-posts from which to proclaim their territory, small willow bushes often suffice.

The first bird may be heard in the first half of April, with 10th April 1981 the earliest in this survey period, but most arrive in early May. A bird ringed in Hertfordshire on 10th May 1979 reached Red Rocks three days later, averaging 80 kilometres per day. Males sing vigorously on arrival but pairing is often rapid and song is then seldom heard, posing problems for census work. Sedge Warblers in Cheshire are single-brooded – although they may replace lost clutches – and most birds leave the breeding-sites in July to congregate in *Phragmites* reed-beds. A few such areas hold large numbers of birds, with Frodsham at the forefront, where, for instance, 125 were ringed on 4th August 1979. They soon move south, and few are seen after early September, although late migrants are seen in October in some years. Sedge Warblers frequent reed-beds in autumn mainly to feed on the plum-reed aphid and fatten up for migration. In good conditions they can accumulate fuel rapidly, like the juvenile which added 70% to its weight in seven days at Woolston in 1981, enough to give sufficient energy for a non-stop flight to Portugal or beyond. This autumn migration strategy means that birds are seldom seen except at a few favoured sites, and, as for the Reed Warbler, the map shows few 'possible' breeding records, almost all of which are attributable to prospecting birds on spring passage.

In 1967 at least 40 pairs were breeding on Bidston Moss, Wirral, but this total had fallen to less than ten pairs by the start of this survey. In the 1970s some 25 pairs annually bred around the Sandbach Flashes as against three in 1984. Knutsford Moor remained the only regular breeding-site in east Cheshire with three or four singing males during the early part of the present survey but even this had become an irregular breeding-site by the 1980s.

In May 1976 there were up to 40 singing males on the Eaton estate at Chester and in 1979 there were 15 singing males in Neston reed-bed and twelve pairs at Handley Covert. 35 males were singing at Moss Side in 1981. This species is outnumbered by the Reed Warbler around Northwich although six pairs held territory at Pickmere in 1981 and five pairs in 1982.

At Woolston, now the species' stronghold, the population was considered to be at least 60 pairs in 1979, and, although not an easy bird to census, 100-125 pairs were estimated in 1982. A more intensive census on 15th May 1983 found a staggering total of 154 singing males in an area of approximately three square kilometres (much of this standing water). Many of these birds apparently lost nests in a severe hail storm on the evening of 7th June as the air was full of song the following morning (an unusually late date for song) when males were presumably encouraging females to re-start the breeding cycle (Norman & Martin 1983). In 1984 a census on 12th May revealed only 71 singing males, increasing to about 95 following an arrival between 16th and 18th May. This represented a 38% decline on the previous year matching a 31% decline based on national CBC figures. The situation at Woolston was complicated by habitat loss. However, there was a major, unexplained reduction on No. 3 bed where there was very little habitat change. The proximity of territories, despite the availability of much deserted habitat, suggested an important social element in the birds' distribution there.

Sharrock[35] suggested an average density corresponding to four pairs per occupied tetrad, but the decline since then may make two pairs per tetrad a more reasonable figure. Allowing for the exceptional concentration at Woolston, the Cheshire population probably fell during the survey from around 650 to below 500 pairs.

REFERENCES

Norman, D. & Martin, B. (1983): Some observations on the effects of a summer hailstorm. *Cheshire Bird Report* pp79-80.

Reed Warbler
Acrocephalus scirpaceus

AS its name suggests, the Reed Warbler is typically a bird of *Phragmites* reed-beds with large concentrations at Frodsham, Rostherne Mere NNR, Moore and Marbury Mere (Great Budworth). However, reedmace supports regular colonies at the Sandbach Flashes, Woolston Deposit Grounds and elsewhere. At Tabley Mere birds sometimes sing from yellow flag and lesser reedmace, and, at Eaton Park near Chester, nesting has been noted in mereside bamboos[7]. Early immigrants have been heard on occasion singing from hedges before the reeds have made sufficient growth, and in early June late arrivals have stayed briefly in rhododendrons and other bushes in gardens, more suitable territories having already been occupied. Small isolated stands of reeds seem to be favoured provided that there are adjacent bushes from which the adult birds can collect food such as insects and caterpillars for their young. Reed Warblers are basically very territorial although, because the necessary breeding habitat is limited, they appear to be colonial.

Coward & Oldham[2] knew the Reed Warbler as an abundant summer resident in the reed-beds fringing the larger meres, adding that it was seldom met with far from large sheets of water. It was scarce in Wirral and absent from the eastern hill areas. A very similar situation was described in the *L&CFC Report* for 1938 which listed 23 nesting localities. Griffiths & Wilson had seen birds at Moreton brick-pit, north Wirral, in the years before 1945. Boyd[5] records birds seen as early as mid-April (17th in 1920, 13th in 1944 and 16th in 1945) although they typically arrive during the first two weeks of May and may have an extended breeding season. Indeed, it is probably one of the last of the warblers to leave in the autumn with odd birds seen into October.

Some expansion in the range of this species has occurred since the mid-1960s. In 1965 several pairs bred in reedy ditches on Frodsham Marsh and in 1967 three or more males were singing, and breeding was proved, at Bidston Moss in Wirral where Brockholes had found birds annually a century earlier prior to drainage. Further colonisation has since occurred in Wirral, notably at the Neston reed-bed where there was an increase from one male in 1972 to nine in 1979 and 15 in 1980. Elsewhere small colonies have been established beside the canals and the River Weaver in mid-Cheshire, with at least 62 territorial males located at 18 sites in the four mid-Cheshire 10-km squares in 1981 (JPG), not including the Frodsham birds: in all perhaps 100 pairs were present in those four 10-km squares that year.

The limited amount of habitat available is easily checked for presence, singing birds being particularly conspicuous – although some people find difficulty in distinguishing the song from the more rapid, scratchy tones of Sedge Warblers. The map must come very close to representing the exact distribution of the species in Cheshire and Wirral at the present time. Suitable habitat is lacking in the eastern hills and the outpost at Redesmere is threatened as remaining reeds are trampled down by anglers. The many sand-quarries in this and other areas of the county might provide future nesting sites if sympathetically managed.

Many occupied sites hold very few pairs but at the Woolston Deposit Grounds, where regular spring warbler counts are now held, there were 20-22 territorial birds in 1982, 23 in 1983 and 20 in 1984. At Rostherne Mere NNR the Reed Warbler has been studied intensively by M. Calvert since 1973. The population there over the twelve seasons 1973-84 averaged 34 pairs, with extremes of 25 in 1974 and 43 in 1984. In all it is estimated that the county population is in the region of 300 pairs. The Lancashire and Cheshire colonies represent the north-western limit of this species' breeding range in Europe.

Exceptionally, nest-building commences in the first few days of May. Eggs in a nest at Marbury Mere (Great Budworth) had hatched by 19th May 1958, so that clutch must have been started by 3rd or 4th May (Trelfa 1959). At Rostherne, in 1977 and 1984, the first egg was laid on 10th May (Calvert *in litt.*) and a nest near Frodsham, in 1981, had five well grown young on 31st May, giving a first-egg date of 5th May (DN). The median egg-laying date at Rostherne over the twelve-year period 1973-84 was 18th May. Nest-building typically continues from the third week of May and throughout June, although in some years the first nests of late arrivals are not begun until early July. Young are found fully fledged in first nests from mid-June to mid-July whilst a second batch of nests sometimes follows in the last week of July and the first week of August (Boyd 1933). At Rostherne early nests are built on old reeds, but new reeds are much preferred and nests are very rarely used for successive clutches (Calvert 1983a, 1983b). The mean clutch-size at Rostherne for the period 1973-84 was 3.93 eggs (sample size 493) and the mean brood-size over the same period was 3.37 chicks (sample size 406).

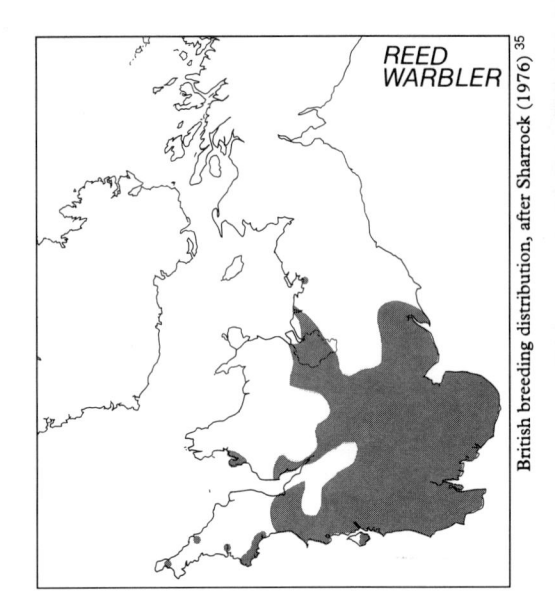

In some parts of the country the Reed Warbler is regularly parasitised by the Cuckoo, but in Cheshire there had been only two known cases up to 1984: one at Marbury Mere (Great Budworth) in 1934, Boyd's only record despite examining over 200 nests up to 1951[5], and one at Rostherne in 1977 (Calvert 1983b). At the Woolston Deposit Grounds and elsewhere, Cuckoos regularly move through the reed-beds and were suspected of cuckolding Reed Warbler nests although there had been no proof of this by 1984. The other main threats to breeding Reed Warblers are damage to exposed reed-beds by strong winds (Calvert 1981, 1984) and roosting Starlings.

REFERENCES

Boyd, A. W. (1933): Notes on the nesting of the Reed Warbler. *British Birds* **26**: 222-223

Calvert, M. (1981): Reed Warbler nest survival during exceptional rainfall. *Cheshire Bird Report* pp70-71.

Calvert, M. (1983a): Height and support of Reed Warbler nests. *Naturalist* **108**: 105-106.

Calvert, M. (1983b): Reed Warbler nests and instances of re-use for further breeding attempts. *The North Western Naturalist* pp8-10.

Calvert, M. (1984): Cheshire Reed Warblers and Weather. *The North Western Naturalist* pp17-20.

Trelfa, G. (1959): Early Nesting of Reed Warbler. *British Birds* **52**: 165-166.

During 1978 to 1984 Reed Warblers were encountered in 101 tetrads (15.07%).

Tetrads with breeding
- confirmed: 55 (54.46%)
- probable: 29 (28.71%)
- possible: 17 (16.83%)

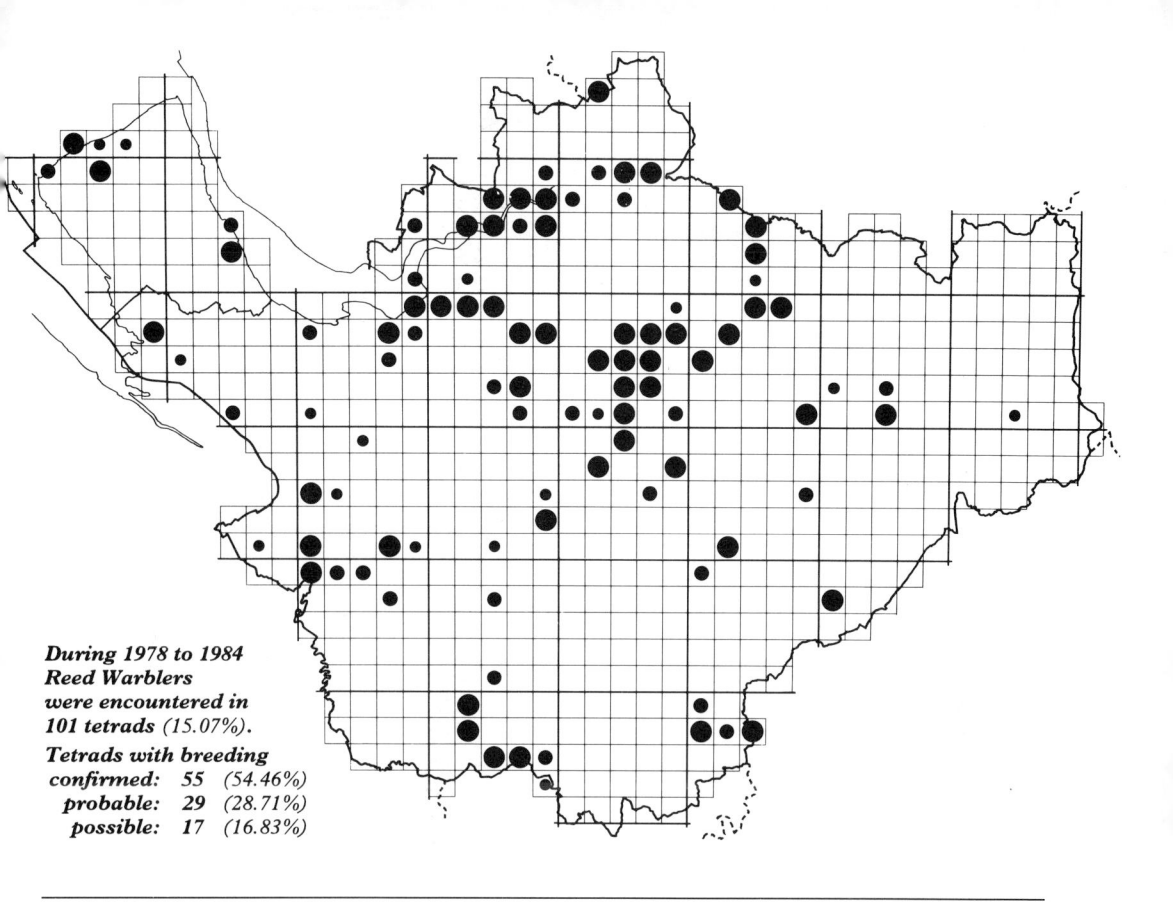

Lesser Whitethroat
Sylvia curruca

THE Lesser Whitethroat frequents tall overgrown thorn hedges or shrubbery of a similar height. It is not an uncommon species in Cheshire, though often regarded as scarce. Certainly it is far more often heard than seen, and, as many bird-watchers are not familiar with the song, recalling a flat, rattling Yellowhammer song, it is doubtless much overlooked. In many recent years a pair has bred in the hedge along the southern side of Elton Hall Flash, unnoticed by the majority of visiting watchers! As well as rural hedgerows, garden hedges and shrubberies are not infrequently occupied. For example birds held territory in most years during the 1970s in gardens near the River Dean at Handforth, singing from topiaried variegated privet and other ornamental shrubs. Newly-arrived birds in April and May can also be found amongst gorse and low brambles. Boyd[6] heard his earliest on 17th April in 1926, with an average arrival date over twenty years of 2nd May. In 1981 birds arrived from 17th April, and from 15th in 1982. Some birds, however, may arrive late – even into June – and start singing then. Most birds have departed by mid-September, at which season they may be seen feeding on elder berries, wild raspberries or on blackberries as in the case of a very late bird at Withington on 12th October 1980.

Song soon ceases once the pair has formed. In many cases where a vigorous songster, present in spring, is thought to have moved on, there may be a resurgence of song some weeks later, after the first brood of young has flown. Lesser Whitethroats seem more secretive than other warblers, perhaps because the males are fully involved, even incubating the eggs, and breeding is not easily confirmed unless adults are seen feeding young, sometimes with caterpillars collected from hedgerow trees. At this stage the presence of the family is often betrayed by the "tic" alarm note of the older birds, sharper and more metallic than comparable calls of other warblers. Almost all of the nests detailed in the BTO Nest Record Card collection were in prickly shrubs or hedges – bramble, hawthorn, rose or blackthorn – which might be thought to give some protection against predation, but in fact they lost significantly more nests to predators than did the other three *Sylvia* warblers, possibly because their nests, generally sited higher off the ground, were more easily found by corvids (Mason 1976).

Coward & Oldham[2] classed the Lesser Whitethroat as a common summer visitor. It was distributed throughout the lowlands, especially in the west, where, around Chester, it was sometimes almost as numerous as the Whitethroat. In the centre and east of the county it was less plentiful, and, while it reached the foot of the hills around Congleton, it was scarce actually in the hills. It was probably overlooked in many areas however. Abbott[10] mapped nine territories in 22 square kilometres over the three seasons of 1919, 1921 and 1922 in the Wilmslow/Alderley Edge area. Up to the 1940s, Hardy[4] regarded it as an uncommon summer visitor in the Liverpool area and Griffiths & Wilson[13] described it as scarce in north Wirral, where it occurred mainly on passage, though a few pairs bred. Raines[18], writing in 1961, similarly referred to small numbers breeding in Wirral, describing it as rare on passage. Boyd[6] regarded it as far less numerous than the Whitethroat around Great Budworth. In May

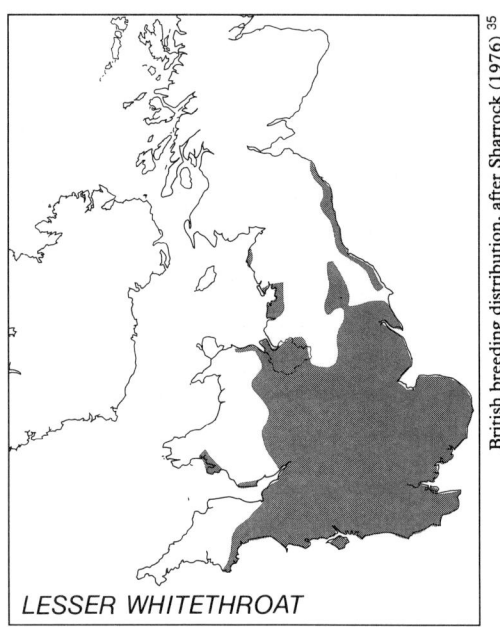

LESSER WHITETHROAT

British breeding distribution, after Sharrock (1976)[35]

1942 he had noted unusual numbers singing in hedgerow and garden, but a comment on wartime agricultural practice in his diary for 16th March 1944 must be of significance for Lesser Whitethroats: "Almost all of the old tall thorn fences have gone – some of them cut down to the ground – to let sun and air have access to the ploughland".

The *L&CFC Report* for 1961-62 recorded that no more than one singing male could be heard in the Nantwich area in any one year. Bell[7] talked of "a marked alteration in status" since Coward & Oldham's day, depicting the species as "scarce and irregular". Very few published records were available, although Bell too acknowledged that it might be overlooked in many districts. No sooner had the species thus been designated scarce than records began to pour in. The 1964 *CBR* lists some ten males in Wirral, with birds at ten other scattered localities. 13 sites were listed in the 1965 *CBR*. 14 males or pairs were reported in the Aldford/Farndon area in 1966, as well as males at seven sites in north Wirral, and birds at thirteen localities elsewhere. In 1968, ten additional sites were recorded from north Wirral, though only one bird was reported from the east of the county. At Rostherne only six records were available from 1963 to 1976 including three in the latter year. Boyd had not recorded the Lesser Whitethroat at Rostherne, though Coward had implied its presence[24]. A relative surge of records over the next few years suggests an

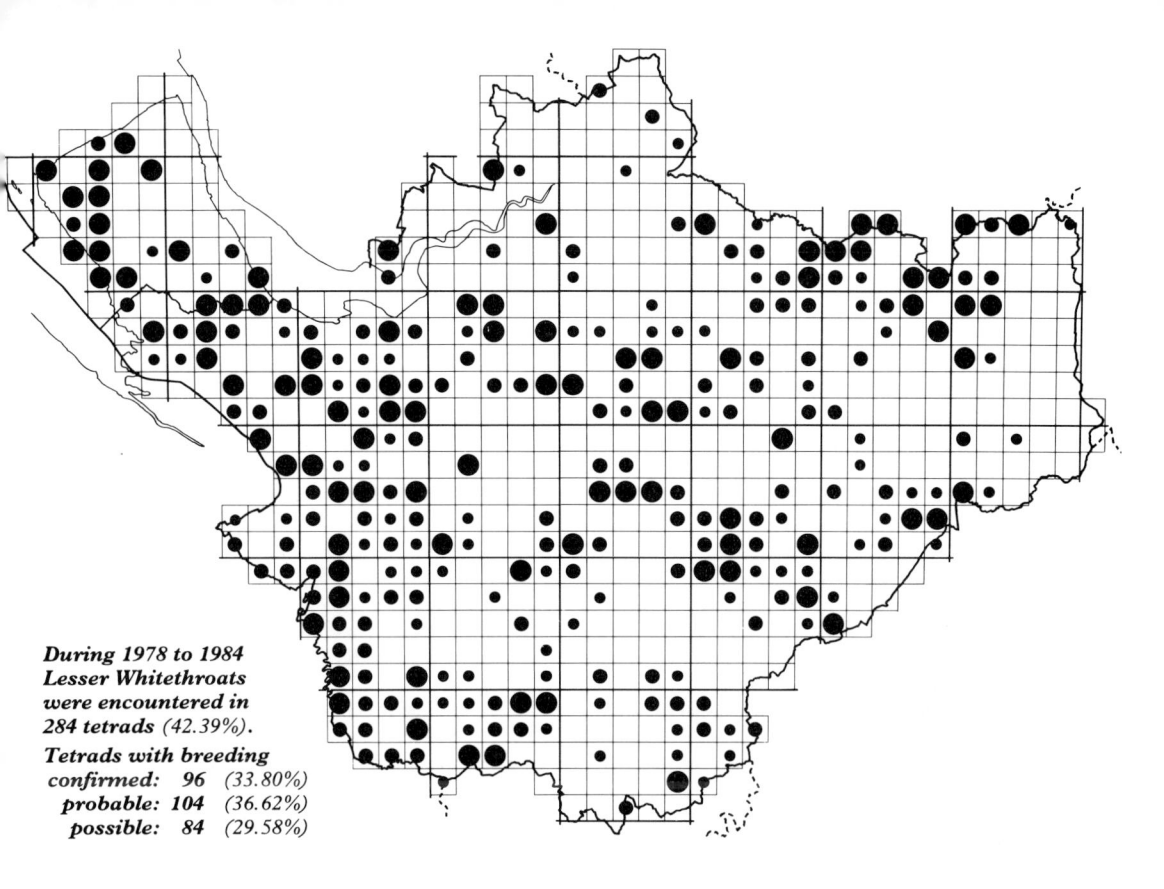

During 1978 to 1984 Lesser Whitethroats were encountered in 284 tetrads (42.39%).

Tetrads with breeding
confirmed: 96 (33.80%)
probable: 104 (36.62%)
possible: 84 (29.58%)

increase at this time. However, JPG's experience is that many observers in the east of the county were just learning to recognise the bird, so how much the increase is due to improved numbers, and how much to improved identification, is not clear. By 1977, twelve pairs were located around the Sandbach Flashes alone, with records from 19 other localities in eastern Cheshire[21].

In a similar vein it is noteworthy that *CBRs* contained records from only four sites in the mid-Cheshire recording area for 1979, and four for 1980, yet a series of visits by JPG to this area in 1981 revealed birds in 20 sites including 14 instances of confirmed breeding. In 1982 records fell back to a more normal level of four sites. Given such discrepancies, it is difficult to assess the validity of apparent localised increases reported in *CBRs*. The Lesser Whitethroat is ancestrally a central European species which spread here after the Ice Age: Cheshire and Lancashire are at the north-western limit of its breeding range, and it is important that its status in the county is carefully monitored in future to detect any possible expansion or contraction of its range. This species might be expected to be sensitive to long-term climatic change, in summer or winter, but the well publicised recent drought in its wintering areas of Sudan and Ethiopia has apparently not so far affected its numbers.

The map shows the species to be widespread but thinly distributed throughout the county, with the densest concentration in west Cheshire and in the Wirral peninsula. It is absent from the eastern hills and scarce in the north. The more intensively agricultural parts of the plain provide few records. A few tetrads hold three or more pairs of Lesser Whitethroats, and 'possible' breeding records may refer in many cases to isolated pairs. With 200 tetrads with at least 'probable' breeding, the county population may lie in the region of 350-400 pairs.

REFERENCE

Mason, C. F. (1976): Breeding biology of the *Sylvia* warblers. *Bird Study* **23**: 213-232.

Whitethroat
Sylvia communis

UNLIKE its retiring cousin, the Lesser Whitethroat, this species has a conspicuous display flight, delivering its scratchy warble whilst dancing over the top of the scrub before plunging back into cover.

Coward & Oldham[2] stated that "throughout Wirral and the lowlands generally, the Whitethroat abounds, being by far the commonest member of the genus. A few pairs nest amongst the scanty vegetation on Hilbre...Even on the bare hills of the east of the county it may be seen in the stunted bushes high up amongst the heather....It is less retiring than many of the warblers, and nests in hedgerows and rank vegetation bordering country lanes, as well as in woods and thickets. It is partial to osier beds where there is growth of young withies". Boyd[5], in his country diary for 11th May 1944, wrote of "common whitethroats in every hedgerow" and Bell[7] referred to them as abundant and widespread.

Over much of Britain the Whitethroat was in fact commoner even than the Willow Warbler[35, 47], although Abbott[10] mapped only 86 territories in just under 22 square kilometres over the three seasons of 1919, 1921 and 1922 in the Wilmslow/Alderley Edge area as against 331 Willow Warbler territories in the same area. Boyd[6] found Willow Warblers more common than Whitethroats around Great Budworth. On Hilbre, where detailed records have been kept since 1957, record numbers were observed on spring passage in 1966, when 60 visited the island on 3rd May, and 1967 when 150 were logged on 14th May.

The 1969 *CBR* however, read "in common with the rest of the British Isles this species is reported to be very scarce this year or in greatly reduced numbers. It is thought that some catastrophe must have occurred either in their winter quarters or during migration". In fact 77% of the 1968 British population failed to return in 1969, a situation equalled over much of western Europe, and by 1974 the level fell to one sixth of that in 1968. Drought in the Sahelian zone of West Africa and the spread of the Sahara were found to be responsible. It is in this region that most Whitethroats spend the winter, and birds had starved due to lack of fruit on the salt-bush, normally their staple diet, and of the associated insects.

CBRs reported a slight recovery annually over the next ten years, but, although no directly comparable figures exist from before the decline, these subjective comments may be due more to bird-watchers becoming familiar with the remaining haunts of this species rather than any real increase in birds. National CBC figures show no overall improvement from 1969 to 1983 and there was a further fall of one third from 1983 to 1984 following yet another particularly dry winter in Africa.

The present breeding distribution of the Whitethroat in Cheshire shows a marked westward contraction (cf. Sedge Warbler). In the eastern hills the species is generally scarce and outnumbered by the Lesser Whitethroat in some areas, as around Disley. In the eastern half of the county birds are concentrated into scrubby areas which offer ideal habitat, for example disused sand- and clay-pits with developing gorse, willow or thorn scrub and rough undergrowth of brambles and willowherb; defunct settling-tanks at sewage works with elder and willow scrub and extensive nettle-beds; and the rough herbage of disused market gardens with regenerating scrub. Westwards from Northwich pairs can be found nesting in hedgerows on farmland also, though generally by an unmown verge. On Hilbre, where Coward & Oldham[2] reported a few pairs nesting, and Hardy[4] said the species nested occasionally, there have been no breeding records in recent decades.

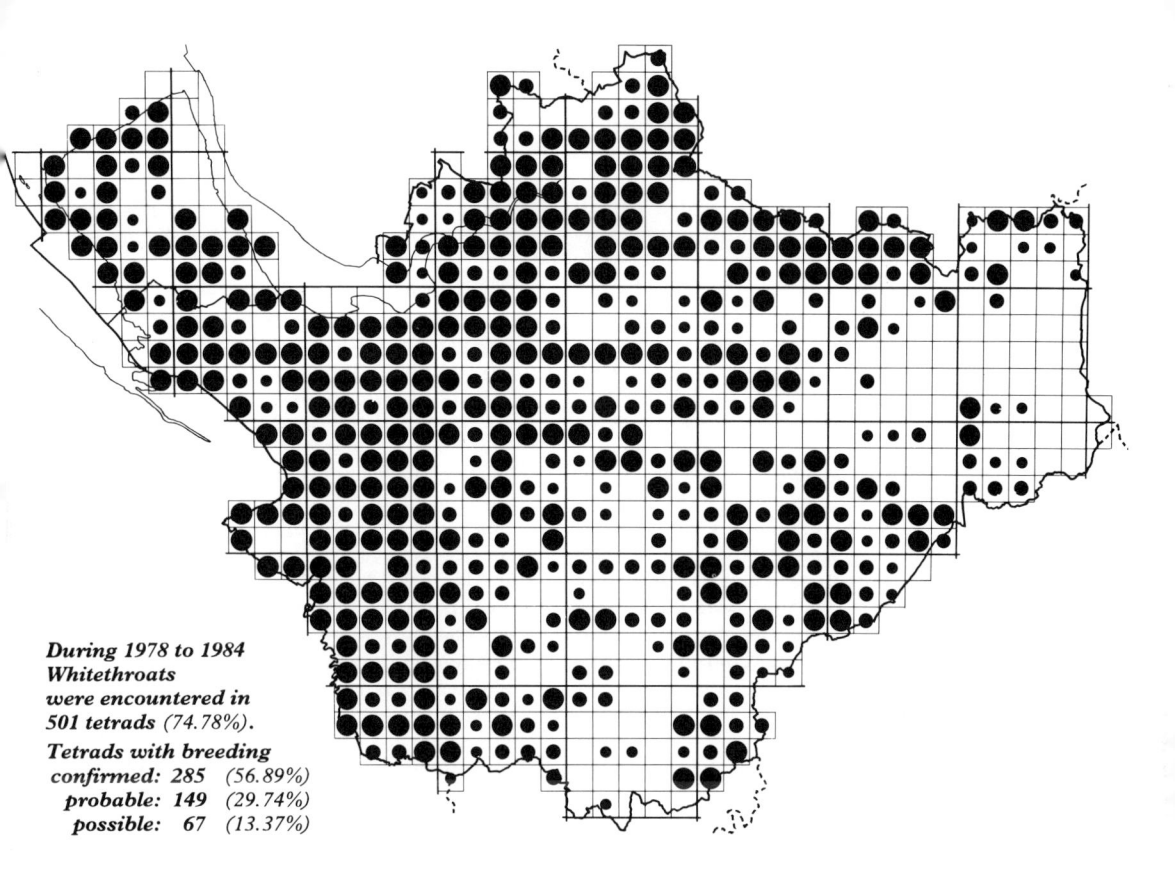

During 1978 to 1984 Whitethroats were encountered in 501 tetrads (74.78%).

Tetrads with breeding
confirmed: 285 *(56.89%)*
probable: 149 *(29.74%)*
possible: 67 *(13.37%)*

Nest-building begins soon after the arrival of the first females, in early or mid-May. Boyd found a nest with six eggs on 13th May 1928, but it is generally into June before the parents can be seen carrying food. Song may be heard into the middle of July, and sometimes a second brood is reared. Boyd had a record of young in the nest on 11th August, and in 1975 a nest at Sandbach held young on 12th August (JPG). When nesting, birds will often scold an intruder with harsh churring notes, and the behaviour of a bird at Prestbury on 23rd August 1983 suggested it still felt defensive towards its recently fledged young.

The close trimming of hedgerows and adjacent verges accounts for the species' scarcity in the south of the county, but a greater threat under present circumstances is the lack of safeguards for scrubland habitats which, by their nature, are only transitional. Osiers are no longer cut on a regular basis, so the few withy-beds in the county have become overgrown and unsuitable. The only concentration in SJ98, at Higher Poynton clay-pits, was threatened by tipping; an important site at Knutsford was built on in the mid-1970s; and the obsession with "tidiness" which afflicts the landscaping sections of too many local authorities led to the obliteration of a rich Whitethroat site at Marbury Country Park, Northwich.

Many counts of singing males are available. For example there were five at Higher Poynton clay-pits in 1983 (perhaps half the population of the Cheshire part of this 10-km square) and six or seven territories in the early 1980s at Prestbury Sewage Works. Six farmland CBC plots in south-east Cheshire held 17 territories in 393 hectares, and counts of singing males around the Sandbach Flashes gave 16 in 1973, 18 in 1978, although only nine in 1984. At Woolston 37 males were counted in 1983; 20 at Moss Side in 1980; 20 pairs at Handley Covert (twelve acres) in 1979; and 15 males at Thurstaston in 1978. At Hogshead Wood, Delamere, ten were singing in 15 hectares of young plantations in May 1982, their territories being mostly centred on brambles along the rides. 20 pairs were located around Aldford in 1983 and eight around Saighton in 1984. Many tetrads in the east and centre of the county held only one or two pairs however. Altogether 501 tetrads held birds at some time during this survey and the county population may have been around 3000 pairs.

Garden Warbler
Sylvia borin

THE Garden Warbler is a bird of tall shrubbery and thickets, from which its song may be heard from late April to early July. Few birds arrive before the end of April in fact, with larger arrivals often from the second week of May. Very early migrants were at Red Rocks, Hoylake, on 11th April 1981 and at Bradley on 17th April 1982. Most have generally departed by mid-September.

To differentiate between the song of this species and that of the Blackcap is regarded as a considerable achievement by many bird-watchers, although full-hearted songs of the two species are quite distinct. That of the Garden Warbler is typically more even, both in intonation and quality, with a more rapid flow of phrases – a sweet, chattery warble.

The full song of the Blackcap on the other hand may start with scratchy, clearly rougher phrases developing into a loud clear whistle with a final flourish. Subdued song of the latter is also more scratchy, less flowing than that of the Garden Warbler. Once glimpsed however, this species is quite distinct, with its uniform pale mousy plumage, broad-based bill and dark eye giving a soft facial expression.

The habitats of the two species differ also, though there is considerable overlap, and it is not unusual to hear a Garden Warbler and a Blackcap singing in the same copse, apparently in competition with each other. Whilst the Blackcap is a bird of tall woodland with some shrub layer, the Garden Warbler will occupy open birch woodland with a thick undergrowth of bramble or gorse, or extensive bramble clumps without taller song-posts. In Delamere Forest unthinned conifer stands some 20 feet tall with their bottom branches not yet "brashed" are used. Overgrown thickets of thorn are also popular.

Any changes in the species' status this century are difficult to assess, for very little has been written on the subject, most authors having concentrated on the relative abundance of this species compared with the Blackcap. Coward & Oldham[2] stated that Garden Warblers occurred throughout Wirral and the plain, though not so commonly as Blackcaps. Around Ince numbers of the two species were reported to be about equal, and at Alderley Edge and in Eaton Park and other localities near Chester the Garden Warbler was the more numerous. It was also abundant in the Dane valley near Wincle, where the Blackcap was rare. Abbott[10] recorded eleven territories in 13 square kilometres near Wilmslow in 1919 – three in Styal Woods and eight in Alderley Woods. On 17th May 1925 he noted eight to ten birds in the Bollin valley at Castle Mill.

Griffiths & Wilson[13] found neither species common in north Wirral, the Garden Warbler being scarcer if anything. Bell[7] added only that since the 1930s it had been recorded at some stage to outnumber the Blackcap around Nantwich, Sandbach and Great Budworth. In fact Boyd[6] implied that this was generally the commoner species in the last-named vicinity. Almost every wood and small covert had a pair in summer, and Garden Warblers were especially numerous at Arley, Tabley, Marbury and Little Leigh. In May 1941, when Boyd[5] visited the wooded grounds of a large house at Caldy, he found, "Blackcaps are more plentiful than Garden Warblers...contrary to my experience at home (Great Budworth)".

The *Lancashire Bird Report*[3] for 1957 mentions a bird at Risley Moss in late May, adding that the species had not been seen in the Leigh and Wigan areas for many years. The map shows that this scarcity north of the Mersey still applies.

It appears that this species has decreased in the Great Budworth area, and possibly also in the hills, where the Blackcap is now more numerous. The 1968 *CBR* reported an increase in Wirral with eight sites holding several pairs each, although by 1982 it was again "markedly uncommon" on the peninsula. In eastern Cheshire however there

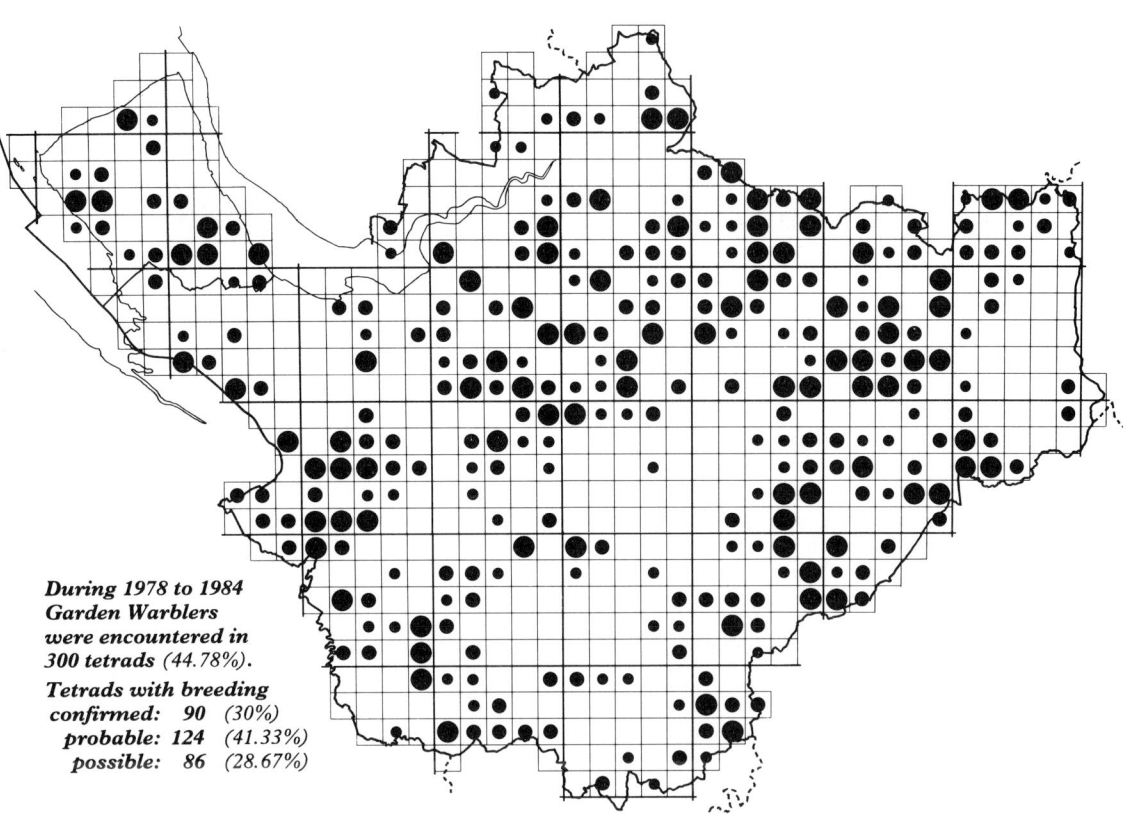

During 1978 to 1984 Garden Warblers were encountered in 300 tetrads (44.78%).

Tetrads with breeding
confirmed: 90 (30%)
probable: 124 (41.33%)
possible: 86 (28.67%)

was a definite increase in reports during the 1970s and early 1980s, with some birds occupying vacated Whitethroat territories, as at Prestbury Sewage Works for example (JPG). They winter far enough south to have avoided most of the problems of the sub-Saharan droughts which have afflicted such species as the Whitethroat and Sedge Warbler.

Where suitable habitat is extensive local concentrations of Garden Warblers will occur: there were seven occupied territories in the Alderley Park CBC plot (13 ha.) in 1979; in 1982 five pairs were located at Risley Moss, and seven males were singing along three kilometres of the Vale Royal Cut; in 1983 nine were singing at Farmwood Pool, Chelford.

Allowing for choice tetrads which may hold five or more pairs, and given that 212 tetrads were found with Garden Warblers probably or definitely breeding, the county population may be around 500 pairs.

Blackcap
Sylvia atricapilla

FROM mid-April the scratchy warble and fluty crescendo of the Blackcap's song is heard in most woods throughout the county. Often the bird remains unseen amongst the newly unfurled leaves until it hops to some new perch where the greyish plumage and glossy black cap may be discernible.

Since the decline of the Whitethroat in 1969, the Blackcap has become the commonest member of its genus over wide areas in the east and centre of the county, though of course the traditional habitats of the two species are quite distinct. However in the 1980s DN found a few Blackcaps as hedgerow birds, possibly as their population increased and spilled over from their usual woodland areas and they found little competition from the reduced density of

Whitethroats. The Blackcap appears to require taller woodland with more mature trees than does the closely related Garden Warbler. A well developed shrub layer is also necessary, and Blackcaps are often closely associated with rhododendrons. In some former mossland birchwoods, where rhododendrons were planted as cover for game and now form a continuous ground layer up to four or five metres high, as at Siddington, Tabley, and Aston-by-Budworth for example, the Blackcap thrives and is the characteristic bird of such woods. At Rostherne Mere NNR a census of 23 hectares of mixed woodland held between twelve and 22 pairs annually from 1976 to 1980, but only six or seven pairs over the next three years as a programme of eradicating rhododendrons came into effect. This association with rhododendrons was remarked upon in the Knutsford OS report for 1974, and enlarged upon by Sharrock[35] who found the correlation strongest in Scotland and especially Ireland, postulating that much of the mature deciduous woodland in those countries lay within old estates where ornamental rhododendrons had been planted. This certainly holds true for Cheshire also.

In Delamere and Macclesfield forests, Blackcaps are restricted to the mature fringe of broad-leaved trees around the conifer plantations, only occasionally moving into stands of older pines where some undergrowth of bramble or underplanted conifers exists. At Dean Row, Wilmslow, a pair nests annually in a small wood of copper beech, merging with the shrubbery of a large garden. Similar territories are found throughout the county.

Where Blackcaps and Garden Warblers occur together in a wood, they react to each other's song. Their territories may overlap however[35], so the apparent replacement of one species in a wood by the other may be due not to competition but to coincidence. For example, in 1982 none was seen in Highlees and Henbury Big Woods where they usually outnumbered the Garden Warbler. The latter species has been increasing in this district however.

Coward & Oldham[2] found the Blackcap to be fairly abundant throughout Wirral and the wooded parts of the plain, and it occurred in a few sheltered valleys amongst the eastern hills. Abbott[10] mapped six territories in 1919, four in the Bollin valley woods and two in Alderley Woods, but found none in non-wooded areas. Brockholes[2] stated that it was abundant in Wirral in the late nineteenth century, but Griffiths & Wilson[13] regarded it as scarce in north Wirral with nesting records only from Caldy and Thornton Hough. Hardy[4] knew it to be much commoner in Wirral than in southern Lancashire (which then included the Warrington district) however, and listed nine nesting sites from the peninsula, not including Caldy.

Sibson[14] recorded the Blackcap only once in the Sandbach area between September 1935 and July 1939, and Boyd[6], while considering it not uncommon, found it less numerous than the Garden Warbler around Northwich. He saw it most often at Arley, Tabley and Marbury in the better wooded parts. Bell[7] stated that he and Samuels had found the Blackcap to be the more numerous species around Wilmslow, probably in the period up to 1955. The *CBR* for 1964 classed the species as a "summer resident", breeding in small numbers, and scarce as a passage migrant.

It would appear that there has been a considerable increase in numbers more recently. By 1973 the *CBR* classed the Blackcap as a common resident, and in 1977 the *CBR* alluded to "indications of increasing numbers". In 1963, Whalley considered the species fairly common around the Sandbach Flashes, though no breeding records were given; Goodwin[27], however, reported that three or more pairs bred annually since at least 1973. The WGOS report for 1970 gave the species nesting in "most local woodlands". In Alderley Woods the breeding population is much higher than in Abbott's day, probably reflecting the closing of the canopy overhead in many parts since the 1920s.

Recent information in *CBRs* includes numerous counts of singing and territorial birds. At Risley Moss there have been between ten and 14 territories in 25 hectares of woodland; 16 were singing around the edge

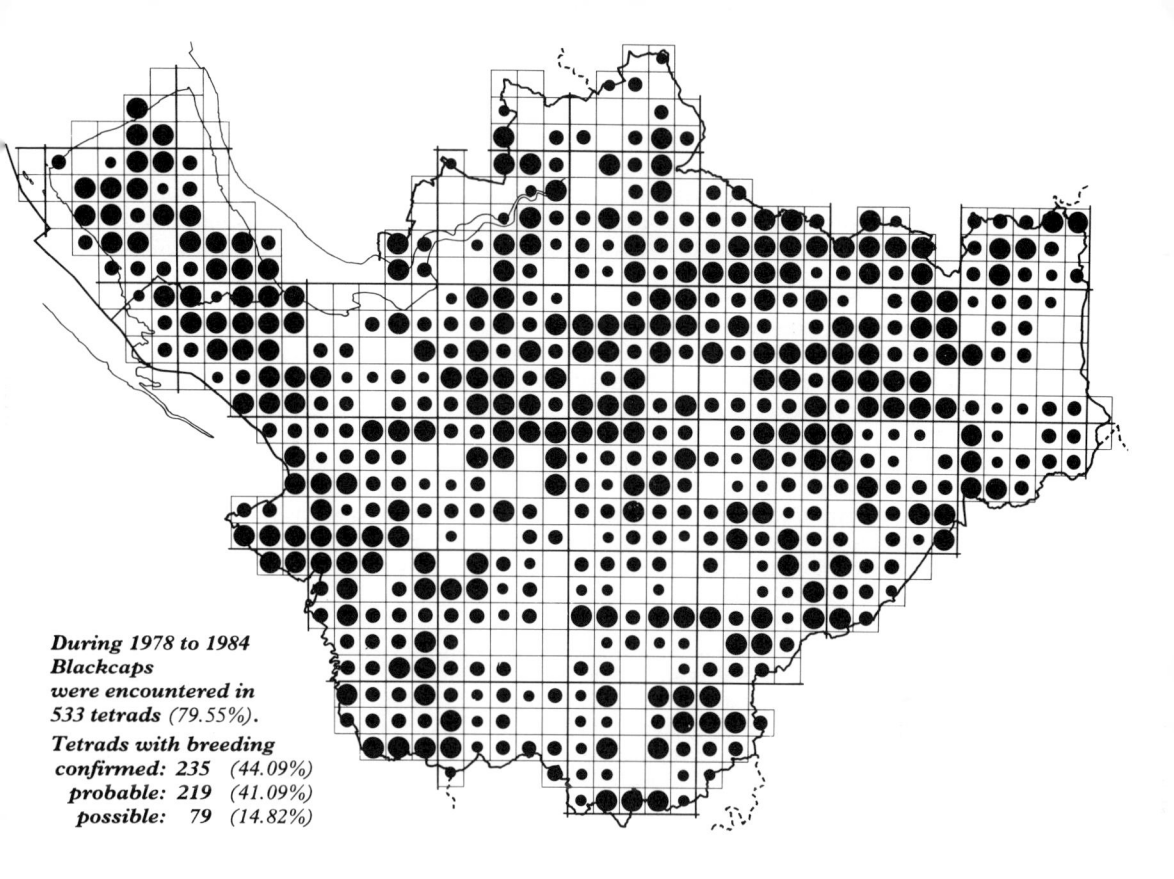

During 1978 to 1984 Blackcaps were encountered in 533 tetrads (79.55%).

Tetrads with breeding
  confirmed: 235 *(44.09%)*
  probable: 219 *(41.09%)*
  possible: 79 *(14.82%)*

of Combermere in May 1984; and four to six pairs at any one site do not appear unusual. Fifteen pairs were found in the Aldford tetrad (SJ45E) in 1984. The county population is probably over 2500 pairs.

Boyd[6] gave 19th-20th April as the average arrival date around Great Budworth, and at Wilmslow, Bell & Samuels[17] gave 23rd April as the average over a period of ten years. Recent records over a ten-year period at Rostherne Mere NNR suggest 17th-18th April for the first arrival, with the last bird generally seen about mid-September. A few earlier spring birds are seen, though these are not reliably separable from the increased number of central European breeders which have spent the winter here in recent decades. In 1974 a bird was in song at Redesmere on 31st March, and in 1977 one at Bromborough was singing on 11th March. Song is heard regularly from April until July and young are seen out of the nest from late May.

Wood Warbler
Phylloscopus sibilatrix

TO the woodland bird-watcher, the hesitant trilling of the Wood Warbler is one of the most evocative sounds of spring. It is often interspersed with a secondary song of "peu...peu...peu..." notes on a descending scale and may be heard from the first birds' arrival in late April through to the end of June or, exceptionally, later. Whilst trilling, the cock bird's whole body vibrates and the dashing display as the bird flies rapidly between the trees, changing direction every few feet, is one of our more breathtaking avian displays.

In Coward & Oldham's day the Wood Warbler occurred locally throughout the county, especially where oaks and beeches predominated. Brockholes[1] had described it as common in Wirral; it was very plentiful in unreclaimed parts of Delamere Forest, outnumbering the Willow Warbler and

Chiffchaff there; abundant in parks such as Tatton where there was plenty of mature timber; plentiful in the beech woods at Alderley but curiously rare on the wooded ridge at Peckforton; and common in the eastern hills from the Dane to the Goyt, being more abundant than any other warbler in the Dane valley above Bosley. It was often absent from plantations and game-coverts of recent origin however. In 1919 Abbott plotted single males in Styal Woods and Carr Wood, Wilmslow, and 14 males in the Alderley Woods.

Hardy[4] regarded it as a common passage migrant in beech woods as at Bidston, but scarce as a nester, and quoted Wirral records only from Storeton in 1938 and Dibbinsdale in 1941. Griffiths & Wilson[13] added a record from Bidston in 1932. Bell[7] knew of only one recent breeding record from Wirral – at Dibbinsdale in 1956.

By the early 1960s the Wood Warbler had become much scarcer[7]. It was still common in Delamere, though surely nowhere near to its former prevalence. Otherwise it then bred occasionally on the Peckforton Hills and sparingly in the east of the county.

Records in subsequent *CBRs* suggest that numbers declined until the early 1970s. This may reflect recording procedures and the enthusiasm of observers rather than its true status although populations may also fluctuate with time. The number of reports has increased considerably since the start of the present survey and 1978 in particular seems to have been a year in which above-average numbers were present: 34 territories were located in eastern Cheshire, including twelve in Alderley Woods and six around Lyme Park – in 1982 these two sites held seven and five territories respectively. On the Peckforton Hills nine males were located in 1978 and seven in 1982. In Wirral three pairs were reported at Caldy and two pairs at Eastham in 1979, and up to three birds in Burton Woods in 1982. A thorough survey of the 10-km square SJ87 in 1983 revealed a total of twelve singing males whilst ten pairs were found in SJ96, eight in SJ97 and eight in SJ98. In 1984 a national census organised by the BTO suggested an increasing population in east Cheshire, with an estimated 90 males holding territory throughout the county although only 80 were actually recorded. Figure 15 shows the results of this survey by 10-km square.

Migrating birds often stop to sing in small woods where they do not usually breed, for example under copper beeches at Dean Row, Wilmslow – the association with beeches in southern England, and formerly in Cheshire as mentioned by Coward & Oldham[2], no longer holds good in the county – and many of the small dots mapped will refer to such migrants. Furthermore a considerable proportion of our population is made up of unmated males, these being the birds which sing late into the summer. Thus, many instances of mapped 'probable' breeding may come from unmated territorial males. The survey map undoubtedly overstates the distribution, particularly away from its east Cheshire stronghold, partly because of the aggregation of seven years' records and partly because males, even when mated, will sing

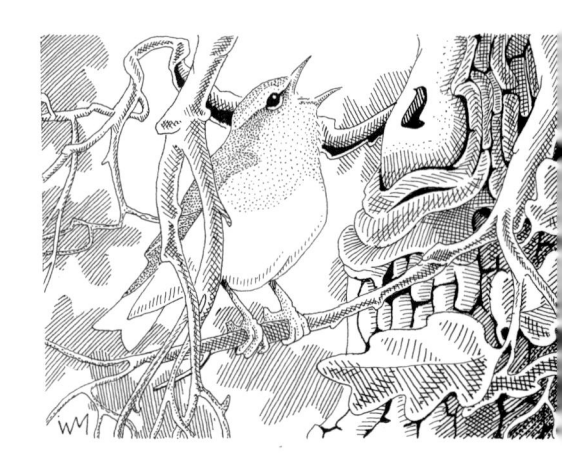

Fig.15. Distribution of Wood Warblers found in the county during the BTO national survey of 1984.

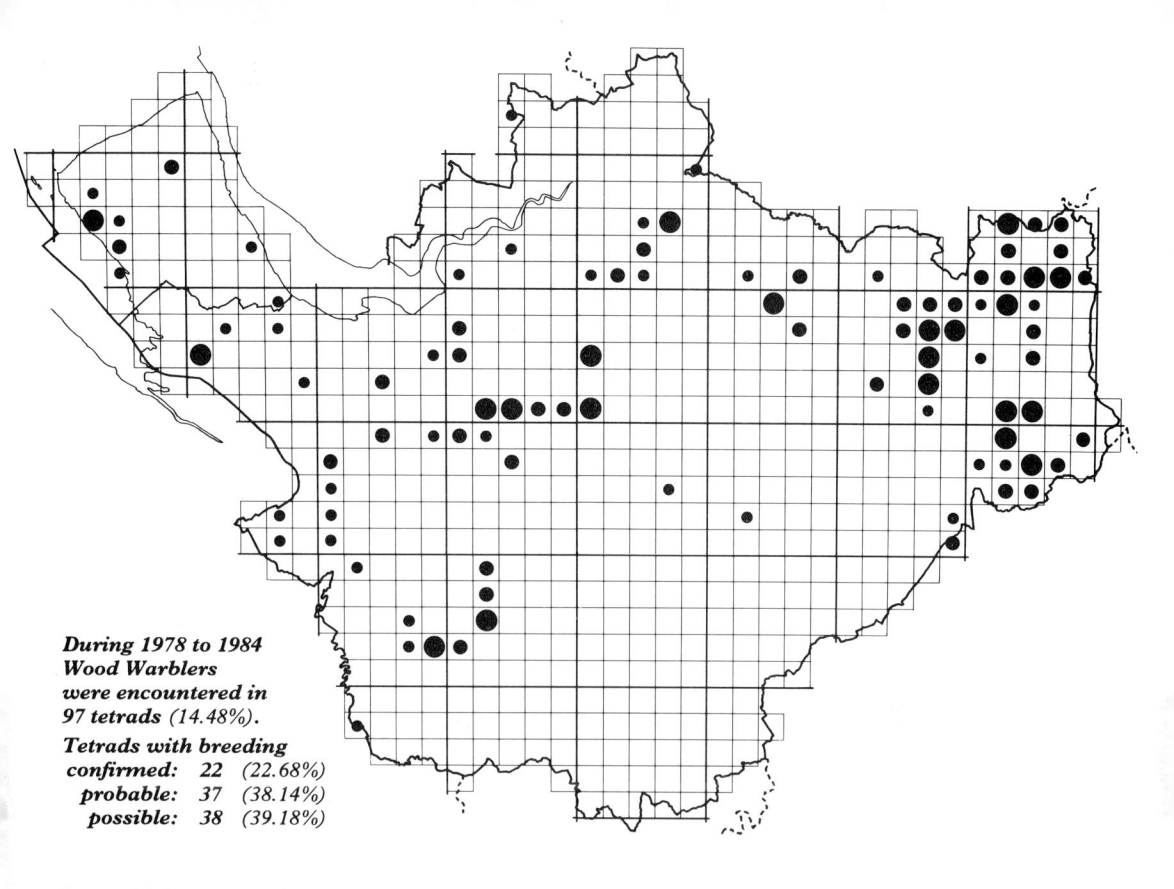

During 1978 to 1984 Wood Warblers were encountered in 97 tetrads (14.48%).

Tetrads with breeding
confirmed: 22 (22.68%)
probable: 37 (38.14%)
possible: 38 (39.18%)

from widely separated areas. Birds have been seen to move half a mile or more, and may be recorded in an adjacent tetrad.

Nowadays, Cheshire birds are most typically associated with oakwoods as at Peckforton, Alderley Edge and the Pennine foothills, though more often than not nesting woods are extensively mixed with other tree species. In Delamere, birds nest beneath the oaks bordering the conifer plantations; at Petty Pool birds sing from isolated red oaks within stands of conifers; on the drained mosses of the east as at Withington, birchwoods are developing into oakwoods and becoming attractive to Wood Warblers; and generally where sycamores are mixed with oaks the warblers will visit the former species to pick aphids from beneath the leaves. In 1983, a bird sang in a larch wood near Prestbury and the pair was subsequently watched feeding young in an adjacent alder carr. Usually birds prefer mature woods with sparse understorey and ultimately will desert a site as secondary growth develops.

The nest is usually situated amongst bracken, bluebells or brambles on the woodland floor, but unless the female is watched building it can be very difficult to find until the parents start feeding the young. As the nest is approached the adults utter a loud alarm call, similar to, but to some ears broader than, the whistle of a Bullfinch. Most records of confirmed breeding were 'FY' or 'NY' categories.

The Delamere population is small, with a maximum of five males found in any year. Some birds have been faithful to the area, three different ringed males having turned up there two years running, but others have changed site from one year to another. One male failed to find a mate in Eastham Woods, Wirral, in 1979 but bred successfully in Delamere in 1981 and another Delamere breeding male, ringed there in 1980, was caught again in 1982 on territory in Shropshire, 66 km SW.

Chiffchaff
Phylloscopus collybita

THE cheerful repetitive song of the Chiffchaff – single high and low notes more or less regularly alternated and interspersed with a grating "heer" – emanating from a blossoming clump of sallow bushes is one of the most welcome sounds of early spring. The Chiffchaff is in fact one of the earliest summer visitors to arrive, the first birds generally being seen in mid-March, and whilst it is not unusual for good numbers to be present by the end of that month, arrivals continue throughout April into May.

In most winters a few birds occur in Cheshire, often of the paler eastern races *abietinus* or *tristis*, and almost invariably frequent patches of willow scrub. These birds obscure the pattern of spring arrival, for they remain inconspicuous until well into March when song begins. Thus, a bird under observation at Foden's Flash, Sandbach from the begining of March 1975, was not heard to sing until 20th March. There is only one known record of winter song in Cheshire – a bird at Winsford on 26th February 1983.

From late March until mid-July song is heard regularly – at Dean Row in 1983 a bird sang almost daily until 18th July. There is then a pause until early September when autumn song begins, this continuing into early October. In 1983 for example all records of autumn song fell between 9th and 29th September. At this season migrant birds may be heard singing in gardens and other areas away from known breeding sites.

As mentioned under Willow Warbler, although the songs of the two species are normally quite distinct, indeed unmistakable, occasional birds are encountered with confusing songs. A bird at Rostherne on 31st March 1946 sang "several chiffchaff and willow-wren phrases"[3], and a Chiffchaff in Alderley Woods in the early 1970s (JPG) had a normal quality to its song, but instead of the usual alternating high and low notes, it strung together descending sequences recalling a Willow Warbler. In 1984 a warbler in Delamere Forest consistently sang "the first few notes of a normal Willow Warbler followed by a few 'chiff-chaff' phrases", and was suspected of being a hybrid between the two species (Norman 1984).

Some Chiffchaffs may linger for a few days in insect-rich willows but many go straight to their favoured breeding habitat. This consists of a variety of mature woodlands, often mixed, but also in large mature gardens and the older stands of conifers in Delamere and Macclesfield forests. In these plantations the ground flora may be limited in variety, but generally comprises a thick layer of brambles and ferns. In Scotland the species' distribution correlates closely with the presence of rhododendrons[34], and many Cheshire coverts containing these shrubs also hold Chiffchaffs. The dense, evergreen nature of the bushes provides shelter during adverse weather in spring.

Most Cheshire Chiffchaffs have young in the nest from the third week of May. Adult birds can be absurdly confiding at this stage, females carrying caterpillars whilst their mates sing or give alarm calls high in the trees. However the final approach to the nest is sufficiently devious to fool most birdwatchers, and the nest itself is usually a foot or two off the ground in dense vegetation, safe from most predators.

Coward & Oldham[2] considered the Chiffchaff to be widely distributed and not uncommon though generally more local and far less abundant than the Willow Warbler. Chiffchaffs did, however, predominate in parts of Delamere Forest. In the eastern hills the bird occurred only sparingly, chiefly in the valleys of the Goyt and the Dane. Around 1920 Abbott found only two singing Chiffchaffs in his mapped area which encompassed parts of Styal and Alderley Woods. In 1919 he commented that he never heard a Chiffchaff on Alderley Edge all summer. More recent authors have reaffirmed the predominance of the Willow Warbler, though given the difference in habitat preference between the two species, local exceptions are to be expected. Thus, Boyd[5] found Chiffchaffs unusually numerous in the wooded grounds of a large house at Caldy, where they outnumbered the commoner species. During the present survey Chiffchaffs and Willow Warblers were found in a ratio of three to one along the Overleigh Drive at Chester. Williams[20], writing in 1971, had reported a similar state of affairs there.

In Delamere current forestry practice provides abundant habitat for Willow

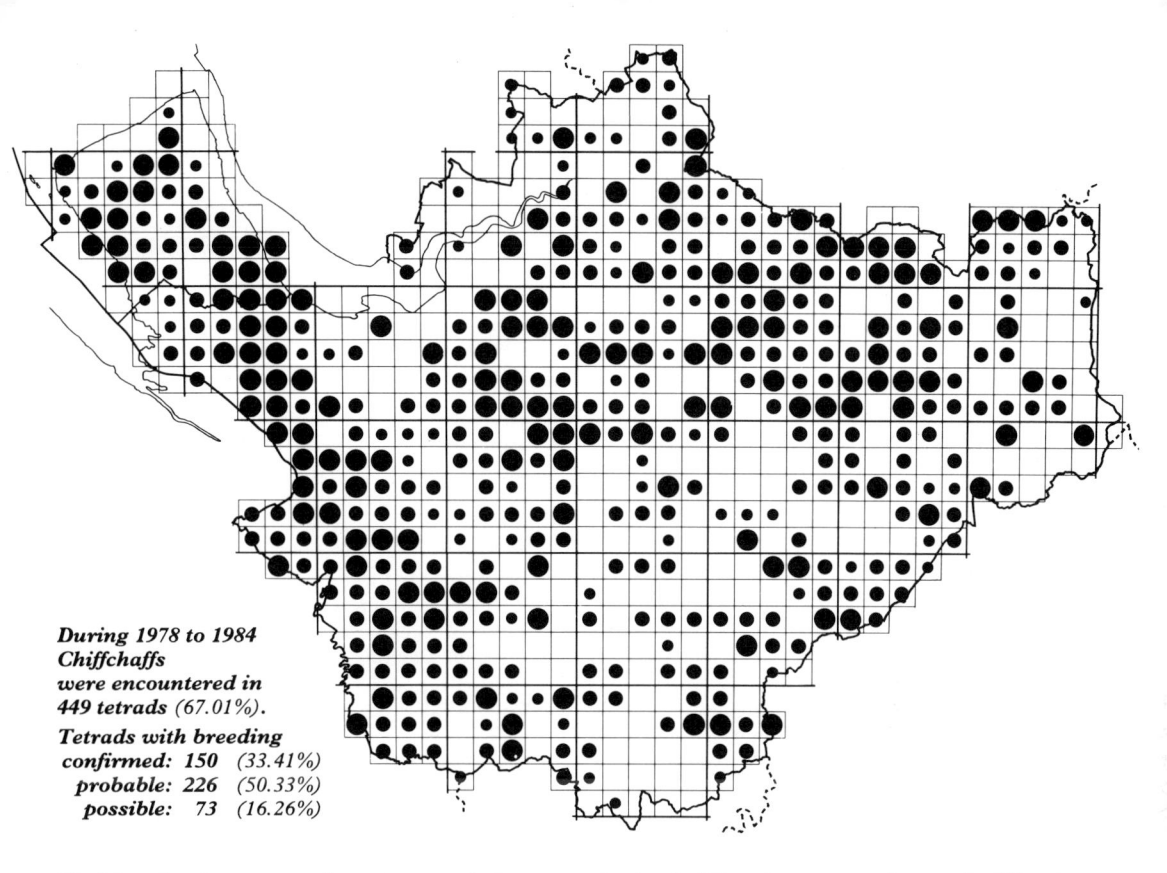

During 1978 to 1984 Chiffchaffs were encountered in 449 tetrads (67.01%).

Tetrads with breeding
confirmed: 150   (33.41%)
probable: 226   (50.33%)
possible: 73   (16.26%)

Warblers in the young plantations and they now outnumber Chiffchaffs generally in the forest. However some areas traditionally attract Chiffchaffs. The first territory to be occupied in Delamere was the same each year from 1981 to 1986, although it was known from ringing that it was a different male each time (DN). The establishment of conifer plantations in Macclesfield Forest and elsewhere this century has allowed some spread by Chiffchaffs into the eastern hills. Neglect of woodlands, and the closing of canopies, has allowed the Chiffchaff to increase, for example at Alderley Edge, in contrast to Abbott's experience in 1919. The national atlas[35] showed avoidance of the Pennine uplands generally, presumably due to lack of trees.

The map demonstrates a clear association with wooded areas. The mosslands north of the Mersey and the southern parts of the plain, both areas lacking in mature woodland, are short of Chiffchaffs. A number of records of 'probable' breeding in the south of the county however referred to males singing from inaccessible fox-coverts, where confirmation of breeding could not be obtained. The clear pre-eminence in Wirral and in the Chester area points to the suitability of habitat in that part of the county. However other migrant species such as the Sedge Warbler and Whitethroat show a western bias.

A census in Harper's Bank Wood and Mere Covert at Rostherne from 1976 to 1983 revealed between four and ten territories each year. Fourteen males were singing around the reserve on 17th April 1977, but this total may include passage birds which did not stay to nest, as might counts of twelve at the Overleigh Drive, Chester on 6th April 1980 and 14 at Combermere and four at Burwardsley. These are probably more realistic figures for well wooded tetrads, though many in the east of the county will hold fewer pairs. Nationally, the mean density in woodland CBC plots was 20 pairs per square kilometre[40], but it is unlikely that any Cheshire tetrads hold as many birds as this. Sharrock[35] estimated an average of 100 pairs per 10-km square, since when CBC results show the population to have fallen by more than one third. The county population may be in the region of 1500 pairs.

REFERENCE

Norman, D. (1984): Possible Hybrid Chiffchaff/Willow Warbler. *Cheshire Bird Report* p89.

Willow Warbler
Phylloscopus trochilus

ONE morning each April, following warm winds from a southerly quarter, sweet, liquid cadences announce the arrival of Willow Warblers, singing from sallows and other bushes. Boyd[6] found that the first birds in the Great Budworth area over 28 years arrived on 9th or 10th April. Bell and Samuels had an average date of 13th April over 14 years in the Wilmslow area[7], and WGOS reports over nine years up to 1977 show an average arrival date of 11th April. In some years a few birds arrive towards the end of March. If birds arrive during unfavourable weather song may be held back for a day or two. One singing at Blacon on 17th and 20th February 1975 was quite exceptional. Females generally arrive a few days later than the males. Waves of migrants continue to pass through the county into May, so that counts of singing birds during April may well exceed the eventual number of breeding territories.

The Willow Warbler, a characteristic bird of pioneer or regenerating woodland, is especially plentiful in birchwoods and willow scrub. Like other leaf warblers it will pick greenflies from beneath the leaves of trees and may be seen flycatching around the flowers of sycamores, which attract many insects. Woods with a closed canopy and little ground cover support few pairs, clearings and woodland edges being much more favoured. Conifer plantations may hold many pairs for a few years until the canopy closes in as the trees reach perhaps four metres in height. Waste ground on the urban fringe, railway cuttings and embankments, and sewage works are other popular habitats, and the species will nest in the shrubberies of large gardens. Tall hedgerows will support pairs, but the low, mechanically trimmed hedges in many agricultural areas are of little value to the species.

Coward & Oldham[2] described the species as abundant throughout Wirral and the lowlands, although less plentiful amongst the hills to the east of Macclesfield. This difference appears now to have largely evened out, as tidying of habitat on the plain has caused some deterioration there, and plantations in the hills have led to some improvement. The Willow Warbler is now undoubtedly the most numerous of our warblers, although prior to the decline of the Whitethroat it may have been outnumbered locally by that species. Griffiths & Wilson[13] suspected that this might be the case in north Wirral. Elsewhere however the former dialect name of "Peggy Whitethroat", applied to the Willow Warbler, is a source of confusion. Abbott's maps show a total of 331 singing Willow Warblers in three areas totalling just over 22 square kilometres around Wilmslow between 1919 and 1922, making this by far the most numerous of the warblers he censused. Statements such as that by Boyd[5], that Willow Warblers were outnumbered by Chiffchaffs in the wooded grounds of a large house at Caldy, reflect the latter species' preference for taller trees rather than scrub.

Song may be heard daily from the birds' arrival in April until the end of June and less often in early July. The species starts to sing again about the end of July or in early August, though this lacks the vitality of spring song. From late August full song becomes less frequent, but from then on into September it often takes the form of a quiet sub-song uttered by yellow juvenile birds. In 1978 one was reported in song at Nantwich on 4th October[23]. A report of one in song at Delamere on 14th December 1980 appears to be without precedent in Britain but unfortunately is poorly documented[8]. Robins not infrequently utter brief snatches of song recalling that of a Willow Warbler.

Birds with poorly developed powers of song are encountered from time to time, and may cause problems of identification (cf. Chiffchaff). A colour-ringed male at Daresbury had a normal song in 1982 and 1983, but on his return in 1984 could manage only the usual first few descending notes followed by a buzzy wheeze. Nevertheless this bird attracted a mate and bred successfully. When it returned to the same territory in 1985 it produced a normal song (Norman

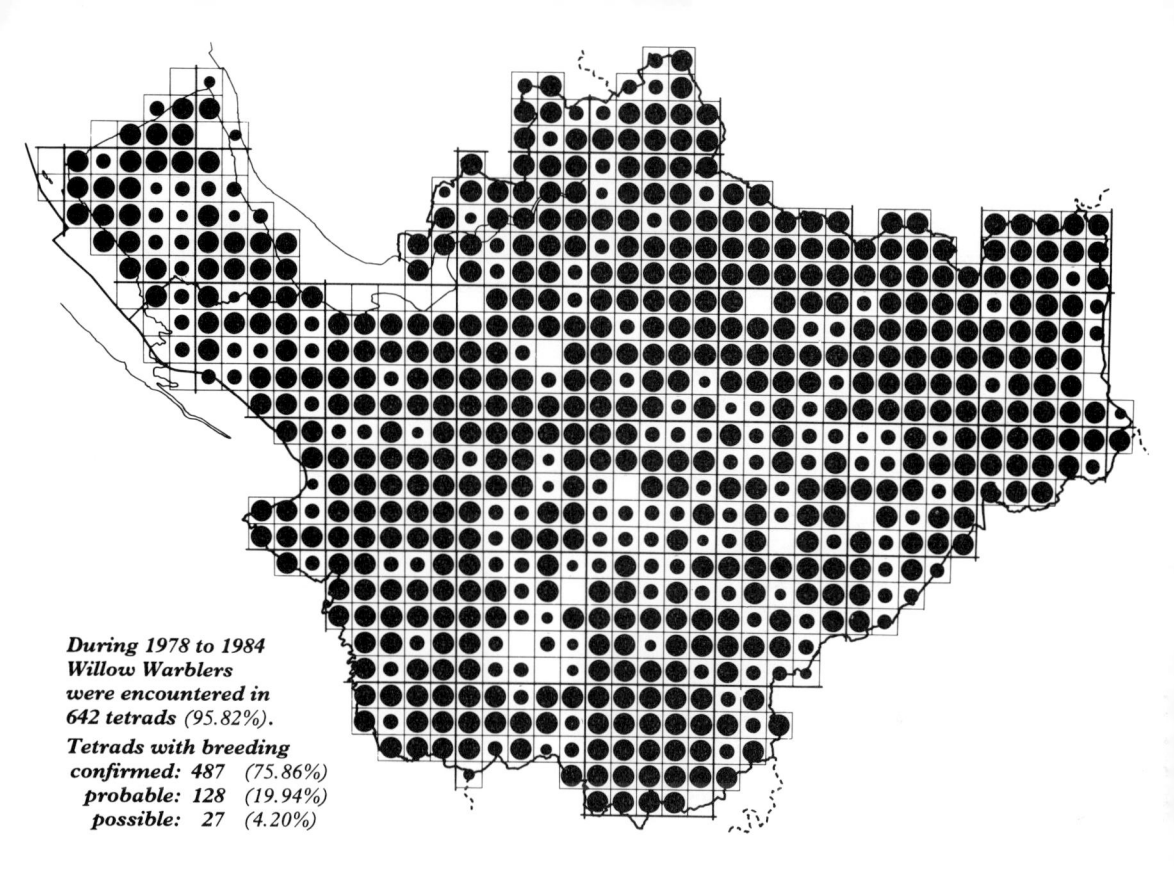

During 1978 to 1984 Willow Warblers were encountered in 642 tetrads (95.82%).

Tetrads with breeding
confirmed: 487 (75.86%)
probable: 128 (19.94%)
possible: 27 (4.20%)

1987). An exceptionally brown bird at Siddington during the present survey had a song approaching the normal intonation of a Willow Warbler but with a dry, tuneless quality somewhat recalling a Chiffchaff (JPG).

In late April and early May a display is often observed in which one or both birds crouch, with wings half spread and quivering, and cheep excitedly rather like young birds begging to be fed. Mating may then follow. As soon as nest-building starts in late April the pair will anxiously call "hooeet" at the approach of a predator and flick their wings in an agitated manner. Boyd[6] had a record of a pair that attacked a dog at a partly-built nest on 4th May. Anxiety displays are usually most intense whilst young are in the nest, from late May in most years, this providing the easiest confirmation of breeding. Some pairs in Cheshire have second broods, and these adults may be seen carrying caterpillars and other food until early August.

Most nests are situated on the ground or on a bank amongst grass and dead leaves, but a small proportion is raised slightly in the low branches of bushes. On 29th May 1948 Boyd found a nest with eggs seven feet up in ivy on a garden wall at Rostherne[3]. At the other extreme Boyd[5] recorded a nest which was so close under a stone that the usual dome was unnecessary.

Sharrock[35] believed that each occupied 10-km square in Britain might hold more than 1000 pairs of Willow Warblers, i.e. 40+ pairs per tetrad. A number of Cheshire tetrads are known to exceed this figure. Six CBC plots on wooded farmland in south-east Cheshire, totalling 393 hectares (one tetrad = 400 hectares), held 58 pairs in 1978. At Risley Moss, 73 hectares, including 25 hectares of woodland, support from 60 to 79 pairs. At Rostherne, 19 hectares of mixed woodland held up to 29 pairs. At Alderley Park, 13 hectares of woodland held between nine and 15 territories from 1978-80. Between Styperson and Higher Poynton, 37 males were singing along 3.5 kilometres of disused railway cuttings and embankments with birch scrub in May 1982, with a further 15 males along a parallel stretch of canal towpath, and, in the same month, 18 males held territory in 7 hectares of young conifer plantations at Hogshead Wood, Delamere.

On the other hand just 47 territories were found in and around Lyme Park in 1982 –

approximately four tetrads of rather better than typical upland habitat, and a thorough survey of the Saighton tetrad (SJ45J) revealed only about thirty pairs. Urban and many intensively farmed areas lack suitable habitat – the records of Birkenhead School Natural History Society for 1963-73 show some withdrawal following housing developments – so an average of rather less than thirty pairs per occupied tetrad may be realistic across the county. This would give a Cheshire population of perhaps 18,000 pairs. CBC results show the breeding total to be remarkably stable from year to year.

REFERENCE
Norman, D. (1987): Willow Warbler with temporarily aberrant song. *British Birds* 80: 578.

Goldcrest
Regulus regulus

ALTHOUGH widely distributed and locally numerous, Goldcrests attract less than their fair share of attention because of their small size and concealing summer habitat, and their high-pitched calls which are beyond the range of hearing for some people. Consequently there is little quantitative information on breeding populations in Cheshire.

Goldcrests are restricted to coniferous or mixed woodland in the breeding season, although churchyard yew trees or mature gardens with ornamental conifers, for example cedar or cypress, form an acceptable substitute. In some years birds take up territory in the spring in other woodlands, for example alder and willow carr at Foden's Flash, Sandbach, and they may linger and sing in such habitats well into April, but there are no records in Cheshire or Wirral of Goldcrests breeding away from conifers.

The tinkling song emanating from high in a fir tree is often the first indication of the bird's presence. Song is most frequent in April and May, but not infrequently may be heard from January onwards (some of the smallest dots on the map may refer to passage birds singing in early spring), and Coward & Oldham[2] had heard it in every month except August. JPG has modern records of song in every month. With luck the nest may be spotted suspended beneath a branch, although occasionally nests may be found supported in ivy on the main trunk of a tree. However, by far the easiest method of confirming breeding during this survey was to encounter noisy parties of young identified by their squeaky "zit...zit..." calls.

The greatly increased area covered by conifers in recent decades has been of enormous benefit to this species. Up until this century there was a general lack of conifers in Cheshire, other than a few stands of Scots pines. The horizontal branching pattern of spruce and larch trees is preferred to pines for nesting. G. B. Hill has noticed that in Alderley Park, Goldcrests prefer western hemlock, but will also hold territory in stands of pine or larch. The Macclesfield and Delamere forests in particular now provide shelter for dozens or probably hundreds of pairs, though a subjective impression is that Goldcrests are more numerous in the latter lower-lying complex. Pinewoods on sandstone outcrops in Wirral are also favoured. Many formerly broad-leaved woodlands have been felled and replanted with conifers, and many old mosslands have been similarly afforested.

Following severe weather early in 1917, only one Goldcrest was seen anywhere in Cheshire that summer[3]. In part this reflects winter mortality and the scarcity of bird-watchers, but primarily shortage of habitat. Abbott noted Goldcrests in the nesting season on only three occasions between 1913 and 1930. In 1919 he kept a watch on Alderley Edge throughout the summer, finally seeing two birds in October.

Being Britain's smallest bird and almost exclusively insectivorous, it may be surprising that this species survives in normal winters, but it is less of a surprise that many succumb to the rigours of severe weather. Following a long cold spell in early 1979 the population in Wirral was much reduced and the results of a CBC at Alderley Park showed that the

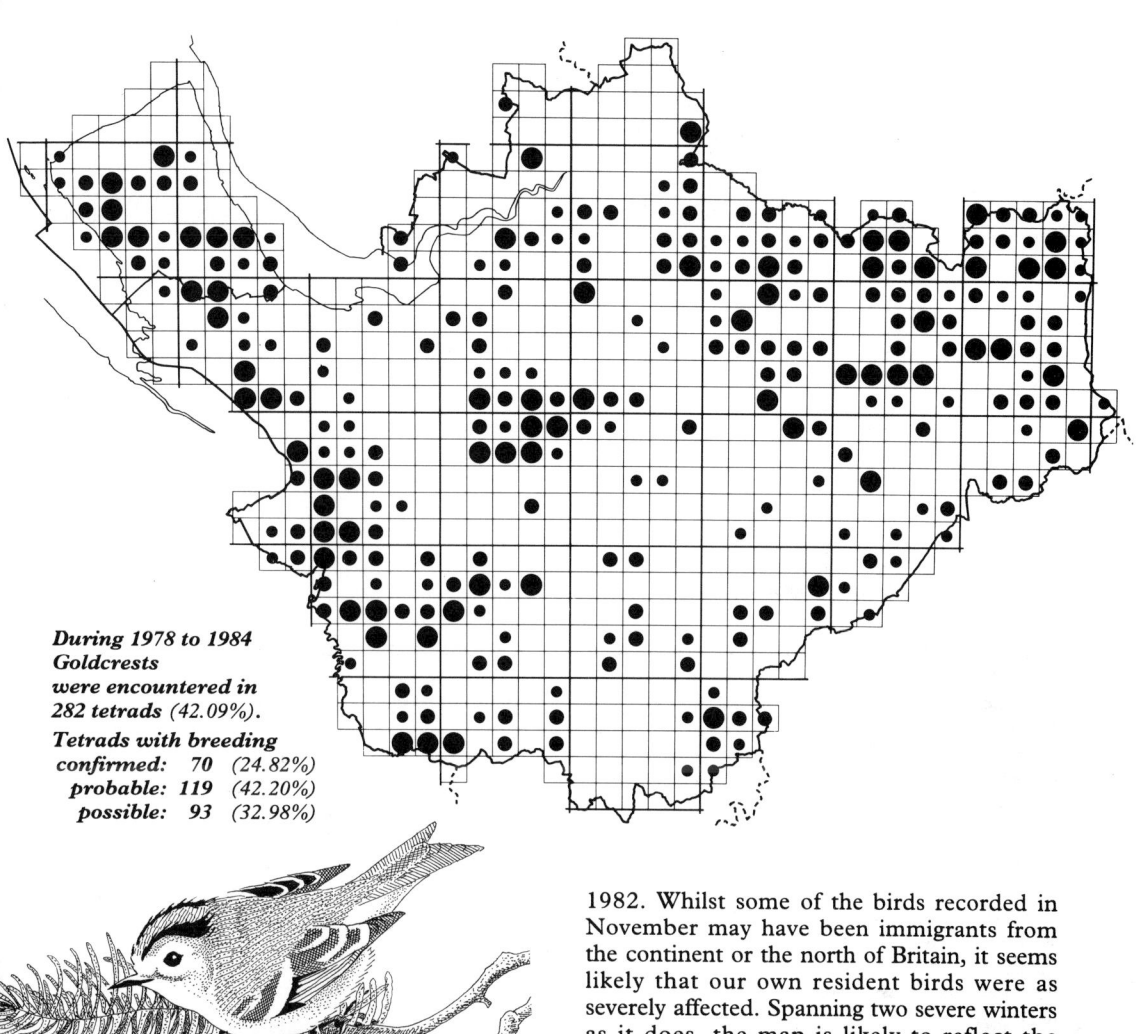

During 1978 to 1984 Goldcrests were encountered in 282 tetrads (42.09%).

Tetrads with breeding
confirmed: 70 (24.82%)
probable: 119 (42.20%)
possible: 93 (32.98%)

population was more than halved, although numbers here recovered to their 1978 level by 1980. A second spell of severe weather in late 1981 also fell within the survey period: on 26th November 1981 a walk through Alderley Woods revealed 39 Goldcrests; the best count following the same route after the bad weather was of just nine birds on 21st January 1982. Whilst some of the birds recorded in November may have been immigrants from the continent or the north of Britain, it seems likely that our own resident birds were as severely affected. Spanning two severe winters as it does, the map is likely to reflect the distribution of the species at a low ebb, although the difficulties of recording this species may also reflect the apparently thin distribution. The absence of records from the Mersey mosslands and parts of the agricultural lowlands reflects the lack of suitable habitat in these areas.

National CBC figures show a peak in Goldcrest numbers from 1973 to 1975 with a crash to little over a quarter of this figure in 1979, the first full year of this survey. The most densely populated tetrads in Cheshire may hold up to 100 pairs, but others are only sparsely occupied. The county population probably lies in the range 1500–3000 pairs.

Firecrest
Regulus ignicapillus

THIS species was first detected within the present recording area in 1968, when a bird was seen at Storeton on 9th March. Since then there have been records in most years although breeding has yet to be confirmed in the county.

In 1975 singles were seen at Thurstaston on 26th April and at Moss Side on 29th April. That same spring there was a report of a bird singing regularly in the mornings at Alderley Edge but this could not be verified. A male was watched singing amongst yews, other conifers and evergreens at Adlington Hall on 17th May 1976. However, a passage bird noted at Hilbre two days previously suggests this bird may also have been a migrant.

Breeding season records during this survey included:

1978 A pair in a wood at Thurstaston on 17th April, the male singing.
1981 A bird watched feeding in sycamores in Macclesfield Forest on 17th April.
1982 Three or more males singing at three sites in the east of the county between 8th and 26th May.
1983 A male singing in a tall thorn hedge with scattered pines at Whirley on 28th April. (A passage bird was at Hilbre two days later.) A male singing on 7th June in one of the same localities as in May 1982.
1984 A male singing in Delamere Forest on 20th April.

In the West Midlands, breeding was first reported in 1975[47], and in Derbyshire in 1981[44], though in neither area has a population become established.

Spotted Flycatcher
Muscicapa striata

THE Spotted Flycatcher lives up to its name, using a branch or other perch with good visibility from which to sally forth and catch its insect prey, frequently returning to the same perch. Because of its specialised manner of feeding this species is one of the last of the regular summer visitors to return from its sub-Saharan wintering grounds, arriving late in spring when winged insect food is usually more numerous. The first birds sometimes sneak in during the last few days of April, but they are seldom present in any numbers until the middle of May, with birds still appearing in new territories in June. In fact, because the song is weak and unobtrusive, their arrival often goes relatively unnoticed compared with other, more vociferous migrants. In many parts of Cheshire, the Spotted Flycatcher is primarily a bird of woodland edges or glades, being equally at home in the cloughs of the eastern hills, in lowland parks or in wooded gardens. On the plain it also breeds commonly around the buildings of dairy farms, where insects associated with cattle provide rich feeding, even though in some areas only a single line of roadside trees may be present.

The nest-sites chosen reflect this range of habitats. Many woodland birds choose open cavities or forks in trees; in parkland areas the

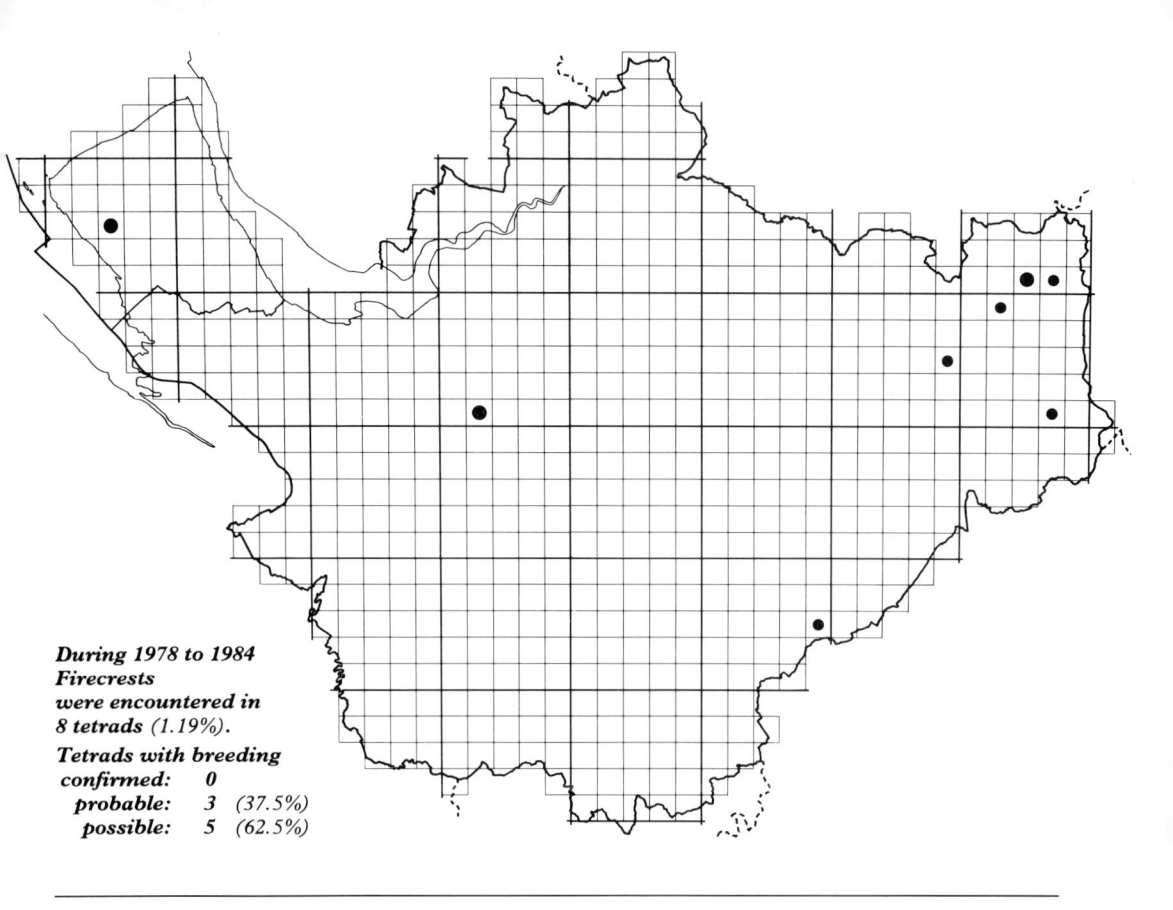

During 1978 to 1984 Firecrests were encountered in 8 tetrads (1.19%).

Tetrads with breeding
confirmed:	0	
probable:	3	(37.5%)
possible:	5	(62.5%)

clusters of epicormic shoots on limes and other trees may be used; and around habitations creeper-covered walls provide suitable sites, hence the old dialect name of "Wall Robin". Boyd[6] listed the big hinges of stable doors or a garden door as a favourite site. He also had a note of a pair occupying a Chaffinch's old nest as soon as the finches had flown, and Coward & Oldham[2] knew of cases where the nests of the House Martin, Swallow and Song Thrush had been adapted. Spotted Flycatchers sometimes build multiple nests on man-made structures where a pattern is repeated as in brickwork or on machinery. In 1948 a pair built two nests at Plover's Moss, Delamere, and on 1st August there were three eggs in one nest and a bird sitting on the other[3]. At Arclid sand-quarry in 1982 a pair nested on a narrow ledge of redundant equipment, the occupied nest adjoining the remains of seven or eight earlier nests[8].

Nesting operations may start within a day or two of the birds' arrival. Usually birds are feeding young of the first brood from the second week of June and of later broods into August. An apparent family party of eight at Dean Row on 8th September 1983 may have moved in from further north.

Early this century it was a common summer resident, widely distributed across the lowlands and in many of the wooded cloughs in the hills. It was considered sacred in some of the country districts, and its nest and eggs were not molested by young egg-collectors[2]. Abbott mapped 28 territories in 22 square kilometres around Wilmslow between 1919 and 1922.

Whilst by no means skulking and quite easy to prove breeding, Spotted Flycatchers are rather quiet, and isolated pairs are easily missed unless a tetrad is thoroughly covered. This factor, coupled with the somewhat shorter recording season for such a late arrival, means birds are likely to have been overlooked, particularly in those areas of central Cheshire which relied heavily on brief visits by outside observers for coverage. A last-ditch effort by JPG in 1984 added records of confirmed breeding in 22 tetrads in the east, centre and south of the county (all four such records in SJ64 for example), and there can be little doubt that this is one of the most under-recorded species during the survey. The absence from the mossland areas of the Mersey valley probably reflects lack of suitable woodland, but even allowing for sparse coverage in the centre of the county, it is clear from the map that the species is thinly if widely distributed.

On 3rd July 1973, JPG found 11 pairs in Tatton Park and estimated that perhaps twenty were present. The previous year five family parties were seen along less than half a mile of the River Croco near Holmes Chapel. During the survey, some tetrads were reported to hold four or five pairs, but there were also many cases where only a single pair could be found. The situation in Cheshire probably exceeds somewhat Sharrock's[35] estimate of 30 or more pairs per 10-km square. A total Cheshire population of about 1000-1250 pairs is suggested.

Pied Flycatcher
Ficedula hypoleuca

ALTHOUGH a numerous breeding species in many Welsh oakwoods, it is only in recent years that the Pied Flycatcher has begun to nest in Cheshire. In his black-and-white summer plumage the male is unmistakeable, and the loud song with its clear intonation is quite distinctive although unfamiliar to many bird-watchers.

Coward & Oldham[2] knew the species only as an occasional bird of passage, chiefly in the spring, and all their listed records are obviously of such birds with the possible exception of a pair at Tintwistle (now GMC/Derbyshire) on 21st May 1898. Boyd, in his country diary of 8th May 1936, found it puzzling that none stayed to nest in the wooded cloughs of the eastern hills which offered just the right kind of habitat – old oak trees with cavities for nesting and a plentiful supply of caterpillars and other insect food. It seems mildly ironic therefore that some of the earliest records of birds summering in the county came from Wirral, which is, however, close to the Welsh strongholds.

In 1941 a male was present during May, June and July at Raby, but no nest was found[13]. Two years later a pair was seen in Wirral, and in 1951 a pair nested in Dibbinsdale – the first breeding record for the peninsula. Since then Wirral records have been few and far between: birds were seen at Noctorum in summer 1963; a pair at Royden

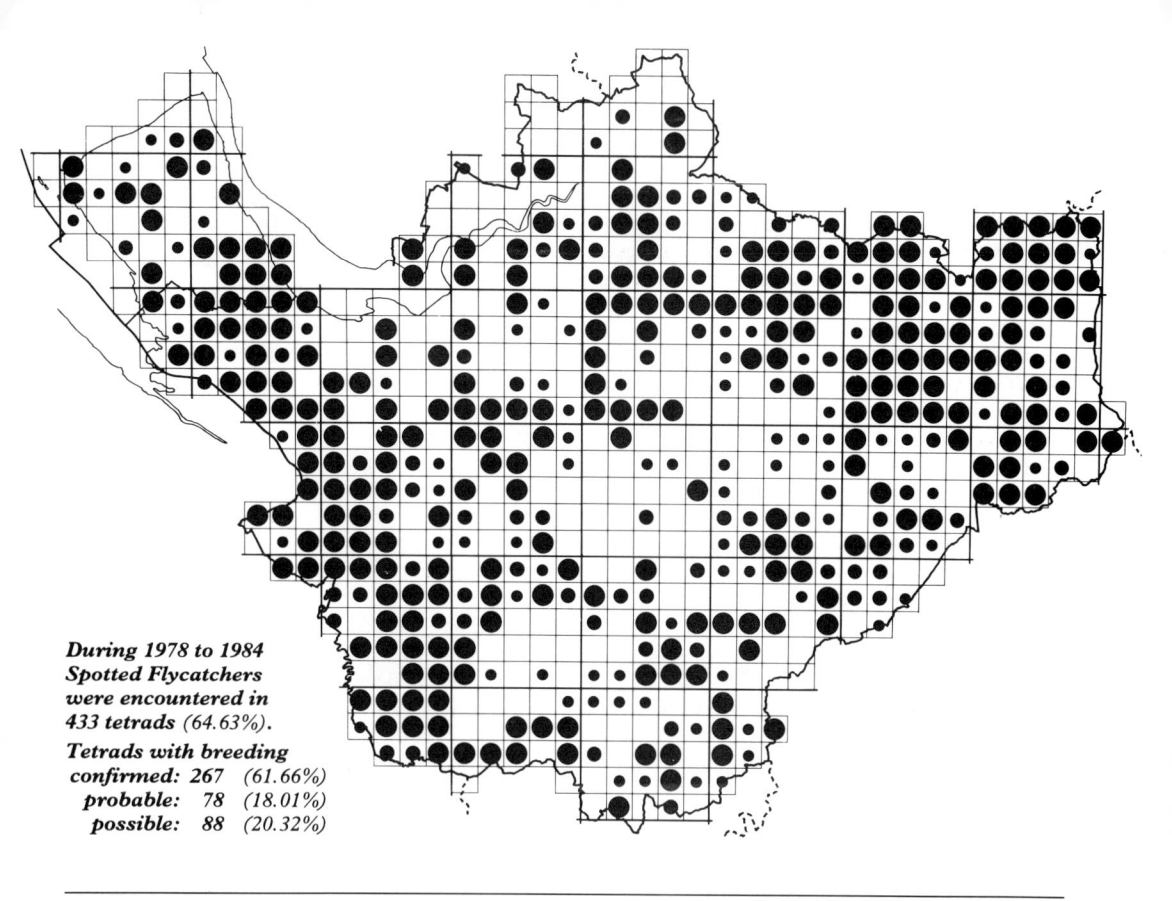

During 1978 to 1984 Spotted Flycatchers were encountered in 433 tetrads (64.63%).

Tetrads with breeding
confirmed: 267 *(61.66%)*
probable: 78 *(18.01%)*
possible: 88 *(20.32%)*

Park in May 1971 did not persevere, and a pair bred at an undisclosed site in 1977.

In 1948 young were watched being fed in a beechwood in north-eastern Cheshire – whether this was within the county as currently constituted is not known. In April and May of the same year a male was watched entering a hole in a tree in Delamere[3]. From 1949 onwards however there started a series of records from the Dane Valley at Wincle, with from a single male to two pairs present annually, and sometimes nesting, until 1955. In the autumn of 1956 tree-felling began at the favoured site, but Boyd had definite evidence of nesting along the Dane in 1957, and Bell[9] still had records up to 1966. *CBRs* mention this locality in 1967, 1969 and 1978, with two singing males "at the usual nesting areas in east Cheshire" in 1977. Elsewhere in the east of the county a pair bred in 1957 at Alderley Edge where the species was also present in 1969. A female was in Lyme Park in late May of 1967, and a pair also bred there in 1969. In 1973, a female lingered around Wilmslow Park for a month from mid-June.

Two other areas have traditions of nesting attempts. In Delamere, site of the 1948 record listed above, a pair nested in 1952 and birds probably did so in most years until 1968, with three pairs in May 1956[7, 8, 9]. In 1962 the favoured nesting trees were felled and the birds moved elsewhere. The *CBRs* for 1969 and 1971 specifically mention absences and no further Delamere records came to light until 1982, when a male stayed for four days in May. The following year a pair bred successfully and two other males were seen briefly in April and May. Finally in 1984, three pairs nested but only one brood fledged. One of these pairs used the same nest-box as in 1983, but it is known from ringing that different birds were involved (DN).

From about 1954, nesting has occurred more or less regularly on the Peckforton Hills, with almost annual mention in *CBRs* except for a gap from 1971 to 1977 – probably more a reflection of editorial policy than of status. Seven birds were at Beeston on 12th May 1957; seven males were counted at Peckforton on 8th May 1960; at least four pairs bred in 1965 and four pairs bred in 1977.

Isolated records come from Cholmondeley where breeding was suspected in 1967 and birds were present in 1969; from Combermere where birds bred in 1970; and from Frodsham Hill where a pair was present in May and June 1975. In 1977, singing birds, presumably on passage, were noted in the Eaton estate and at Tatton.

Spring migrants undoubtedly cloud the picture and probably account for most of the small dots on the map, but given the rather erratic history of breeding in the county, many of the passage birds must be regarded as potential nesters, especially when males linger for a week or two in mature woodland as happened at Redesmere in 1980 and Styal in 1983. A small number of other records refer to wandering birds in mid-summer, such as those at Rixton in July and August 1980 and Rostherne in July 1982. A record of a young bird ringed in Clwyd in 1968 and found dead at Frankby, Wirral in July of that year points to the origins of some of these wandering birds.

Although the typical nest-site is in a small hole in a tree trunk, nest-boxes are readily accepted and probably account for the majority of Cheshire nests. A pair in Lyme Park in 1982 bred in a hole 20 feet up a birch tree. Competition with tits and other hole-nesting species may be severe, particularly as

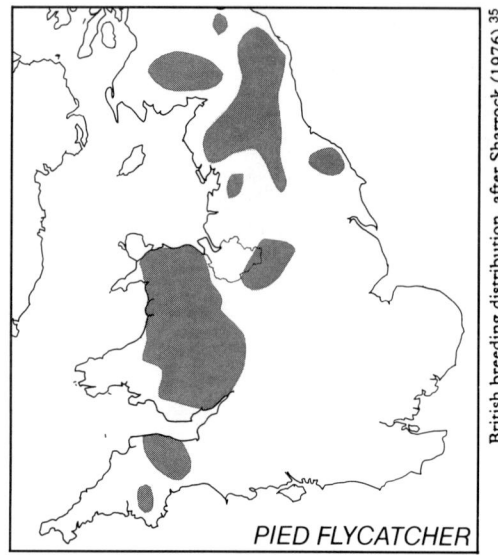

PIED FLYCATCHER

British breeding distribution, after Sharrock (1976)

Pied Flycatchers generally arrive on territory in late April or May when the tits are already firmly ensconced. Nest-box schemes might well encourage the occupation of suitable woods in the eastern hills and elsewhere. With a maximum of six pairs reported at the Peckforton stronghold during this survey, and erratic numbers elsewhere, the county population may currently stand at 10-15 pairs – the highest level since records began.

Bearded Tit
Panurus biarmicus

A pair of birds was shot during the middle of the nineteenth century at Whitley Reed, then a wild, unreclaimed bog. This suggests at least the possibility that the species may have bred there.

Birds have been seen in the county in autumn and winter, particularly in the 1960s and 1970s when large post-breeding flocks were irrupting from the Netherlands, but few seemed to stay long and there were no breeding season records.

On 3rd April 1981 a pair was seen briefly in the *Phragmites* reed-bed at Knutsford Moor. That same spring, on 10th May, JPG saw a male bird at Woolston and later the same day a different male and other birds calling in a second, larger reed-bed site. Two pairs were subsequently located at the latter site where they remained for several weeks. In 1982, a male frequented Neston reed-bed from 12th April until 18th May, and one was seen at Gatewarth sewage farm on 4th August. The following year two males and a female were noted at a reed-bed site on 8th May; and in 1984 two males and a female again frequented a suitable breeding site from 16th to 24th April.

It appears to be only the lack of good quality habitat that is preventing the Bearded Tit from nesting in Cheshire.

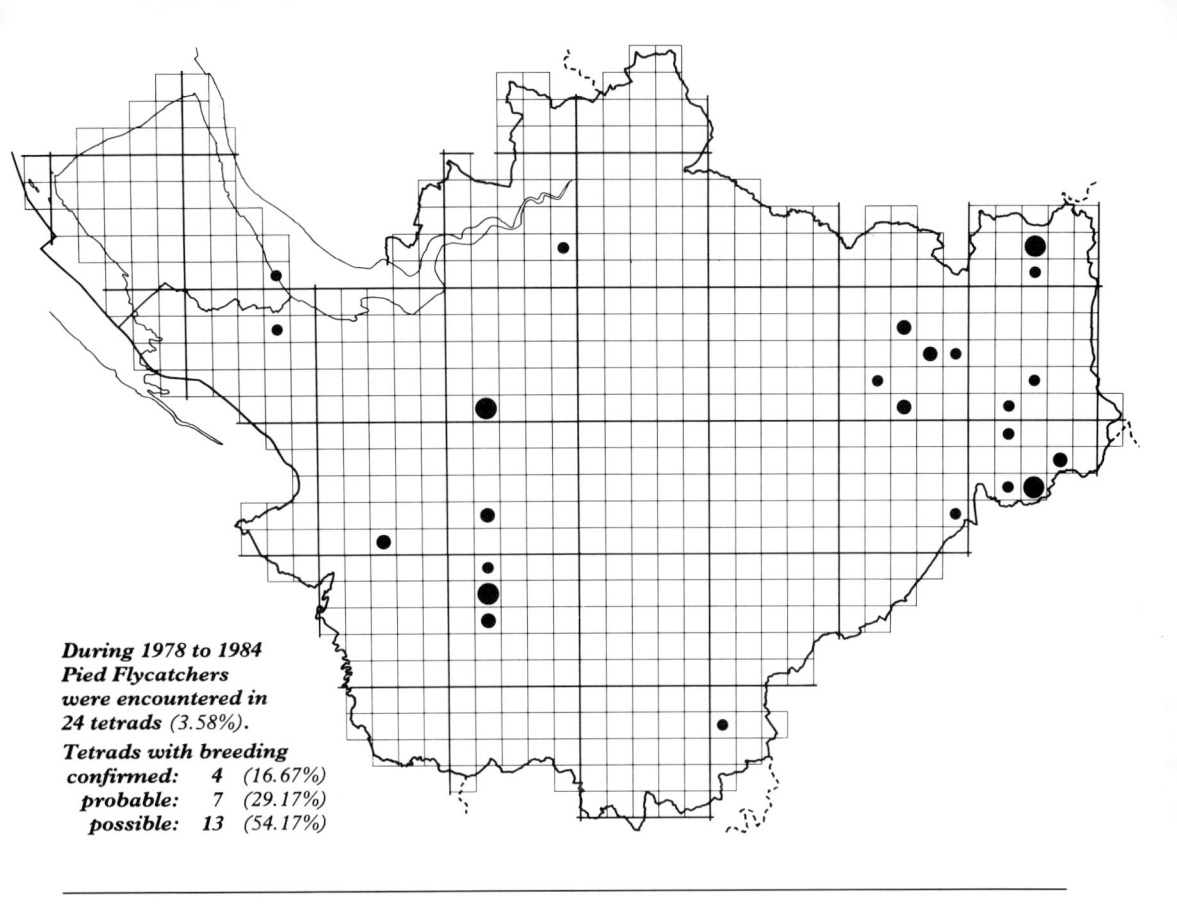

During 1978 to 1984 Pied Flycatchers were encountered in 24 tetrads (3.58%).

Tetrads with breeding
confirmed:	4	(16.67%)
probable:	7	(29.17%)
possible:	13	(54.17%)

Long-tailed Tit
Aegithalos caudatus

UNLIKE the true tits, *Parus*, the Long-tailed Tit does not nest in holes, and so it is not necessarily restricted to woodland in the breeding season although woodland does play a considerable part in the species' distribution. At Prestbury sewage farm, for example, with plenty of scrub but few natural cavities, it is the only tit species which breeds regularly. Nests are typically built in dense prickly bushes, with gorse, wild rose, bramble, holly and young conifers being preferred. A completely different type of nest-site is not uncommonly used, high up in the fork of a tree.

The domed nest, which gave rise to the old dialect names of "churn", "bottle-tit" and "two-fingered tit" (two fingers being all that could be poked through the opening by young egg-collectors), is typically built of moss and spiders' webs. Lichens are listed in many text books as nest material, but until recently the air in Cheshire was too polluted to allow the appropriate leafy and shrubby lichens to grow, although nests containing *Hypogymnia physodes* have been found during the present survey in the eastern hills. As air pollution at ground level has subsided with the installation of taller chimneys and industrial decline, various lichen species (including *Evernia prunastri*, much used for nest building in other parts of the country) are starting to reappear in sheltered sites, and nests in such areas should be watched in the future for signs that the Long-tailed Tit is again using this lichen. Boyd[5] describes a pair visiting a nest of grey lichens in a blackthorn bush in May 1936. However, if this was in Cheshire, it seems improbable given the pollution levels at this time unless the site was in a sheltered valley in the eastern hills. It is uncertain just how beneficial the use of lichens is in nest building, although probably they give some rigidity to what may otherwise be a rather flimsy structure.

The scarcity of Long-tailed Tits north-eastward from Ellesmere Port, Stanlow and Runcorn/Widnes may further be correlated with air quality (cf. Chaffinch): it may be difficult for the birds to find the appropriate moss species for binding the nest together, these also being sensitive to high levels of pollution. Boyd in his country diary of 2nd August 1943 detailed the 1269 feathers lining one particular nest: 117 Woodpigeon's, 15 Starling's, 1136 poultry and a tail feather probably from a Robin, plus rabbit fur, grass and moss – the nearest farm was over 300 yards away.

Nest building begins early, sometimes in late February, although such early nests may not be completed until a month or so later, especially if there are inclement weather conditions, and may be left until April before the lining of feathers is added prior to egg-laying. A nest in a gorse bush near Sandbach in 1975 was started by 26th February and still being lined on 24th March (JPG). Nests built in March and April are particularly exposed because of the lack of leaf cover at this time. Many nests are destroyed by predators (for example all of 13 at Marbury Country Park in 1980), so nests left intact during the period prior to lining may be more likely to survive until the young are ready to fly. When a nest is robbed of its eggs or young, the bereaved birds do not necessarily start a replacement clutch, but may join a neighbouring pair. In all cases studied, birds helped out at a brother's nest, and an average of 70% more young fledged from broods with helpers (Glen & Perrins 1988). Sometimes Long-tailed Tits may even help feed at a nest of another species[39]. On some occasions extra birds may be recorded at a nest right from the beginning of the breeding season. At Clive Green, Winsford, in 1981 at least three adults were watched carrying food to one nest (JPG). Similarly, four adults attended a nest in Delamere in 1983 (DN). Newly fledged young will often huddle together along a twig awaiting parental attention, a habit shared by roosting birds in winter. The earliest fledged broods may be seen away from the nest in the first few days of May, with others appearing into July.

Adults and young will remain together as a family party often throughout autumn and winter. They are normally sedentary, roaming around the flock's territory of 25 hectares or so[39]. Surprisingly however they may occasionally move considerable distances, apparently often as a flock, for example five birds ringed together at Meols, Wirral, on 7th October 1975 were caught together again 26 days later in Nottinghamshire, some 135 kilometres to the east, and three juveniles ringed in Warwickshire on 21st June 1981 were retrapped together on 14th February 1982 at Caldy, Wirral (142 km NW). Although often found in mixed tit flocks Long-tailed Tits have only short stubby bills unsuited to dealing with large seeds such as beechmast or peanuts. Consequently they suffer greatly during severe winters when insect food is scarce. *L&CFC Reports* record scarcities following the severe winters of 1916/17, the early 1940s and 1961/62/63. When Abbott saw half a dozen between Mouldsworth and Hatchmere in November 1920 he commented that they were "the first I have seen in Cheshire for quite three years".

As the map shows it is easy to confirm breeding for this species. 'Probable' breeding records include those of displaying birds: on 8th March 1982, three birds were seen to indulge in repeated short display flights from the top of an elder bush, rising some six feet in the air on each occasion before plummeting back into the bush (JGP).

Where suitable scrub is present several pairs may nest within a few hundred square yards. However, many occupied tetrads, particularly in farming areas with neatly trimmed hedgerows, held only one or two pairs. This scarcity in open farmland areas reflects the species' predominantly woodland distribution and Long-tailed Tits are also absent from built-up areas. An estimate of the county population is 1000 to 1500 pairs.

REFERENCE

Glen, N.W. & Perrins, C.M. (1988): Co-operative breeding by Long-tailed Tits. *British Birds* 81: 630.

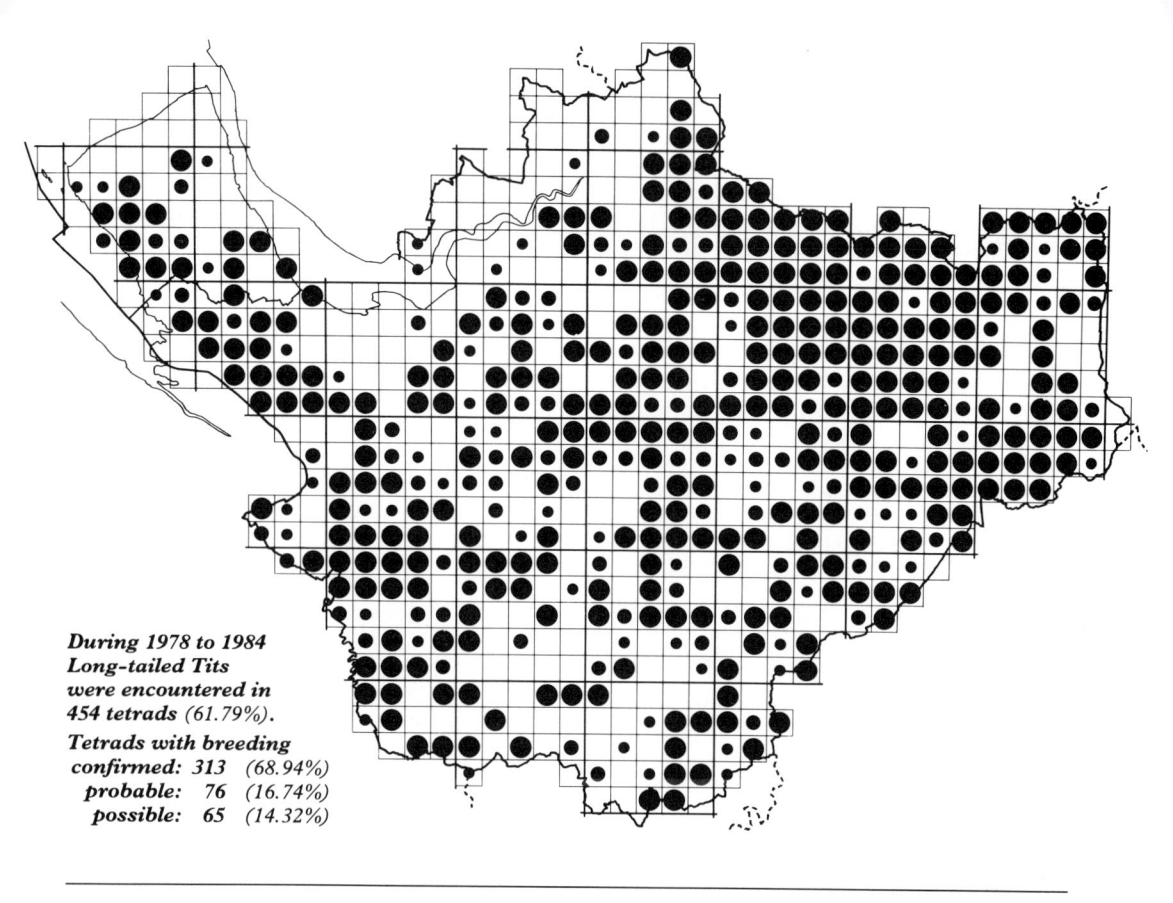

During 1978 to 1984 Long-tailed Tits were encountered in 454 tetrads (61.79%).

Tetrads with breeding
confirmed: 313 (68.94%)
probable: 76 (16.74%)
possible: 65 (14.32%)

Marsh Tit
Parus palustris

UNTIL the Willow Tit was recognised as a separate British species in 1897 all black-capped tits were referred to as "Marsh Titmice". What we now know as the Marsh Tit is very much a species of mature broad-leaved woodlands and less typical of wet woods which are preferred by the Willow Tit. The Marsh Tit's distribution in eastern Cheshire has been correlated with river valleys (JPG/ [21]), and the map shows this to be largely true also along the Weaver valley further west. Many of these river valleys and their tributary cloughs contain some of the oldest woods in the county. Wandering birds are occasionally met with outside the breeding season following the courses of rivers and streams away from known nesting woods, and have been seen as far east as the River Goyt at Disley (SJ98S).

Other mature woods are also occupied, for example derelict coppices, fox- and game-coverts and the planted woods of large estates. As such plantations have matured this century, it seems likely that Marsh Tits will have increased, although past confusion with Willow Tits renders historical information confusing. More efficient drainage has lowered the water-table in the Cheshire mosslands, and many former birchwoods are developing into oakwoods more suitable for this species (for example Withington Mosses, SJ87G). Conifer plantations are avoided. It is suggested (Snow 1954) that Marsh Tits are to some extent prevented from spreading out of broad-leaved woodland by competition with Willow Tits. Of the Cheshire tetrads with confirmed or 'probable' breeding by Marsh Tits, half are shared with Willow Tits, and half are occupied by Marsh Tits alone.

The food and feeding habits of the tits have been intensively studied[36]. Perhaps surprisingly, those of the Marsh Tit resemble those of the Great Tit more closely than any of the other *Parus* species: it takes much larger insect prey than the Coal or Blue Tit, and is more likely than any other tit to be found feeding on the ground. Its beak is stouter than the Willow Tit's, and it can open beechmast and other strong nuts and seeds in winter.

Separation of Marsh from Willow Tits in the field has never been easy, and a small proportion of records submitted to this atlas may be erroneous. The abrupt "pitchou" note of the former is diagnostic and is often the first indication that the species is present. The song "pee-chew-pee-chew-pee-chew" and variants are equally diagnostic once learnt. Song is most frequent in winter and early spring with records spanning the period 31st October to 28th April (JPG). Whilst the Marsh Tit does not actually excavate its own nest-hole (unlike the Willow Tit) it will enlarge the entrance of existing holes. It prefers a low nest-site, and will occasionally use nest-boxes. Proof of breeding in most cases came from sightings of family parties or adults carrying food for their young.

Griffiths & Wilson[13] had seldom seen this species in Wirral, where most sightings were during winter. One at Bidston on 15th April 1938 was considered worthy of mention. Boyd[6] found this the least common of the tits around Great Budworth, although one or two were occasionally seen in the woods at Marbury, Arley and Tabley, and Whalley[19] knew of only a single record around the Sandbach Flashes. These observations apply equally well today, although there have been a few winter records in the Sandbach area. The 1966 *CBR* described this species as common in south-west Cheshire, especially in the Eaton Park area, but the map shows its distribution to be unexceptional here. This is the most sedentary of all the tits[39] and many of the 'possible' and 'probable' breeding records will relate to actual breeding pairs. Their sedentary nature is born out by the lack of movements of Cheshire-ringed birds, Marsh Tits being retrapped only in the localities where they were ringed.

Although not obvious from the map this is another species which is only scarcely found north of the Mersey. The South-West Lancashire RG, which covers Liverpool and south-west Lancashire, has ringed only two birds in 21 years (1964-84) whereas the MRG, which covers Wirral, north & mid-Cheshire and north-east Clwyd has ringed 116 in the same period. There are only two confirmed, two 'probable' and three 'possible' breeding records in the area covered by the Greater Manchester Atlas[46]. Marsh Tits are only rarely found on the Wirral peninsula.

Four broods of young were seen on three visits to the Weaver valley between Dutton and Kingsley in 1981 (SJ57) and it was thought that the species was more numerous there than anywhere further east (JPG). Marsh Tits are thinly scattered in Cheshire and possibly no more than 20 pairs are present in any 10-km square – far short of the national estimate of 50-100 pairs per occupied 10-km square[35]. Where the habitat is suitable, this species probably breeds at a somewhat higher density than the Willow Tit. The county population probably lies somewhere between 150 and 200 pairs.

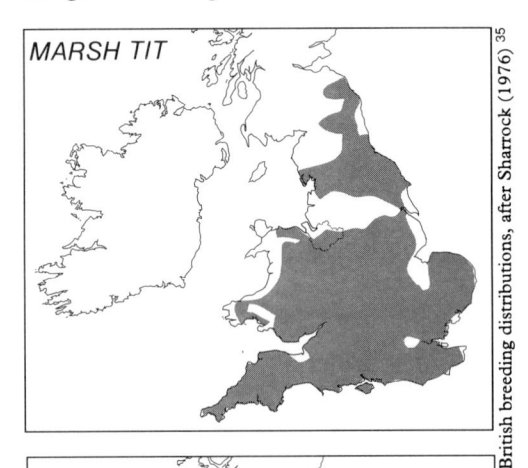

REFERENCE

Snow, D. W. (1954): The Habitats of Eurasian Tits (*Parus* spp.) *Ibis* **96**: 565-85.

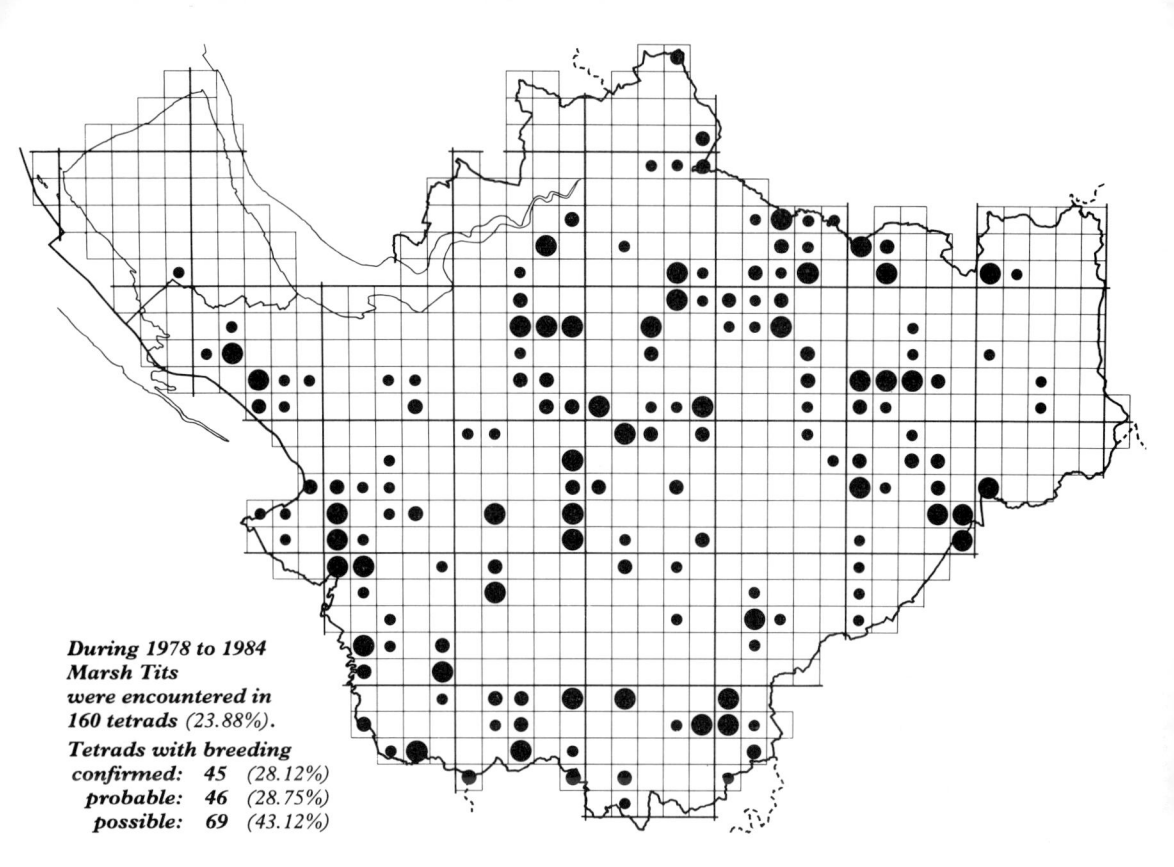

*During 1978 to 1984
Marsh Tits
were encountered in
160 tetrads (23.88%).*

Tetrads with breeding
confirmed: 45   *(28.12%)*
probable: 46   *(28.75%)*
possible: 69   *(43.12%)*

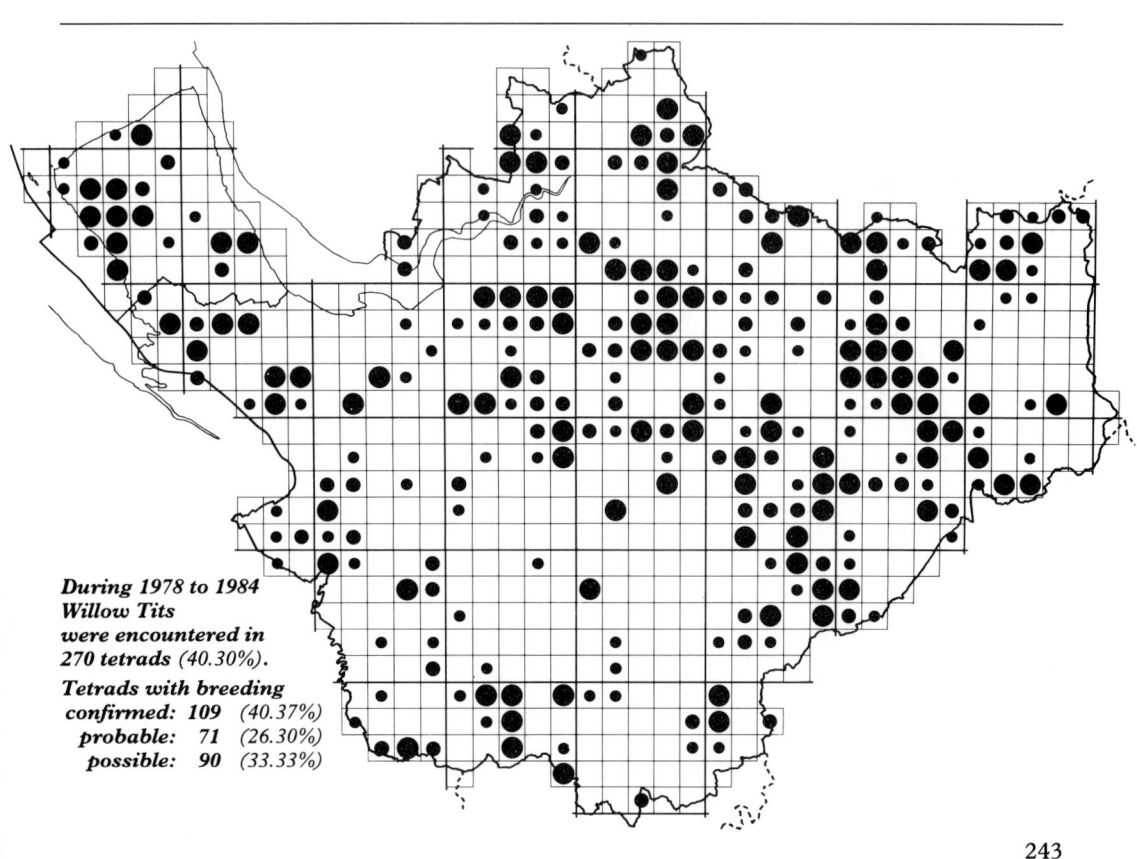

*During 1978 to 1984
Willow Tits
were encountered in
270 tetrads (40.30%).*

Tetrads with breeding
confirmed: 109   *(40.37%)*
probable: 71   *(26.30%)*
possible: 90   *(33.33%)*

Willow Tit
Parus montanus

THE Willow Tit was not separated as a species from the Marsh Tit until the end of the nineteenth century. In the first edition of *British Birds* magazine (June 1907) a plea was made to discover more of the species' status in Britain. Information was not long in coming (see Alexander 1974) and on 17th April 1912 Coward identified the first Cheshire bird at Rostherne. In the following spring he found a pair breeding at the same site and since then it has been found nesting in most parts of the county including Wirral, being more widespread than the Marsh Tit at the time of this survey except in the area south of Chester and around Peckforton.

Willow Tits, which excavate their own nest-holes, frequent scrubby woodland, particularly where soft, rotten wood is plentiful. Thus, river- or stream-side alder or willow carrs, mossland birchwoods and elder thickets are all popular. Both Willow and Marsh Tits occur together in a number of localities where mature woodland merges into wet carr, as at Dog Wood, Tatton Park. Willow Tits will breed in conifer plantations, although the lack of rotten wood for nesting may restrict the number of breeding pairs. In 1981 a nest was found in a horizontal dead branch of an oak tree at the edge of a homogeneous stand of young pines at Whatcroft (SJ67), and in Delamere pairs nest in the stumps of alders fringing the boggy hollows. Stands of pine between 15 feet and 25 feet in height appear most favoured for feeding. Birds will sometimes use nest-boxes packed with a material (such as polystyrene) for them to excavate. The placing of a rotten stump in a hedgerow has been successful in providing a nest-site. Less typically a pair bred in a stone wall at Holmes Chapel in 1979. Unlike the Marsh Tit this species is not uncommon north of the Mersey where birch and alder woods are widespread.

The normal song of the Willow Tit, heard at almost any time of the year, is a repeated descending "seer..seer..seer..", vaguely recalling a Chiffchaff to some ears though lacking the staccato quality or variation in pitch. JPG has records of such song from late December to mid-May, with a slight resurgence in late August and September. A peculiar warbling song is also heard at times, perhaps most frequently in the spring. Most records of confirmed breeding for this survey were sightings of adults carrying food for their young, or from observers finding nests, although these can be difficult to locate owing to the boggy nature of many territories. Family parties appear from the first week of June until mid-August. This species is less sedentary than the Marsh Tit and may wander some distance after breeding, for example a bird ringed at Ashton-in-Makerfield (Greater Manchester) on 27th June 1982 was caught at Rostherne thirteen days later, some 24 kilometres south-east.

In 1972 and 1978, there were six territories in the Risley Moss CBC, increasing to eleven in 1979 and nine in 1980. In 1981, five territories were found in 25 hectares of birch woodland there with eight pairs in 1983. Densities may be higher in wet carr, with two to three pairs annually in six hectares at Foden's Flash, Sandbach. At Rostherne, one pair is present in most years at Shaw Green Willows (4 ha.), and up to four pairs in Mere Covert (10.5 ha. of mixed woodland fringed by willows and reed-bed).

In the east of the county Willow Tits occur sparingly in wooded areas near Macclesfield, but there has clearly been something of a decline in this area since the mid-1970s and recent scarcity has been noted at Prestbury, Adlington, Dean Row, Alderley Edge and Styal.

The map (previous page) may contain a few errors because of possible confusion between this species and the Marsh Tit. At the time of this survey there were probably some 350-400 pairs in the county.

REFERENCE

Alexander, H. G. (1974): *Seventy Years of Birdwatching*, T. & A. D. Poyser, Berkhamsted.

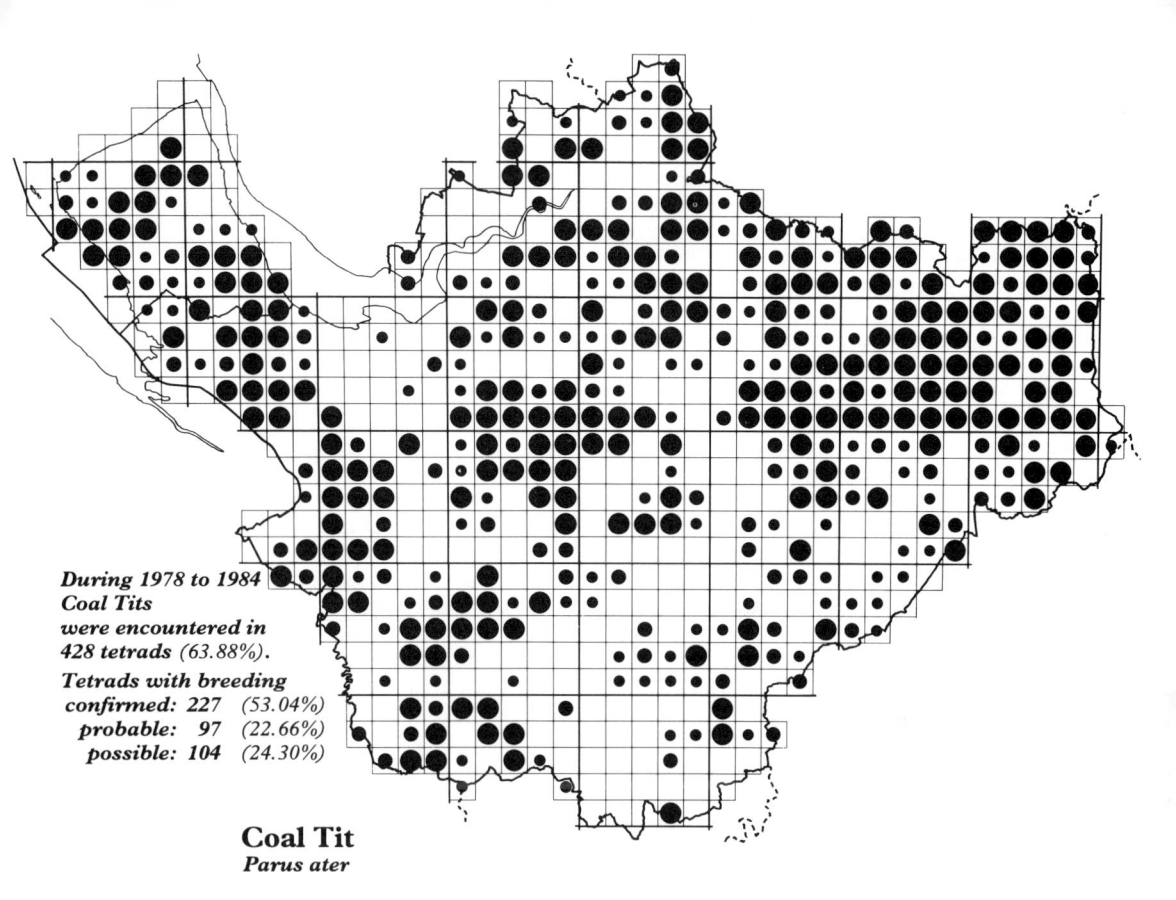

During 1978 to 1984 Coal Tits were encountered in 428 tetrads (63.88%).

Tetrads with breeding
confirmed: 227 (53.04%)
probable: 97 (22.66%)
possible: 104 (24.30%)

Coal Tit
Parus ater

THE Coal Tit is distinctive in having a white stripe down the back of its black-capped head and, as its name suggests, has a rather dusky appearance. It is the smallest of the British tits and appears to be more or less exclusively associated with conifers, whether in pure commercial stands, in mixed woodlands, or in mature gardens with a few ornamental trees. It is particularly common in the mixed Scots pine and oak woodland at Alderley Edge, and in an oakwood near Tarporley a pair was found nesting beneath the only pine in the wood.

Afforestation with conifers at Delamere and Langley (Macclesfield) has undoubtedly been of benefit to this species, although Coal Tits are not usually present until stands of trees reach 15-20 feet in height. In mature stands it may be the most numerous species, vying with the Goldcrest for this status. Certainly this is the most numerous tit species in tall plantations, its finer bill making it better adapted than the others to life in conifers[39]. It hovers a great deal more than other tit species when feeding, this being particularly advantageous for feeding along twigs bristling with conifer needles, the undersides of which may be accessible only with difficulty to other tits. This ability to hover also enables Coal Tits to take beechmast before it falls.

Shortage of nest-sites is undoubtedly a limiting factor for several bird species in conifer plantations because the trees are felled before suitable cavities develop. Coal Tits overcome this problem by nesting in mouse-holes or other cavities beneath the tree-roots, and at Alderley Edge have been found nesting in a hole in a sandstone face. Standard nest-boxes are used only infrequently, this species preferring boxes with very small holes or narrow entrance slits. A bird was caught at Newchurch Common on one occasion during June as it flew to holes at a Sand Martin colony, this being an indication of their inquisitiveness for any type of hole. In this case there was no evidence that the bird was breeding.

Before mass planting of conifers became widespread the Coal Tit was much scarcer. Coward & Oldham[2] declared it less plentiful than the Great and Blue Tit everywhere, and around Chester it was far less numerous than the "Marsh" Tit (in those days "Marsh" Tit would have included the Willow Tit also). Griffiths & Wilson[13] recorded it as resident and common in Wirral in suitable localities, but seen more often in winter than summer. Boyd[6] had no evidence of its nesting around Great Budworth, there being no conifer woods of any size in the area at that time. Harrison & Rogers[24] thought that few stayed to nest at Rostherne because of the lack of conifers and although song has been heard in spring the first confirmed breeding record here was in 1974 when a pair bred in the garden of the warden's house. In 1978 breeding was confirmed for the first time on the reserve.

Territorial birds are easily located during the spring by their song, a loud piping "seetoo seetoo...". Breeding was confirmed in many cases by watching the parents carrying caterpillars to the nest. Breeding is timed to coincide with the main feeding period of the few caterpillar species which feed in conifers, and most broods fledge synchronously around the last week in June. There is a resurgence of song in September and October, this being more pronounced in warm autumns.

The only figures relevant to population density include one to six territories in Mere Covert, Rostherne (10.5 ha. of mixed woodland with scattered pines) between 1976 and 1983, between seven and nine territories in the Alderley Park CBC (13 ha.) and only one or two territories in the Risley Moss CBC (25 ha.) both between 1978 and 1980. Many tetrads with confirmed breeding will hold ten or more pairs, with perhaps only one or two pairs in other occupied tetrads. This would give a county total of around 1600 pairs.

The map shows clearly the concentration of Coal Tits in those areas with conifers. The plantations in the eastern hills and the many parkland woods of the north-east, the Delamere plantations, the Eaton estate and the wooded heaths and parks of Wirral form major strongholds. Conifers are relatively scarce in the south-east of the county and on the low-lying land north of the Mersey. In this latter area chronic air-pollution early this century killed whole stands of trees, especially pines (e.g. Coward 1903).

REFERENCE
Coward, T. A. (1903): *Picturesque Cheshire*. Sherratt & Hughes, London and Manchester.

Blue Tit
Parus caeruleus

THE Blue Tit is so generally distributed and numerous throughout Cheshire and Wirral that little has been written of it. Whilst ancestrally a woodland species, the Blue Tit is perhaps most familiar as a garden bird, adopting nest-boxes readily, sometimes within days of their being erected. It is also widespread on farmland where hedgerow trees or small copses provide nest-sites, and on four such CBC plots in south-eastern Cheshire it was the third most numerous species. In conifer plantations the relative shortage of insect food and the difficulty of extracting it discourages Blue Tits, which are not so well adapted to such habitat as are Coal Tits. The absence of natural cavities for nesting is a further drawback. Birch woodland too may have a limited supply of nesting holes, and at Risley Moss most of the 27 pairs present in 1983 were using boxes.

Song is often resumed in November and becomes more frequent in mid- to late December. By early February the gliding display flight can be seen. Nest-boxes are often visited from January onwards, although whether these are birds prospecting for future nest-sites, merely feeding or looking for roost holes is not clear. Boyd[6] found a bird on

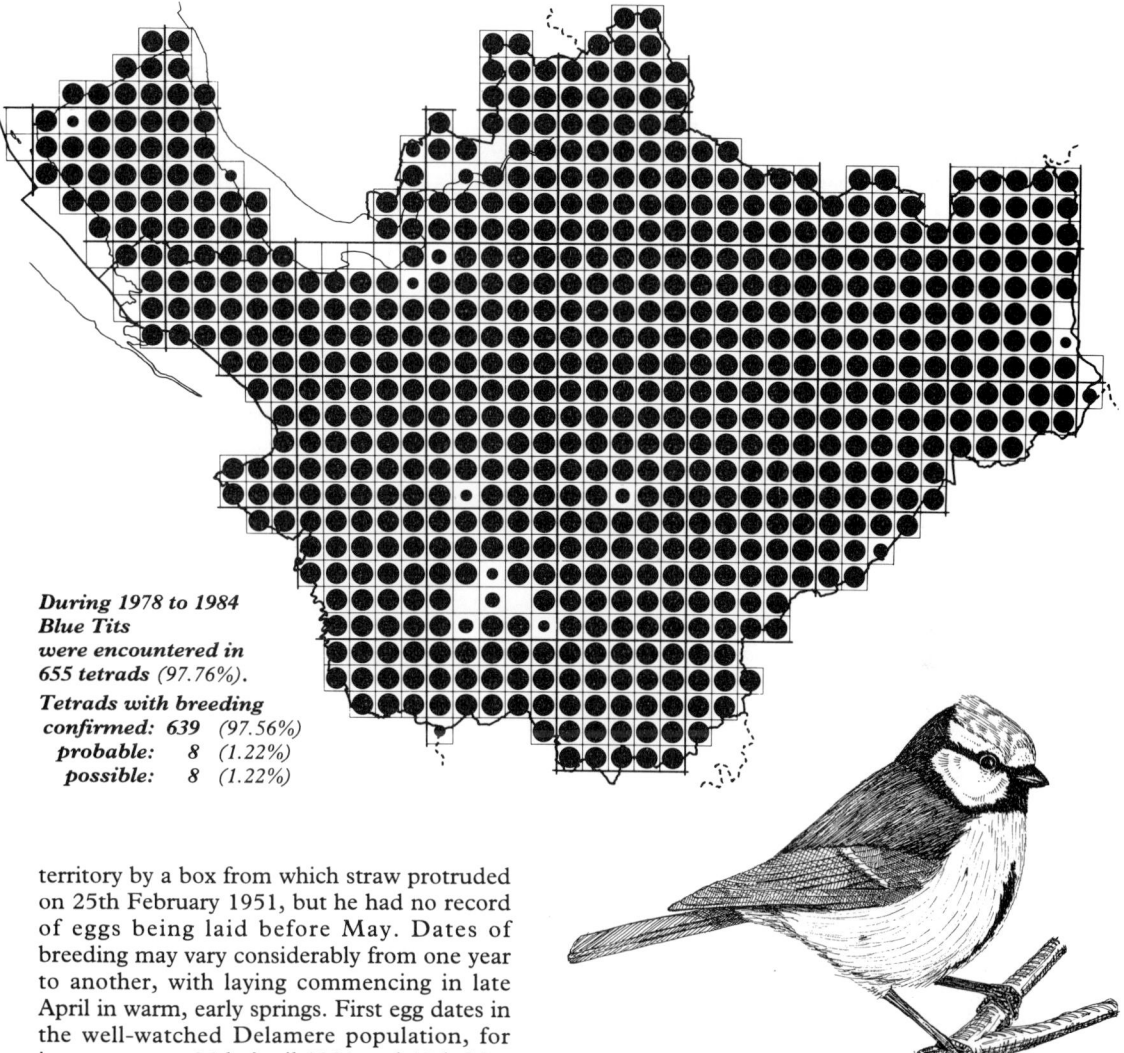

*During 1978 to 1984
Blue Tits
were encountered in
655 tetrads (97.76%).*

Tetrads with breeding
confirmed: 639 *(97.56%)*
probable: 8 *(1.22%)*
possible: 8 *(1.22%)*

territory by a box from which straw protruded on 25th February 1951, but he had no record of eggs being laid before May. Dates of breeding may vary considerably from one year to another, with laying commencing in late April in warm, early springs. First egg dates in the well-watched Delamere population, for instance, were 26th April 1982 and 10th May 1983.

Once the female is incubating, the cock bird may perch for long periods near the nest-hole, uttering a three- or four-note, shortened version of the song with monotonous regularity whilst swivelling from side to side in display. The sitting bird when threatened will hiss suddenly at intruders. After the eggs hatch, both parents bring their young copious quantities of caterpillars, and their breeding success depends critically on the coincident timing of broods with the major flush of caterpillars, the winter moth being especially important in oakwoods. Adults with large broods (which may contain up to 14 or more chicks) may bring food, on average, as frequently as once a minute from dawn till dusk during each of the 19 days of the typical fledging period. This prodigious work-rate indicates how readily caterpillars can be found. The young leave the nest from early June onwards. Fledging dates appear up to a fortnight later at Higher Disley (230m above sea level) than at Wilmslow (70m above sea level). Such differences may not always be altitudinal, however, since the birds in nest-boxes at Eastham and Royden Park, Wirral, appear up to a week later than those at Delamere, and birds breeding in gardens are usually earlier than those in woodland[39]. The source of a record of a family of fledged young at Wilmslow on 30th April 1975[8] has not been traced and must be regarded as highly dubious.

The map shows the Blue Tit to be one of Cheshire's most widespread breeding species. It is absent from some tetrads in the highest part of the eastern hills, the Dee saltmarshes and Hilbre islands, and was not proved to breed on the treeless Frodsham Marsh. The only other squares without confirmation of breeding were most likely the result of under-recording. This applies to some rural tetrads, particularly in SJ55 (although woodland is scarce in parts of this area), and also to a handful of urban tetrads (Meols, Port Sunlight and the Widnes area). Even the most built-up parts will hold one or two pairs of Blue Tits, but it may be difficult to find a nest, possibly in a garden nest-box or other artificial site such as a lamp post, a bus-stop sign or a hole in a wall.

18 territories were located in 19.3 hectares of mixed, mainly broad-leaved woodland at Rostherne in 1976 but numbers varied between fourteen and eight during the period 1978-1983 possibly as a result of rhododendron and sycamore clearance during the same period. In the Alderley Park CBC up to 15 territories were recorded between 1978 and 1980 and at Risley Moss (25 ha.) 14 were recorded in 1978 although only eight were recorded in 1979 and 1980. Additionally there were from 56 to 66 territories on four farmland CBC plots (totalling 393 hectares) between 1979 and 1981. A number of farmland CBC plots in adjacent counties give an average density of 60 pairs per tetrad[42], and national woodland CBC figures for 1980 ranged from 90 to 300 pairs per tetrad[40]. The best broad-leaved woodland may hold up to 1000 pairs of Blue Tits per tetrad[36]. However, some urban tetrads will have held few birds. Assuming the density lies in the range 50-100 pairs per tetrad across the county, the total population would lie between 32,000 and 65,000 pairs.

Great Tit
Parus major

FROM early winter the bell-like double note of the Great Tit's song acts as a reminder that spring will eventually follow. At closer quarters a more metallic, slightly rasping quality is audible, giving rise to the old dialect names of Sawfinch and Saw-filer in the days when saw blades were sharpened rather than thrown away and replaced. Most males have a repertoire of variations on the usual song, all sounding basically disyllabic. Witherby et al.[30] gave January to early June as the main song period, less frequently to early July, with some resurgence, after the moult, from late August to early October. There are a few records of Cheshire birds singing in late October and November. A few birds start to sing annually around Christmas, at about the same date as many Blue Tits.

The distribution map is very similar to that of the Blue Tit. Great Tits avoid the barren areas of the eastern hills, the Dee saltmarshes and Frodsham Marsh. The same comments as for the Blue Tit apply to the agricultural southern and urban northern parts of the county, although the Great Tit is much more of a woodland species than the Blue Tit, and Great Tits are found at considerably lower densities in farmland and surburbia: this is reflected in the somewhat lower proportion of confirmed breeding records. In suburban and even urban areas birds are found nesting in boxes and other artificial cavities. Whilst holes in trees or walls form typical nest-sites, Great Tits also take readily to cavities in metalwork. For example a pair reared several young in a disused downspout at Wilmslow in 1970, and in 1983 five nests were found in metal gate posts in a small area around Appleton Reservoir. Broken lamp posts are a favourite site on an industrial estate near Runcorn. They have also been found down a pipe some three feet out of the ground at Lower Peover and in a broken drainpipe in Holmes Chapel. Great Tits tend to nest lower down than Blue Tits,

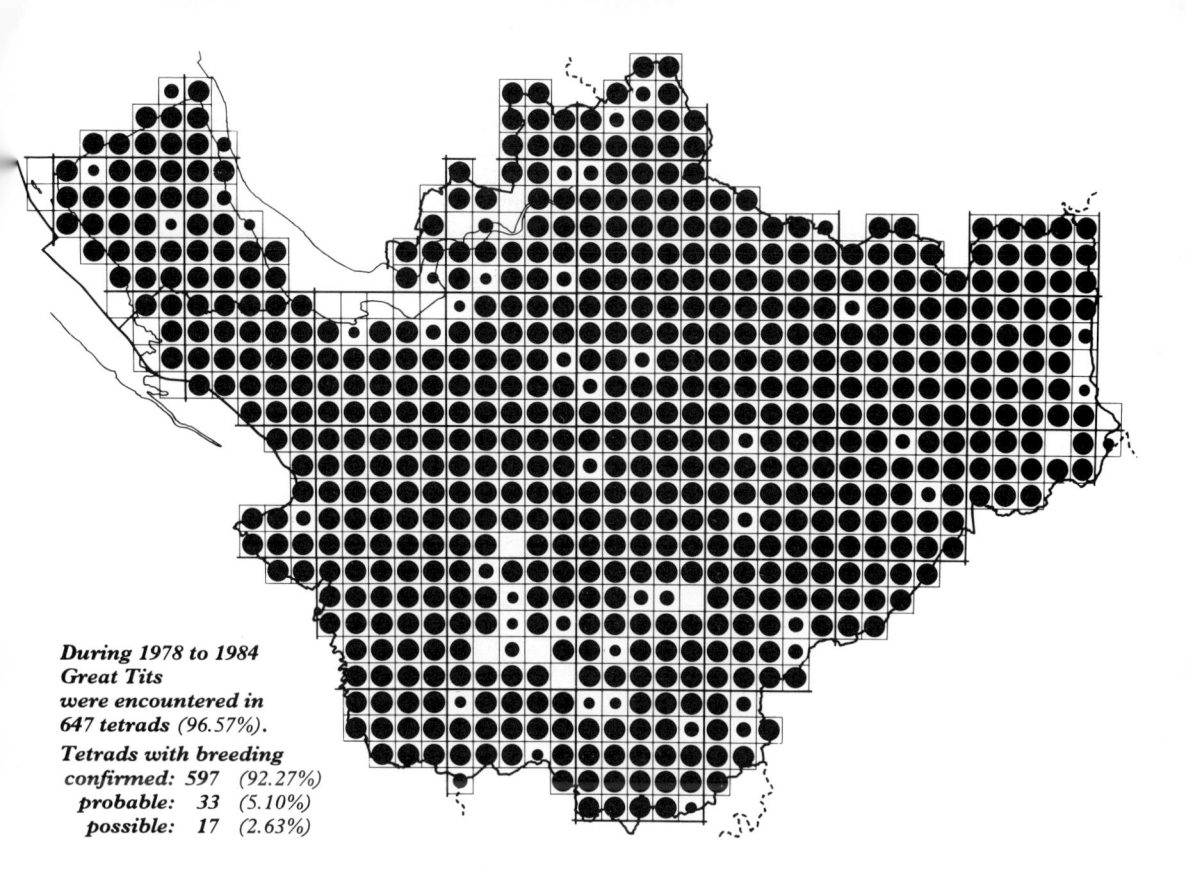

During 1978 to 1984 Great Tits were encountered in 647 tetrads (96.57%).

Tetrads with breeding
confirmed: 597 (92.27%)
probable: 33 (5.10%)
possible: 17 (2.63%)

and a nest-box sited within easy reach of the ground is much more likely to be occupied by the larger species.

Nest-building is at its peak in late April and early May. Clutches of as many as ten or twelve eggs are not unusual. Coward & Oldham[2] mentioned a nest with 15 eggs at High Legh, where they suspected two females had laid in the one nest; and in 1983 a female in Delamere sat on ten Blue Tit's eggs and eight of her own before eventually deserting. The sitting bird can be very reluctant to leave her eggs, hissing angrily at any intruder. Broods of noisy young leave the nest from early June. Parties of yellowish juveniles, lacking the white cheeks and glossy caps of their parents, can then be seen begging for food. These family parties often mingle with parties of other tits and warblers, especially as the young birds mature. A family of young Great Tits being fed out of the nest at Alderley Edge on 17th May 1969 was considered to be exceptionally early, especially in view of the late spring that year[21].

Nationally, this species is said to have become "rather less common" since about 1966[39]: this might be attributed to a change in hedgerow structure, with the removal of some trees and mechanised flailing of hedges[42]. Information suggesting such a change in Cheshire is not available, but cannot be discounted. However, published information shows no major difference between the present and 1910 when Coward & Oldham considered it the most abundant tit in some of the parks and woodlands, though in most districts it was outnumbered by the Blue Tit, and in the hills east of Macclesfield it was less common than the Coal Tit also. It is doubtful whether the Great Tit is now commoner than the Blue anywhere in the county. Annual ringing totals of the MRG for birds caught in a wide variety of habitats indicate that Great Tits are just under half as common as Blue Tits throughout the year. Numbers of Great Tits in the eastern hills have apparently increased, but they may still be locally outnumbered by the Coal Tit in conifer plantations, as at Delamere, Macclesfield Forest and the Dane Valley near Wincle, and possibly also in mixed woodland with plentiful pine and larch, as at Alderley Woods.

Quantitative information is limited, as with so many of the smaller birds in Cheshire. However, some 23 hectares of mixed woodland at Rostherne Mere NNR held between nine and fourteen Great Tit territories from 1977 to 1983; five or six pairs in 25 hectares of birch woodland is typical at Risley Moss; the Tatton CBC plot held 14 territories in 1982 and twelve in 1983. Three farmland plots in south-east Cheshire held from 20 to 22 territories between 1980 and 1983; and some 40 hectares of farmland at Rostherne held three to six pairs between 1977 and 1983. Between 15 and 20 pairs per tetrad were found around Aldford[28]. The suitability of farmland varies according to the supply of old hedgerow trees for nesting. Tetrads in the more intensively farmed areas can be expected to hold far fewer pairs. CBC figures for several adjacent counties give a mean farmland density of 30 pairs per tetrad[42] and, nationally, woodland CBC figures range from fewer than 100 to more than 300 pairs per tetrad[40]. With 25-50 pairs per tetrad the county may hold between 15,000 and 30,000 pairs of Great Tits.

Nuthatch
Sitta europaea

THE Nuthatch is generally a very sedentary bird and, as its winter diet consists largely of such tree-fruits as acorns, beechmast and sweet chestnuts, its distribution follows closely the availability of mature broad-leaved woods, parkland often being ideal. Conifer plantations are felled before suitable nesting cavities can develop, and Nuthatches are absent from the Macclesfield Forest. In Delamere, however, they breed in the lines of broad-leaved trees at the edges of plantations, foraging for food amongst the conifers also.

The ability of this species to climb in all directions, up and down trees, is unique in this country and its distinctive colouring of chestnut underparts and flanks, blue-grey crown and upperparts, white cheeks and throat and bold black streak through the eye make it one of the most attractive members of our avifauna.

The territorial calls of the male may be heard from mid-November onwards through the winter, although more frequently from early March until May – either a drawn-out high-pitched trill or a fluty, almost thrush-like, song of "tleeoo" and similar notes delivered with the bill pointed skywards. Birds become quieter during incubation and brooding. The nest may be in a cavity or hole in a tree or in a nest-box. Whatever nest-site is chosen the entrance hole is plastered up with mud to fit the bird exactly and deny access to larger predators or rivals. The nest itself is quite distinctive, but the birds can be very secretive until the young hatch, when the adults can be watched carrying food back to the nest. DN has a record of fledged young on 1st June in 1982, and family parties can still be seen into early September.

At the turn of the century, Coward & Oldham[1,2] described the Nuthatch as "a very local resident... practically confined to the south-west of the county". It bred regularly around Chester, was abundant at Edge Hall near Malpas, and several pairs bred annually in Oulton Park. It was absent from Wirral[2] and the eastern hills, and very scarce on the Cheshire plain, although odd pairs may have bred in Tatton Park. It had "disappeared ... from many places in England".

During the 1930s and 1940s Nuthatches began to spread back through the county, aided in part by the maturation of nineteenth-century hardwood plantations. In Wirral, Hardy[4] noted it was a regular breeding bird at Dibbinsdale since 1931. Elsewhere in Wirral early breeding records came from Burton Wood (1931) and Thornton Hough (1934). Griffiths & Wilson[13] noted breeding at Arrowe

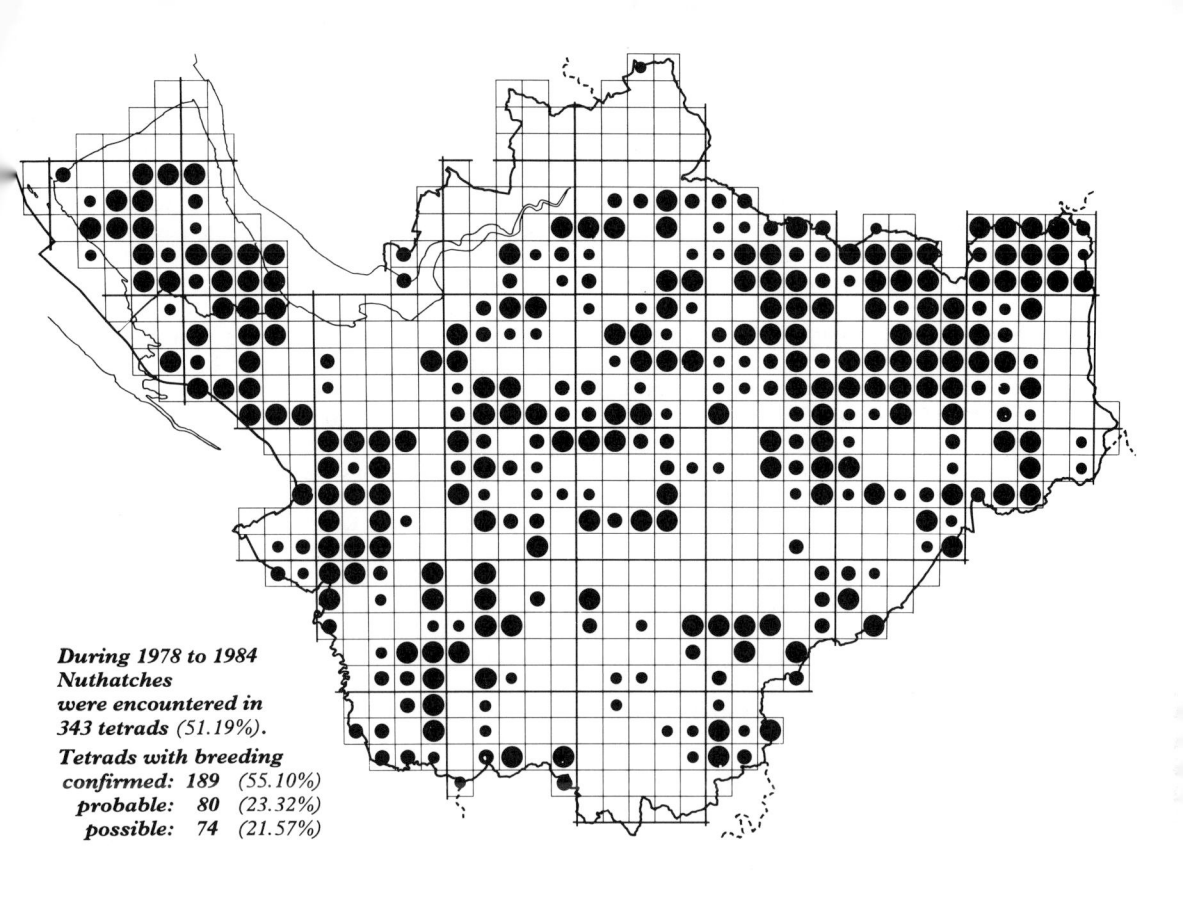

During 1978 to 1984 Nuthatches were encountered in 343 tetrads (51.19%).

Tetrads with breeding
confirmed: 189 *(55.10%)*
probable: 80 *(23.32%)*
possible: 74 *(21.57%)*

Park in 1932 and at Bromborough. There were further records from Arrowe Park, Upton, Burton and Mollington between 1944 and 1949[3].

Elsewhere in Cheshire, Boyd[5,6] watched a pair nesting at Marbury Big Wood, Great Budworth in 1938, and considered that a bird at Frodsham in March 1945 was "further evidence of this bird's remarkable extension of range...in the last few years". *L&CFC Reports* referred to records from Pettypool and Hartford, Doddington and Cholmondeley since 1926, a return of birds to Tatton Park in 1933-34, and increases at Crewe and Nantwich in 1945-46. The 1949 report also listed records from fourteen sites in north and east Cheshire with one from the hills, at Goyt's Bridge (now Derbyshire), in 1949.

Bell[7,9] catalogues further spread into eastern Cheshire during the 1940s and 1950s including the first records from Lyme Park and Prestbury in 1964. By 1966, it was found to be widespread throughout the county with five or more pairs in Tatton Park, 20 or more in Eaton Park (Chester), and at least twelve pairs breeding in Wirral. At Lymm, birds had started nesting in 1957, but Nuthatches were still scarce in the north of the county[8]. In Alderley Woods, Nuthatches seemed to increase about 1968 (A. Booth pers. comm.). At Rostherne, Harrison & Rogers[24] reported that it had been present since 1946 and was well established although numbers had been much reduced for a few years by the severe winter of 1962-63.

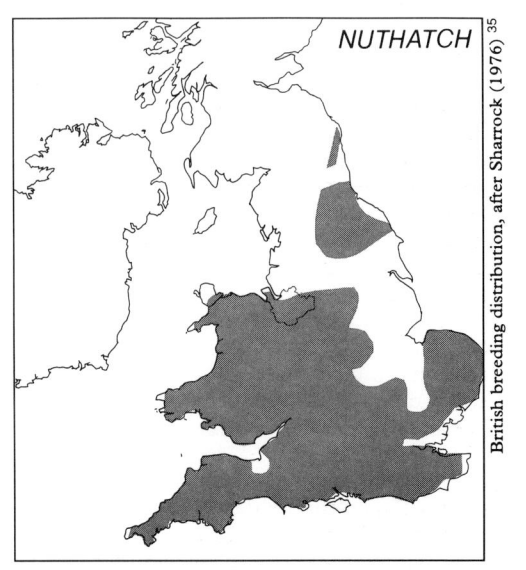

Although adults may be very sedentary, juveniles undoubtedly move out of natal areas and this must have been so for the species' distribution to have spread so much this century. Records from the MRG show that a ringed nestling from Eastham Woods bred the following year in Brotherton Park (2 km) and another Eastham chick was taken by a cat in a Moreton garden (14 km). Some adults do move from time to time, as evidenced by an adult male, ringed at Eastham on 15th June 1982 and found dead at Bebington (7 km NW) on 26th January 1983.

The map shows continued expansion northwards by the time of this survey, but the belt of birchwoods along the Mersey valley mosslands provides poor winter feeding and has proved a barrier to further spread, as was clearly shown in the BTO Atlas[35]. The records of the South-West Lancashire RG, which cover the areas of Liverpool, south-west Lancashire and north of the Mersey, show that only four birds have been caught in the years 1964-84, while the MRG, working south of the Mersey and in Wirral, have ringed almost 400 birds in the same period.

Elsewhere in the county, the distribution reflects that of suitable woodland, although odd pairs may have been missed in the private fox-coverts of south Cheshire. Because of the largely sedentary nature of the Nuthatch, most of the 'possible' and 'probable' breeding records are likely to refer to breeding birds. However, bearing in mind it is not a difficult species to prove breeding and its continuing spread, some of these records may reflect the presence of wandering birds in suitable areas but not yet paired or breeding.

Eight males were singing around Tatton Mere in March 1983 and there may now be more than 20 pairs in the park. Seventeen were counted in Alderley Woods during three hours on 1st March 1982. Most occupied tetrads hold fewer than ten pairs, and, taking an average of three, the county probably holds between 700 and 850 pairs.

Treecreeper
Certhia familiaris

COWARD & Oldham[2] described the Treecreeper as "a not uncommon resident" and "generally distributed in the lowlands". It also occurred in the wooded valleys of the upper Goyt and Dane. Griffiths & Wilson[13] stated that it nested in small numbers in most of the wooded parts of Wirral although it was less frequently found in the coastal districts. Boyd[6] described it as "not really common" in the Great Budworth area but "often... seen in the Marbury, Arley and Tabley woods and, less frequently, in hedgerow trees and small coverts at Frandley and Antrobus".

Today the Treecreeper is a familiar woodland bird throughout the county, nesting even in small isolated stands of trees and hunting the trunks of nearby hedgerow trees. Most types of woodland are inhabited, the chief criteria evidently being the roughness of the bark and the availability of cavities for nesting. Thus willow carr is ideal, the rough mossy bark of crack willows harbouring many of the invertebrates which the Treecreeper seeks using its tweezer-like bill, and nests may be placed between loose bark and the wood beneath, or even in a crack in a split trunk.

Oak woodland and stands of mature pines in Delamere are equally favoured, but young trees of whatever species tend to have smooth bark as do mature beech and many sycamores which consequently provide a poorer supply of food. Young plantations of mixed broad-leaved species at Prestbury sewage farm containing little wood with a girth of more than two to three inches were seldom visited by Treecreepers, and briefly at that – the frequency of visits was expected to increase as the trees grow.

Unlike the Nuthatch which climbs up and down trees, the Treecreeper always climbs from the bottom upwards usually spiralling around the trunk and flying to the bottom of the next tree after exploring the first. Exceptionally birds will feed on the ground at the base of trees. Other suitable substrates may be searched for insects: it is not unusual to see birds examining fence posts, particularly of split oak with the bark intact, and in the eastern hills drystone walls provide

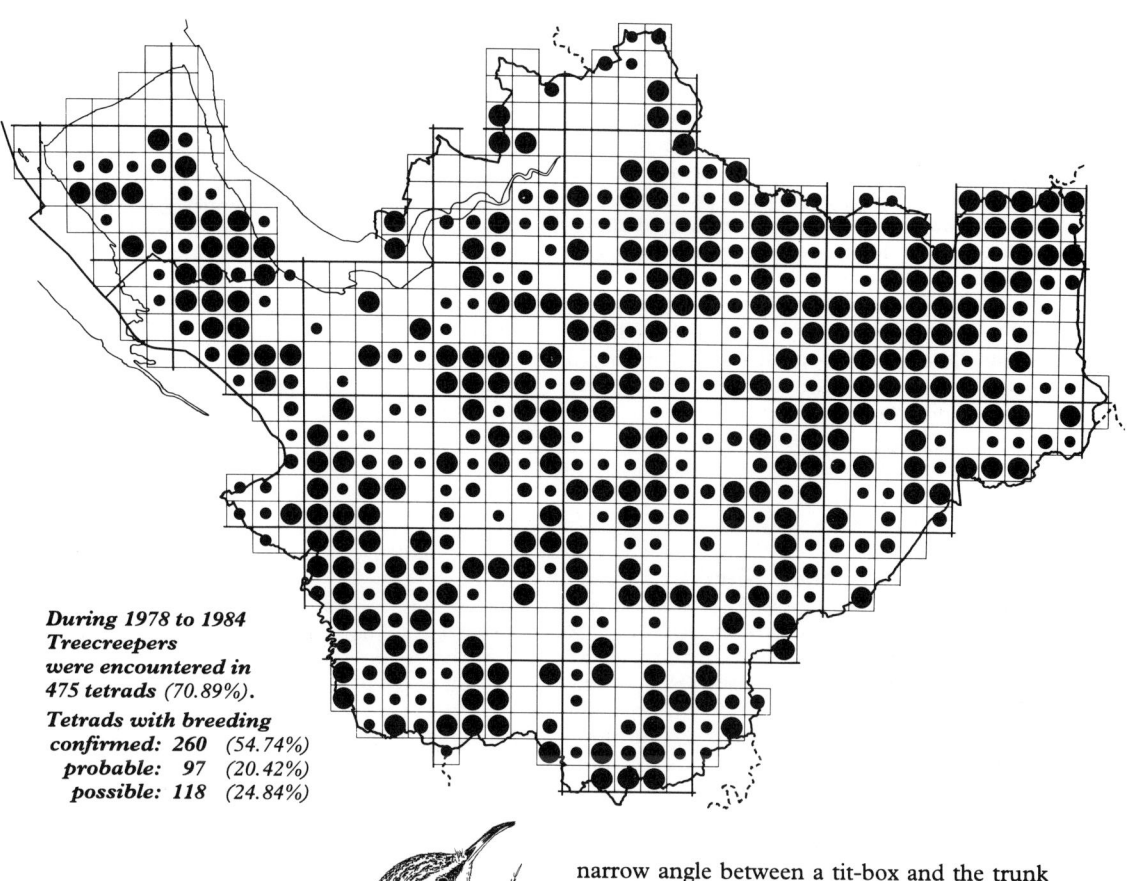

During 1978 to 1984 Treecreepers were encountered in 475 tetrads (70.89%).

Tetrads with breeding
confirmed: 260 *(54.74%)*
probable: 97 *(20.42%)*
possible: 118 *(24.84%)*

a rich hunting ground and perhaps not infrequently nest-sites. On 7th June 1983, a pair was feeding young in a nest in such a wall in Lyme Park[8] and there have been several further reports of birds examining cavities in spring. In 1948, a pair raised a brood in a nest in a hole in a plank at the top of a wooden shed at Comberbach[6]. In 1982, a pair nested in a conventional tit-box at Delamere (DN), though elsewhere pairs have built in the narrow angle between a tit-box and the trunk of a tree.

Both the descending song with its final flourish and the rapid courtship chases – the birds flying up and down around the trunk of a tree within inches of the bark – provide early evidence of breeding intent. JPG has noted song infrequently in September to November, more regularly from early December through to February, depending on the mildness of the weather, then frequently into early summer. Nest-building generally starts in the last few days of March, continuing for late nests until the end of May. By the beginning of May in some years the parents can be watched carrying food to the young, often unexpectedly sidling out of sight into some cavity where the nest is concealed. The male will pause to sing even during such operations as this, his voice not being audibly affected despite his bill being full of wool during nest-building, or, later, of insects. Young Treecreepers recently out of the nest will huddle together against the trunk (as with the Long-tailed Tit, this is a habit shared by roosting adults) and when disturbed during their early life they will often freeze close to the bark relying on their superb camouflage for protection. Second broods may not leave the nest until the second week in July.

As with other insectivorous species the Treecreeper suffers during severe winters, particularly if snow becomes frozen to the bark of trees. Thus three CBC plots in south-eastern Cheshire which between them held six territories in 1978, held only one in 1979, building up again to three in 1980.

In 1965, three pairs were present in 20 acres of mixed woodland on Bidston Hill and in the spring of 1983 eight males were heard singing along two kilometres of the Dane valley near Wincle. In Mere Covert (10.5 ha.) at Rostherne between three and six pairs were located annually between 1976 and 1979 although there were only one or two pairs over the next four years following clearance of rhododendron and sycamore. It is not unusual to see ten or so birds during an afternoon's watching in many wooded areas, and if an average of five pairs is taken for each tetrad in which breeding was confirmed and one pair in all other occupied squares, the total population may be around 1500 pairs.

Red-backed Shrike
Lanius collurio

DESCRIBED as a rare and irregular summer visitor in the early years of this century, there are some nine or ten breeding attempts on record. Coward & Oldham[2] listed: Claughton (1863); Alderley (about 1869); Bidston (young reared prior to 1874); Ince (1886) and Leasowe (1892). Bell[7] adds records from Appleton (1901 & 1902); Arrowe (1913); Wilmslow (1926) and Wildboarclough, where one was thought to be feeding young on 7th June 1936. The Wilmslow record referred to a male seen by E. Cohen at Lindow Common on 8th August, subsequently confirmed by Abbott who saw a pair with two young on 15th and 19th August. Abbott also watched a male at Hatchmere on 5th June 1920.

Since 1936 there have been no recorded nesting attempts nor any suspicion of breeding. Early in this survey, a female was at Red Rocks on 19th and 20th May 1979, typical dates for an overshooting spring migrant.

The breeding range of Red-backed Shrikes has been contracting to south-east England, where they are now on the verge of extinction, although in recent years very small numbers of birds from Scandinavian stock have nested in Scotland. During the years of this survey the numbers of British counties holding breeding pairs declined from eleven to four, and the total number of pairs from 37 to ten[43].

Jay
Garrulus glandarius

THOUGH severely persecuted during the nineteenth century, the Jay was able to survive by means of its secretive behaviour and retiring habits during the breeding season. Byerley, writing in 1854, described it as formerly common in Wirral but scarce by that date, and Brockholes[2] referred to it as "a much persecuted resident". Nevertheless, the Jay persisted in the wooded parts of Wirral and Coward & Oldham[2] regarded it as common in most parts of Cheshire. It was indeed "an abundant resident in the parks and coverts" in many areas, and "nowhere...so plentiful as in Delamere Forest". Jays were absent from the treeless hills of the east and, although quite numerous in the upper Dane valley, they were scarce even in the well-wooded areas of Middlewood and Disley. Despite persecution from gamekeepers, they owed their existence largely to the privacy of game-coverts.

During the thirty years up to 1951, Jays

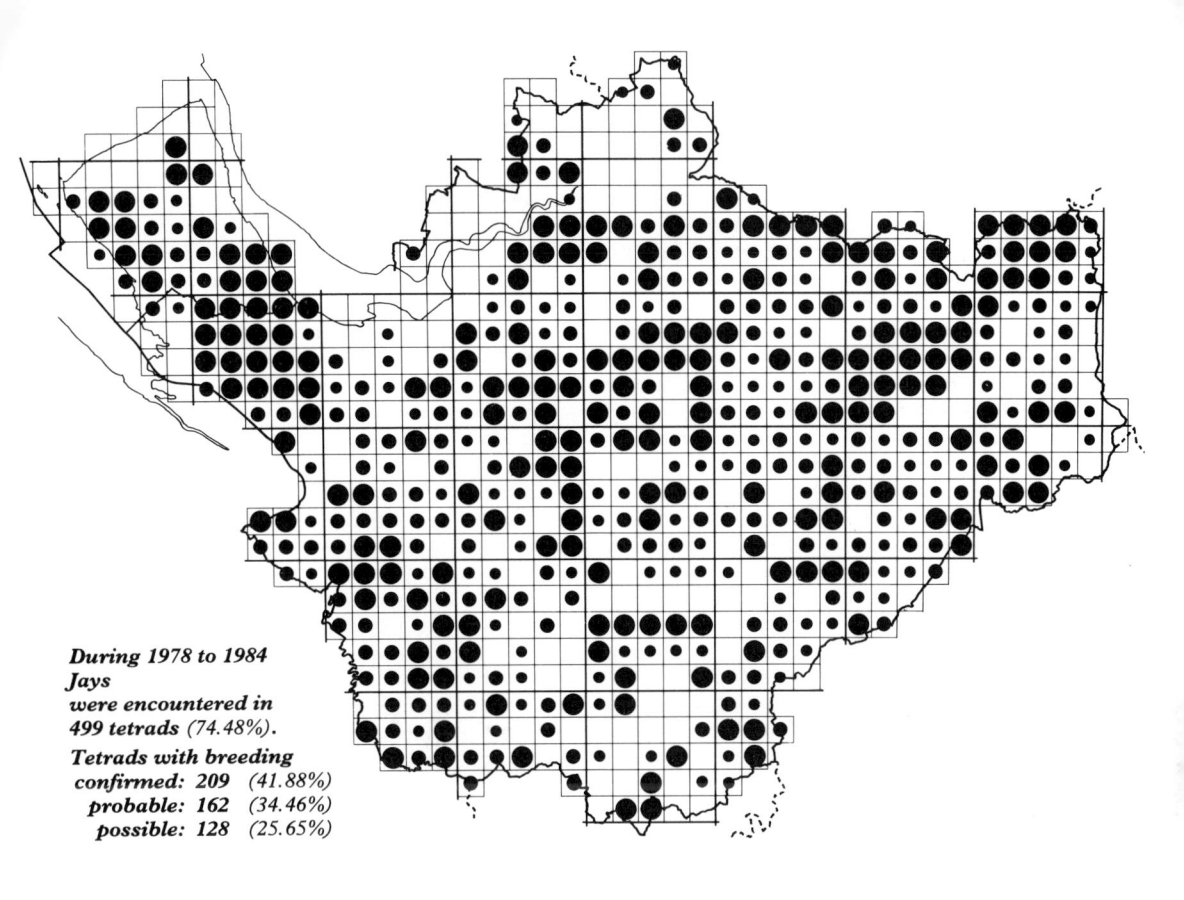

During 1978 to 1984 Jays were encountered in 499 tetrads (74.48%).

Tetrads with breeding
- confirmed: 209 (41.88%)
- probable: 162 (34.46%)
- possible: 128 (25.65%)

"increased greatly" around Great Budworth and were by then present in all woods and coverts[6]. The increase had been especially marked since 1940 when the scarcity of cartridges in wartime had led to reduced persecution. Similarly, in north Wirral, Griffiths & Wilson[13] noted an increase since 1938. Boyd[16] included the Jay among breeding species at Rostherne whereas Coward[11] had not mentioned it, and this "may reflect a relaxation of pressure by game-keepers"[24]. By 1962, Bell could describe it as common and widespread throughout the county, including the wooded valleys of the hills, and particularly in such areas as Delamere and Styal Woods.

Subsequently there is almost no mention in *CBRs* until the late 1970s, but records since then point to a continuing increase and spread. Birds were seen regularly at Woolston in 1984, where previously they had been scarce, and a party of twelve at Rostherne in April of that year prompted the comment that they appeared to have increased there since ten years earlier when two or three pairs at the most had been considered to breed[24]. The species is now common in the area of Lyme Park.

In March and April noisy displaying parties of up to half-a-dozen birds may be seen and at this time a variety of peculiar calls is heard, notably a soft "tic-tiddle-wurrh" noted by Boyd from a displaying bird at Marbury in April 1928, and more recently by JPG in March 1983 in the eastern hills. At Rostherne in 1982 a bird gave a strange "kewick" call like a subdued Tawny Owl. At Withington in March 1982 a bird gave a remarkable rendering of the call of a displaying Kestrel and in March 1983 two birds in flight at Dean Row, Wilmslow uttered a screeching "shree-kaaa" (JPG). Some Jays are good mimics, especially of raucous species such as the Magpie, Crow and Heron.

During the actual nesting period Jays can be very secretive, and – particularly as the species is sedentary and seldom leaves its favoured woodland – many 'possible' and 'probable' breeding records on the map will represent pairs which successfully evaded observers. Often the only sign of a Jay's presence came when an observer entered a wood, to be greeted by a strident "squaark", followed by views of a white rump as the bird flapped away. Nests can be difficult to find, and the female sits tight during incubation, which starts with the first egg, as with all corvids and almost all predatory birds. Jays feed their young on invertebrates and eggs and chicks plundered from the nests of songbirds, often in the first hour of daylight. JPG's earliest note of birds feeding young in the nest was on 4th June 1974, and of fledged young on 22nd June 1973. It is not unusual to see Cheshire birds still feeding young in the nest in mid-July, and family parties often pillage fruit and vegetable gardens in late July and August.

The map shows Jays to be absent from those parts of the county where oak woodland is sparse – the eastern hills, the southern plain, the north Wirral coast, the Gowy/Frodsham marshes, and the Mersey valley. The availability of acorns, a favourite winter food, may largely determine the breeding distribution of this sedentary species. Jays will breed in some built-up areas, in parks or large gardens, as at Birkenhead. This spread into towns is a fairly recent event, probably within the last 20 years, although it is not documented for Cheshire.

Jays can breed at quite high densities. Prince[29] found up to four pairs in the 18.2-hectare Bidston Hill CBC, and national CBC figures average around five pairs per square kilometre of woodland. Some tetrads in SJ37 were considered to hold in excess of six pairs, but many with confirmed breeding held only two or three. Allowing three pairs per tetrad with 'probable' or confirmed breeding and one pair per tetrad for those with 'possible' breeding would give an estimate of some 1200 pairs.

Magpie
Pica pica

IN the days when every small farm kept a few poultry, Magpies were customarily shot by farmers as potential egg thieves, and also by keepers protecting their stock of gamebirds. With the advent of larger, more specialised farms in recent years this persecution has largely ceased, and while Magpies still find their way on to the few remaining gibbets, and nests are destroyed in certain areas, their numbers have increased steadily. Numbers shot are insignificant compared to the total population: a winter roost near Sandbach held 470 birds in February 1981 and several other such roosts regularly hold 100 or more birds.

Coward & Oldham[2] noted the Mersey valley above Warrington and the hills around Rainow as areas where the bird abounded. It was also common around Audlem and Doddington and nowhere more plentiful than in the Delamere Forest. Elsewhere scattered pairs existed throughout the county. Brockholes[2] had found it fairly abundant in Wirral. Abbott noted birds on only seven dates between 1913 and 1930, notably "a number on the tops" around Bakestonedale and Billinge Hill on 27th June 1920, and many throughout Delamere Forest from Mouldsworth to Hatchmere on 19th November 1920.

Hardy[4] described the Magpie as a very common nester in suburban Liverpool parks and occurring wherever it was not shot by gamekeepers. He too mentioned its prevalence in Delamere where the keeper at Oakmere had fifty corpses on a gibbet. In Wirral it was common south of a line from Burton to Hooton but less so further north. Griffiths & Wilson[13] noted an increase along the north coast of the peninsula since 1938.

Boyd[5] makes repeated reference to increasing numbers from 1937 onwards, and

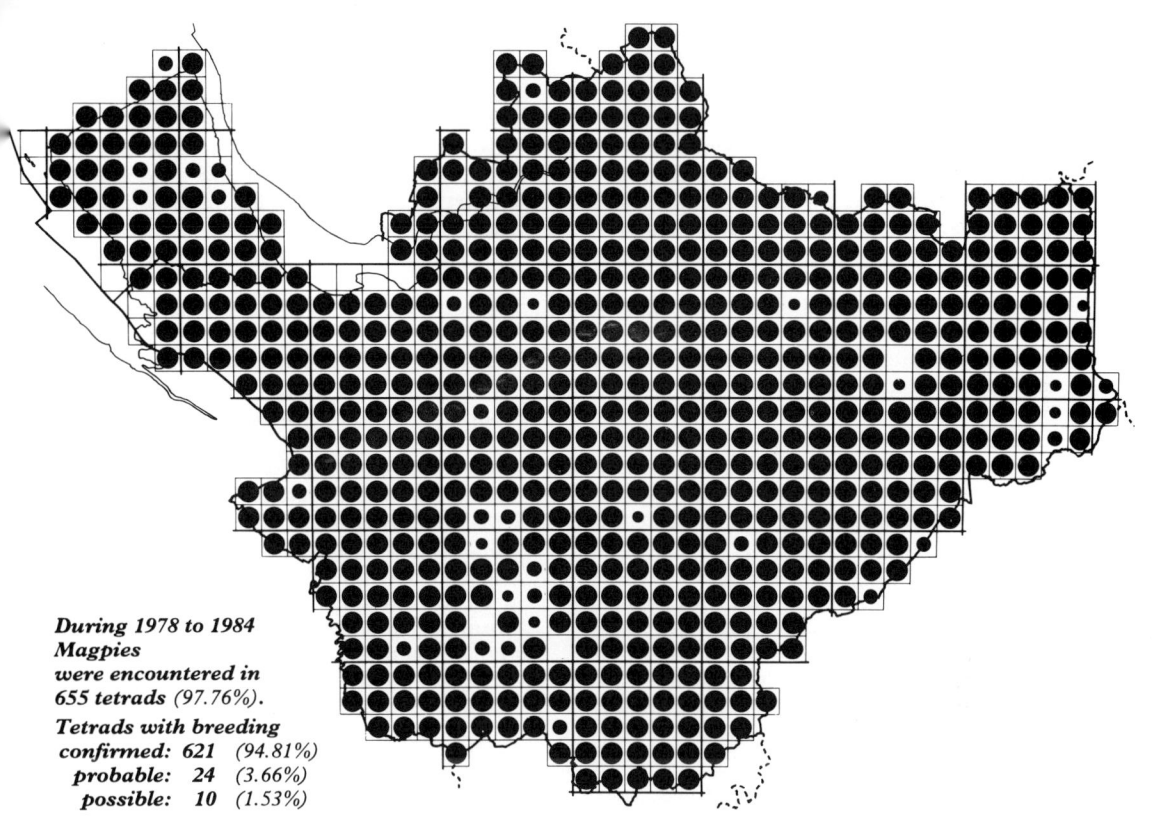

During 1978 to 1984 Magpies were encountered in 655 tetrads (97.76%).

Tetrads with breeding
confirmed: 621 *(94.81%)*
probable: 24 *(3.66%)*
possible: 10 *(1.53%)*

particularly during the wartime years when cartridges were scarce and the birds bred unmolested. This increase has continued ever since, accompanied by a spread into suburban and urban areas. Magpies systematically comb the hedgerows in spring preying on eggs and nestlings, and are a frequent cause of concern to garden bird watchers, nests of small birds in neatly clipped hedges being particularly prone to predation. However this effect may be somewhat overstated since Tatner (1983) found only 5% of Magpie nestlings in his Manchester study to have eggs or chicks in their gizzards: the main food was invertebrates, predominantly leatherjackets. DN's experience at Sutton Weaver is that small birds' nesting activities were badly hit in the first year or two when Magpies summered in a new area, but, after that, they managed to co-exist and breed successfully.

One of a pair nesting by Elton Hall Flash in 1974 was seen to swoop at and kill a fully-grown Starling. On other occasions it was seen to pursue House Sparrows. A breeding pair at Wilmslow killed a grey squirrel.

Pairs may start to renovate old nests in December or January, but more usually from February onwards when breeding pairs will leave the winter flocks to roost by their future home. The bulky domed nest is both distinctive and, in spring and early summer, conspicuous, and those few tetrads where birds were not recorded were probably not visited early enough in the year. Of 167 nests noted in eastern and mid-Cheshire in 1983, 69 (41%) were in hawthorn bushes, 22 (13%) in oak, and 19 (11%) in pear (JPG). A further

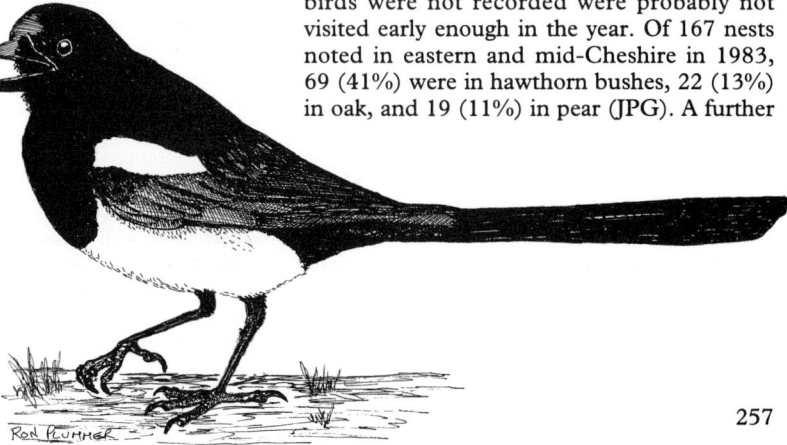

16 tree species were also used, with a preference for those with dense thicket-like branches: Magpie eggs are taken by Crows, and the Magpie's long, rudder-like tail helps it out-manoeuvre the Crow amongst denser branching. One exceptional nest in a larch lacked a dome but held a single young bird on 2nd June. Thorn bushes are particularly favoured in the eastern hill areas where taller trees are largely unsuitable and scarce. The curious preference for old, tall pear trees, often within feet of farm buildings, was noted by Boyd[6]. In 1971 a pair built a dome over a disused Crow nest at Dean Row. The majority of the mapped records refer to nests, although many observers were also able to find parties of noisy, short-tailed 'FL' young birds.

The extent of the population explosion is shown by national CBC results, with breeding Magpies about twice as numerous in 1984 as in 1966. This species also has a large non-breeding population, possibly up to half of the spring total (Holyoak 1974). Most British Magpies are very sedentary, but some of the non-breeding birds may wander. A chick from Bidston, ringed on 5th June 1980, became the national record holder by moving 50 kilometres east to be found in Sale, Greater Manchester, on 11th March 1982.

The map shows almost complete occupation of the county. The apparent scarcity around Cholmondeley and Peckforton (SJ55) perhaps reflects the activities of gamekeepers, although coverage in that area was patchy, and shooting no doubt explains the absence of breeding birds in a few other tetrads. Three farmland CBC plots in south-east Cheshire totalling 167 hectares held on average 14 pairs between 1978 and 1980. The Bidston Hill CBC plot (18.2 ha.) held three or four pairs every year[29]. Densities derived from national CBC figures in 1980[40] averaged twelve pairs per square kilometre in woodland and five pairs per square kilometre on farmland. Taking an average of 25 pairs per tetrad the county total is around 16,000 breeding pairs.

REFERENCES
Holyoak, D. (1974): Territorial and Feeding Behaviour of the Magpie. *Bird Study* **21**: 117-128.
Tatner, P. (1983): The diet of urban Magpies. *Ibis* **125**: 90-107.

Jackdaw
Corvus monedula

THE Jackdaw is so familiar as a bird of open grassland, daintily picking insect prey from the surface, that it may come as a surprise to find birds in spring rather clumsily searching the woodland canopy for caterpillars. However, this is evidence of the adaptability which has helped the species to spread and increase in numbers during this century[34]. Jackdaws are colonial nesters, regularly using the same sites from one year to the next, and may be found throughout the county albeit locally in many parts. Their distribution is probably restricted only by the availability of nest-sites and suitable feeding areas.

In parkland, where mature trees and grassland for feeding occur side by side, pairs nest in cavities in the trees, old beeches being particularly favoured. Occasional pairs may nest in woodland. In the eastern hills and in Wirral large numbers of birds nest in chimney-pots and in spring they may be seen lining the roof-tops in Rainow, Kettleshulme and elsewhere. Even in lowland tetrads where the species is relatively scarce an examination of large, old houses with plentiful chimney-pots will often result in the location of a pair or two. Similarly, quarry-faces and church towers, as at Oughtrington, may support colonies. Coward & Oldham[2] had records of eggs laid in old Magpie nests and reported that open nests were built in many places on the broad, flat branches of Scots pines. There is no record of either of these types of nest in recent years.

Where Jackdaws are scarce they are often first detected by their harsh calls, and birds with swollen crops can be watched flying back to the nest, although, as with Rooks, Jackdaws will feed well away from the nest and quite often in different tetrads from the actual nest-site. Nevertheless, by following flight lines, the nesting area can often be pin-pointed. Many of the mapped 'possible' breeding records

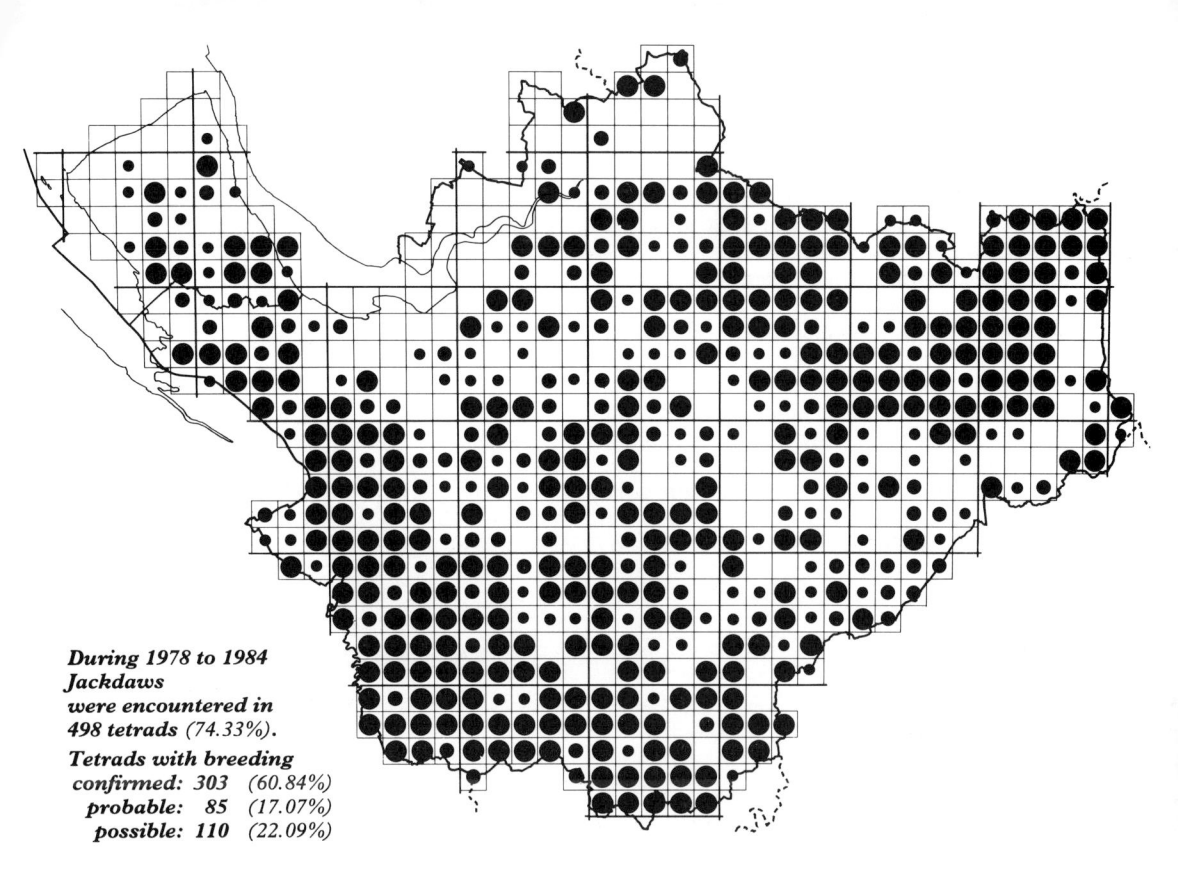

During 1978 to 1984 Jackdaws were encountered in 498 tetrads (74.33%).

Tetrads with breeding
confirmed: 303 *(60.84%)*
probable: 85 *(17.07%)*
possible: 110 *(22.09%)*

may refer to off-nest feeding birds from adjacent tetrads or to early wandering juveniles.

Jackdaws' nesting starts later than Rooks'. Whereas the latter leave the winter roosts during February, Jackdaws continue to occupy them until late March or April. However, even in December observation of birds flying to roost shows most of them to be in pairs, and as early as January some of these will inspect potential nest-sites. Building may start in March – in the eastern hills birds can be seen pulling wool from sheeps' backs at this time – but eggs are not normally laid before April. Boyd[6] found one nest which "held two young Starlings and two young Jackdaws but later only one fledged Starling and a Jackdaw not so much advanced". Young birds squawk exceptionally loudly when visited by the parents – an easy way to confirm breeding. The young have usually left the nest by the end of June, but Boyd watched a parent feeding a recently fledged youngster by the Dee at Chester in early October 1939.

At Rostherne "there is frequent competition for nest-sites with kestrels, the jackdaws usually winning through force of numbers"[24]. The two species will compete especially for cavities in trees or for nest-boxes, but otherwise there is little overlap in preferred sites, and indeed their feeding habitats are quite distinct. The map shows a concentration of Jackdaws in the north-east of the county, where parkland trees and the chimneys of terraced cottages form nest-sites, and also along the south-western fringes, from the undulating land of the Ellesmere moraine up to south Wirral. The lowest-lying land of the Mersey and Gowy valleys is largely avoided, whereas the Dee valley, with plentiful old trees, is well-populated. Along the Mersey valley suitable trees are scarce and competition with other hole-nesting species, notably the Kestrel and Tawny Owl, may be a significant factor. Jackdaws are thinly spread over the intensive dairying areas of the southern plain, here nesting mostly in chimneys. The species' fortunes have been shown to correlate with the area of mown grassland in agricultural rotations[42] so this is one species towards which silage cutting may be beneficial. North Wirral has an abundance of suitable buildings which could hold nests, but here a shortage of pastures for feeding may be critical.

National CBC plots hold widely varying densities of breeding Jackdaws, with many around two to three pairs per square kilometre but others, presumably with good colonies, over ten pairs per square kilometre. The lower figure (eight pairs per tetrad) would give a Cheshire total of 2500-3000 pairs.

Rook
Corvus frugilegus

THE reoccupation of rookeries may be a welcome sign of spring nowadays but it was not always so. In 1544 a Barnton man appeared before Halton Court for allowing "crows" to nest in his woods and, between 1699 and 1703, the Great Budworth churchwardens paid out a farthing each for 4470 "crow heds" [sic], 1552 of these in 1703[5, 6]. This alone suggests a very considerable Rook population in the area. The general hostility displayed by countrymen towards Rooks led to birds concentrating to nest in secure private estates where they remained free from persecution, and parties of Rooks flying home to their colonies or roosts were nicknamed according to their landowning benefactor, for example, "Lord Egerton's Pigeons" at Tatton. Persecution is less intense and more sporadic nowadays, and many rookeries exist in copses on open farmland, although there is still a tendency for clusters of rookeries to develop in secure estates, as at Alderley Park.

Coward & Oldham[2] stated that rookeries could be found throughout Cheshire. Some of them were very large and they specifically mentioned one at Ashton Hayes near Mouldsworth. While there is still nowhere within the county which is beyond the comfortable foraging range of birds from one rookery or another, the number of rookeries has certainly declined since then, and some contraction of range has occurred as suburban development spread in Wirral (Henderson 1968) and to the south of Manchester[46].

A national census in 1944-45 revealed 439 rookeries within the present county boundary and Wirral, containing some 15,866 nests (Boyd 1954). A repeat survey in 1975 gave 346 rookeries holding 8824 nests (Ankers & Elphick 1981). The latter survey included areas to the north of the Mersey, not included in the 1944-45 census, holding eight rookeries with 298 nests. However the area between Bunbury and Crewe was not adequately covered in the earlier survey, so the results are broadly comparable and show a decline of around 45% over the intervening 30 years. A sample survey of six 10-km squares in 1980 suggested a partial recovery from 1556 nests in 1975 to 2519 nests.

A number of more local studies adds to the picture. In Wirral, the population showed only a 3% decline between 1929 and 1944, but over the next five years fell by 32% (Henderson 1946, 1953 & 1964). A comparable 36% decline occurred in a west Cheshire study area over the same five years (Henderson 1968), and a more limited survey carried out by Boyd[6] in the Northwich area showed a 33% decline between 1931 and 1945.

Between 1949 and 1952 there was something of a resurgence in Wirral and west Cheshire with increases of 28% and 42%

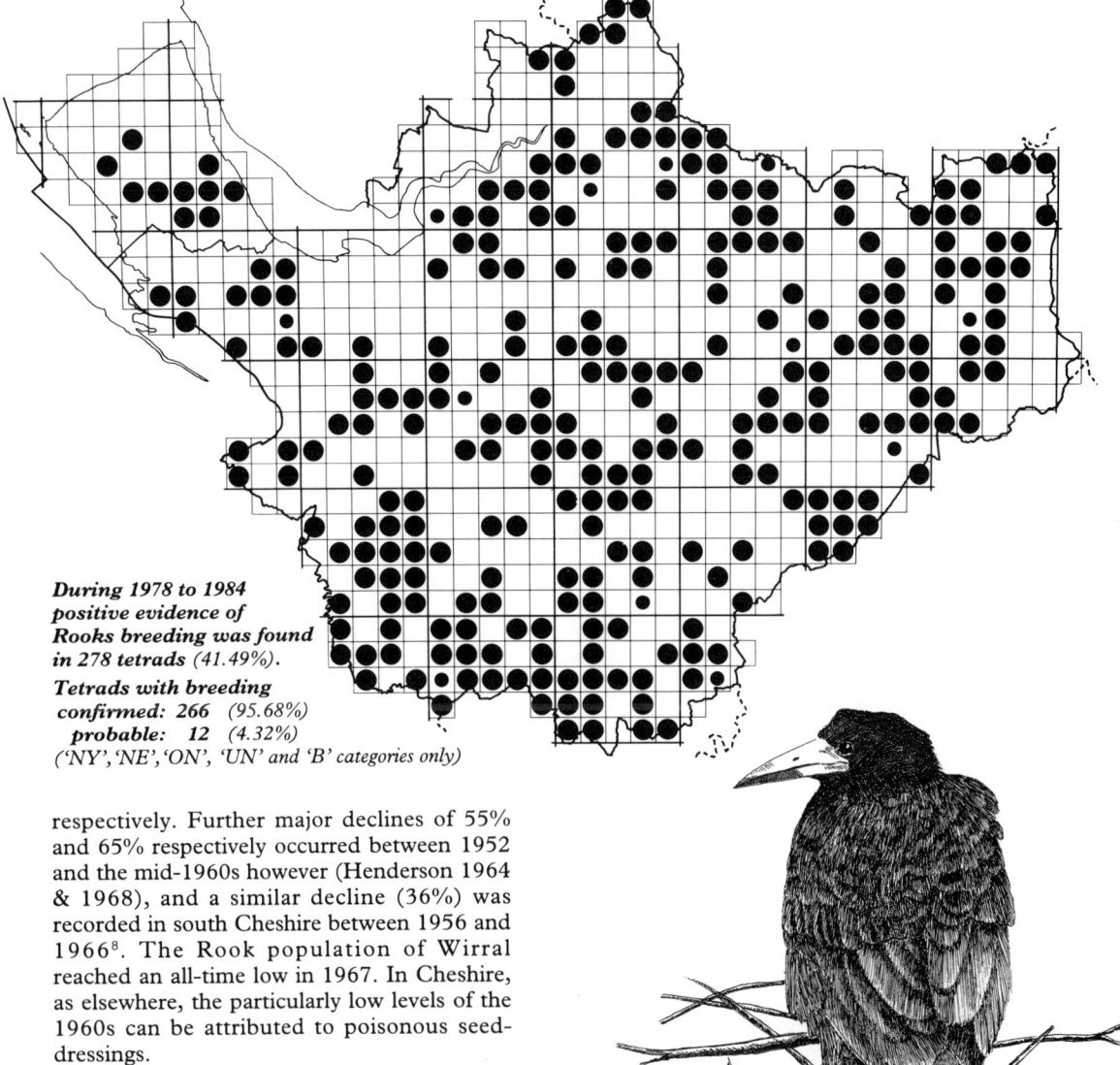

During 1978 to 1984 positive evidence of Rooks breeding was found in 278 tetrads (41.49%).

Tetrads with breeding
confirmed: 266 (95.68%)
probable: 12   (4.32%)
('NY', 'NE', 'ON', 'UN' and 'B' categories only)

respectively. Further major declines of 55% and 65% respectively occurred between 1952 and the mid-1960s however (Henderson 1964 & 1968), and a similar decline (36%) was recorded in south Cheshire between 1956 and 1966[8]. The Rook population of Wirral reached an all-time low in 1967. In Cheshire, as elsewhere, the particularly low levels of the 1960s can be attributed to poisonous seed-dressings.

Agricultural practice is likely to be the key factor affecting Rook populations, this being one of the most economically important farmland birds[42]. It eats largely grain and grassland invertebrates, with earthworms particularly important throughout the year. Cold or dry summers, when worms are difficult to find, lead to heavy mortality. Optimum habitat has been shown to be a roughly equal mix of grassland and arable, whether cereals, roots or vegetables (Brenchley 1984) and any shift from this balance, towards more specialised farming of one sort or another, reduces the Rook population.

The upward trend in Cheshire may have continued since 1980. Five rookeries south of Crewe which held 117 nests in 1980 grew to 169 in 1983; four rookeries near Comberbach increased from 107 nests in 1975 to 225 in 1983; the Alderley Park rookeries increased from 69 in 1978 to 210 in 1982 but then declined; and eight rookeries in the Holmes Chapel—Middlewich area increased from 249 in 1980 to 262 in 1984 (+5%). Drakelow rookery, in this last area, runs against this trend. It held 253 nests in 1931, declined to 142 in 1945[6], 57 in 1980 and further to 36 by 1984. Counts of small groups of rookeries such as these may not reflect the changes in the county as a whole however, since local movements of birds from one colony to another between seasons are commonplace. Isolated counts add little to our knowledge.

The map shows that Rooks avoid most of the major urban areas such as Birkenhead, Chester, Widnes and Northwich, and also the highest land in the eastern hills and the lowest-lying parts of the county, along the north Wirral coast and the valleys of the Gowy and Mersey. This altitudinal preference is probably explained by the lack of suitable nest-sites.

Unfortunately, recent surveys have paid little attention to the species of trees in which rookeries are located. Henderson (1946) recorded oak, ash, elm, sycamore and horse chestnut as the main tree species used in Wirral, whilst pines, fir, larch and other conifers were also used. Similarly, Boyd[5, 6] reported the majority of nests in his study area to be in oak, ash, elm, sycamore and beech which were then the most prevalent large trees. He also found many in Scots pine, alder, horse chestnut and poplar, with some in cherry, lime, elder, birch, larch and hawthorn. The South-East Cheshire Ornithological Society survey of 1966 recorded 44% of nests in oak, 34% in beech, sycamore and elm, with use also being made of eight other species of trees[8]. Scattered observations suggest little change from Boyd's time. Elm is not a particularly important nesting tree in Cheshire, and Dutch elm disease has had no noticeable effect on the fortunes of the county's Rooks.

Although birds may visit rookeries at almost any time from November onwards, the main return movement occurs during February as the winter roosts break up and the immigrants go back to their continental breeding areas. Thus, in spring 1983 a roost at Mottram Hall woods decreased from 739 on 8th February to 24 by 2nd March. Nest-building is very easily observed, and once a rookery has been located it is a simple matter to return at a later date to confirm breeding. Occasionally, single nests may be found in isolation. There is probably no tetrad in the recording area where Rooks do not appear at some stage during the breeding season, and they may even be found scavenging in rubbish bins behind suburban supermarkets several miles from any rookery. For this reason, only 'NY', 'NE', 'ON', 'UN' and 'B' categories are plotted on the map. It is now known that a number of rookeries were missed during the 1975 survey and it is estimated that, by 1984, the county population may have stood at 12,000 or more nests.

REFERENCES

Ankers, J. A. & Elphick, D. (1981): BTO Rookery Surveys 1975 & 1980 – Cheshire and Wirral Results. *Cheshire Bird Report* pp59-64.

Boyd, A. W. (1954): Cheshire Rookeries. *L&CFC 30th Report*, 1954: 24-38.

Brenchley, A. (1984): The use of birds as indicators of change in agriculture. In *Agriculture and the Environment*, ed. D. Jenkins pp123-8. (Cambridge: Institute of Terrestrial Ecology).

Henderson, M. (1946): Rookeries Census, Wirral and West Cheshire. *Proc. of the Liverpool Nat. Field Club*, pp27-32.

Henderson, M. (1953): Some Observations on the Rook Population of West Cheshire. *Proc. of the Liverpool Nat. Field Club*, pp14-15

Henderson, M. (1964): Observations on the Rook Population of Wirral, Cheshire. *Proc. of the Liverpool Nat. Field Club*, 1961-63 pp15-17.

Henderson, M. (1968): The Rook population of a part of West Cheshire 1944-1968. *Bird Study* 15: 206-208.

Crow
Corvus corone

PRIOR to 1865, Crows bred commonly in the neighbourhood of the Dee marshes and probably across much of the county. With the advent of game-preservation however they were heavily persecuted and by 1910 Coward & Oldham described the species as a scarce resident, thinly distributed throughout the county, and rare in many of

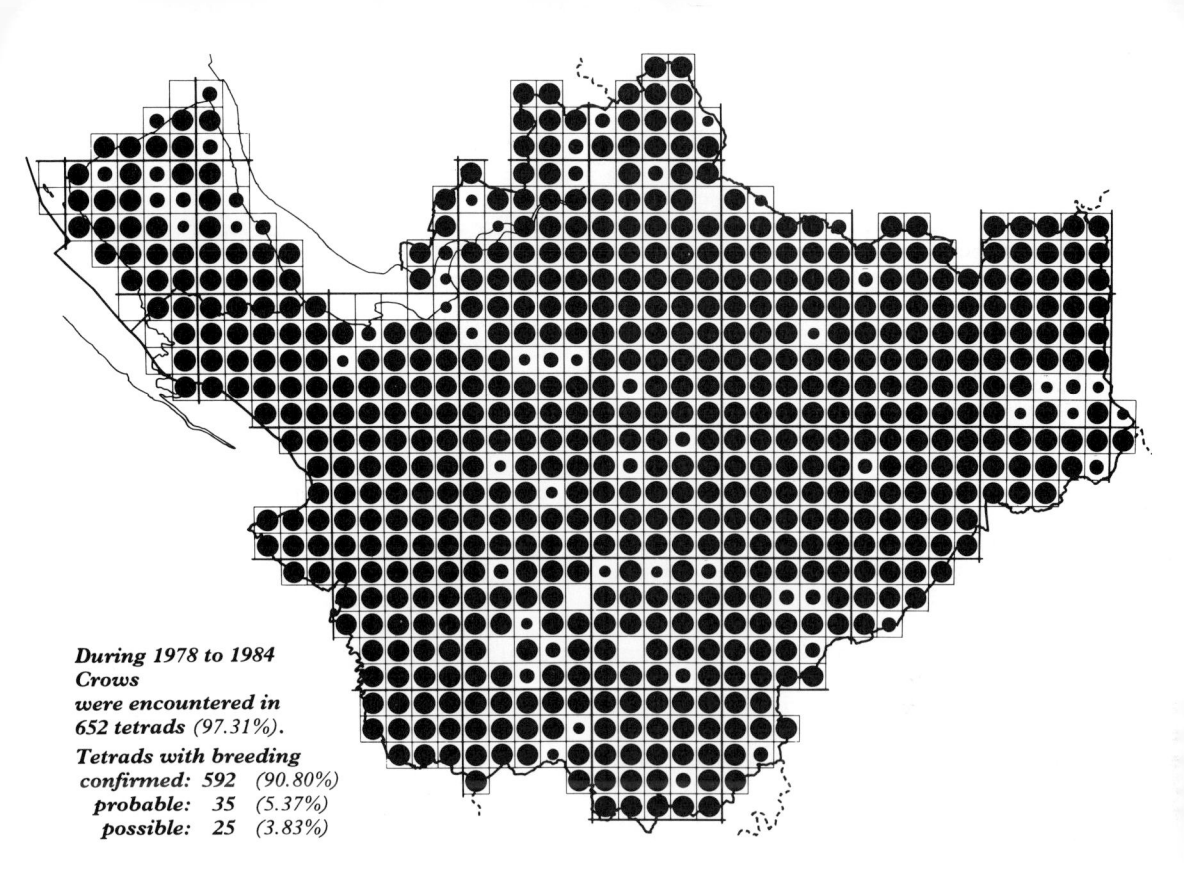

During 1978 to 1984
Crows
were encountered in
652 tetrads (97.31%).
Tetrads with breeding
confirmed: 592 (90.80%)
probable: 35 (5.37%)
possible: 25 (3.83%)

the eastern hill districts. The only breeding season records mentioned in Abbott's diaries between 1913 and 1930 were of a pair at Marbury Park, Great Budworth, in March 1923 and April 1925. Gamekeeping had lapsed during the 1914-18 war, but where keepers were still active few Crows survived. In 1929 the keepers at Arley considered there were only three pairs in 5000 acres – much scarcer than elsewhere in that district[6]. Since then numbers have increased steadily.

In 1941, Hardy described the Crow as a widespread nester in southern parts of Wirral and adjacent areas of Cheshire. In north Wirral and around Warrington it was still chiefly a winter visitor but increasing as a breeding species. Griffiths & Wilson[13] similarly knew of only small numbers in north Wirral although some increase had occurred. This was probably due to a further reduction in game-preservation during the 1939-45 war. By 1949, the species was abundant in much of the Cheshire plain and common in Wirral[3]. Boyd[6] described it as numerous and regretted that "the division of land into small farms restricts any organised killing of these destructive birds".

With the more recent demise of small mixed farms and the arrival of battery methods for poultry rearing, fewer farmers now bother to shoot birds nesting on their land, and the efforts of those who do try to keep numbers down amount to little when flocks of 100 or more Crows are increasingly met with even in summer. At Prestbury Sewage Works in 1983 there were counts of up to 355 in April and 240 in June. These are subordinate birds, incapable of breeding without a territory, but which may rob the nests of territorial pairs. When a territory becomes vacant a new pair will leave such a flock to take up residence. The Crow population is to an extent self-regulating, for when the number of non-breeding birds is high, breeding success is proportionately lower (Coombs 1978).

Crows have a considerable repertoire in addition to the familiar cawing notes. JPG has notes of a gulping "kowp-kowp-kowp" call from October to April, less often in September and May. A curious loud rattling call, not mentioned by Witherby, often deceives bird-watchers into seeking a drumming woodpecker. This call is given especially towards dusk at communal roosts (which may number several hundred birds) from February to April. A subdued version was heard from below a nest at Adlington where a bird was sitting on 18th May 1977.

During March and April birds can be seen snapping twigs off trees for nesting material, and whilst a new nest may be completed within a day or so, old nests are frequently renovated. One nest in a copper beech at Dean Row, Wilmslow, was used in at least nine successive seasons, with birds sitting by 15th March 1981 and 18th March 1974, although incubation usually starts in April. Of 33 nests noted in eastern Cheshire in 1983, ten were in oak, six in ash, four each in alder, hawthorn and sycamore, three in beech and one each in lime and crack willow (JPG). The low-growing hawthorn and the wind-resistant sycamore are particularly popular in the hill areas, where choice of sites is restricted. Mature willows are used at Woolston and Scots pines in Delamere. Double telegraph poles are another favoured site, and in 1955 a pair nested on the survey mast at Hilbre[25]. In 1966, young were reared from a nest near Frodsham situated in a gorse bush only four feet from the ground, a site recalling that on a lopped hedge near Kelsall[3] where clutches were laid without success in 1948 and 1949. In 1981 eggs were laid but later disappeared from a nest on the ground in a hayfield near Crewe[8].

Persecution by man, the Crow's only real enemy, has largely ceased. Indeed the brown-plumaged juveniles, which may be very tame, are sometimes taken as pets. CBC figures show an enormous increase, the national breeding population in 1984 being over twice the 1966 level. Two other factors may have contributed to the recent population increase. One is a recovery from the effects of pesticides, Crows having suffered some eggshell thinning caused by ingestion of organochlorine seed-dressings in the 1950s and early 1960s. The other is the increased availability of sheep-carrion in hill districts[42] leading to an increased production of young Crows which may emigrate to the lowlands.

The map shows Crows to be very widespread. The few tetrads without 'probable' or confirmed breeding are mainly in the most built-up areas (Birkenhead docks, the Widnes area and Warrington), the hills above Macclesfield (with no proven breeding in Macclesfield Forest), and under-recorded parts of the south of the county. Absences from other tetrads are difficult to explain, however, – for example Lymm village, mid-Wirral, and the well-watched Frodsham/Kingsley/Acton Bridge area.

Densities on CBC plots nationally averaged five pairs per square kilometre on farmland and eleven pairs per square kilometre in woodland in 1980[40]. Sharrock[35] suggested a figure of ten pairs per tetrad, since when the population has increased substantially. Some areas hold locally high densities such as three pairs in 18.2 hectares at Bidston[29] and 33 pairs in 393 hectares in six CBC plots in south-east Cheshire in 1978. Other tetrads, however, have few suitable nest-sites and may hold only one or two pairs of Crows. All crow species are still noticeably less numerous in keepered areas, as at Withington and Aston near Frodsham, although such local variations are not detectable on the map. An average of only twelve pairs per tetrad (three pairs per square kilometre) would give a county total of 7200 pairs.

REFERENCE

Coombs, F. (1978): *The Crows – A Study of the Corvids of Europe.* Batsford, London.

Raven
Corvus corax

THE Raven was widespread as a breeding species throughout Britain until the early nineteenth century when its range contracted because of persecution. It is now mostly found in coastal and upland districts of the north and west, frequenting rocky crags, open moorland and steep sea cliffs[35].

In Cheshire, King, in his "Vale-Royall" of 1656, mentions Ravens as well as crows in the forests of "Delamere and Maxfield" in 1617. At the beginning of the eighteenth century Stockport churchwardens paid only one penny per head for their destruction; an indication of their abundance at that time, although it has been suggested that the word "raven" might then have been loosely applied to other crow species.

Byerley[2] noted that Ravens occurred "occasionally in Wirral", and Brockholes[2] referred to them as formerly abundant in winter on the Dee marshes, though he had not seen any since about 1866. A pair nested at Hilbre in 1857 but were driven away before the eggs hatched[2].

Bell[7,9] listed only nine records this century. Some of these were considered to be escaped pets. Since 1980 there has been some resurgence, although there was no sign of nesting during this present survey. The nearest breeding sites are in the Clwydian hills and it may only be a matter of time before Ravens once more breed in Cheshire. Certainly there is no shortage of carrion in the eastern hills.

Starling
Sturnus vulgaris

"THERE are people who dislike shepsters just because their nests are dirty and untidy and their appetites good". Boyd's[6] words seem to have such general application that little seems to have been written on the bird's breeding habits within the county since. It is at once one of our most and least familiar birds.

Prior to about 1830 the Starling was surprisingly scarce in much of northern and western Britain[35], so a reference in Leland's itinerary, around the 1530s, to "a great reedy pool (near Spurstow), whither an innumerable sight of stares resort at night" (Harrison 1910) is of interest in that it implies at least that Starlings were visiting, if not actually nesting in the county in those distant days. Little change in status is evident from available writings this century although there appears to have been an increase in the numbers nesting in the woods at Rostherne in recent decades[24]. In 1940 Starlings and House Sparrows were the only birds breeding in parts of Birkenhead, the same perhaps being true of other Cheshire towns at that time[5].

Starlings nest commonly in cavities around buildings and in holes in trees. Nest-boxes are used, and woodpecker holes, woodpeckers or Nuthatches quite frequently being dispossessed. At Wilmslow a bird was watched enlarging the entrance to a Lesser Spotted Woodpecker's old hole in a rotten alder stump. Hardy[4] noted that it "often usurps occupied nests of Green Woodpecker at Oakmere and Spotted Woodpecker at Norton". Boyd[6] once found an open nest, resembling a large, untidy Blackbird nest, in a creeper on a house wall. At Hatherton near Nantwich in 1942, a pair nested in the same hollow bough as a Little Owl[3].

In early March the birds return to their nest-sites, and birds may then be heard singing from tree-tops in woods from which they have largely been absent all winter. The song is a spluttery wheeze, more or less elaborated with fluty whistles and mimickry. In January 1942 Boyd[5] noted one "giving splendid imitations of a mistle-thrush" and "quacking like a mallard and whistling like a redshank". The bubbling trill of a Curlew and the song of the Yellowhammer are other frequently adopted phrases. Nowadays the sound of a telephone may be added to their repertoire.

By mid-May broods of noisy youngsters are easily heard in woods and buildings alike, and the identity of the birds is confirmed by the white-washed entrance to the nest. Often the first species noted on a recorder's form when entering a new tetrad would be a Starling flying to its nest with a beakful of food. In fact the Starling produced more tetrads with confirmed breeding than any other species, showing both its ubiquity and the ease of proving breeding. First broods leave the nest by the end of May and the brown-plumaged young gather to feed in noisy flocks in the pastures. For example 200 such birds were in a field at Kirkleyditch (SJ87) on 25th May 1978 and 100 or more near Prestbury on 27th May 1982. On 11th June 1921 Abbott watched flocks of young birds feeding in oak and birch trees on Alderley Edge that had been stripped of leaves by caterpillars. Boyd[6] ringed 209 broods: of 172 marked in May, the average brood size was 4.26, and of 37 second or replacement broods marked between 6th June and 9th July the average was three young. JPG has extreme records of a fully-grown youngster caught by a Magpie near Sandbach on 2nd May 1974 and two recently fledged youngsters at Frodsham Marsh on 3rd September 1984. An early nest at Eaton Hall held four eggs on 29th March 1943[3].

Between 1963 and 1969 seven to ten pairs bred on Hilbre[25] where the Starling is one of the few regularly breeding species, and resident birds have flocked to a total of 90. Elsewhere in our area the species is generally so numerous that no breeding census appears to have been taken. Starlings defend only a small distance, up to five or ten metres, next to their nest, and share feeding grounds. In this way the species can attain very high densities in the best areas. Concentrations of up to 500 nests per tetrad may be found in older suburbs with extensive gardens[31], with similar figures in parkland (Feare 1984). Cheshire and Wirral have no shortage of such habitats. However agricultural land carries much lower densities, farmland CBC densities averaging 19 pairs per tetrad in a number of adjacent counties[42]. Sharrock's[35] estimate of an average of 40-80 pairs per tetrad probably applies to our area, giving a Cheshire total of some 25,000-50,000 pairs.

Unusually for a passerine, there is a large pool of non-breeding Starlings. Most females breed in their first year, but many males do not try until their second summer (Feare 1984). It is probably these non-breeders, with some off-duty breeding males, that make up the large summer roosts. Some birds spent the night on the Runcorn–Widnes bridge in every month of this survey, and at Woolston from 1980-84 the roost formed in late March or early April and held 5000-10,000 birds until late May when numbers increased greatly with the influx of locally-bred juveniles.

REFERENCES

Feare, C. (1984): *The Starling*. OUP. Oxford.
Harrison, W. (1910): *Leland's Itinerary* in *Trans. of the Lancs. and Ches. Ant. Soc.*

House Sparrow
Passer domesticus

THE House Sparrow shares with the Starling the dubious distinction of being one of our most familiar yet ignored species. It seems never to occur far from the haunts of man and is largely dependent on people for winter feeding and for nest-sites. On Hilbre, House Sparrows bred only so long as the island's warden kept poultry, and the provision of grain to these poultry was clearly an important source of food for the sparrows. In the absence of poultry from 1956 onwards, the sparrows died out, only to recolonise in

During 1978 to 1984 Starlings were encountered in 663 tetrads (98.96%).

Tetrads with breeding
confirmed: 657 (99.10%)
probable: 3 (0.45%)
possible: 3 (0.45%)

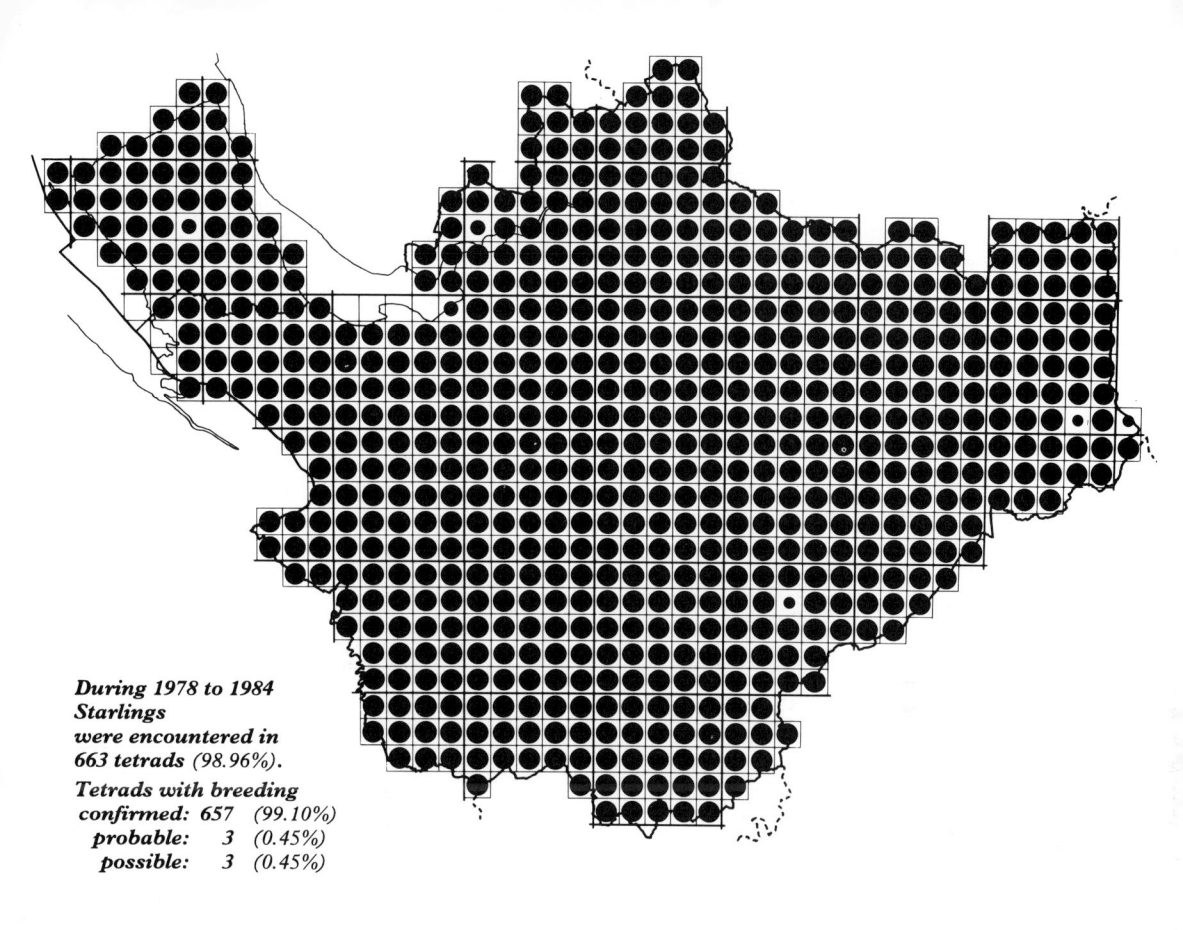

1971, poultry and other livestock having been brought back in by the new warden in 1968[25]. Similarly, there are several localities in the county where it is known that few sparrows nest, but which are visited by large flocks in the autumn and winter. Thus, at Woolston, where only a handful of pairs nests, a flock of 130 was present in late autumn 1981 and 70 fed in willow scrub in September 1982. From July and August onwards flocks visit cereal fields to feed on ripening grain or stubble, and congregations of 100-300 are not unusual in such situations.

Nest-building has been recorded in most if not all months of the year. At Wilmslow, in the late 1970s, nests were constructed in several winters between November and January inside the ridge of a garage workshop, presumably encouraged by the artificial (albeit poor) heating. There was no evidence however that eggs were ever laid, and it is far more likely that the birds were improving their roosting quarters. Boyd[6] had watched sparrows building on 27th November, 25th January and 2nd March but he, too, doubted whether these were serious nesting attempts. A pair carrying feathers into a barn at Disley on 5th March 1983 during mild weather would appear to have got at least as far as lining their nest, although birds were seen collecting feathers in February 1985 in extremely cold conditions when a breeding attempt would appear to be highly improbable. However, in 1948 a pair was nesting in the rafters of a corrugated-iron garage at Audlem, the brood flying around Christmas[3]. In early spring, prior to egg-laying, hen sparrows in particular may be seen eating the nutritious spore-capsules of mosses on roof-tops.

Most nests are situated in cavities in buildings, particularly under eaves, but nests in tall thorn hedges near settlements are not uncommon, and it is by no means unusual for

House Martin nests to be taken over. In 1982 a nest at Langley (SJ97) was placed on top of batteries inside a farm tractor. Boyd[6] had records of nests in poplars, pear trees, yews and other trees, and once in a hole in a beech tree. Once a newly-built Song Thrush nest was occupied and adapted, once a Greenfinch nest and on several occasions nests of Swallows and Magpies, and at Anderton (Northwich) those of Sand Martins. Boyd had also seen a nest a furlong away from the nearest building, such a nest being quite exceptional. Copulating pairs are frequently seen on roof-tops, overhead wires or even in the middle of a road. Eggs are laid in April and on into late summer, and young birds being fed by their parents are a common sight in gardens from June onwards.

House Sparrows were proved to breed in all but a handful of tetrads, with the only areas from which they are truly absent being the saltmarsh tetrads of the Dee and Mersey and the highest land in the eastern hills. Sharrock[35] suggested a mean density of 10-20 pairs per square kilometre (40-80 pairs per tetrad) in Britain, but his figure averages over the whole country including the sparsely populated areas. A considerably higher density must be expected in Cheshire, perhaps as many as 100-200 pairs per tetrad, which would give a county population of 65,000-130,000 pairs. An estimate close to this upper level is obtained using the novel formula quoted by Summers-Smith[41] of approximately one pair of breeding House Sparrows for every five members of the human population. Sample censuses in a variety of habitats would make an interesting project for the county's bird-watchers.

Tree Sparrow
Passer montanus

COWARD & Oldham[2] noted that in the 1870s the occurrence of the Tree Sparrow in Cheshire was hardly recognised, and Brockholes writing in 1874 had known of none nesting closer to Wirral than at Bache House, Chester. By 1894 a colony was established in the old sandstone quarry at Burton Point, nesting in holes in the rockface, and by the early years of this century it was not uncommon on the peninsula and around Chester. Indeed it then occurred in all lowland parts of the county including the Frodsham and Peckforton Hills, but was not common on the higher ground in the east. Griffiths & Wilson[13] still regarded it as very local in north Wirral however.

Bell[7] could add little, although in his Supplement of 1967[9] he emphasised the local nature of its distribution by listing recent breeding sites only eleven of which fell within the present recording area. It seems at least probable that this scarcity of information reflected lack of interest on the part of observers, and that, as both Coward & Oldham and Boyd suggested, the bird was even then more widely distributed than was generally believed.

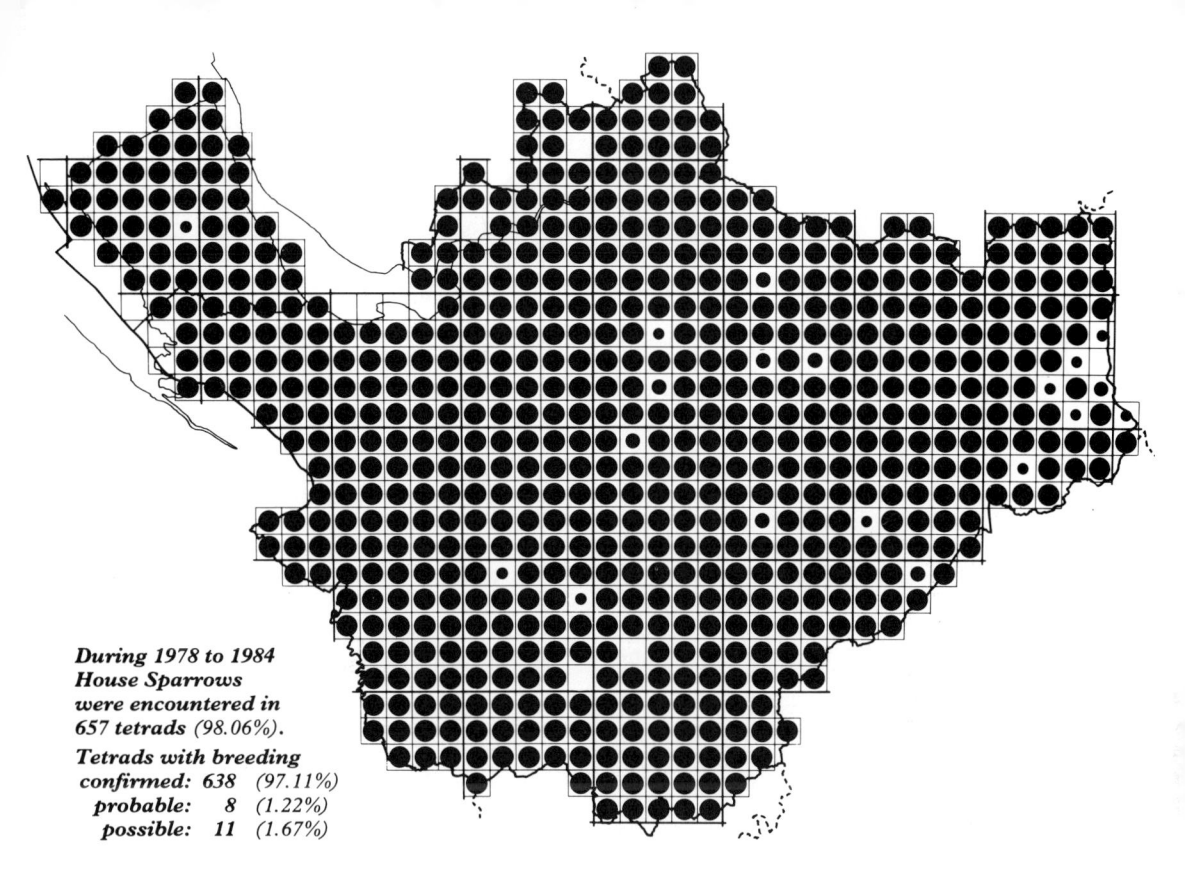

During 1978 to 1984
House Sparrows
were encountered in
657 tetrads (98.06%).

Tetrads with breeding
 confirmed: 638 (97.11%)
 probable: 8 (1.22%)
 possible: 11 (1.67%)

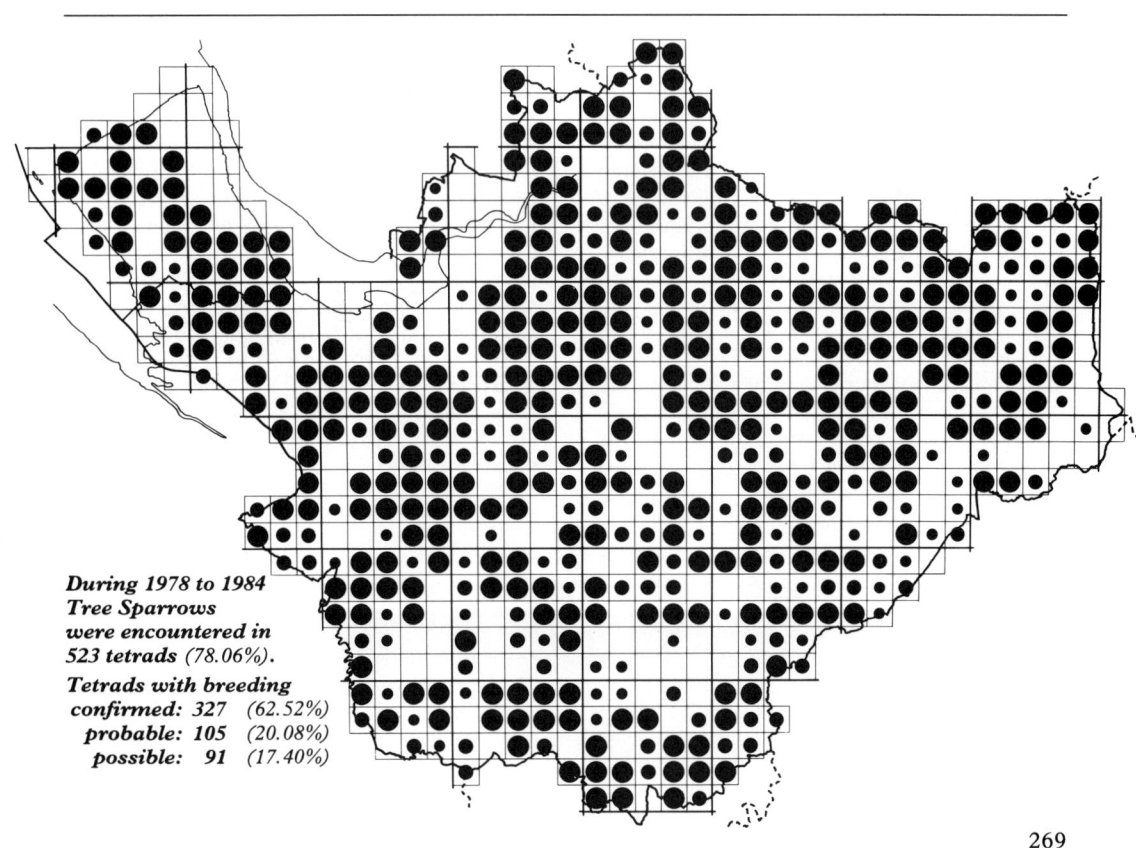

During 1978 to 1984
Tree Sparrows
were encountered in
523 tetrads (78.06%).

Tetrads with breeding
 confirmed: 327 (62.52%)
 probable: 105 (20.08%)
 possible: 91 (17.40%)

Nationally the Tree Sparrow has undergone large fluctuations in numbers and range. Populations are thought to have decreased through most of the first half of this century to a low in the 1950s but they increased very rapidly in the 1960s [34]. The cause of these changes is not known, although the use of organochlorine seed-dressings in the 1950s could not have helped this species which, at least in winter, feeds mainly on weed seeds and grain. Numbers were at their peak for a few years each side of 1970, but have since declined again, along with those of other seed-eating birds of farmland such as Linnets and Corn Buntings, with the trend to "cleaner" arable agriculture clearly implicated[41]. Along with their usual diet of insects, chicks were fed on elm seeds until recently, and Dutch elm disease may have had an adverse effect on some colonies. Cheshire is one of the westernmost counties in which Tree Sparrows breed in substantial numbers, and must have shared these nationally observed changes, although they were apparently not recorded here. Boyd's comment[5] that "few of my neighbours realise that there are two kinds of sparrows" unfortunately still applies to many bird-watchers.

The most typical nest-site is in a cavity in a hedgerow tree, the circular holes where branches have fallen from the trunks of oak or ash being especially favoured, and old riverside willows often support a pair or two. Nest-boxes are often occupied: Boyd[5] had a record of a pair which built inside the skeleton of a Stock Dove in one; and the bulky nests of Rooks and Herons sometimes harbour sparrows' nests built into the undersides. Boyd also knew of instances where Magpies' nests were adopted, and Tree Sparrows will nest in Sand Martins' burrows. A pair or two were nesting in a barn at Little Leigh in April 1981, and in the same year two pairs nested under the eaves of Risley Moss visitor centre.

At Rostherne, birds returned to a nest-box colony in late March or early April, and pairs reared three broods in some years[24]. Boyd[6] studied another colony, again in nest-boxes, at Frandley and ringed some 300 broods between 1924 and 1939. Annual average brood-size varied from 3.14 to 4.0 young. Some birds occupied boxes as early as the last week in February with others returning as late as mid-April. Eggs were first seen on 8th May, and the boxes were deserted by the end of August. Provision of additional boxes boosted the colony from two to at most 27 pairs, suggesting that availability of nest-sites may regulate populations.

Tree Sparrows have never really been urban birds, and this survey map reflects this. They are absent from most of the larger towns, including Birkenhead and the east Wirral conurbation, Chester, Ellesmere Port, Runcorn, Widnes, Warrington and Crewe. However, they will penetrate suburbs to nest in large gardens, and find parkland ideal, as at Tatton and in the Wirral peninsula. Despite their English name, they are not woodland birds, and are scarce in areas like Delamere and Peckforton. Also belying their scientific name, they are mainly a lowland species and at present the bird remains scarce in the eastern hills. Perhaps the absence of arable land is crucial here, for weed seed and grain are important parts of the diet, and ringing has suggested that local breeders are sedentary even in winter.

Four CBC plots on farmland in south-east Cheshire held from ten to 16 territories between 1979 and 1981. National CBC densities vary widely, from less than one to over 30 pairs per square kilometre. Many Cheshire tetrads held only a pair or two, but others with good colonies may have had 20 or more. An average of four to six pairs per tetrad with 'probable' or confirmed breeding would give a county population in the range 1700-2500 pairs, somewhat over half of Sharrock's[35] estimate of 150 pairs per 10-km square during the 1968-72 BTO Atlas period, since when the population has fallen considerably.

During 1978 to 1984 Chaffinches were encountered in 638 tetrads (95.22%).

Tetrads with breeding
confirmed: 552 *(86.52%)*
probable: 72 *(11.29%)*
possible: 14 *(2.19%)*

Chaffinch
Fringilla coelebs

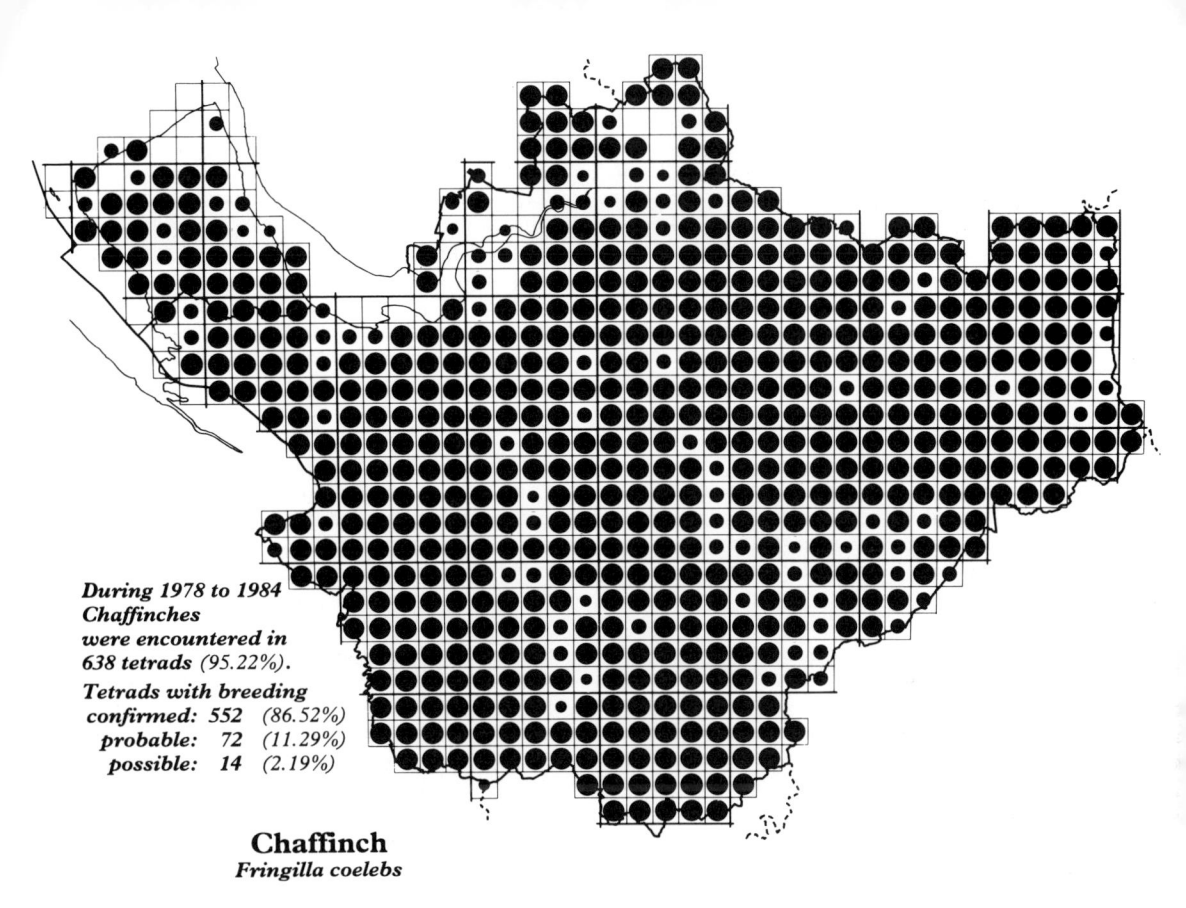

OUR most widespread finch and one of our commonest song-birds, the Chaffinch breeds throughout the county wherever low scrub or hedges provide nest-sites with nearby trees for song-posts. They even manage without trees on Frodsham Marsh, declaring their territories by singing from the tops of clipped hedges. The Chaffinch is one of few species which have adapted readily to large stands of conifers, breeding in quantity in the mature pines of Delamere and the spruces of Macclesfield Forest, although densities are much higher in deciduous woods than in spruce, and higher in spruce than in pine[33]. The only areas from which Chaffinches are absent are the treeless parts of the eastern hills, built-up north-eastern Wirral and the inner Mersey valley.

Song is heard from the middle of February until the end of June, and exceptionally outside this period. Boyd[6] had a record of song on 10th August, and in his experience song started generally in the third or fourth week of February, but sometimes not until March. Whether birds begin to sing earlier nowadays is doubtful: the greater number of early records is proportionate to the increased number of bird-watchers, although there is clear evidence that they breed earlier in warm springs[33]. The male bird frequently utters a repeated "zree" note during the song period, and nesting birds often give a high-pitched squeak as an intruder approaches. Another territorial call, heard only in the breeding season, is a disyllabic "hooeet", similar to but rather harsher than the familiar alarm call of Willow Warblers.

Nest-building begins in April, resident birds commencing their breeding duties while continental winter visitors are still present in flocks. The typical nest-site is low down in a hedge, and many nests are torn out by Magpies. Coward had records of nests as low as two feet and as high as 40 feet off the ground. Cheshire nests are typically of moss and hair but nests from Macclesfield Forest and Higher Disley have been found to contain fragments of lichens (*Hypogymnia physodes* and *Parmelia saxatilis*) in their structure. Coward & Oldham[2] mention the use of incongruous materials such as paper, and a nest in Antrobus village in 1931 was covered in red, white and blue confetti, much of it built into the nest[6]. A nest at Hatherton in 1944 was half made of owl castings[3].

By early June the young are being fed in the nest, and it is then that breeding is most easily confirmed, insect food (such as caterpillars) being carried in the bill and not regurgitated, this being the main feature distinguishing the fringilline finches (Chaffinch and Brambling) from most finch species. This behaviour facilitates confirmation of breeding and the map shows a much higher proportion of confirmed breeding records than for other finches. They normally have only one brood and are one of the longer-lived passerines.

In Coward's day the Chaffinch was the commonest passerine around the farms in the eastern hills, far outnumbering even the House Sparrow, but numbers have declined nationally since the days when corn was threshed in the farmyard and finches gathered to glean amongst the chaff. The advent of herbicides means there are generally fewer weeds to produce seed in the fields. Comments in *CBR* for 1965 show that in north Wirral that year, "it was never noted in some areas where previously common", and it was very uncommon in many parts of mid-Cheshire that summer. The following year it was stated to be outnumbered by the Greenfinch in parts of mid-Cheshire, although still far more numerous in the south-west. The *L&CFC Report* for 1966 stated that the bird was still decreasing except in the eastern hills, and that it would soon "apparently have to be regarded as an uncommon breeding bird in Wirral and much of the Cheshire Plain". In 1975 the decline was summarised as having "affected largely suburban and agricultural land, less in woodland and least in upland and particularly forestry commission land, where there may even be some increase". In suburban gardens it seemed to have been replaced by the Goldfinch[8]. National figures suggest a peak around 1950, with a steady decrease from then to 1963[34] possibly caused by organochlorine seed-dressings[33], and CBC surveys have shown little or no change in numbers from 1966 to 1984.

The absence from the inner Mersey valley came as one of the main surprises of this survey. This area was characterised until recently by very severe air pollution which prohibits the growth of the epiphytic mosses which Chaffinches collect for nest-building, so lack of nest-material may be an important factor. A similar scarcity is found for breeding Long-tailed Tits, which also have moss-built nests. Insect food may also be limited by pollution. Chaffinches in summer feed on defoliating caterpillars (especially winter moths) and aphids, the adults frequently eating the aphids themselves and saving the more nutritious caterpillars for their brood[33].

Chaffinches are among the country's most numerous birds, still common in lowland woods and hedgerows. CBC figures average 20 pairs per square kilometre on farmland in a number of adjoining counties[42] and may exceed 120 pairs per square kilometre in deciduous woodland[40]. However, their density is lower in suburban and urban areas and a mean figure of 15-20 pairs per square kilometre (60-80 pairs per tetrad) would give a county total of 35,000-50,000 pairs.

Brambling
Fringilla montifringilla

RECORDS of this winter visitor straying into the first few days of May are not exceptional. At this season birds resort to stands of pine and other conifers, and males may be heard singing. In 1956 two males and a female were seen in Alderley Woods on 30th May[7]. On 11th June 1966 a male was seen in Delamere Forest[9], and on 6th June 1982 one was singing in a Hoylake garden[8].

Hardy[4], writing in 1941, noted that aviary escapes were sometimes evident in summer. However fewer finches are caged nowadays and there have been several summer records of singing birds in north Derbyshire[45].

During 1978 to 1984 Greenfinches were encountered in 581 tetrads (86.72%).
Tetrads with breeding
confirmed: 295 (50.77%)
probable: 217 (37.35%)
possible: 69 (11.88%)

Greenfinch
Carduelis chloris

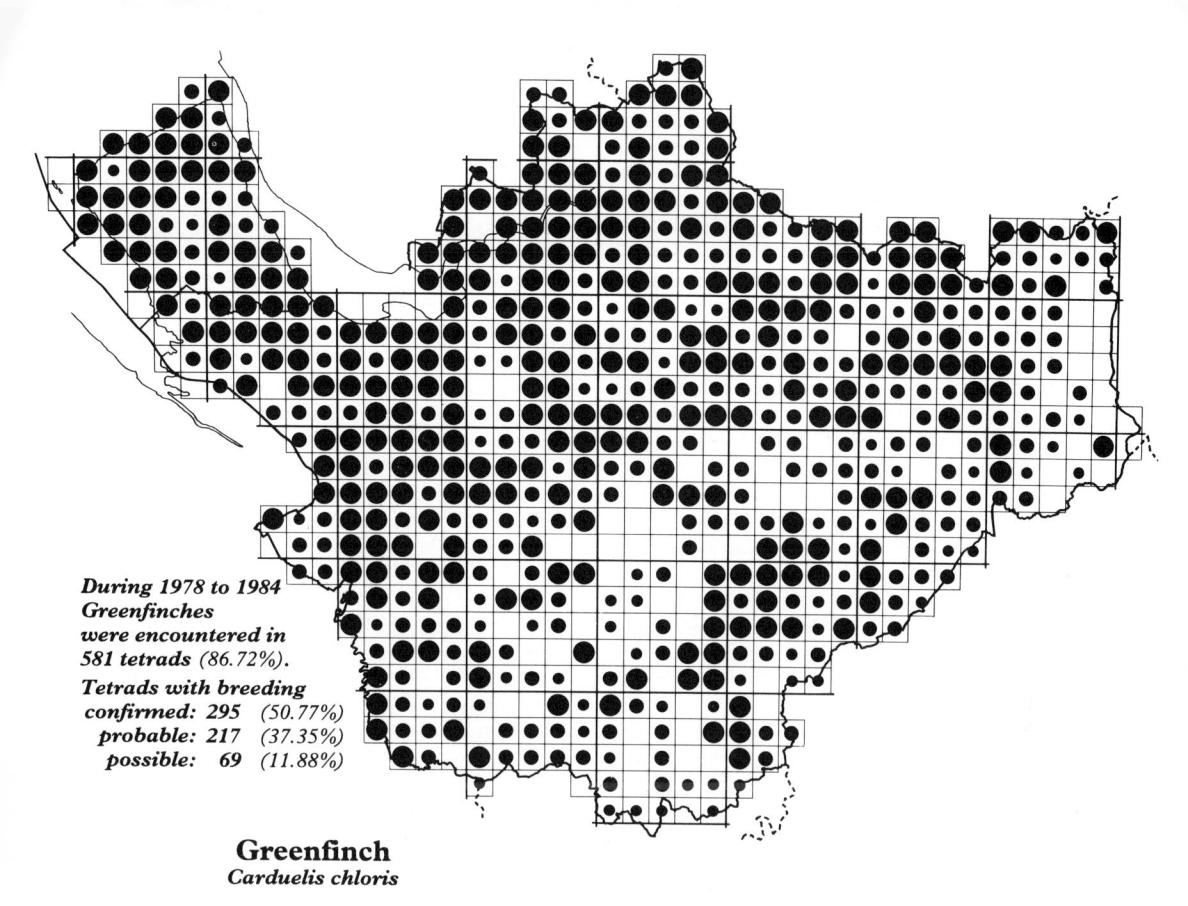

A familiar garden bird, whether taking peanuts from a feeder or seeds from a sunflower, the Greenfinch also nests commonly near habitation. In spring and summer the male has a buzzing, "wheezing" call, heard sometimes in February during bright weather but more often from March onwards. The song, a more tuneful elaboration of the usual twittering calls, is given either from a prominent perch or in flight. At its most pronounced, the song-flight of the male consists in his flying in an irregular circle with slow bat-like wingbeats. Song is heard from mid-March to July, occasionally during August and October into early November.

Apart from gardens, favoured habitats include shrubberies in parks, young plantations and the edge of woodland. The nest is typically in a tall thick hedge, holly and other evergreens being particularly favoured although nests in hawthorn are frequent given the prevalence of this hedging shrub. Of 32 nests on which Boyd[6] had notes eleven were in "quickset hedges" (hawthorn), six in yew, seven in holly and the remainder in privet, cypress and other bushes.

The breeding season often starts late, but may last well into "autumn". Boyd's[6] earliest recorded nest held two eggs on 17th April, with nesting continuing into August, May being "the month of most nests". A nest in his orchard hedge contained three eggs on 24th August 1934; only one of these hatched and the nestling died on 4th September. Forty years later a nest in a holly hedge at Sandbach held young on 27th April (JPG). At Wybunbury, in 1980, three young were still being fed by their parents as late as 20th October[8].

Coward & Oldham[2] knew the Greenfinch as an abundant resident and partial migrant, with most birds arriving in spring and leaving again in the autumn. It occurred at considerable elevations in the eastern hills provided there were thorn hedges or bushes to nest in. In market garden districts of northern Cheshire it was persecuted because of its habit of pulling up young turnips, radishes and sprout plants, but where it was little molested the species was sociable in the breeding season with many nests being built close together. In 1943 five nests were occupied in one hedge at Gorstage: a distinctively marked female visited three of these and was driven off before eventually settling on one[3].

Griffiths & Wilson[13], writing in 1945, found Greenfinches plentiful in Wirral, the only change in status from Coward & Oldham's time being that birds were by then present in winter also. Boyd[6] similarly regarded them as one of the most plentiful birds in the Great Budworth district throughout the year. He made a special study of this species showing the early dispersal of juveniles, the faithfulness of breeding birds to an established site from year to year, and the tendency for juveniles to undertake longer-distance movements than adults (Boyd 1931). More recently there is rather little information on the movements of our breeding birds. However, some birds from Cheshire and adjacent counties have been found nesting in milder climes in north Wales and Ireland.

The present survey shows the Greenfinch to be more plentiful in the west and north of the county, but with local concentrations in suburban areas such as Crewe, Nantwich and Sandbach, Wilmslow, Alderley Edge and Macclesfield. The availability of cultivated ground and associated weed seeds, whether in gardens or arable fields, appears important, for the species is scarce in the grazing land of the eastern hills and in the southern half of the plain where silage and pasture fields predominate. The neatly clipped hedges and scarcity of evergreen shrubbery in the latter area may also restrict the choice of nest-sites.

In the suburban area around Sutton Weaver, DN has suggested that recent increases in breeding Greenfinches may result from the provision of artificial food (peanuts) into mid-May, allowing more birds to survive their peak mortality period in April when weed seeds are scarce (Mead 1974). O'Connor & Shrubb's[42] suggestion, that a great increase nationally since the 1962-63 hard winter is due to increased cultivation of cereal and oilseed rape cultivation, is likely to have limited applicability in pastoral Cheshire. On the contrary the loss of elm seeds following Dutch elm disease has removed a significant food for nestlings.

Sharrock[35] estimated an average of 300-600 pairs per occupied 10-km square. National average CBC densities in 1980 were seven pairs per square kilometre on farmland and 20 pairs per square kilometre in woodland. Four CBC plots in south-east Cheshire totalling 167 hectares held ten territories in 1979, thirteen in 1980 and ten in 1981. The Bidston Hill CBC in Wirral, comprising 18.2 hectares of mixed, deciduous woodland, heath and scrub surrounded by urban and suburban districts, held between four and ten pairs (average 6.7) between 1975 and 1984[29]. On the other hand the species is scarce in the breeding season at Rostherne Mere NNR and at Alderley Park – both well-watched areas. Some Greenfinch colonies seem to move from one year to the next, and Newton[33] also states that many birds will move some distance (possibly into an adjacent tetrad) for further breeding attempts in any one season, so not all mapped tetrads may have contained birds each year. In 1965 W. T. C. Rankin found eight pairs in 240 acres at Noctorum where he had only found one pair in 1964.

Greenfinches do not carry food in their bills, and their swollen crops cannot be seen from any distance. Most of the 'probable' breeding records will have referred to nesting pairs. Most tetrads with 'probable' or confirmed breeding in the better populated north and west of the county probably hold between ten and 15 pairs. Elsewhere numbers are much smaller however and the county population may be around 4000 pairs.

REFERENCES

Boyd, A. W. (1931): On Some Results of Ringing Greenfinches. *British Birds* 24: 329-337.
Mead, C. J. (1974): *Bird Ringing*. BTO, Tring.

During 1978 to 1984 Goldfinches were encountered in 560 tetrads (83.58%).

Tetrads with breeding
  confirmed: 281 (50.18%)
  probable: 193 (34.46%)
  possible: 86 (15.36%)

Goldfinch
Carduelis carduelis

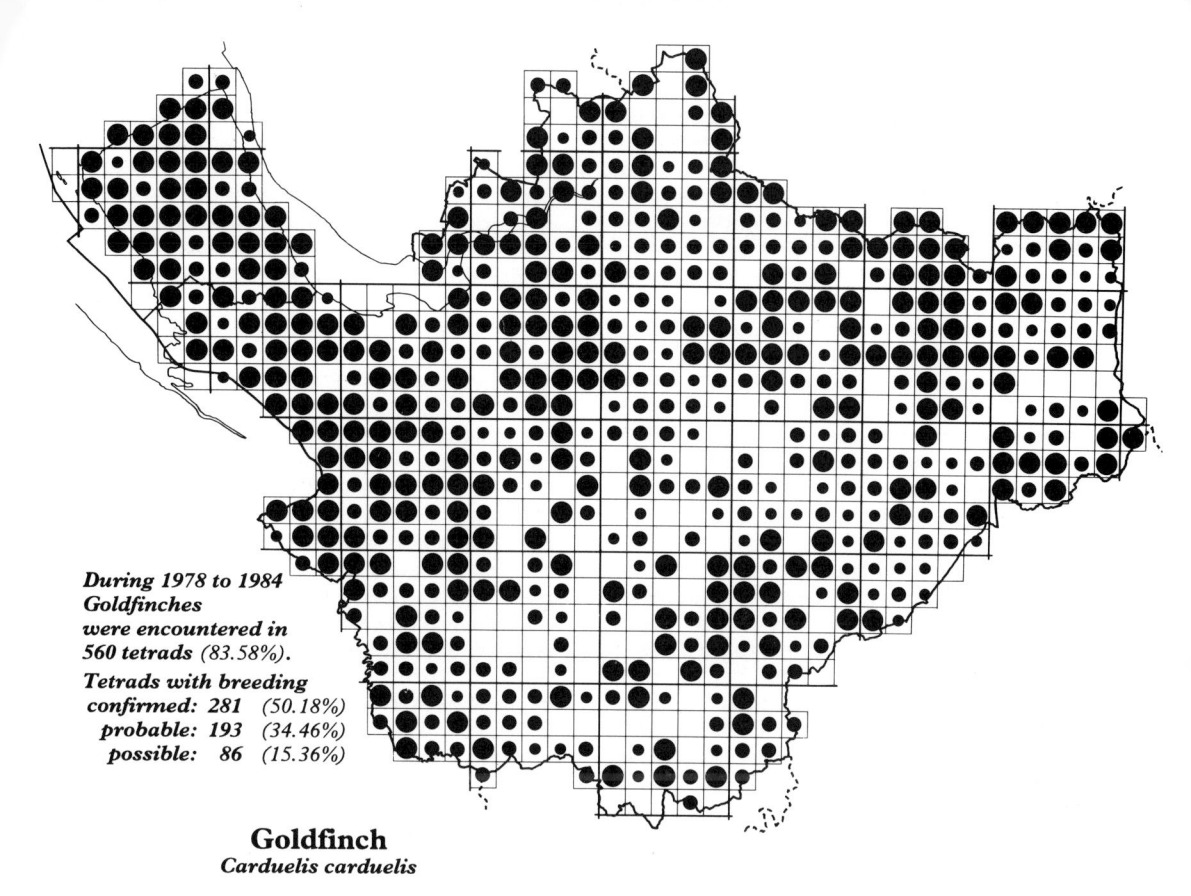

BEING one of our most attractive songbirds, Goldfinches formerly suffered severely from the activities of bird-catchers. Saving the Goldfinch was one of the first tasks declared by the newly formed Society for Protection of Birds (later to become the RSPB) at the end of the last century. Coward & Oldham[2] attributed the species' decline to the reclamation of waste-land with trapping being "instrumental in further reducing its numbers". It had then been rare in the county for many years and even exterminated in some districts. In the Liverpool district and in Wirral it had been scarce as early as 1850, and a nest at Bidston in 1864 was considered worthy of note. By the turn of the century no recent nesting records were known from the peninsula. In the county as a whole only odd pairs or small groups, sometimes under local protection, nested sporadically in perhaps a dozen sites. These were mostly in the west and south, but with isolated instances at Mobberley in 1890, Holmes Chapel in 1892 and Henbury in 1907. There were no records whatsoever of birds nesting in the eastern hills.

A general increase took place in the first 40 years of this century, Goldfinches deriving benefit from the spread of thistles as a result of deterioration in farm husbandry (Alexander & Lack 1944). It appears that a decline in catching during the war years of 1939-45 helped the species very greatly. Hardy[4] stated that odd, tame specimens near towns were generally escaped cage-birds, but by 1945 Griffiths & Wilson could report some recovery

275

in north Wirral, with nesting in seven named areas. Indeed between 1943 and 1949 there were many records of nests and of birds seen during the summer in other parts of Cheshire[3]. At Chester for example, where Williams[20] had known it as a rare nesting species prior to 1939, there was a marked increase by 1942, and birds were nesting in and around the city. The increase continued into the 1950s. By 1951 however, Boyd still knew the bird almost exclusively as a winter visitor to the Northwich area, but with summer records for the first time in 1950. During the 1950s it became much commoner in Wirral and nesting began to occur in the east, although it still remained scarce in the eastern hills[7].

Subsequently the species has continued to increase and by the early 1970s its prevalence was being noted in suburban areas throughout the county, to some extent replacing the Chaffinch which had by then declined in such habitats. Indeed areas on the urban fringe, now largely (but not entirely) devoid of bird-catchers, form ideal habitat. Goldfinches feed on the abundance of seed from thistles and other weeds of derelict farmland in this zone and even in urban waste-ground. Furthermore tall shrubs and ornamental trees in gardens provide ideal nest-sites. Flowering cherries are as popular with Goldfinches as they are with local authorities.

In winter the bulk of Goldfinch records are from the estuaries and coastal strip. Inland sightings are relatively few until April when there may be a marked easterly passage across the plain. From late April onwards birds appear on territory, the males singing from some prominent position, often on a tree top. JPG has notes of song between 4th March and 6th August, including a snatch of poor song on 28th July 1983 from a juvenile bird, still lacking its red face, in a larch tree that had held a nest. In Chester, Williams noted that during the period of expansion in the 1950s, birds had arrived at their breeding sites earlier, for example on 20th March 1957. In 1946 there were records of nest-building as early as 14th April at Kelsall and 19th April at Warford, and a nest with five eggs at West Kirby by 9th May. Young birds leave the nest from mid-June, although there is an exceptional record of three seen beside Poynton Pool on 12th May.

In common with other cardueline finches, the Goldfinch feeds its young by regurgitation rather than by carrying food in its bill. Consequently proof of breeding may be difficult to obtain until the young leave the nest. The male will often sing directly above the nest however, and pairs of finches visiting trees used earlier as song-posts may indicate the presence of a brood of young. The nest is typically built in a fork of twigs up to 20 feet or more from the ground, and nests in exposed sites are vulnerable to high winds. However, once leaves have fallen in the autumn, unsuspected nests, of moss and grasses and lined with hair and feathers, are often revealed.

No accurate counts of breeding birds are available, and, as the species is largely migratory (a juvenile ringed at Red Rocks on 15th July 1980 was retrapped at Meols on 7th September and subsequently caught in Spain (San Sebastian) on 3rd April 1981), flocks of dozens or even hundreds of birds seen in the autumn cannot be taken to reflect local breeding success. In the Chester area the population has been estimated at five pairs per occupied tetrad[28]. National CBC results for farmland average two to three pairs pairs per square kilometre[40] which is much too high for Cheshire. Sharrock's[35] suggestion of a density of 100 pairs per 10-km square (four pairs per tetrad) seems more likely to apply, giving a county total of 1500-2000 pairs.

REFERENCE
Alexander, W. B. & Lack, D. (1944): Changes in Status Among British Breeding Birds. *British Birds* **38**: 42-45, 62-69, 82-88.

Siskin
Carduelis spinus

UP until the early years of this century the Siskin was an irregular winter visitor to Cheshire, exceedingly scarce in Wirral. Griffiths & Wilson[13] knew of only two records from north Wirral; 12 at Thornton Hough on 13th March 1937 and a dead bird found at West Kirby in 1935. Even in 1962, its status was somewhat unpredictable, being abundant and widespread in some winters, scarce or even absent in others[7].

In 1976, Sharrock[35] referred to "a real increase and extension of range....during the last 30 years, especially since 1960". This increase was aided largely by the wholesale planting of conifers in the Scottish highlands and elsewhere, for conifer woods form typical nesting habitat. It seems likely that numbers visiting Cheshire increased in parallel over this same period: Coward & Oldham[2] quoted records spanning the period from 12th November to 29th March only, yet Bell[7] had records from 16th October to 16th May. The Siskin is now common as a Cheshire wintering bird, with large numbers on spring passage and smaller numbers in autumn. Ringing recoveries show some of these to be Scottish-bred birds, for example an adult male ringed at Bidston on 2nd April 1981 was controlled at Kiltarlity, Highland, in May of that year; a bird at Bidston on 30th April 1983 had been ringed in the nest at Lairg, Highland, on 21st May 1980; and other recoveries link Cheshire and Wirral with breeding areas elsewhere in Highland, Tayside, Strathclyde, and Dumfries and Galloway. Some of these birds may be returning by a northerly route to breeding grounds in Scandinavian spruce forests.

Sharrock[35] also suggested that it is probably through wintering Siskins lingering on in suitable haunts into spring that new sites are colonised. Thus the newly maturing plantations at Delamere and Macclesfield Forest began to offer suitable conditions around the middle of this century. A bird in song in Delamere on 11th May 1940 was considered to be possibly breeding[5], but it was not until 1962 that the first confirmation of breeding occurred. On 12th May of that year P. H. G. Wolstenholme saw a male bird feed a female on the nest in Delamere Forest, but the nest was later found to have been deserted[9].

In 1964 seven pairs were located in stands of mature pines in this forest, remaining throughout April. On 3rd May only three males could be found, but, on 4th July, R. P. Cockbain saw a male and a female with three young birds[8].

Subsequently there has been only one further breeding record reported. The 1976 *CBR* mentions a pair feeding young at Willaston, Wirral in June, but gives no detail. During this survey there was a sprinkling of records in April and May, and a bird seen at Kettleshulme on 1st August 1981.

As spring approaches, Siskin flocks, which have spent the winter in alder carrs, move increasingly into conifer plantations, attracted by the opening cones and accessible seeds of larch trees. Towards the end of March the buzzing, twittering song becomes more frequent, and by early April the cocks are seen to indulge in bat-like display flights around the tree-tops. Most birds disappear during April however, but the few that remain into May, generally amongst pines by this time, may well be breeding. The acreage of conifers in Delamere and Macclesfield Forest is huge, and few birds are present, but clarification of the breeding status of the Siskin, like the Crossbill, deserves more attention.

Linnet
Carduelis cannabina

AS with the Goldfinch, the activities of bird-catchers and the reclamation of waste-ground had reduced the status of the Linnet to that of a scarce breeder by the early years of this century. It was exceedingly local on the plain, but nested commonly around Chester and in some numbers on the gorse-covered parts of the Peckforton Hills and in open parts of Delamere Forest. Elsewhere on the plain small numbers bred where suitable habitat existed, for example the gorse scrub on Knutsford Heath, and sites between Nantwich and Audlem. The strongholds at that time were in the eastern hills and along the coast of Wirral where, however, it had probably decreased since Brockholes described it as abundant in 1872[2].

There had evidently been some increase by 1951 for Boyd[6] regarded Linnets as "common but of patchy distribution" around Great Budworth. A similar situation still prevails across much of the county with small scattered colonies in many areas. The burning of gorse is an increasing hazard in the eastern hills where numbers have probably declined in recent years.

Although mention of nesting Linnets often summons up visions of gorse-clad heathland, in Cheshire more pairs nest in neatly trimmed hedgerows than in any other habitat. Much former heathland is now planted with conifers, as in Delamere, or has become overgrown with birch scrub, as at Little Budworth and Lindow Commons. Nevertheless in Delamere young pine plantations perhaps six to ten feet in height hold many nesting pairs. Boyd[6] made notes on 33 nests: eight in gorse, eight in hawthorn, seven in holly, three in privet, three in laurel, etc.

Except around the estuaries and a few small roosts inland, Linnets are absent from the county in winter, returning to nesting areas in late March or April. In eastern parts of the county there is a marked eastward passage at this time in some springs. Eggs are laid from late April until June. They may breed into August, or even September on occasions. In 1969 a clutch of nine eggs was found on Bidston Hill, but five eggs is more usual.

As with other small finch species, confirmation of breeding is difficult unless young birds are seen away from the nest, although initial location of singing birds is relatively easy. From January to March occasional snatches of song are heard from birds detached from flocks, either perched or in flight, but as soon as flocks arrive on the breeding grounds a constant twittering is maintained. Linnets often nest colonially, with groupings of 5-10 pairs typical, as at Frodsham and Woolston (DN).

On Hilbre, where Griffiths & Wilson[13] recorded breeding in 1943 and 1944, a pair bred in 1960. Numbers have increased to 6-8 pairs in recent years, probably encouraged by tree-planting. Elsewhere six farmland CBC plots in south-east Cheshire held a total of 22 territories in 393 hectares in 1978. There are usually four pairs present in 73 hectares at Risley Moss. From one to three pairs breed each year on an 18-hectare census plot at Bidston Hill. It thus seems likely that many tetrads with confirmed breeding will hold at least ten pairs, giving a total population in Cheshire and Wirral of around 3500 pairs.

Elsewhere in Britain, the Linnet underwent a much-publicised decline from the mid-1960s, national CBC figures showing a 50% drop over 20 years, attributed to reduced supplies of weed seeds following greater use of herbicides. However, detailed habitat analysis shows the decrease to have occurred on arable land whilst populations

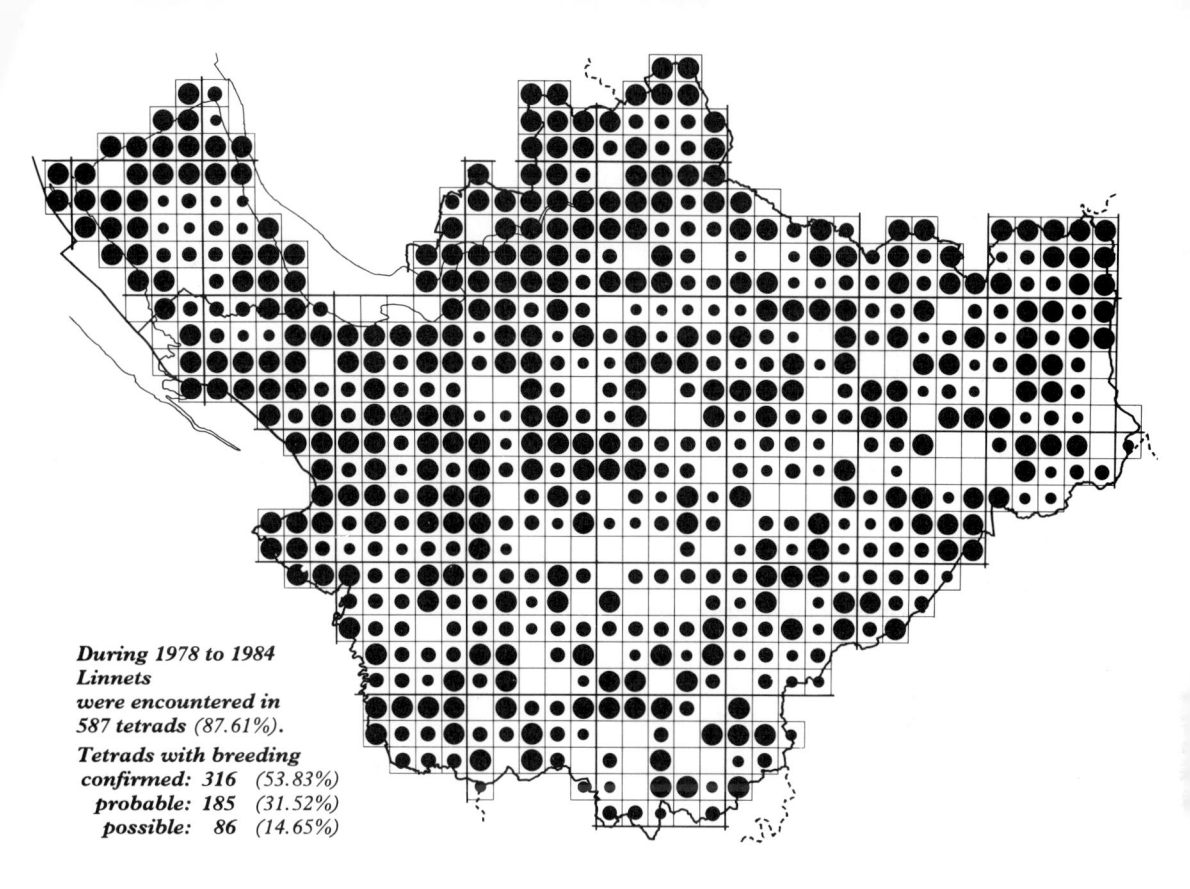

During 1978 to 1984 Linnets were encountered in 587 tetrads (87.61%).

Tetrads with breeding
- confirmed: 316 (53.83%)
- probable: 185 (31.52%)
- possible: 86 (14.65%)

were stable in grassland areas[42]. Cheshire may have escaped so far and, other than in the eastern hills, there has been little evidence of a drop in the county total. In any case, long-term population trends may be difficult to follow because birds move around from one year to the next. For instance, as a patch of gorse is killed off by hard weather a colony of Linnets will move to other nest-sites nearby. The species has benefited from one agricultural change, the increased acreage of oilseed rape. The large flocks that gather in rape-fields from mid-June are mainly first-brood juveniles but also include some adults gathering seed to feed their young.

Twite
Carduelis flavirostris

DESPITE the excellent work which Orford (1973) did to establish the breeding distribution and habitat preferences of the Twite, the species remains an exceedingly elusive and enigmatic member of the moorland fauna, although this is perhaps less true of the few birds breeding in the saltmarsh of the Dee estuary. It is such a drab bird, and so easy to confuse with its relatives at a glance, that there has been a strong suspicion that it has been overlooked in the moorland areas by surveys such as the present one. Its colonial behaviour, too, means that the species is more easily missed than one whose population is evenly spread through the preferred habitat.

In the current survey there were no definite breeding records from the moorland areas within the county boundary; only three tetrads with 'probable' and seven with 'possible' breeding. However there were two records, one confirmed breeding and one 'possible' breeding, from the Dee estuary where it has been a regular winter visitor since

TWITE

British breeding distribution, after Sharrock (1976)[35]

the mid-1950s. There is a regular easterly passage of birds across the county during April and the record from SJ64 was almost certainly such a bird.

The moorland records suggest a lower population than reported by Orford, who specifically mentions Whetstone Ridge and the Cat & Fiddle areas as breeding localities. Moreover in the Peak District generally he examined 128 nests in the six years 1964-1969 which seems far more than one could expect to find now. In a series of transects across moorland in which all birds were counted, Twites contributed only five out of 950 territories in 1980-81 (Yalden 1984). There is then a temptation to conclude that the species is less common now than it was 15 years ago. Against this, quite large flocks were sometimes recorded on the moors at the end of the breeding season: although these could have moved in from elsewhere, they suggest an overlooked, or at least underestimated, local population.

Orford pointed out that the species relies on moorland for nest cover, but does most of its feeding on nearby pastures. These might be 1.5 kilometres away, and when there are young in the nests there is a constant procession of adults back and forth along well defined flight-lines. This ought to make breeding birds conspicuous, but implies that birds seen in suitable habitat during the breeding season are possibly only feeding, with nests some distance away. Thus the temptation is to suppose that only the 'probable' breeding records refer to nesting birds, implying a population of only three pairs and certainly no more than ten pairs.

At the turn of the century the species was already confined to the hill country, though previously it had nested on Carrington Moss (now Greater Manchester) and the Lancashire mosses[2]. Throughout the period from the 1920s to the 1950s, however, the species was barely recorded, either as a breeding bird or as a winter visitor to the saltmarshes of the Dee estuary[7]. Then both wintering flocks on the coast and breeding birds on the hills became more common or at least more commonly recorded. There is some evidence to believe, moreover, that this was a genuine increase, although it certainly owes something to the activity of Noel Orford at this time[9]. The population seems to have increased through the 1960s and into the 1970s. By way of confirmation, similar changes were recorded in neighbouring counties, both for breeding areas and for wintering sites[45, 47]. More recently, at least to judge from county bird reports, there seems to have been a decline in both Derbyshire and Staffordshire, and this may well be reflected in the sparsity of records obtained for this survey. This species is certainly one where specific attention from bird-watchers is desirable.

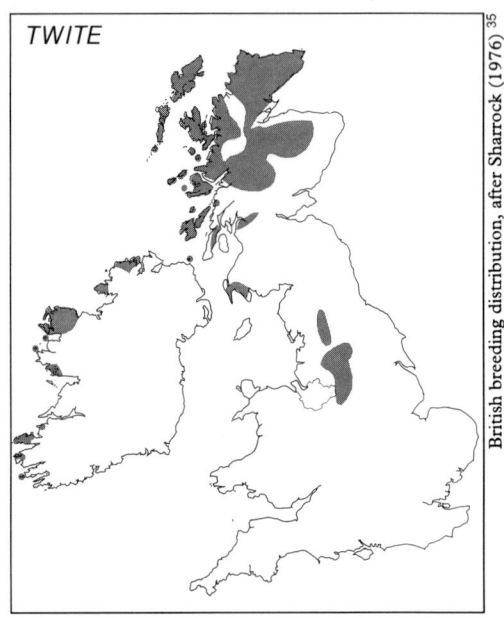

In the saltmarsh of the Dee estuary a pair was watched and a nest with three eggs found by J. R. Mullins on 13th May 1967 in the same tetrad as that shown on the map. Subsequent observations revealed five eggs with one adult present on 20th May but an empty nest on 27th. It was thought that the nest, which was six inches above the ground in a clump of sea purslane, had possibly been flooded by a high tide. In 1979, B. Barnacal

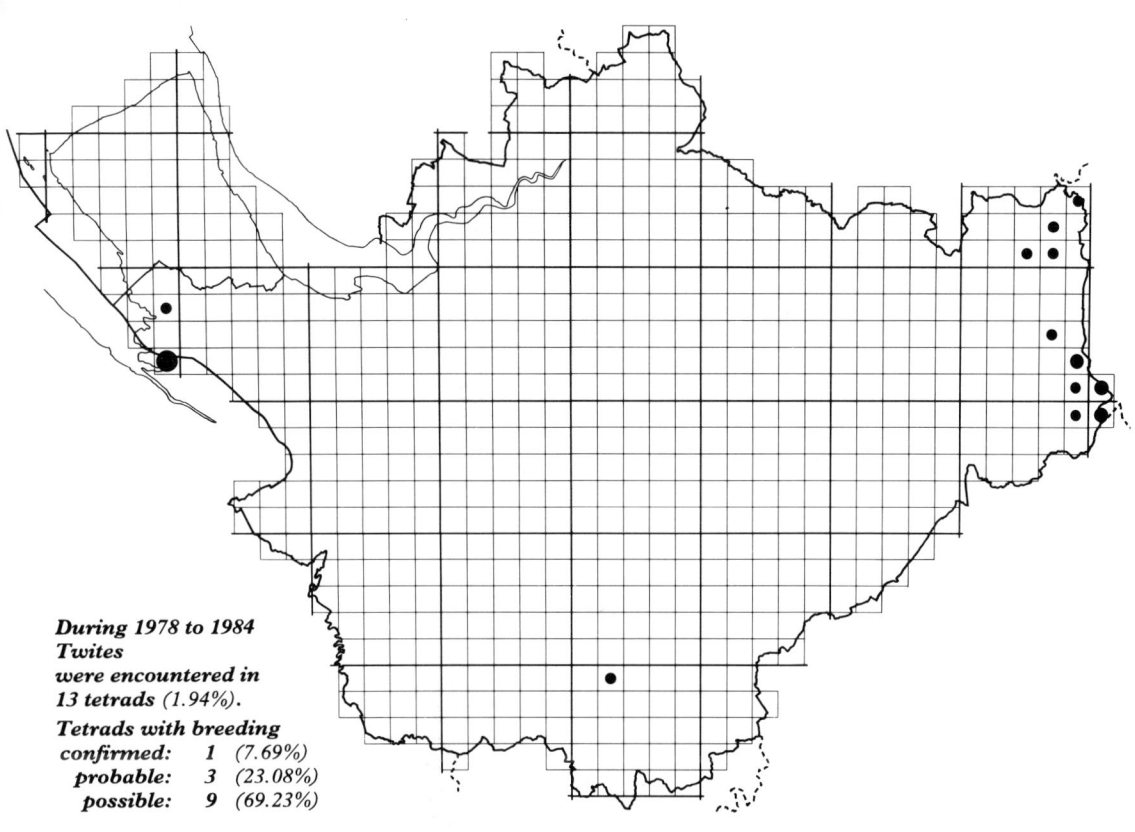

During 1978 to 1984 Twites were encountered in 13 tetrads (1.94%).

Tetrads with breeding
confirmed: 1   (7.69%)
 probable: 3 (23.08%)
 possible: 9 (69.23%)

saw a pair of adults with three recently fledged young at the edge of the *Spartina* marsh off Burton Point, in the same area as the 1967 record. In 1980, a single adult was seen in mid-June by E. J. Abraham on the *Spartina* marsh off Neston.

There was a series of records from beside the Mersey estuary at New Ferry and Frodsham Marsh in the mid-1970s. At New Ferry, six birds in August 1974 were thought to be a family party, a pair was seen with four recently fledged young in 1975, and birds were again present in June 1976. One was seen at Eastham on 11th May 1978. At Frodsham, where a bird had been seen on 10th July 1968, up to 30 or 40 were present at either end of 1975 and birds were present in summer, local breeding perhaps accounting for the winter flocks. In 1976, birds were reported from mid-April with 18 in July, "almost certainly breeding birds and their young"[8].

                                                                                            DWY

REFERENCE
    Orford, N. (1973): Breeding Distribution of the Twite in Central Britain. *Bird Study* **20**: 51-62, 121-126.

Redpoll
Carduelis flammea

FOR a time in the 1970s Redpolls increased to the point where they were undoubtedly commoner as a breeding species, in Cheshire as well as nationally, than at any time in the past hundred years[35]. While still inhabiting their traditional mossland birchwoods, Redpolls have spread this century into new conifer plantations where stands of trees between six and 15 feet tall are preferred.

Coward & Oldham[2] classed them as a "not uncommon resident", breeding in many parts of Wirral, throughout the Cheshire lowlands, and in the valleys and on the lower hills in the east of the county. At Wincle, Bosley and in other parts of the hill country the species was very plentiful. Abbott saw a flock of 50 at Alderley Edge on 10th May 1919. Nationally, there was "a widespread decline, leading to the species' complete disappearance from many lowland areas in the 1920s" [34], but it is not known what happened in Cheshire at that time.

Coward & Oldham also quoted from Byerley who stated that the Redpoll occurred around New Brighton in the summer months, and Brockholes who had found it to be a rather scarce resident in Wirral. In 1945, Griffiths & Wilson[13] also considered it somewhat local on the peninsula, although it was fairly plentiful in those districts where it did occur. Hardy[4] gives nesting records from Wirral at Arrowe Park, Irby, Burton and elsewhere. He linked its considerable increase as a breeding species this century, particularly in south Lancashire (now largely north Cheshire) to the spread of birch trees, on whose seeds the Redpoll feeds, as the mosslands dried out due to drainage. The *L&CFC Report* for 1949 reported breeding in urban locations at Queen's Park, Chester in 1943, at Prenton in 1944 and at West Kirby, where eggs hatched on 12th August 1948.

Bell[7] found the species to have a patchy distribution, including such suburban areas as those listed above plus Hartford. It nested in plantations in the lower eastern hills, but appeared to have deserted some localities during the previous decade. The *L&CFC Report* for 1963 recorded absence from the Nantwich district.

The 1966 and 1967 *CBRs* listed eight sites in Wirral holding 16 pairs, and in the following year reported that the Redpoll was increasing quite rapidly in the east of the county as well as in Wirral – in the Birkenhead area it had become quite a common garden bird. By 1970 it was considered common in several wooded areas in eastern Cheshire and from 1972 bred for several years at Dean Row, Wilmslow[21]. In 1975 many pairs were displaying in the birchwoods at Plumley lime-beds, and around Sandmere, Allostock[22]. Around the Sandbach Flashes, where Sibson recorded the species as being widely distributed in the 1930s, Whalley[19] could give no recent records, yet three or more pairs bred annually from 1973 to 1976[27].

By 1977 there was an impression of reduced numbers in eastern Cheshire, and two years later very few summering birds were reported in that area[8]. Sites in willow scrub along railway embankments and at sewage works which had held birds a few years previously were deserted, and the Redpoll had reverted to its former, patchy distribution centred on birch woods. The CBC plot on Bidston Hill, which supported eleven pairs in 1977 and 1978, held just three pairs in 1984. Also in 1984 absence was noted from gardens at Nantwich where the species had bred regularly in the past. However a pair bred on Hilbre for the first time in 1981.

It seems that many of our breeding Redpolls may be summer visitors. Breeding sites are occupied during late April and May, when vociferous flocks of displaying birds appear. This ties in with a marked passage at Hilbre at this time, as well as numerous, scattered records of northward- or eastward-bound migrants inland. For example, in 1981 DN noted eastward passage at Woolston from 10th April to 4th May. Similarly birds ringed there in August and September have been found to move quickly to south-eastern England, perhaps on their way to the continent.

Song is often delivered in flight, sometimes apparently while the birds are going about their routine business and without any obvious territorial meaning. At other times it is given from a perch, or the males may circle slowly around the tops of a stand of trees in display, singing as they go. This display gives some indication of nesting intent, though confirmation is difficult to obtain and generally dependent on encountering broods of newly-fledged young.

Few population figures are available. Many occupied tetrads probably held only one or two pairs. On the other hand the species will nest sociably with several nests close together. During this survey Redpolls

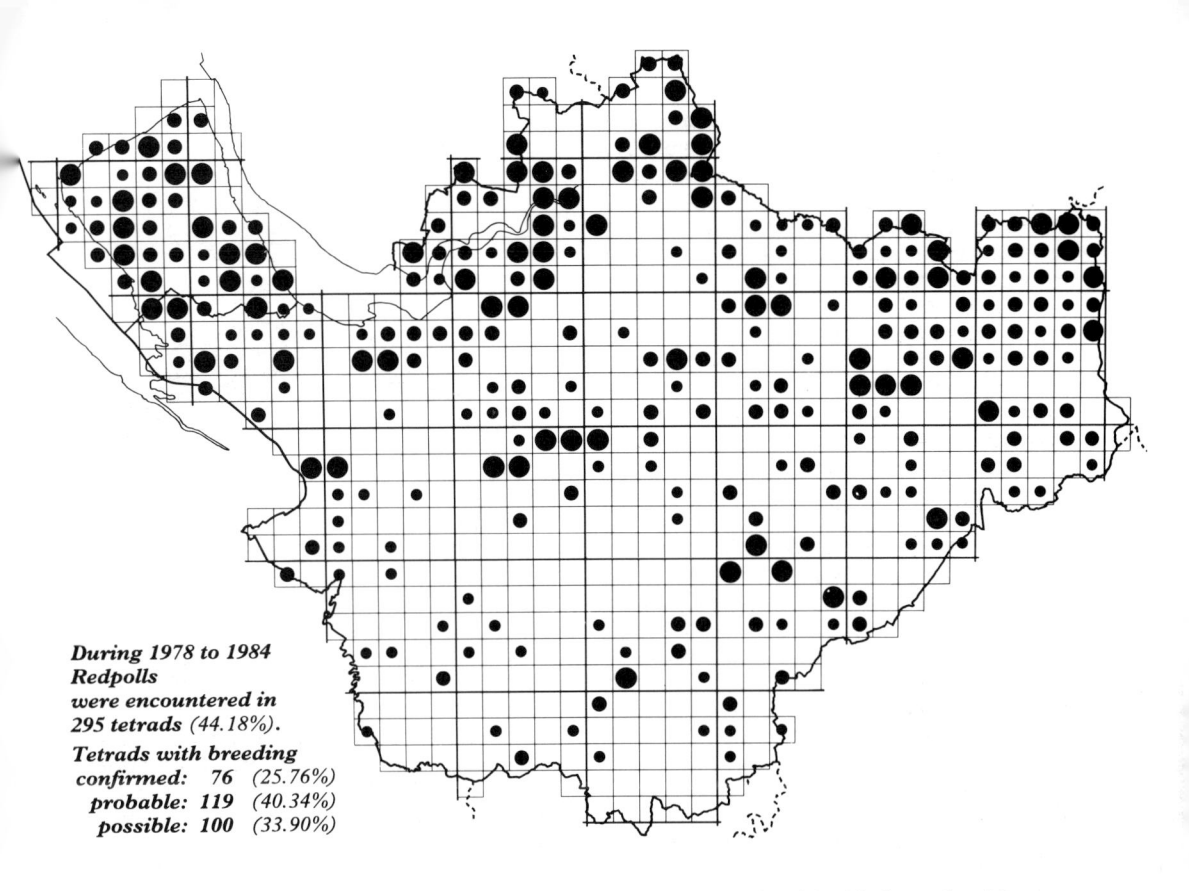

During 1978 to 1984 Redpolls were encountered in 295 tetrads (44.18%).

Tetrads with breeding
 confirmed: 76 (25.76%)
  probable: 119 (40.34%)
  possible: 100 (33.90%)

thrived in the alders and birches planted around the new towns of Runcorn and Warrington. Thus some 15 pairs bred at Woolston Park in 1983. Nearby, Risley Moss held four small colonies in 1982. Perhaps six or seven pairs bred in birch scrub with gorse at Alderley in 1983. The 293 occupied tetrads in Cheshire and Wirral may have supported between 500 and 800 pairs of Redpolls at the time of the survey.

Crossbill
Loxia curvirostra

ALTHOUGH Scots pines have been planted on Cheshire heaths since the eighteenth century, it was not until the 1920s that conifers were extensively planted in Cheshire, notably in the Macclesfield and Delamere forests. As these and other plantations in northern England and Wales matured, immigrant Crossbills from northern Europe settled on occasion to nest, and at present it is possible that a small population may be resident in the county.

Coward & Oldham[2] made no mention whatsoever of breeding, although parties of birds had been seen in the 1830s and again from the 1880s onwards. Even then records came most frequently from the Delamere area. A large irruption of continental birds occurred in the summer of 1909, and Bell[7] refers to evidence of breeding at Delamere in May 1910. A nest with eggs was found in a yew tree at Oulton Park in 1914 or 1915, and birds may have bred in 1917 at Delamere. In 1928 up to 30 birds were at Belgrave Moat Farm south of Chester between 6th April and 3rd May, and song was heard (T. S. Williams). Again at Delamere, nest-building was observed in late March 1931.

Subsequently infrequent waves of immigrants seem to have fizzled out without any further evidence of breeding until 1964 when up to 100 birds were present in Delamere in mid-April and R. P. Cockbain found a female feeding a fledged youngster in mid-May. In 1967 J. R. Mullins found seven or more pairs and five nests at a locality in north Wirral. Building was observed between 4th March and 16th April. In 1968 a bird was seen carrying nesting material in these same woods on 25th May. Nesting may also have occurred at Delamere in 1973.

No birds whatsoever were reported in Cheshire in 1976, but in June and July 1977 an irruption brought new blood into the county, and birds were noted in every year during this survey period. The only positive evidence of breeding came from south Wirral and from Macclesfield Forest in 1981, although small numbers were present at Oakmere in early spring 1984 and a flock of 46 there in May of that year contained young birds. It must be remembered however that Crossbills are often most active in early spring and the recording season did not officially begin until 1st April in each year, by which time birds may well have been sitting inconspicuously on eggs, and by the time most observers got into the field in May and June young birds may already have flown.

A flock was found in Scots pine at Burton, Wirral, in May 1981. Five birds ringed there on 19th May by D. Cross and A. Ormond included a female with an active brood-patch and two recently fledged juveniles with their bills only just starting to cross. In Macclesfield Forest in 1981 song was heard from early February, birds were seen carrying grasses in early March, and recently fledged young were seen on 17th April (JPG). The chief attraction here appears to be the seeds of larch which are particularly easily obtained in March and April as the cones open. In Delamere birds have been seen feeding from pine-cones between late April and late June.

Bullfinch
Pyrrhula pyrrhula

BETWEEN 1654 and 1674, church-wardens' accounts from Wilmslow, Goostrey and Rostherne show that a penny a head was paid in reward for Bullfinches[2] presumably because of their destructive habit of nipping off the buds from fruit trees, a habit that continues to this day: it is for its spring-time incursions to gardens that this retiring species is best known. In winter small parties may be encountered eating the seeds of birch trees, or feeding on hawthorn buds, though nettle and dock seeds or shrivelled blackberries are also sought, and the typical habitat at all times of year is scrubby woodland, shrubberies in gardens, or the overgrown hedgerows and thickets of small fields.

Coward & Oldham[2] knew Bullfinches as fairly common residents in the lowlands, but scarcer in the eastern hills where they occurred mainly in autumn and winter. However Abbott noted the species only three times between 1913 and 1930: one near Wilmslow on 16th October 1915, one at Soss Moss on 22nd May 1921, and two in Fulshaw Park on 12th June 1921. Boyd[6] regarded it as only thinly distributed around Great Budworth and, by 1951, had seen birds around Frandley on only nine occasions during many years of observation. Hardy[4] recorded an increase since the law against bird-catching in 1933, but though it nested in many woods in Wirral, as at Dibbinsdale, Irby, Storeton and Shotwick for example, it was not abundant. Griffiths & Wilson[13] also noted a considerable increase over the twenty years to 1945, and considered it then to be fairly common in north Wirral. Bell[7] was of

During 1978 to 1984 Crossbills were encountered in 3 tetrads (0.45%).

Tetrads with breeding
confirmed: 1 (33.33%)
probable: 0
possible: 2 (66.67%)

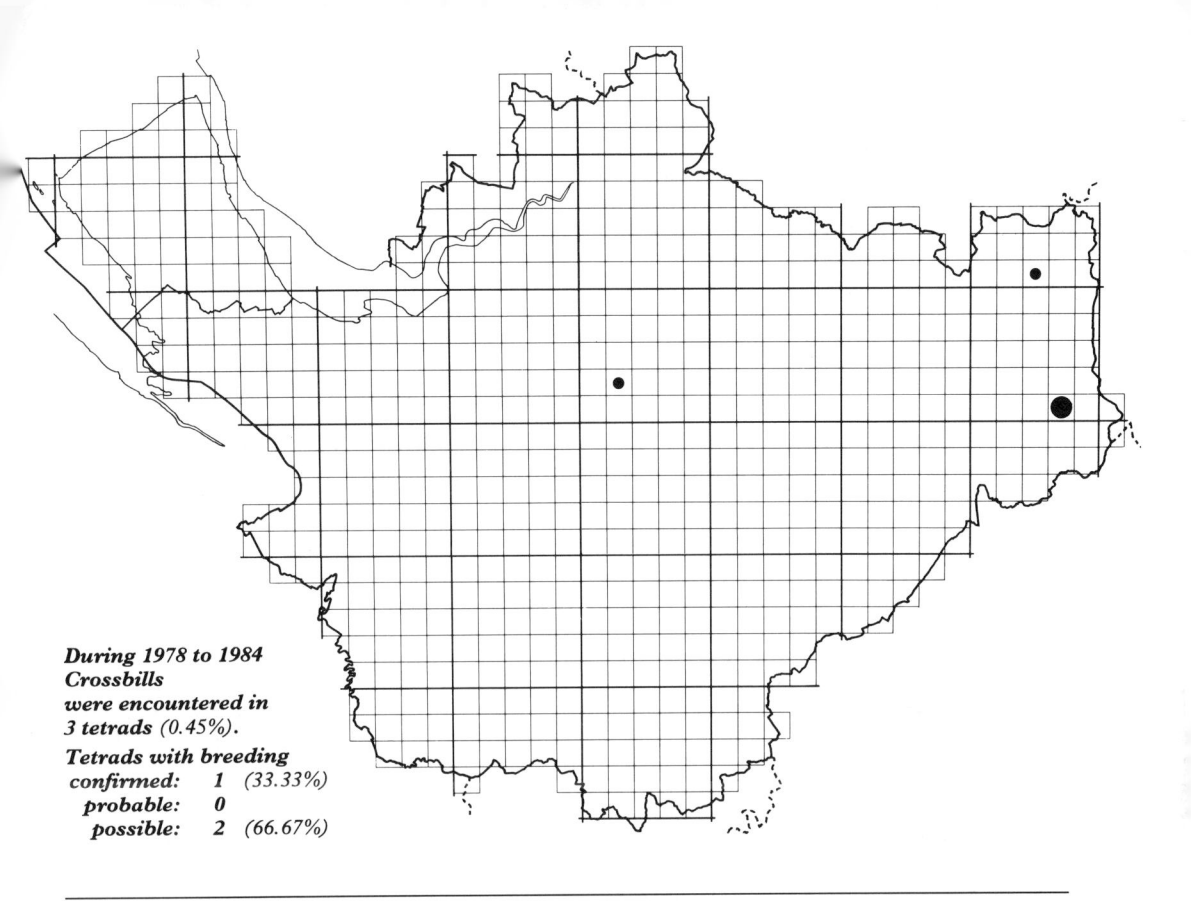

the impression that it had increased over the 15 years to 1962, and in his Supplement[9] refers to a marked increase in the Delamere area during 1961 and 1962. The species was by then widespread in Wirral. This increase occurred throughout Britain, with Bullfinches spreading into new habitats, especially gardens and town parks. In many rural areas they were increasingly noted in more open habitats and were no longer confined to thick cover[33].

Subsequently it appears that Bullfinches have increased in the eastern hills, helped in part by the planting of conifers. In June 1978 four males were uttering their song, a short undistinguished warble, from the tops of conifers in Wildboarclough, and on 25th February 1985 a similar performance was given by two males atop spruces in Macclesfield Forest. In both cases dense branches remained down to ground level, the trees having not yet been "brashed". In certain of the river valleys, as for example the Goyt at Disley, where thorn scrub is plentiful, the species now breeds regularly in small numbers.

Courtship displays may be seen at almost any time of year, Bullfinches appearing to pair for life. A pair ringed at Bidston on 24th April 1976 was retrapped there four times, the latest date – still together – being 15th July 1980. Tall thorn bushes are a favourite nest-site, and repeated visits by a pair to likely bushes will often indicate the location of a nest. Some males will bring food to their mate on the nest. Generally much difficulty is experienced in confirming breeding however, until young birds appear from early June, with fledglings from later broods still lacking their black caps into mid-September.

There is little information on the population of this unobtrusive species. Sharrock's[35] average figure of 200 pairs per 10-km square (8 pairs per tetrad) is unlikely to apply to many parts of Cheshire. The neatly tended hedgerows of the central plain provide little suitable cover for Bullfinches, and the species is particularly scarce from Winsford and Tarporley southwards to Audlem. The only figure available for breeding numbers is three pairs nesting at the Woolston Eyes in 1981. (24 birds ringed there between 15th July and 29th August 1981 were believed to have moved in from gardens in Grappenhall to the south.) In particularly favourable tetrads, for example the derelict market garden and scrubland areas around Fodens Flash, Sandbach, between five and ten pairs or more may be present, but many others probably hold only two to four pairs. The county population thus probably lies between 1500 and 2000 pairs.

Hawfinch
Coccothraustes coccothraustes

AROUND 1880 the Hawfinch was rare in Cheshire, but subsequently it increased such that Coward & Oldham[2] could describe it as "a resident, increasing in numbers; nowhere very common, but widely distributed". Since 1860, when it was first recorded as a Cheshire bird, it had become so generally distributed that a detailed account of its range was considered unnecessary. In Wirral, it nested annually in many localities and, though absent from treeless parts of the eastern hills, it occurred in some of the wooded valleys – at Wincle it was troublesome in gardens, presumably eating peas or destroying tree-buds. On the plain, game-coverts formed an invaluable habitat. Abbott's diaries contain references to a pair feeding young in a nest in Wilmslow Park on 24th May 1919 and one feeding a bird, probably its mate, in a nest about 30 feet from the ground at the end of a branch of a fir [sic] in Alderley Woods.

In the 1930s and 1940s, records came from localities spread throughout the county, though nowhere was it regarded as anything but scarce. In January and February 1943, L. P. Samuels found a party of 40 in Carrs Wood, Wilmslow, and, in mid-summer 1947, Boyd found a party of twelve at Over Tabley though generally numbers were much smaller. By 1949 it was "well distributed in Cheshire" and a nest with five eggs was found at Aldford near Chester in May 1945[3]. Nevertheless the species remained scarce into the 1960s[7] and has subsequently decreased considerably. Four young were reared in a nest in a dense beech tree at Lymm in 1970, leaving the nest on 3rd July. Otherwise there were no other breeding records until the start of this survey. Indeed there were very few records in any one year, and none at all for the first time in 1977.

This survey confirmed the scarcity of the Hawfinch in Cheshire. The stronghold in the Chester area furnished records from six tetrads. In 1983 up to 14 birds were present around Overleigh Cemetery in March and April, with display seen and song heard. Two adults were carrying nest material in late May, but with no further evidence of breeding. In the same year a pair reared three young at Eccleston, where two large young accompanied a female in 1984. Elsewhere a bird was seen in Ness Gardens on 14th April 1979, and in central Cheshire an adult and a fledged youngster were seen at Cuddington on 20th July 1978.

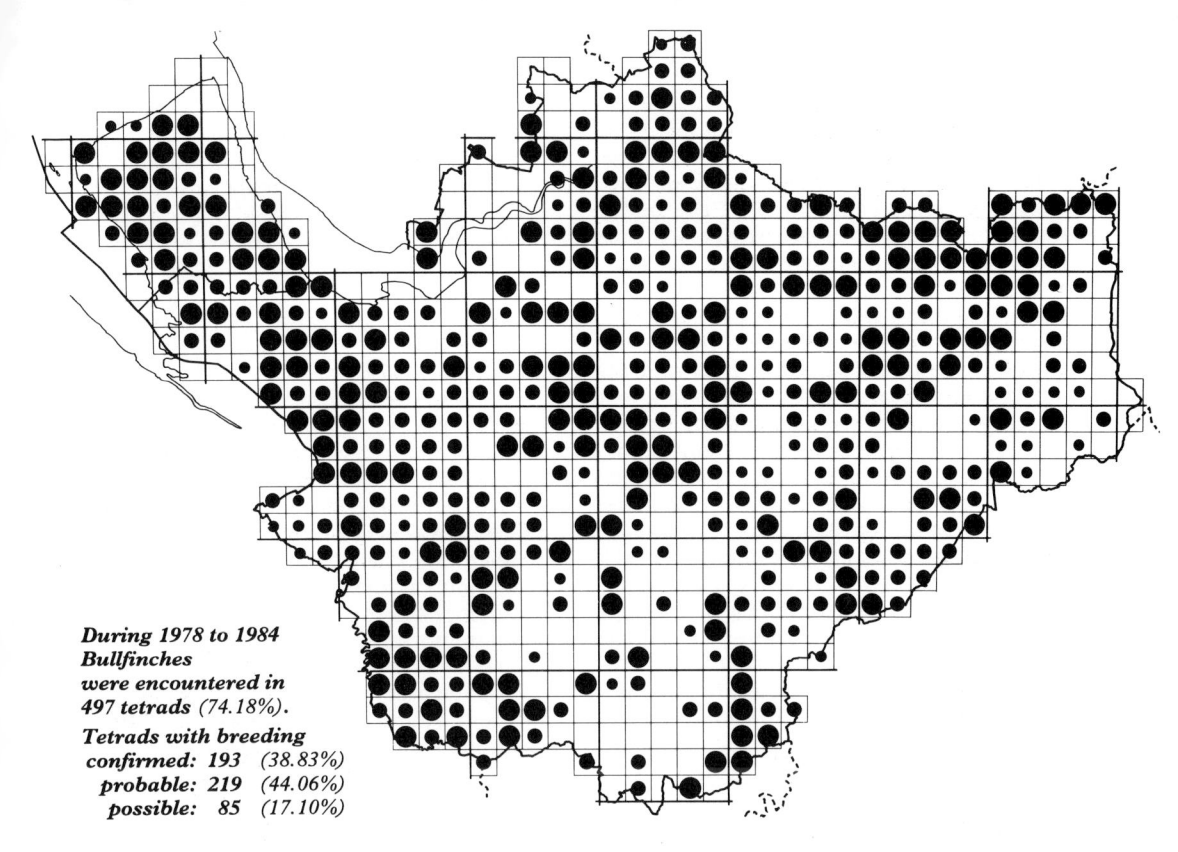

During 1978 to 1984
Bullfinches
were encountered in
497 tetrads (74.18%).

Tetrads with breeding
 confirmed: 193 (38.83%)
 probable: 219 (44.06%)
 possible: 85 (17.10%)

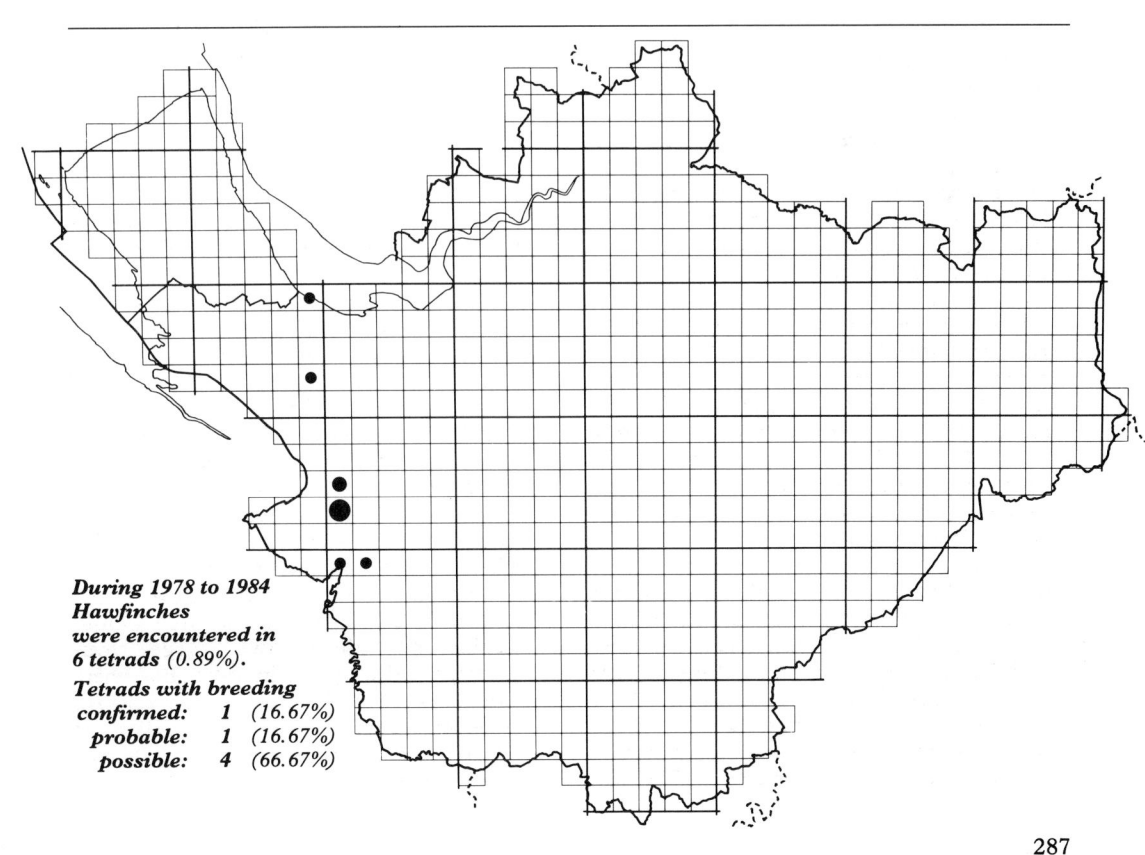

During 1978 to 1984
Hawfinches
were encountered in
6 tetrads (0.89%).

Tetrads with breeding
 confirmed: 1 (16.67%)
 probable: 1 (16.67%)
 possible: 4 (66.67%)

Yellowhammer
Emberiza citrinella

COWARD & Oldham knew the Yellowhammer as "one of the most familiar of Cheshire birds, being plentiful everywhere in the lowlands and on the hills up to the edge of the moors". In 1941, Hardy[4] regarded it as a common nester, mentioning the species' preference for bracken beds on heaths in Wirral and at Norton Priory, as well as its predilection for arable land. It also bred around the urban fringe and on the Wirral marshes, although it was less numerous on the peninsula than in the south Lancashire area, which then included Warrington. Griffiths & Wilson[13] however considered it to be not nearly so plentiful as formerly in north Wirral. "The spread of house-building, the regular lopping of hedges, and the frequent gorse-fires at such places as Caldy and Thurstaston" were thought to be probable causes of the decline. In mid-Cheshire, Boyd[6] found it to be "abundant at all seasons, especially on arable farms": in his country diary for February 1936 he had qualified this by stating that in a district such as Great Budworth, of small mixed farms, Yellowhammers were as abundant and widespread as any bird. In 1962, Bell[7] echoed almost exactly Coward & Oldham's description of the bird's distribution.

Subsequently no information on status was included in any county bird report from 1964 to 1972. In 1973 however the editor stated that it "seems to be declining still, it is now almost entirely missing from parts of central Wirral, where it was a common breeding bird a decade ago." The report of the Birkenhead School NHS for 1973 showed a decrease from ten territories on their census plot in 1963 to just one in 1973 as the area became increasingly built up. Craggs[25] recorded a noticeable decrease in sightings at Hilbre since 1970.

The map shows the species' avoidance of urban areas, this being particularly noticeable in north Wirral, but also around Chester, Runcorn, Warrington and Crewe. There are large gaps in the south of the county where pastoral farms with low, trimmed hedgerows characterise the landscape; and large parts of the eastern hills are now devoid of Yellowhammers. By 1983 the species was reported to have disappeared from an area to the south of Alderley Edge where until recently it had been common, and a marked reduction was evident in the Dean valley east of Wilmslow where formerly it had been one of the most numerous farmland species. At Rostherne, where several pairs were present annually up until 1976, the species is now seldom recorded[24].

A comparable decline has been noted in adjacent counties. In Derbyshire, Frost[45] mentioned some decrease "in the past two decades", perhaps attributable in part to destruction of hedgerows and tidying-up of farmland. In Greater Manchester, Holland et al.[46] reported "a serious contraction in distribution in the last 30-50 years" which they put down to urbanisation and hedge-clearance, although they added that rural tetrads to the north-east of the city are now without Yellowhammers. Neither urbanisation nor removal of hedgerows can be blamed for the current decline in eastern Cheshire. Although burning of gorse may have had some local impact, many areas which formerly held the species still appear eminently suitable. That stretch of the Goyt valley around Disley for example contains plenty of tall hedges but almost no Yellowhammers.

Early this century it was noted that the species was absent from the hills around Buglawton in winter, returning in mid-March[2], and into the early years of the present survey a small flock would arrive each spring at Lyme Handley. It may be that, with numbers in the lowlands reduced as a consequence of more intensive agriculture, birds are no longer "spilling over" into upland districts. The national atlas[35] shows an avoidance of upland parts of Scotland and the northern Pennines. Lowland birds appear to be relatively sedentary. Of approximately 1000 which Boyd ringed at Frandley, 192 were retrapped, all but three within one-and-

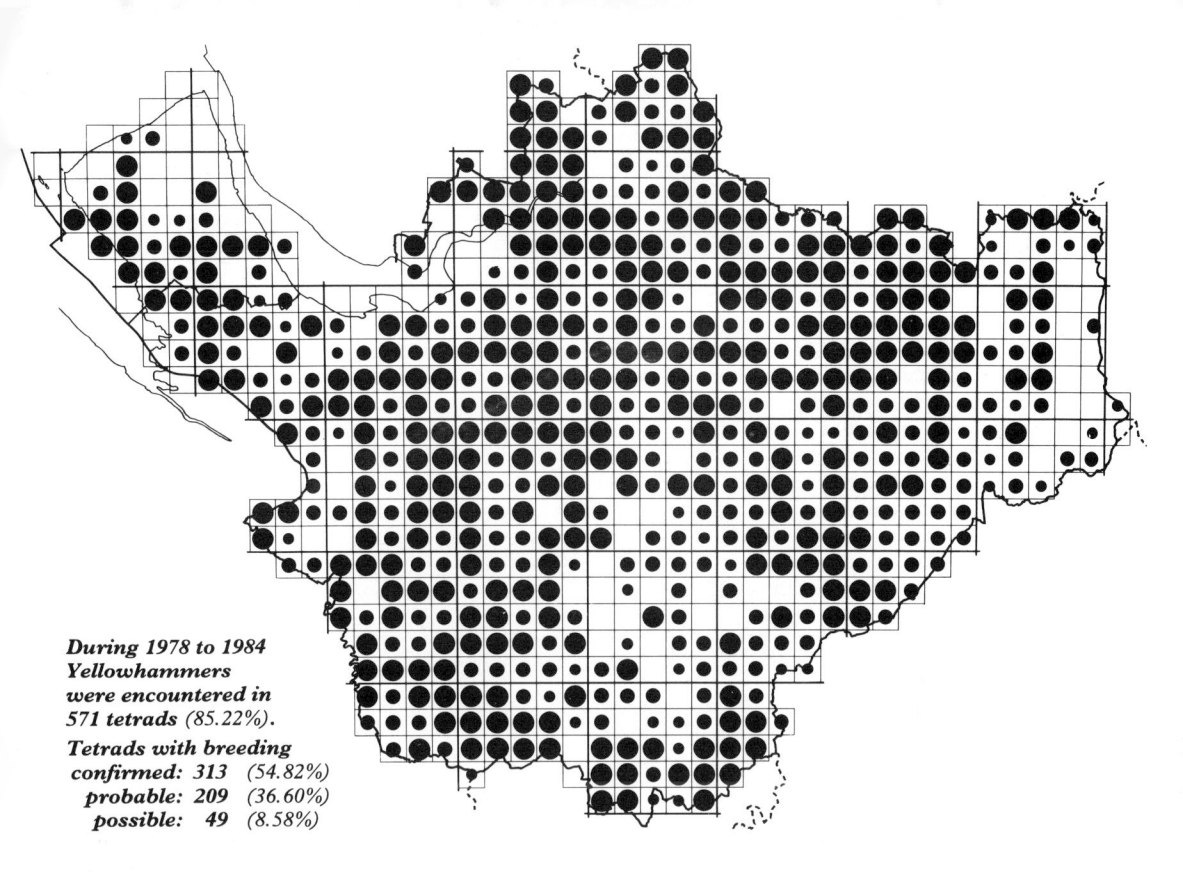

During 1978 to 1984 Yellowhammers were encountered in 571 tetrads (85.22%).

Tetrads with breeding
confirmed: 313 (54.82%)
probable: 209 (36.60%)
possible: 49 (8.58%)

a-half miles. The farthest movement was only eight miles. The oldest was "within a few months of at least 10 years old."

The cheery song of the Yellowhammer is one of the best known of all bird-songs, commonly written as, "a little bit of bread and no cheese", although early singers often leave off the "cheese". Both Coward & Oldham[2] and Boyd[6], who wrote extensively on this species, stated that it was first heard in most years about the second or third week of February, the latter author adding that it was sometimes delayed until early March or even later by severe weather. 3rd February was given as the earliest recorded date[2]. More recently the onset of song appears to have crept back by nearly a month. At Rostherne, song started in the third week of March, and, around the Sandbach Flashes, 7th March was given as the earliest song date[27]. JPG has several records from the first week of March, although in some years none is heard until the end of that month.

Birds often sing into August and the nesting season is also prolonged. Coward & Oldham had records of young in the nest on 22nd April and 5th September, although May and June are the chief nesting months[6]. Breeding is most easily proved between early June and August by watching adults carrying caterpillars and other food for their young. On 21st March 1943 an early nest was begun at Gorstage[3] and on 12th March 1978 a nest was said to be occupied at Mobberley (per A. C. Usher). On 9th September 1973 a bird by Watch Lane Flash, Sandbach, was carrying nesting material (JPG).

The male bird, when displaying to the female, will raise his yellow crest and flicker his wings and tail[6]. Later, when eggs or young are in the nest, the parent will feign injury in much the same fashion as a Reed Bunting.

Most nests are on or near the ground, in the bottom of hedges, on hedge-banks and roadside verges, or low down in gorse or other bushes, although a nest ten feet up in a copper beech was reported in 1946[3]. In Delamere, Yellowhammers nest commonly in young conifer plantations, although the species is relatively scarce in similar plantations in the east of the county. Clutches are small: of 36 nests examined by Boyd none held more than four eggs, with two or three quite usual. 22 broods averaged 2.27 young, with none larger than three. A number of local dialect names, such as "Scribbling Bunting" or "Writing Master" refer to the dark squiggles on typical eggs. Not all eggs are so marked however. Boyd recorded a clutch of two pale brown eggs and another of four pure white eggs. A nest with three pure white eggs at Mobberley in July 1974 was later found deserted[21].

In late February 1936, Boyd referred in his diary to birds singing "in every hedgerow" but, whilst there were many tetrads during the present survey where several males could be heard singing simultaneously, it seems doubtful whether such a statement could be justified anywhere in the county at present. Sharrock[35] estimated a national average of 300+ pairs per occupied 10-km square, equivalent to twelve pairs per tetrad. In 1984 just eight pairs were counted in the survey of a Saighton tetrad, and the CADOS bunting survey of 1978 concluded that a density of only four pairs per occupied tetrad was likely in the west of Cheshire. The six CBC plots in south-east Cheshire, amounting to 393 hectares, held 68 territories in 1978, five times Sharrock's estimated density, and 73 hectares of Risley Moss (including 25 hectares of woodland) held from eight to twelve pairs between 1978 and 1982. In Delamere, eight males were singing around the shores of Nunsmere in June 1984, and ten males in 15 hectares of young plantations at Hogshead Wood in 1982. The national farmland CBC index has remained remarkably constant from 1964 to 1984, with a mean density equivalent to 55 pairs per tetrad. Neither the constancy nor this density is typical of Cheshire. Allowing just five pairs for each tetrad with breeding confirmed and one or two pairs in tetrads with poorer evidence would give a county population of 1800-2050 pairs.

Cirl Bunting
Emberiza cirlus

COWARD & Oldham[2] had an August record from Flintshire within three and a half miles of the Cheshire boundary, but stated that the bird had never been seen within the county in the breeding season. Hardy[4] reported that, "a nest the Warrington Field Club noted at Rows Wood, Hatton, 1910, is in Warrington Museum". He added that the *Lancashire Naturalist* for 1917 stated that, "it occurs in low-lying lands between Warrington and Hale". However the authenticity of the account is brought into question by the later inclusion of a pair identified by their "rufous rumps" – unquestionably characteristic of Yellowhammers.

Reed Bunting
Emberiza schoeniclus

THE Reed Bunting occurs in summer throughout the county, almost wherever there is standing or running water. Riverbanks, lakesides, marl-pits, sewage works and shallow-sided reservoirs are all frequented, provided at least some small patch of rank vegetation exists in which to build the nest. Birds nest in some numbers along the saltmarshes of the Dee, and locally in rushy hollows in the eastern hills, though the species remains scarce on the sandy soils of the Delamere Ridge. Inland it is primarily a summer visitor, with birds filtering back from February in some years, but usually during March and April. Gradual reoccupation of territories is the rule, rather than a sudden arrival "en masse". However, migrant parties of up to a dozen brightly plumaged males may appear, with pipits and wagtails, on short turf

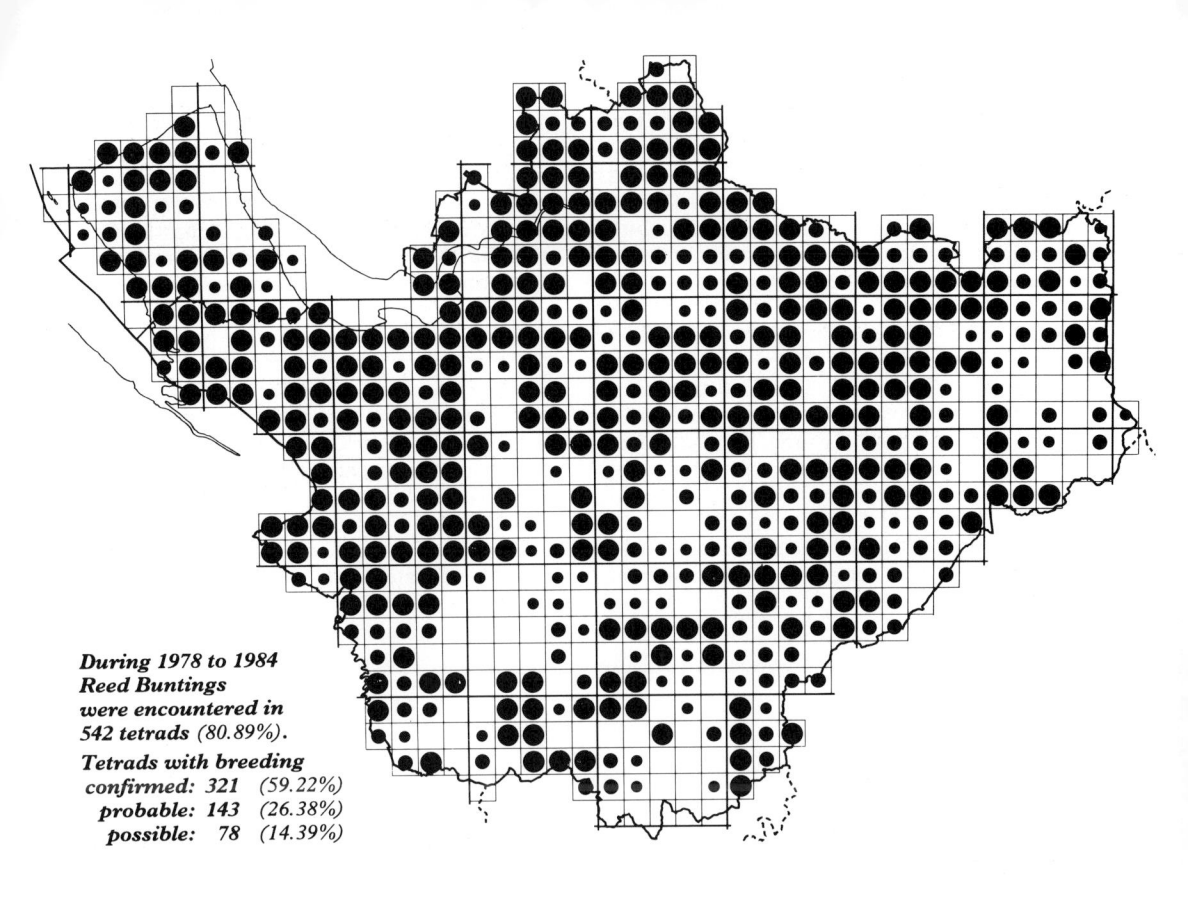

During 1978 to 1984 Reed Buntings were encountered in 542 tetrads (80.89%).

Tetrads with breeding
confirmed: 321 (59.22%)
probable: 143 (26.38%)
possible: 78 (14.39%)

beside the meres and flashes at this season.

Since the 1930s an increasing trend has developed in other parts of the country where birds will nest in drier habitats away from water[35]. Increased competition within the species due to the reduction of wetlands by drainage has been suggested as one cause of the spread: on the other hand reduced competition with the Yellowhammer, whose numbers have declined in recent years, may have allowed adaptation to this new habitat. Both changes may well also apply in Cheshire. Coward & Oldham[2] regarded "the reedy margins of the meres and the coarse herbage and undergrowth which fringe many of the marl-pits" as favourite nest-sites of the Reed Bunting. While such sites are still occupied, birds have nested since the early 1970s or earlier in farmland sites which would previously have been regarded as Yellowhammer territory, for example at Dean Row and Newton in 1972 (JPG) and at Mobberley in 1974[22]. Such hedgerow sites are generally adjacent to moist, unmown verges or damp hollows. However the occupation of young conifer plantations marks a more extreme stage in the relaxation of the Reed Bunting's habitat requirements: in 1979 birds held territory in two young plantations at Alderley Park, and on 20th May 1982 three males were singing, with Yellowhammers, in a plantation of pines some six to ten feet tall, on a sandy heath in Delamere.

Song is first heard towards the end of February (the earliest from two at Redesmere on 19th February 1983), and from then on the rather metallic stammer may be heard with increasing frequency as territories are occupied. From late March onwards, the males are to be seen chasing the females in display, though it is generally late April before nest-building begins. The nest is usually situated low down amongst reeds, sedges, rushes or rough grass, though others have been found four feet up in a bramble[13], in a small pile of thorn cuttings, in a triangle of briars, and twice on cinders by the Northwich Flashes[6]. A nest in short grass at a pathside at Woolston in 1980 was predated soon after the young had hatched.

Often the proximity of a nest is betrayed by the female, fluttering weakly over the sedges as though injured, or by the high-pitched squeaks of the nestlings. In mid-May and into June the adults are seen carrying beakfuls of flies to their brood, and from late

May onwards the young leave the nest. The pair may then rear a second brood, and song sometimes continues into July. A bird at Rostherne in 1973 sang until 29th July and was considered to be unmated. In 1977 a bird uttered a particularly disjointed song at Fodens Flash on 17th September.

By far the largest concentration in the county breeds in the shallowly flooded sludge-beds at Woolston. 107 males were singing there in 1983, and 89 in 1984. Fourteen pairs bred at Risley Moss in 1983, one more than in the previous year. Around the Sandbach Flashes 26 pairs were located in 1974; and 393 hectares of farmland CBC plots in south-east Cheshire held 22 pairs in 1978. Eight pairs were found in a tetrad at Saighton in 1984 and 14 males sang around Pickmere on 17th June 1981. The 1976 *CBR* included an estimate of 50 pairs around the Neston reed-bed – though a count of 16 territorial pairs there in 1980 must be considered more accurate. Numbers breeding along the Dee marshes were halved by severe weather in December 1981, but information from other areas is sparse as to the winter's effects. Many of the county's birds, particularly the smaller females, try to avoid problems with severe weather by migrating, either to the coast or to the south-west, and there are records of locally-ringed birds wintering in Dyfed, Somerset, Devon and Dorset. Many tetrads will have held five to ten pairs, though some will have held fewer than this. Allowing for local concentrations, the county population may be estimated at 2500-3000 pairs.

Corn Bunting
Miliaria calandra

A characteristic bird of arable farmland, the Corn Bunting may be encountered in many lowland areas where potatoes or cereals, especially barley, are grown. The males sing conspicuously from posts or wires.

Song may be heard during the winter months when the birds are still in flocks. JPG has noted song occasionally from mid-December until February, then more frequently from March until early August. Many Corn Buntings time their breeding to coincide with the late summer abundance of seed and grain. Indeed, many observers finished their season's recording before these buntings began to feed their young, and, as nest-sites in or around cornfields are not accessible for searching, the low proportion of records of confirmed breeding is not surprising. Most records of 'probable' or confirmed breeding will represent nesting pairs or groups however. This species tends to nest in small groups and some males are polygamous.

The Corn Bunting is never known to have bred in the eastern hill-pastures but Coward & Oldham[2] considered its absence there surprising as it then nested plentifully in similar areas of northern Derbyshire from which it is now absent. In fact the parts of Derbyshire then favoured, between Brough and Tideswell, were all on limestone rather than the grit towards the Cheshire border[45]. In Cheshire its strongholds were along the Mersey valley eastwards from Warrington, around Wallasey in Wirral, and along the enclosed parts of the Dee marshes from Heswall to Saltney. It was present in one district in the eastern part of the plain. Boyd, writing in 1936, described it as "one of the least common birds" on the plain, although it was frequently met with nearer the coast. Hardy[4] said it nested widely over the mosslands around Warrington, specifically naming Houghton Green, Croft and Woolston.

A colony became established at Byley (SJ76) in the 1950s and this may have marked the start of an upsurge in the species' fortunes although in 1962 Bell[7] stated that it remained scarce on the plain, had disappeared from areas in the north-east of the county (presumably the Mersey valley, now in

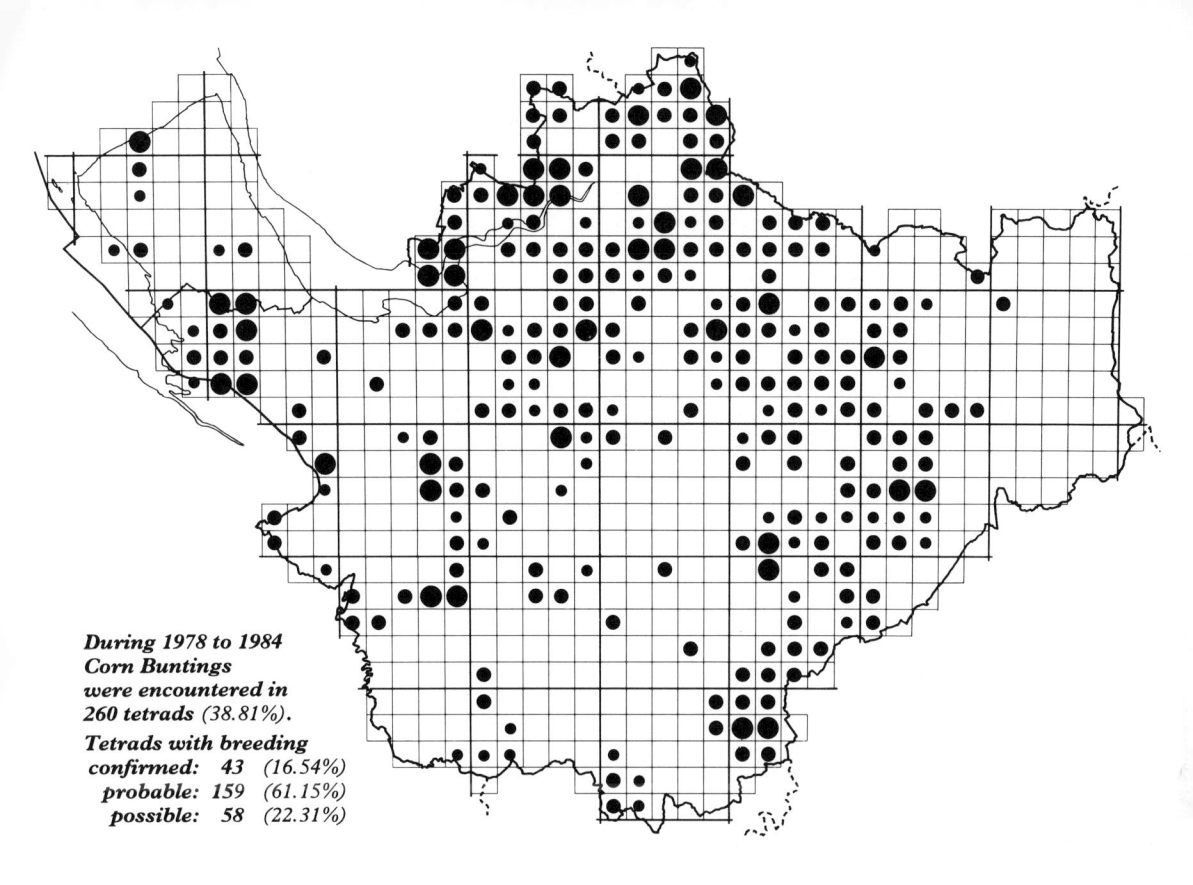

During 1978 to 1984 Corn Buntings were encountered in 260 tetrads (38.81%).

Tetrads with breeding
confirmed: 43 (16.54%)
probable: 159 (61.15%)
possible: 58 (22.31%)

Greater Manchester) where Coward & Oldham had found it abundant, and then had its main stronghold on the Mersey marshes around Frodsham and Runcorn. Just one year later, the *L&CFC Report* clarified the situation somewhat, describing two main stocks in the county. Firstly, birds bred from Mount Manisty through Frodsham and up the Weaver valley to Dutton, and along the Mersey valley through Runcorn to Altrincham; and secondly on the reclaimed Dee marshes around Shotton, Puddington and Burton. Elsewhere there were reports that year from several sites around Northwich, from Knutsford, Ashley, and on one day only at Prestbury. 1966 and 1967 *CBRs* detail a considerable extension of the species' range in the eastern half of the plain and into south Cheshire. In 1966 P. Schofield found over 20 pairs in the Antrobus area where Boyd[6] had seen only two birds in many years.

More recently there is evidence of a reversal of this trend in the east of the county where territories south-east of Wilmslow and near Danes Moss have been deserted during this survey, and winter flocks appear to be dwindling. In the Byley/Middlewich area DE found birds frequently in the late 1970s but seldom since. Use of chemical sprays on crops may have reduced the supply of weed seeds and insect food for the young, or alternatively earlier harvesting of cornfields may be interrupting the breeding process.

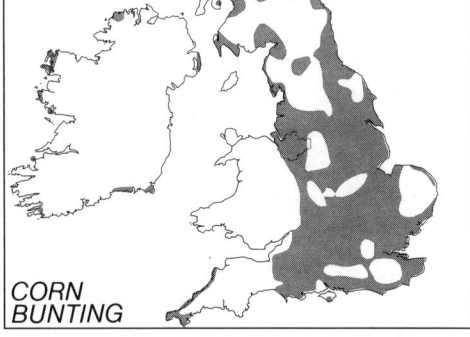

The map shows clearly the distribution of arable farming in the county, with Corn Buntings being absent from the hill-pastures in the east and from the intensive dairying districts of the southern plain. The bunting survey organised by CADOS in 1978 revealed no more than one pair per occupied tetrad. Averages for the county as a whole may be slightly higher than this, and a few tetrads may have held five or more territorial males, but altogether the county population is unlikely to exceed 450 pairs.

CORN BUNTING

British breeding distribution, after Sharrock (1976)[35]

APPENDIX I

SCIENTIFIC NAMES OF ANIMALS MENTIONED IN THE TEXT

Mollusca
- Lamellibranchia:
  - family: Mytilidae — mussels
  - family: Cardiidae — cockles
- Gastropoda:
  - family: Helicidae
    - *Cepaea hortensis* — Garden Snail

Annelida: Oligochaeta
- family: Lumbricidae — earthworms

Arthropoda: Insecta
- Ephemeroptera:
  - family: Ephemeridae — mayflies
- Plecoptera: — stoneflies
- Orthoptera:
  - family: Gryllidae — crickets
- Hemiptera:
  - family: Aphididae — greenflies; aphids
    - *Hyalopterus pruni* — Plum-reed Aphis
- Coleoptera: — beetles
  - family: Carabidae — ground beetles
  - family: Curculionidae — weevils
- Trichoptera: — caddisflies
- Lepidoptera: — butterflies; moths; (caterpillars)
  - family: Geometridae (Larentiinae)
    - *Operophtera brumata* — Winter Moth
- Diptera: — flies
  - family: Chironomidae: midges
  - family: Tipulidae — craneflies; (leather-jackets)
  - family: Syrphidae — hoverflies
- Hymenoptera:
  - family: Formicidae — ants
  - family: Tenthredinidae — sawflies

Chordata: (Vertebrata):

Pisces — fishes
- Salmoniformes:
  - family: Esocidae
    - *Esox lucius* — Pike

Amphibia — amphibians
- family: Ranidae — frogs

Aves — birds
(Additional to those species given main treatment in the text.)
- Procellariiformes:
  - family: Procellariidae
    - *Fulmarus glacialis* — Fulmar
- Anseriformes:
  - family: Anatidae
    - *Mergus merganser* — Goosander
- Galliformes:
  - family: Phasianidae
    - *Gallus gallus* — Red Jungle Fowl (farmyard fowl; chicken; "hen")
    - *Pavo cristatus* — Blue Peafowl ("Peacock")
- Charadriiformes:
  - family: Scolopacidae
    - *Tringa ochruros* — Green Sandpiper
  - family: Sternidae
    - *Chlidonias niger* — Black Tern

Mammalia — mammals
- Insectivora:
  - family: Soricidae
    - *Sorex araneus* — Common Shrew

Mammalia (continued)
    family: Talpidae
        Talpa europaea   Mole
Chiroptera: bats
Primates:
    family: Hominidae
        Homo sapiens   Human ("Man")
Lagomorpha:
    family: Leporidae
        Oryctolagus cuniculus   Rabbit
Rodentia:
    family: Sciuridae
        Sciurus carolinensis   Grey Squirrel
    family: Cricetidae
        Microtus agrestis   Field Vole ("Short-tailed Vole")
        Clethrionomys glareolus   Bank Vole
    family: Muridae
        Apodemus sylvaticus   Wood Mouse
        Mus musculus   House Mouse
        Rattus spp.   rats
Carnivora:
    family: Canidae
        Canis familiaris   Dog
        Vulpes vulpes   Fox
    family: Mustelidae
        Mustela erminea   Stoat
        M. vison   Mink
Artiodactyla:
    family: Suidae
        Sus scrofa var.   Pig
    family: Bovidae
        Ovis aries   Sheep
        Bos taurus   Cattle

APPENDIX II

SCIENTIFIC NAMES OF PLANTS MENTIONED IN THE TEXT

Lichenes   lichens
    Hypogymnia physodes
    Evernia prunastri
    Parmelia saxatilis

Fungi   fungi
    Ceratocystis ulmi   Dutch Elm Disease Fungus
    Piptoporus betulinus   Razor-strop Fungus
    Laetiporus sulphureus   Sulphur Polypore

Bryophyta: Musci   mosses

Pteridophyta: Pteropsida   ferns
    Filicales:
        family: Hypolepidaceae (Dennstaedtiaceae)
            Pteridium aquilinum   Bracken
        family: Dryopteridaceae (Aspidiaceae)
            Dryopteris spp.   buckler ferns

Spermatophyta: Angiospermae
    Dicotyledones:
        family: Ranunculaceae
            Ranunculus auricomus   Goldilocks Buttercup
        family: Fumariaceae
            Fumaria officinalis   Fumitory
        family: Cruciferae
            Brassica oleracea   Cabbage; Brussels Sprout
            B. napus   Oilseed Rape
            B. rapa   Turnip
            Raphanus sativus   Radish

Dicotyledones: (continued)
    family: Tiliaceae
        Tilia x *vulgaris* — Common Lime
        T. cordata — Small-leaved Lime
    family: Aquifoliaceae
        Ilex aquifolium — Holly
    family: Aceraceae
        Acer pseudoplatanus — Sycamore
        A. campestre — Field Maple
    family: Papilionaceae
        Ulex spp. — gorse
        Trifolium spp. — clovers; trefoils
        Lotus spp. — trefoils
        Arachis hypogea — Groundnut ("Peanut")
        Pisum spp. — peas
    family: Rosaceae
        Prunus laurocerasus — Laurel
        P. spinosa — Blackthorn
        P. avium — Wild Cherry
        Rubus fruticosus — Bramble; Blackberry
        Rosa spp. — wild roses; briars
        Sorbus aucuparia — Rowan (Mountain Ash)
        Pyrus communis — Pear
        Malus sylvestris — Crab Apple
        M. domestica — Cultivated Apple
        Crataegus monogyna — Hawthorn
    family: Grossulariaceae
        Ribes uva-crispa — Gooseberry
    family: Onagraceae
        Epilobium spp. — willowherbs
    family: Umbelliferae
        Oenanthe crocata — Hemlock Water Dropwort
    family: Araliaceae
        Hedera helix — Ivy
    family: Caprifoliaceae
        Sambucus nigra — Elder
    family: Rubiaceae
        Galium aparine — Goosegrass
    family: Compositae
        Helianthus annuus — Sunflower
        Petasites hybridus — Butterbur
        Senecio spp. — ragworts
        Carduus spp. — thistles
        Cirsium spp. — thistles
    family: Ericaceae
        Vaccinium myrtillus — Bilberry
        V. oxycoccus — Cranberry
        Calluna vulgaris — Ling/Heather
        Erica spp. — heathers
        Rhododendron ponticum — Rhododendron
    family: Empetraceae
        Empetrum nigrum — Crowberry
    family: Primulaceae
        Primula veris — Cowslip
    family: Oleaceae
        Fraxinus excelsior — Ash
        Ligustrum ovalifolium — Privet
    family: Solanaceae
        Solanum tuberosum — Potato
    family: Scrophulariaceae
        Melampyrum pratense — Common Cow-wheat
    family: Chenopodiaceae
        Chenopodium spp. — goosefoots
        Beta vulgaris — Sugar Beet
        Salvadora persica — Salt-bush
        Halimione portulacoides — Sea Purslane
    family: Polygonaceae
        Rumex spp. — docks
        Rheum rhaponticum — Rhubarb
    family: Euphorbiaceae
        Mercurialis perennis — Dog's Mercury

Dicotyledones: (continued)
 family: Ulmaceae
  Ulmus glabra — Wych Elm
  U. procera — English Elm
 family: Betulaceae
  Betula pendula — Silver Birch
  B. pubescens — Downy Birch
  Alnus glutinosa — Alder
 family: Corylaceae
  Corylus avellana — Hazel
 family: Hippocastanaceae
  Aesculus hippocastanum — Horse Chestnut
 family: Fagaceae
  Quercus robur — Pedunculate Oak
  Q. petraea — Sessile Oak
  Q. rubra — Red Oak
  Q. cerris — Turkey Oak
  Castanea sativa — Sweet or Spanish Chestnut
  Fagus sylvatica — Beech
  F. s. purpurea — Copper Beech
 family: Salicaceae
  Salix spp. — willows; sallows
  S. fragilis — Crack Willow
  S. viminalis — Common Osier
  S. cinerea — Common Sallow
  S. caprea — Goat Willow or Great Sallow
  Populus nigra betulifolia — Black Poplar

Monocotyledones:
 family: Iridaceae
  Iris pseudacorus — Yellow Flag
 family: Liliaceae
  Hyacinthoides non-scripta — Bluebell
 family: Juncaceae
  Juncus spp. — rushes
 family: Typhaceae
  Typha latifolia — Reedmace; "Bulrush"
  T. angustifolia — Lesser Reedmace
 family: Cyperaceae
  Eriophorum angustifolium — Common Cotton-grass
  E. vaginatum — Hare's-tail Cotton-grass
  Carex spp. — sedges
 family: Gramineae
  Spartina anglica — Common Cord-grass
  Molinia caerulea — Purple Moor Grass
  Phragmites australis — Common Reed
  Triticum sativum — Wheat
  Hordeum murinum — Barley
  Bambusoideae — bamboos

Spermatophyta: Gymnospermae — conifers
 family: Pinaceae
  Picea abies — Norwegian Spruce
  P. sitchensis — Sitka Spruce
  Larix spp. — larches
  L. decidua — European Larch
  Pinus sylvestris — Scots Pine
  P. nigra — Corsican Pine
  Tsuga heterophylla — Western Hemlock
 family: Taxaceae
  Taxus baccata — Yew
 family: Cupressaceae
  Cupressus spp. — cypresses
  x *Cupressocyparis* spp. — cypresses
  Chamaecyparis spp. — cypresses
  Cedrus spp. — cedars

REFERENCES

Clapham, A. R., Tutin, T. G. & Warburg, E.F. (1958): *Flora of the British Isles.* Cambridge Univ. Press.
Stace, C. A. (1991): *New Flora of the British Isles.* Cambridge Univ. Press.

APPENDIX III

a.) GAZETTEER OF PLACE NAMES IN CHESHIRE AND WIRRAL

THERE follows a comprehensive list of Cheshire and Wirral place-names with a tetrad designation for each.

Many names will be found on Ordnance Survey maps; other names - those of parishes for instance - are given, being commonly used in older books about Cheshire's wildlife. The larger places are referred to by a tetrad near to the centre. Most Cheshire meres, pools and reservoirs are listed, and also many woods, parks, farms and country houses where these have either ornithological significance or are sited in isolation from better-known places on the map, or because their names are duplicated at widely different localities.

A description of the atlas survey's Recording Area is given on page 31; places within Cheshire but outside the Recording Area are given in brackets, other place-names within the Recording Area but just over the boundary in neighbouring counties are followed by the name of that county.

This gazetteer was compiled by Alan Hunter, who acknowledges the co-operation of the 10-km square co-ordinators (page 36) and of Steve Barber.

The 100-km reference is SJ unless otherwise stated.

Place	Ref	Place	Ref	Place	Ref
Abbot's Meads	36Y	Aston, nr Runcorn	57P	Betchton Heath	76Q
Abbot's Moss	56Z	Aston-by-Budworth	67Z	Betley Mere (part Staffs.)	74N
Acre Nook SQ (Mosses SQ)	87A/B/F/G	Aston Heath	57U	Bewsey	58Z
Acre Nook Wood (Mosses Wood)	87G	Aston juxta Mondrum	65N	Bewsey New Hall	58Z
Acton, nr Nantwich	65G	Audlem	64L	Bewsey Old Hall	58Z
Acton Bridge	57X	Austerton	64P	Bexton	77N
Adder's Moss	87T			Bickerton	55B
Adlington	98A	Bache	46E	Bickerton Hill	45W
Agden, nr Lymm	78C	Bache House	65C	Bickley	54J
Agden, nr Malpas	54C	Backford	37V	Bickley Moss	54P
Agden Hall, nr Lymm	78C	Backford Cross	37W	Bickley Town	54J
Agden Hall, nr Malpas	54C	Baddiley	65A	Bickleywood	54I
Alderley Edge	87P	Baddiley Mere Reservoir	55V	Bidston	29V
Alderley Park	87M	Baddington	64P	Bidston Dock	39A
Aldersey	45N	Bag Mere	76X	Bidston Hill	28Z
Aldersey Green	45T	Bakestonedale	97P	Bidston Moss	29V
Aldford	45E	Balderton	36R	Big Wood	87R
Aldford Brook	45J	Ball o' Ditton	48Y	Billinge Flashes	67V
Allgreave	96T	Balterley (Staffs.)	75Q	Billinge Green	67Q
Allostock	77K	Balterley Green (Staffs.)	75Q	Billinge Hill	97N
Alpraham	55Z	Balterley Heath (Staffs.)	75K	Birchenough Hill	96Y
Alsager	85C	Bank, The	85N	Birch Heath	56K
Alsager Heath	75Y	Bank Quay	58Y	Birchwood	69K
Alvanley	47X	Barbridge	65D	Birkenhead	38E
Alvaston	65S	Bar Mere	54I	Birkinheath Covert	78R
Anderton	67M	Barnett Brook	64H	Birtles Lake	87M
Anker's Knowl	97K	Barnhill	45X	Blackden Heath	77V
Antrobus	67P	Barnston	28W	Black Hill	98W
Appleton, nr Warrington	68G	Barnton	67H	Black Lake	57F
Appleton (Widnes)	58D	Barrett's Green	55Z	Black Rocks	98W
Appleton Moss	68L	Barrow	46U	Blacon	36Y
Appleton Reservoir		Barrows Green	65Z	Blakelow	65V
(formerly Walton Reservoir)	68C	Barrow's Green	58I	Blakemere	57K
Arclid	76V	Barthomley	75R	Blakemere Moss	57K
Arclid Green	76V	Bartington Common or Heath	57Y/67D	Blakenhall	74I
Arclid Sand Quarry	76R	Barton	45M	Blakenhall Moss	74J
Arley	68Q	Basford	75B	Boden Hall	85E
Arley Hall	68Q	Bate Heath	67Z	Bold Colliery (Merseyside)	59L
Arpley Dredging Deposit Grounds	58Y	Batemill Sand Quarry	87B	Bolesworth Castle	45Y
Arrowe Hill	28T	Batherton	64U	Bollington	97I
Arrowe Park	28T	Bath Vale	86R	Bollington Cross	97I
Arthill	78H	Baxendale (now Bakestonedale)	97P	Bollington, Little	78I
Ashbank	65Y	Beachins	45N	Bollinhurst Reservoir	98R
Ashley	78S	Beacons, The	28Q	Bollin Point	68Z
Ashton	56E	Beambridge	65L	Booston Wood	37Z
Ashton's Flash	67S	Bebington	38H	Booth Bank	78H
Astbury Marsh	86L	Beckett's Wood	57P	Booth Green	98F
Astbury Sand Quarry	86L	Beechwood	58F	Boothlane Head	76F
Astle	87G	Beeston	55P	Boothsdale	56I
Astle Hall	87B	Beeston Castle	55J	Booth's Mere	77U
Astle Pool	87B	Beeston Moss	55P	Boots Green	77L
Astmoor	58G	Belgrave	36V	Bosley	96C
Astmoor Saltmarsh	58G/H	Bell o' th' Hill	54H	Bosley Cloud	96B
Aston, nr Nantwich	64D	Betchton	75Z	Bosley Minn	96I

298

Name	Ref	Name	Ref	Name	Ref
Bosley Reservoir	96H	Burtonwood Airfield	59K/Q	Cloud Side	86W
Bostock	66U	Burtonwood Brewery	59L	Clutton	45S
Bostock Green	66U	Burwardsley	55D	Cocksmoss Wood	86N
Bottom-of-the-Oven	97W	Butley Town	97D	Coddington	45M
Bottoms Reservoir, nr Langley	97K	Butterfinch Bridge	67S	Collins Green	59M
Boughton	46D	Butter Hill	47A	Comberbach	67N
Boughton Heath	46I	Butt Green	65Q	Comber Mere	54X
Bowstonegate	98Q	Butt Lane (Staffs.)	85H	Combermere Park	54X
Bradeley Green	54H	Byley	76J	Common Side (Little Budworth)	56Y
Bradfield Green	65Z			Commonside, nr Kingsley	57L
Bradley	54C	Caldecott	45F	Commonside, nr Helsby	57C
Bradley Common	54C	Caldy	28H	Congleton	86L
Bradley Green	54C	Caldy Blacks	28H	Congleton Edge	86Q
Bradley Mount	97D	Caldy Brook	46H	Cooksongreen	57S
Bradwall	76L	Caldy Hill	28H	Coole Pilate	64N
Bradwall Green	76L	Calrofold	97M	Coomb Dale	55C
Brassey Green	56F	Calveley	55Z	Coppenhall	75D
Brereton	76X	Capenhurst	37R	Coppenhall Moss	75E
Brereton Green	76S	Capesthorne Hall	87L	Copthorne	64L
Brereton Heath	86C	Capesthorne Park	87L	Cotebrook	56S
Brereton Pool	76S	Carden	45R	Cotteril Clough (GMC)	88B
Bretton (Clwyd)	36L	Carlett Park	38Q	Cotton Abbots	46S
Brickhouses	76Q	Castle Cob	57G	Cotton Edmunds	46S
Bridgemere	74C	Castlefields	58G	Coxbank	64K
Bridgemere Park	74C	Castle Hill (GMC)	88B	Crabmill Flash	76A
Bridge Trafford	47K	Castle Park	57D	Cracow Moss (Staffs.)	74N
Bridgewater Lock	48W	Castle, The	46C	Crag Hall (Wildboarclough)	96Z
Brimstage	38W	Castletown	45F	Cranage	76P
Brindley	55W	Cat-and-Fiddle Inn	SK07A	Cranberry Moss	86S
Broad Hill	86E	Catchpenny Lane	87A	Cranshaw Hall	58E
Broken Cross (Macclesfield)	87W	Catchpenny Sand Quarry (Pool)	87A	Crewe	75C
Broken Cross (Northwich)	67W	Cats Tor	97X	Crewe	75H
Bromborough	38L	Caughall	47A	Crewe, nr Farndon	45G
Bromborough Pool	38M	Cessbank Common	96J	Crewe Green	75H
Brookhouse	97M	Chapel End	64R	Crewe Hall	75G
Brookhouse Green	86A	Chapel Mere (Cholmondeley)	55K	Croft	69G
Brookhouse Moss	86A	Charles Head	97U	Croker Hill	96J
Brookhurst	38K	Cheaveley Hall	46A	Cromwell's Bank	58M
Brook Place	58G	Checkley	74I	Crook of Dee	46F
Brookvale	58K	Checkley-cum-Wrinehill	74H	Cross Lane	66Q
Broomedge	78C	Checkley Green	74H	Cross o' th' Hill	44Y
Broomhall	64I	Chelford	87C	Cross Town (Knutsford)	77P
Broomhall Green	64I	Chester	46D	Croughton	47B
Broomhill	46U	Chester Castle	46C	Crowley	68R
Broseley Hall	69N	Chester Zoo	47A	Crow Mere	57H
Broughton (Clwyd)	36L	Chidlow	44X	Crowton	57S
Brownedge	76R	Childer Thornton	37T	Crow Wood	58I
Brown Heath	46M	Cholmondeley	55K	Croxton Green	55L
Brown Knoll	45W	Cholmondeley Castle	55F	Cuckoo Rocks	97V
Brownlow	86F	Cholmondeley Park	55K	Cuddington, nr Malpas	44N
Brownlow Heath	86F	Cholmondeston	65J	Cuddington, nr Northwich	57V
Brownmoss	64X	Chorley, nr Alderley Edge	87J	Cuddington Heath	44T
Brown's Bank	64L	Chorley, nr Nantwich	55Q	Cuerdley	58N
Broxton	45X	Chorlton, nr Crewe	75F	Cuerdley Cross	58N
Bruen Stapleford	46X	Chorlton, nr Malpas	44T	Cuerdley Green	58N
Bruera	46F	Chorlton-by-Backford	47B	Cuerdley Marsh	58M
Bryn	67B	Chorlton Hall	44U	Culcheth	69M
Bucklow Hill	78G	Chorlton Lane	44N	Culcheth Carrs	69N
Buckoak	57B	Chowley	45T	Cumberland Brook	96Z
Budworth Heath	67U	Christleton	46M	Cut-thorn Hill	SK06E
Budworth Mere	67N	Church Hill	66M		
Budworth Pool	56X	Church Hulme	76T	Dairyhouse Farm Sand Quarry	87A
Buerton, nr Audlem	64W	Church Lawton	85H	Dale Brow	87Y
Buerton, nr Chester	46F	Church Minshull	66Q	Dales, The	28Q
Buerton Moss	64X	Church Shocklach	44P	Dales Green (Staffs.)	85N
Buglawton	86R	Churton	45D	Dale Top	98Q
Buglawton Hall	86X	Churton-by-Aldford	45D	Dallam	59Y
Bulkeley	55H	Churton-by-Farndon	45D	Danebank	98S
Bulkeley Hall	55H	Churton Heath	46F	Danebower	SK07A
Bulkeley Hill	55H	Cinnamon Brow	69F	(Danebridge	96S)
Bunbury	55U	Clark Green	97J	Dane-in-Shaw	86V
Bunbury Heath	55P	Clatterbridge Hospital	38B	Dane-in-Shaw Brook	86R
Bunsley Bank	64R	Claughton	38E	Dane's Moss	97A
Burford	65G	Claverton	46B	Dane Viaduct	86X
Burland	65B	Cleulow Cross	96N	Daresbury	58R
Burleydam	64B	Clifton (Runcorn)	57J	Daresbury Delph	58R
Burntcliff Top	96Y	Clive	66S	Daresbury Hall	58W
Burton, nr Neston, Wirral	37C	Clive Green	66S	Darley Hall	66C
Burton, nr Tarporley	56B	Clotton	56G	Darnhall	66L
Burton Marsh	27W	Clotton Common	56H	Davenham	67K
Burton Point	37B	Clotton Hoofield	56G	Davenport Golf Club	98G/H
Burtonwood	59R	Cloud, The (Bosley)	96B	Davenport Green (Wilmslow)	87J

299

Dawpool	28M	Farmwood Pool Sand Quarry	87B	Hale	48R
Dawpool Bank	28G	Farndon	45C	Hale Bank	48W
Day Green	75T	Farnworth	58D	Halebarns (GMC)	78X
Dean Hill	75U	Fearnhead	69F	Hale Decoy	48R
Dean Row	88Q	Feldy	67Z	Hale Hall	48Q
Deansgreen	68X	Fiddler's Ferry	58N	Hale Head	48Q
Decca Pools	27X	Fiddler's Ferry Lagoons	58M	Hall Green	85I
Deer Park Mere (Cholmondeley)	55K	Fields Farm Flash	76A/B	Hall Marsh	48W
Delamere	56U	Finch's Plantation (Merseyside)	58P	Hallowsgate	56I
Delamere Forest	57K	Finney Green	88L	Halton, nr Runcorn	58F
Denhall (Lane)	27X	Fivecrosses	57I	Halton (Widnes)	58I
Derbyshire Bridge (Derbys.)	SK07A	Flaxmere	57L	Halton Brook	58G
Derbyshire Hill (Merseyside)	59M	Foden's Flash	76F	Halton Moss	58S
Dibbinsdale	38G/L	Forge Pool	67S	Hammerton Knowl	96T
Dingle, The	66R	Foulk Stapleford	46W	Hampton	54E
Dingle Bank Sand Quarry	87A	Four Lane Ends	97T	Hampton Green	54E
Disley	98S	Fourlanes End	85E	Hampton Heath	44Z
Ditton	48S	Fourways Sand Quarry	56U	Handbridge	46C
Ditton Marsh	48X	Fowley Common	69T	Handforth	88L
Dodcott-cum-Wilkesley	64C	Fox Howl	57F	Hand Green	56K
Doddington	74D	Foxwist Green	66J	Handley	45T
Doddington Park	74D	Frandley	67J	Handleyfoot	98V
Doddington Pool	74D	Frankby	28N	Hankelow	64S
Dodd's Green	64B	Frodsham	57D	Hapsford	47S
Dodleston	36Q	Frodsham Hill	57D	Harden Park	87P
Doe Green	58N	Frodsham Marsh	57E	Hardings Wood	85H
Dones Green	67D	Frodsham Score	47Z	Hare Hill	87T
Dove Point	29F	Fullers Moor	45X	Hargrave	46W
Duckington	45V	Fulshaw Park	88K	Harrock Wood	28S
Duddon	56C			Harrop	97U
Duddon Common	56H	Gallantry Bank	55B	Harrop Brook	97P
Duddon Heath	56C	Gallowsclough Hill	57Q	Harrow Hill	56N
Dudlow's Green	68H	Gateshealth	46Q	Hartford	67G
Dungeon, The	28L	Gatewarth Sewage Farm	58Y	Hartfordbeach	67G
Dungeon Bank	48Q	Gaunton's Bank	54T	Harthill	55C
Dunham Hall (GMC)	78I	Gawsworth	86Z	Hart Hill	57Q
Dunham Massey (GMC)	78I	Gawsworth Common	96J	Haslington	75I
Dunham-on-the-Hill	47R	Gayton	28Q	Hassall	75T
Dunham Park (GMC)	78I	Gayton Sands	(27P)/28K	Hassall Green	75U
Dunkirk	37W	Gibb Hill	67P	Hatch Mere	57K
Dutton	57U	Ginclough	97N	Hatherton	64Y
Dutton Hall	57Y	Glazebrook	69W	Hatton, nr Daresbury	58W
		Glazebury	69T	Hatton, nr Tattenhall	46Q
Earlestown (Merseyside)	59S	Gleadsmoss	86J	Hatton Heath	46K
Earl's Eye	46D	Golborne Bellow	45U	Hatton's Hey	57T
Eastham	38K	Golborne David	45P	Haughton	55Y
Eastham Ferry	38Q	Goostrey	77Q	Haughton Moss	55T
Eastham Locks	38Q	Gorsley Green	86P	Havannah	86S
(Eastham Sands	38V)	Gorstage	67B	Hawkhurst Head (Derbys.)	98V
Eastham Woods Country Park	38Q	Gorstella	36L	Haymoor Green	65V
East Hoyle Bank	(29A)/F	Gorstyhill	75K	Heald Green (GMC)	88M
Eaton, nr Chester	46A	Goyt Forest (part) (Derbys.)	97X	Heath	58A
Eaton, nr Congleton	86S	Gradeley Heath	55W	(Heatley	78E)
Eaton, nr Tarporley	56R	Grafton	45K	Hebden Green	66H
Eaton Park or Estate	45E/46A	Grange (Hoylake)	28I	Heild Wood	96T
Eaton Sand Quarry	86S	Grange (Runcorn)	58F	Helsby	47X
Eaves Brow	69L	Grange, nr Warrington	69K	Helsby Hill	47X
Ebnal	44Z	Grappenhall	68I	Helsby Marsh	47Y
Eccleston	46B	Grappenhall Heys	68H	Hempstones Point	58H
Eccleston Hill	46B	Gravel	66T	Henbury	87W
Eddisbury Hall	97G	Greasby	28N	Henbury Hall	87R
Eddisbury Hill, nr Delamere	56P	Great Barrow	46U	Henbury Moss	87K
Eddisbury Hill, nr Macclesfield	97G	Great Broughton	46I	Henhull	65G
Edge	45V	Great Budworth	67T	Hermitage Green	69B
Edge Green	45V	Great Moreton Hall	85P	Heronbridge	46C
Edgerley	45I	Great Sankey	58U	Heswall	28Q
Edleston	65F	Great Saughall	37Q	Heswall-cum-Oldfield	28K
Egerton	55F	Great Sutton	37S	Hetherson Green	54J
Egerton Green	55G	Great Warford	87D	Heyhead (GMC)	88H
Egremont	39B	Great Woolden Hall (GMC)	69W	Highbirch Wood	87V
Ellesmere Port	47D	Great Woolden Moss (GMC)	69W	Higher Ballgreave	97S
Elton, nr Helsby	47M	Greendale	87Y	Higher Bebington	38C
Elton, nr Sandbach	75J	Greenlooms	46R	Higher Burwardsley	55I
Elton Green	47M	Gresty Green	75B	Higher Disley	98S
Elton Hall Flash	75J	Greystone Heath	58U	Higher Hurdsfield	97H
Elworth	76K	Griffiths Road Lagoons	67W	Higher Poynton	98L
Englesea-brook	75K	Guilden Sutton	46P	Higher Runcorn	58B
Ettiley Heath	76F	Gurnett	97F	Higher Shurlach	67R
Eyes, The (Woolston Eyes)	68P			Higher Walton	58X
		Hack Green	64P	Higher Whitley	68A
Faddiley	55W	Haddon Lane	37C	Higher Wincham	67Y
Fanshawe	87K	Hague Bar (Derbys.)	98X	Higher Wych	44W

300

Highfields	64Q	Kingsley	57M	Lowcross Hill	45Q
Highlane	86Z	Kingsley Sludge Pool	57N/T	Lower Ballgreave	97S
High Lane (GMC)	98M	Kings Marsh	45H	Lower Bebington	38G
Highlees Wood	87S	Kingstreet Hall	66Z	Lower Bunbury	55T
High Legh	68X/78C	Kingsway	58C	Lower Carden	45R
High Moor	97Q	Kinsey Heath	64R	Lower Heath	86M
Hightown	86R	Kirkleyditch	87U	Lower Heswall	28Q
High Warren	68C	(Knar	SK06D)	Lowerhouse	97I
Hilbre Island	18Y/Z	Knathole (Derbys.)	98X	Lower Kinnerton	36L
Hilbre Point	28E	Knight's Low	98Q	Lower Peover	77M
Hill Cliffe	68C	Knight's Pool	97G	Lower Stretton	68F
Hinderton	37E	Knolls Green	87E	Lower Threapwood	44M
Hobb Hill	45Q	Knotbury (Staffs.)	SK06E	Lower Walton	68C
Hockenhull	46X	Knutsford	77P	Lower Whitley	67E
Hockenhull Platts	46S	Knutsford Moor	77P	Lower Withington	87A
Hockley	98G			Lower Wych	44X
Hodgehill	86H	Lacey Green	88L	Lugs Dale	58H
Hogshead Wood	56Z	Lach Dennis	77B	Lunts Heath	58E
Holcroft Moss	69W	Lache	36X	Lyme Hall	98R
Holford Moss	77C	Laches, The	97R	Lyme Handley	98Q
Hollinfare	69V	Ladbitch Wood (Derbys.)	97Y	Lyme Park	98R
Hollinlane	88H	Lady Brook (Ladybrook Valley)	98C	Lymm	68Y
Hollins	97G	Lake, The	66G	Lymm Bongs	68S
Hollinsgreen	76G	Lamaload Reservoir	97S	Lymm Dam	68T/Y
Hollins Green	69V	Landican	28X	Lymm Golf Course	68U
Hollowmoor Heath	46Z	Lane Ends (Coppenhall Moss)	75E	Lymm Wood	68S
Holmes Chapel	76T	Lane Ends (Higher Disley)	98W		
Hoofield Hall	56B	Langley	97K	Macclesfield	97B
Hoo Green	78B	Langley Reservoir	97K	Macclesfield Forest	97Q
Hoole	46I	Lantern Wood	98R	Macclesfield Forest Church	97R
Hoole Bank	46J	Larden Green	55V	Macefen	54D
Hoole Village	46J	Larkton	55A	Madam's Wood	86S
Hooton	37P	Larton	28I	Maiden Castle	45W
Hope Green	98B	Latchford	68I	Malkins Bank	75U
Horse Coppice Reservoir	98R	Lavister (Clwyd)	35U	Malpas	44Y
Horton by Malpas	44J	Lawton-gate	85D	Manchester Airport (GMC)	88H
Horton-cum-Peel	46Z	Lawton Hall Pool	85H	Manisty	37Z
Horton Green	44P	Lawton Heath	85D	Manley	57A
Hough, nr Alderley Edge	87P	Lawton Heath End	75Y	Manley Common	57G
Hough, nr Crewe	75A	Lea	74E	Marbury, nr Combermere	54S
Hough Green	48Y	Lea-by-Backford	37V	Marbury, nr Northwich	67N
Houghton Green	69F	Lea Forge	74E	Marbury Big Mere	54M
Howbeck Bank	64Z	Lea Hall	45J	Marbury Country Park	67N
Hoylake	28E	Leahurst	37D	Marbury No. 1 Lagoon	67M
Hulme	69A	Lea Newbold	45J	Marbury Small Mere	54S
Hulme Walfield	86M	Leasowe	29Q	Marbury with Quoisley	54S
Hulseheath	78G	Ledsham	37M	Marleston-cum-Lache	36W
Hunger Hill	57K	Leftwich	67Q	Marley Green	54X
Hunsterson	64X	Leighton	65T	Marshfield Bank	65S
Huntington	46G	Lightwood Green	64G	Marsh Green	???
Huntington Water Works Lagoons	46C	Lime Wharf	18Y/28C	Marsh Green (Frodsham)	57D
Hurdsfield	97H	Lindow End	87E	Marston	67S
Hurleston	65C	Lindow Moss	88F	Marston Flashes	67S
Hurleston Reservoir	65H	Lingley Green	58P	Marthall	77X
Huxley	56A	Linmere Moss	57K	Martinscroft	68P
		Liscard	39B	Martin's Moss	87A
Iddinshall	56G	Liscard Park	39B	Marton, nr Congleton	86P
Illidge Green	76W	Little Barrow	47Q	Marton, nr Winsford	66E
Ilse Pool	76F	Little Bollington	78I	Marton Green	66D
Ince	47N	Little Budworth	56X	Martonheath Wood	86U
Ince Banks	47P	Little Eye	18Y	Marton Hole	66D
Ince Marshes	47T	Little Heath	64S	Mary Dendy Hospitals	87D
Inner Marsh Farm	38B	Little Hilbre Island	18Y	Maw Green	74D
Irby	28M	Little Leigh	67C	Meadowbank	66P
Irby Hill	28M	Little Moreton Hall	85J	Melchett Mere	78K
		Little Neston	27Y	Meols	29F
Jenkin Chapel	97Y	Littler	66I	Mere	78F
Jodrell Bank	77V	Little Saughall	36U	Mere, The	78F
		Little Stanney	47C	mere	
Keckwick	58R	Little Sutton	37T	Baddiley	55V
Kelsall	56J	Littleton	46N	Bag	76V
Kelsborrow Castle	56I	Little Town	69M	Bar	54I
Kent Green	85I	Little Warford	87D	Betley	74N
Kenyon (GMC)	69H	Liverpool Docks (Merseyside)	39F	Birtles (Lake)	87M
Kermincham Hall	76Y	Loach Brook	86C	Blake	57K
Kerridge	97I	Locking Stumps	69K	Booth's	77U
Kerridge-end	97M	Long Green	47Q	Budworth (adjacent to Marbury	
Kerridge Hill	97N	Longmore	87X	Country Park, was	
Kettleshulme	97Z	Longside (Plantation)	98W	formerly often referred	
Key Green	86W	Lordship Marsh	47Y	to as Marbury Mere)	67N
Kidnal	44U	Lostock Gralam	67X	Capesthorne	87L
Kidsgrove (Staffs.)	85H	Lostock Green	67W	Chapel (Cholmondeley)	55K

301

Comber	54X	Nether Peover	77G	Parkgate (Lyme Park)	98S		
Deer Park (Cholmondeley)	55K	Netherton	57D	Parkgate (Peover)	77R		
Flax	57L	Neumann's Flash	67S	Parkgate (Wirral)	27U		
Hatch	57K	Newbold Astbury	86K	Park Moor	98Q		
Marbury Big	54M	New Brighton	39B	Park Moss	68K		
Marbury Small	54S	Newchurch Common	66E	Parrah Green	74C		
Melchett	78K	New Ferry	38M	Parr Moss (Merseyside)	59M		
Norbury	54P	Newhall	64C	Peckforton	55I		
Nuns	56Z	New Lane End	69H	Peckforton Castle	55J		
Oak	56T	New Mills (Derbys.)	98X	Peckforton Hills	55I		
Oulton Park	56X	New Mills, nr Mobberley	78Q/R/V	Peckforton Mere	55N		
Peckforton	55N	New Platt Lane Quarry Pool	77K	Peckforton Moss	55M		
Pick	67Y	New Rocklands	38A	Penketh	58T		
Quoisley	54M	Newsbank	86I	Pensby	28R		
Radnor	87M	Newton (Chester)	46E	Peover	77W		
Redes	87K	Newton, nr Frodsham	57H	Peover Hall	77R		
Rostherne	78L/M	Newton (Hoylake)	27I	Peover Heath	78W		
Shakerley	77F	Newton, nr Tattenhall	55E	Peover Inferior	77M		
Tabley	77I	Newton, nr Woodford	88Q	Peover Superior	77R		
Tatton	78K	Newton by Malpas	44S	Perch Rock	39B		
Tax	76R	Newton Cross	58V	Petty Pool	67A		
The	78F	Newtown (Frodsham)	57J	Pettypool	66E		
Westlow	86M	Newtown, nr Nantwich	64I	Pettypool Wood	67A		
Mere Heath	67Q	Newtown (New Mills) (Derbys.)	98X	Pex Hill (Merseyside)	58E		
Meremoor Moss	75L	Newtown (Poynton)	98G	Pexhill	87Q		
Mere Moss	86J	Noctorum	28Y	Pickering's Pasture	48W		
Mickle Trafford	46P	No Man's Heath	54D	Pick Mere	67Y		
Middlewich	98M	No Man's Land	48W	Pickmere	67Y		
Midge Brook	86D	Norbury, nr Whitchurch	54N	Picton	47F		
Midway	98B	Norbury Common	54P	Picton Gorse	46J		
Mill House Brow	69G	Norbury Hollow	98H	Piggford Moor	96U		
Millington	78H	Norbury Meres	54P	Pike Low	97T		
Milners Heath	46R	Norbury Moor (GMC)	98C	Pinsley Green	54Y		
Milton Green	45U	Norcott Brook	68A	Piper's Ash	46I		
Minshull Vernon	66V	Norley	57R	Plemstall	47K		
Mobberley	77Z	North Rode	86Y	Plover's Moss	56U		
Moblake	64R	Northwich	67L	Plumley (formerly Plumbley)	77C		
Mockbeggar Wharf	29R	Norton	58K	Plumley Moor	77H		
Mollington	37V	Norton Marsh	58M	Poll Hill	28R		
Monks Coppenhall	75D	Norton Priory Park	58L	pool			
Monk's Heath	87M	Nova Scotia	66E	Brereton	76S		
Moore	58S	Nunsmere	56Z	Bromborough	35M		
Moorfields	65W			Budworth	56X		
Moorside (Neston)	27Y	Oakenclough	96U	Catchpenny (SQ)	87A		
Moreton	29Q	Oakgrove	96E	Decca	27X		
Moreton Clay Pit	29K	Oakhanger	75S	Doddington	74D		
Moreton-cum-Alcumlow	85P	Oakhanger Moss	75S	Farmwood (SQ)	87B		
Morley	88G	Oak Mere	56T	Forge	67S		
Morley Green	88F	Oakmere	56U	Ilse	76F		
Moss Bank	58H	Oat Hill	44Z	Knight's	97G		
Mossend	86B	Occlestone Green	66W	Knutsford Moor	77P		
Mosses (Acre Nook) Sand Quarry	87A/B/F/G	Odd Rode	85I	Lawton Hall	85H		
Mosses Wood (Acre Nook Wood)	87G	Oldcastle	44S	Lymm (Dam)	68T/Y		
Moss Houses	87V	Oldcastle Heath	44S	Mount Farm Ponds	96E		
Moss Lane	97A	Oldgate Nick	97Y	New Platt Lane Quarry	77K		
Mossley	86Q	Old Man of Mow	85N	Petty	67A		
Moss Nook (GMC)	88H	Old Moat House	59L	Poynton	98H		
Moss Side, nr Hollins Green	69Q	Old Woodhouse (Salop)	54W	Rode	85D		
Moss Side, nr Moore	58S	Ollerton	77T	Rookery	67F		
Moston, nr Chester	47A	Orford	69A	Ryle's	97B		
Moston, nr Sandbach	76F	Oscroft	56D	Styperson	97J		
Moston Green	76F	Oughtrington	68Y	Thorneycroft	87Q		
Mottram Cross	87Z	Oulton Lake	56X	Whirley Sand Quarry	87X		
Mottram Hall	87Z	Oulton Mill	56X	Wincham	67S		
Mottram St Andrew	87U	Oulton Park	56X	Winterley	75N		
Mouldsworth	57A	Over	66I	Withington Hall	87A		
Moulton	66P	Over Alderley	87T	Poole	65H		
Mount Farm Ponds	96E	Over Knutsford	77U	Poole Hall Sands	37Z		
Mount Manisty	37Z	Overleigh Drive, nr Chester	46B	Port Sunlight	38M		
Mount Pleasant (Mow Cop)	85N	Overpool	37Y	Pott Shrigley	97P		
Mount Pleasant (Wildboarclough)	96U	Over Tabley	78F	Poulton (Bebington)	38G		
Mousley Bottom (Derbys.)	98X	Overton (Frodsham)	57I	Poulton, nr Pulford	35Z		
Mow Cop (Staffs.)	85N	Overton, nr Malpas	44U	Poulton (Wallasey)	39A		
Murdishaw Wood	58K	Overton Green	76V/86A	Poulton, nr Warrington	69F		
		Oxton	28Y	Poulton with Fearnhead	69F		
Nab End	97S			Pownall Park	88F		
Nab Head	97J	Paddington	68J	Poynton	98G		
Nantwich	65L	Paddock Hill	87E	Poynton Pool	98H		
Ness	37D	Padgate	68J	Poynton with Worth	98G		
Nessholt	37D	Palacefields	58K	Prenton	38D		
Neston	27Y	Pale Heights	56P	Prestbury	97D		
Nether Alderley	87N	Pallotti Hall	87Q	Prestbury Sewage Farm or Works	87Z		

Preston Brook	58Q	Rowe, The	69L	Sproston	76I
Preston on the Hill	58Q	Rowleyhill	45H	Sproston Green	76I
Primrose Hill	56J	Row-of-Trees	87J	Spurstow	55N
Prince Hill	74H	Row's Wood	58W	Spurstow Spa	55S
Prince's Wood	98H	Rowton	46M	Spy Hill	57F
Priory Cross	58K	Rowton Moor	46M	Stamford Bridge	46T
Pryor's Hayes (Prior's Heys)	56D	Royal's Green	64G	Stanlow	47H
Puddinglake	76J	Royden Park	28M	Stanlow Banks	47E/J
Puddington	37G	Rudheath	67V	Stanlow Bay	47J
Pulford	35U	Rudheath Woods and Pools	77K	Stanlow Point	47I
Pumphouse Flash	75J	Ruloe	57W	Stanner Nab	55I
Pym Chair	97Y	Runcorn	58B	Stanthorne	66Y
		Runcorn Gap	58B	Stapeley	64U
Quaker's Coppice	75H	Runcorn Hill	58A	Stapledon Wood	28H
Quarrybank	56M	Rushgreen	68Y	Statham	68T
Quarry Bank Mill (Styal)	88G	Rushton	56X	Stoak	47G
Queen's Park	46C	Ryle's Pool	97B	Stockton	44S
Quoisley	54M			Stockton Heath	68D
Quoisley Meres	54M	Saighton	46L	Stoke, nr Chester	47G
		Saltersford Hall, nr Holmes Chapel	76T	Stoke, nr Nantwich	65I
Raby	37E	Saltersford Hall, nr Kettleshulme	97Y	Stoke Hall	65I
Raby Hall	38F	Salterswall	66I	Stonehouse	57A
Raby Vale	38A	Sandbach	76K	Stoneley Green	65A
Radbroke Hall	77S	Sandbach Flashes	75E/J & 76A/B/F	Storeton	38C
Radmore Green	55X			Storeton Hill	38C
Radnor Bridge	86H	Sandbach Heath	76Q	Stretton, nr Farndon	45L
Radnor Mere	87M	Sandiway	67A	Stretton, nr Warrington	68B
Radway Green	75S	Sandle Bridge	87D	Stretton Airfield	68L
Railway Flash	75E	Sandlow Green	76X	Stretton Moss	68G
Rainow	97N	Sankey Bridges	58Y	Stud Green	76G
Rainow Hill	86W	Sankey Valley Park	58Z	Styal	88G
Rainowlow	97N	Saughall	36U	Styperson Pool	97J
Ravenshall (Staffs.)	74N	Saughall Massie	28P	Sutton, nr Macclesfield	96P/97F
Ravensmoor	65F	Scholar Green	85I	Sutton, nr Runcorn	57J
Raw Head	55C	School Green	66M	Sutton Common	96I
Rease Heath	65M	Score Bank	48Q/V	Sutton Green	37S
Red Bull	85H	Seacombe	39F	Sutton Lane Ends	97F
Reddish Hall, nr Grappenhall	68M	Sevenoaks	67P	Sutton Moss (Merseyside)	59L
Reddish Hall, nr Lymm	68Z	Seven Springs	98X	Sutton Reservoir	97A
Redesmere	87K	Shadow Moss (GMC)	88H	Sutton Weaver	57P
Red Rocks	28E	Shakerley Mere	77F	Swanbach	64L
Reed Hill	97U	Shavington	65V	Swan Green	77G
Reeking Hole	56Z	Shavington-cum-Gresty	76B	Swanwick Green	54N
reservoir		Shaw Heath (Knutsford)	77U	Swettenham	86D
Appleton (formerly Walton)	68C	Shell Brook	96M	Sworton Heath	68X
Baddiley	55V	Shell Green	58I	Sydney	75I
Bollinhurst	98R	Shellow Wood	86Z		
Bosley	96H	Shemmy Moss	56Z	Table Rock, The	55I
Bottoms, nr Langley	97K	Shining Tor	97W	Tabley	77J
Horse Coppice	98R	Shocklach	44J	Tabley Inferior	77D
Hurleston	65H	Shocklach Green	44J	Tabley Mere	77I
Lamaload	97S	Shocklach Oviatt	44J	Tabley Superior	77J
Langley	97K	Shotwick	37F	Tagsclough Hill	96Y
Lowerhouse	97I	Shotwick Lodge	37K	Tanskey Rocks	28C
Ridgegate	97K	Shotwick Park	37K	Tarporley	56L
Sutton	97A	Shutlingsloe (Shutlings Low)	96U	Tarvin	46Y
Tegg's Nose	97K	Siddington	87K	Tarvin Sands	46Y
Trentabank	97Q	Siddington Heath	87F	Tattenhall	45Z
Walton (Appleton)	68C	Simm's Cross	58C	Tatton	78K
Rhuddall Heath	56L	Sink Moss	68R	Tatton Dale	78L
Richmond Bank	58T	Skellorn Green	98F	Tatton Hall	78K
Ridgegate Reservoir	97K	Slaughter Hill	75H	Tatton Mere	78K
Ridley	55M	Sloyne, The	38I	Tatton Park	78K
Risley	69L	Smallwood	86A	Taxal Edge (Derbys.)	97Z
Risley Moss	69Q	Smethwick Green	86B	Taxmere	76R
Rixton Clay Pits	69V	Smith's Green, nr Crewe	75L	Taylor's Wood	28Z
Rixton Halls	68Z	Smith's Green, nr Withington	87A	Tegg's Nose	97L
Rixton Moss	69Q	Smithy Brow	69G	Tegg's Nose Reservoir	97K
Rixton-with-Glazebrook	69V	Smithy Green	77M	Tetton (later Moston)	76G
Rock Ferry	38I	Snelson	87C	Thelwall	68N
Rock Savage (Rocksavage)	58F	Somerford	86C	Thelwall Viaduct	68U
Rode Heath	85E	Somerford Booths	86H	Thingwall	28S
Rodeheath	97U	Somerford Park	86C	Thorns Green (GMC)	78X
Rode Pool	85D	Soss Moss Wood	87I	Thornton Hough	38A
Roe Park	85P	Sound	64J	Thornton-le-Moors	47M
Roodee, The	36X	Sound Heath	64E	Thornton Manor	28V
Rookery Bridge	76F	Southgate	58F	Thornycroft Pools	87Q
Rookery Pool	67F	South Park, Macclesfield	97B	Threapwood	44M
Rope	65W	Spen Green	86A	Three Shire Heads	SK06E
Rostherne	78L	Spike Island	58C/H	Thurlwood	85D
Rostherne Mere	78L/M	Spital	38L	Thurstaston	28M
Roughhill	36W	Sponds Hill	98Q	Thurstaston Common	28M

303

Thurstaston Hill	28M	Weaver Bend	57E	Wincham (Pool)	67S
Tidnock Wood	86U	Weaverham	67C	Wincle	96N
Tilston	45K	Weaver Sluices	48V	Wincle Minn	96N
Tilstone Bank	55U	Weetwood Common	56I	Windgather Rocks	97Z
Tilstone Fearnall	56Q	Wells Green	65W	Windle Hill	37D
Timbersbrook	86W	Wervin	47F	Windmill Hill	58L
Tiverton	56K	West Bank	58B	Windyharbour	87F
Toft	77N	West Heath	86L	Winnington	67M
Toot Hill	97Q	(West Hoyle Bank	18I/J -	Winsford	66N
Tower Hill	97M	T/U)		Winsford Bottom Flash	66S
Townfield Lands	45E	West Kirby	28D	Winsford Top Flash	66S
Town Fields	66M	Westlow Mere	86M	Winterley	75N
Trafalgar	38G	Weston, nr Crewe	75G	Winterley Pool	75N
Tranmere	38I	Weston (Macclesfield)	87W	Winwick	69B
Trap Street	86J	Weston (Runcorn)	58A	Winwick Hospital	59W
Trentabank Reservoir	97Q	Weston Marsh	57E	Winwick Quay	59V
Turk's Head	96J	Weston Point	48V	Wirswall	54M
Turnerheath	97I	West Parkgate (Lyme Park)	98K	Wistaston	65W
Tushingham-cum-Grindley	54H	Westy	68J	Wistaston Green	65S
Tushingham Hall	54H	Wettenhall	66F	Withenshaw	96P
Twemlow	76Z	Wettenhall Green	66F	Withington (Lower)	86E
Twemlow Green	76Z	Whaley Moor	98W	Withington Green	87A
Twiss Green	69M	Wharton	66T	Withington Hall	87B
Two Mills	37L	Wharton Green	66T	Withington Hall Pool	87A
Tytherington	97C	Whatcroft	66Z	Withington Park	87B
		Whatcroft Hall	66Z	Withington Sand Quarries	87A
Unwinpool	98K	Wheelock	75P	Witton Flashes	67M/S
Upper Threapwood	44M	Wheelock Heath	75N	Wolverham	47C
Upton (Chester)	46E	Whetstone Ridge	SK07A	Woodbank	37L
Upton (Widnes)	58D	Whirley Grove	87X	Woodchurch	28T
Upton, Wirral	28U	Whirley Sand Quarry (Lake)	87X	Woodcott (or Woodcote)	64E
Upton Heath	46E	Whisterfield	87F	Woodend, nr Disley	98S
Utkinton	56M	Whitby	37X	Woodend (Widnes)	58C
		Whitby Heath	37W	Woodford Aerodrome (GMC)	88V/(W)
Vale Royal Park	66J	Whitegate	66J	Woodhey	38H
Valley, The (Crewe)	65X	Whiteley Green	97J	Woodhey Green	55R
Verdin's Cut	66N	Whitemoor Hill	96D	Woodhouses	57C
Vicarscross	46N	Whitemore (Staffs.)	86V	Wood Lanes	98F
Victoria Park, Warrington	68D	White Moss	75S	Wood Moss	96Z
Vulcan Village (Merseyside)	59W	White Nancy	97I	Woodside, nr Delamere Forest	57F
		Whitfield Common	28R	Woodside, nr Wettenhall	66F
Wades Green	66K	Whitley	67E	Woodworth Green	55T
Walgherton	64Z	Whitley Reed	68K	Woolfall	64X
Walker Barn	97L	Whittle Hall	58U	Woolstanwood	65T
Wallasey	39A	Widnes	58C	Woolstencroft	78D
Walley's Green	66V	Widnes Wharf	58H	Woolston	68P
Walton	58X	Wigg Island	58H	Woolston Eyes	68P
Walton Dredging Deposit Grounds	58Y	Wigland	44X	Woolston Moss	69K
Walton (= Appleton) Reservoir	68C	Wigwam Wood	98B	Worleston	65N
Warburton (GMC)	68Z	Wildboarclough (formerly Crag)	96Z	Wornish Nook	86H
Warburton Green (GMC)	78X	Wilderspool	68D	Worthington's Flash	67S
Warburton's Wood	57N	Wilkesley	64F	Wrenbury	54Y
Warburton Toll Bridge (GMC)	69V	Willaston, nr Nantwich	65R	Wrenbury-cum-Frith	54Y
Wardle	65D	Willaston, nr Neston	37I	Wrenbury Heath	64E
Wardle Bank	65E	Willey Moor	54I	Wrinehill (Staffs.)	74N
Wardsend	98G	Willington	56I	Wybunbury	64Z
Warmingham	76A	Willington Corner	56I	Wybunbury Sand Quarries	74E
Warmingham Flash	76A	Willow Green	67D	Wychough	44X
Warren	87V	Wilmslow	88K		
Warrington	68E	Wilmslow Park	88K	Yarnshaw Hill	97V
Watch Lane Flash	76F	Wilmslow Sewage Farm or Works	88L	Yatehouse Green	76E
Waverton	46R	Wimboldsley	66W	Yearns Low	97S
Waverton Gorse	46R	Wimboldsley Hall	66W	Yeld, The	56J
Way's Green	66M	Wimbolds Trafford	47L		

b.) ADDITIONAL CHESHIRE PLACE NAMES - PRE-1974

AT the time of the boundary changes of 1974, Cheshire lost north-eastern areas with many place-names familiar to readers of the works of Coward & Oldham, of Boyd and of Bell, as well as to those who know the earlier *Cheshire Bird Reports*.

As Figs. 3 and 4 (page 5) show, the eastern extremity of Longdendale is almost twenty kilometres from the present county boundary.

The place-names which follow, with tetrad designations, are all outside the recording scope of this atlas, but will have relevance to students of Cheshire ornithology.

The 100-km reference is SJ unless otherwise stated.

Adswood	88Y	Hyde	99M
Altrincham	78T/U	Lyne Edge	99T
Arnfield Reservoir	SK09D	Marple	89P
Ashton upon Mersey	79R	Marple Bridge	89U
Black Hill	SE00S	Marple Dale	89P
Bottoms Reservoir, Longdendale (part Derbys.)	SK09I	Marpleridge	89T
Bowdon	78N	Mellor	89Z
Bramall Hall	88Y	Mellor Moor	89Y
Bramhall	88X	Millbrook	99Z
Bramhall Moor	98D	Mossbrow	78E
Bramhall Park	88Y	Mottram in Longdendale	99X
Bredbury	99F	Newton (Hyde)	99M
Brinnington	99B	Offerton	98J
Broadbottom	99W	Partington	79A
Broadheath	78U	Pikenaze Moor	SE10A
Brooklands	79V	Reddish Vale (part Lancs.)	99B
Buckton Vale	SD90V	reservoir	
Carrbrook	SD90V	Arnfield	SK09D
Carrington	79L	Bottoms, Longdendale (part Derbys.)	SK09I
Carrington Moss	79K	Errwood	SK07C
Chadkirk	89I	Fernilee	SK07D
Cheadle	88P	Rhodeswood	SK09P
Cheadle Heath	88U	Swineshaw	SK09E
Cheadle Hulme	88T	Toddbrook	SK08A
Compstall	99Q	Torside	SK09U
Crowden	SK09U	Valehouse	SK09I
Davenport	88Y	Walkerwood	99Z
Davenport Green, nr Hale	88D	Woodhead	SK09Z
Dukinfield	99N	Rhodeswood Reservoir (part Derbys.)	SK09P
Dunham Massey	78P	Ringway	88C
Dunham Town	78N	Romiley	99K
Dunham Woodhouses	78J	Sale	79V/W
Edgeley	88Z	Sale Ees (Water Park)	89B
Errwood Hall	SK07C	Salter's Brook Bridge	SE10F
Errwood Reservoir	SK07C	Shaw Heath (Stockport)	88Z
Etherow Country Park	99Q	Shell Pools	79G
Etrop Green	88C	Sinderland Green	79F
Fernilee Reservoir (part Derbys.)	SK07D	Sinderland Sewage Farm	79K
Furness Vale	SK08B	Stalybridge	99U
Gatley	88P	Staly Brushes	99Z
Gee Cross	99L	Stepping Hill	98D
Godley	99S	Stockport (part Lancs.)	89V
Goyt's Bridge	SK07C	Strines	89T
Goyt's Moss (part Derbys.)	SK07B	Swineshaw Reservoirs (part Lancs.)	SK09E
Hale	78T	Taxal	SK07E
Halebarns	78X	Timperley	78Z
Harridge Pike	99Z	Tintwistle	SK09I
Harrop Edge	99Y	Toddbrook Reservoir	SK08A
Hartington Upper Quarter (part Derbys.)	SK07C	Torside Reservoir (part Derbys.)	SK09U
Hattersley	99S/X	Valehouse Reservoir (part Derbys.)	SK09I
Hawk Green	89N	Walkerwood Reservoir	99Z
Hazel Grove	89I	Werneth Low	99L
Heaviley	98E	Whaley Bridge	SK08A
Heyden Moor	SE00W	Woodhead	SK09Z
Heyrod	99U	Woodhead Pass	SE10F
Hollingworth	SK09D	Woodhead Reservoir (part Derbys.)	SK09Z
Hollingworthhall Moor	99Y/Z	Woodley	99G
Holme Moss (part Yorks. W. Riding)	SE00X	Woodsmoor	98D
Hoo Moor	SK07D	Yeardsley Hall	SK08B
Hough Hill	99T		

INDEX

Page numbers in bold refer to maps

Accipiter (Accipitridae)
    gentilis 77
    nisus 78, **79**
Acrocephalus (Sylviidae)
    schoenobaenus 213, **213**
    scirpaceus 215, **217**
Actitis (Scolopacidae)
    hypoleucos 122, **123**
Aegithalos (Aegithalidae)
    caudatus 239, **241**
Aix (Anatidae)
    galericulata 56, **57**
Alauda (Alaudidae)
    arvensis 165, **167**
Alcedo (Alcedinidae)
    atthis 156, **157**
Alectoris (Phasianidae)
    rufa 88, **89**
Anas (Anatidae)
    acuta 63, **65**
    clypeata 66, **67**
    crecca 60, **61**
    penelope 57, **59**
    platyrhynchos 61, **63**
    querquedula 64, **65**
    strepera 58, **59**
Anser (Anatidae)
    anser 49, **49**
Anthus (Motacillidae)
    petrosus 178
    pratensis 176, **177**
    trivialis 174, **175**
Apus (Apodidae)
    apus 153, **155**
Ardea (Ardeidae)
    cinerea 44, **45**
Asio (Strigidae)
    flammeus 150, **151**
    otus 149
Athene (Strigidae)
    noctua 145, **145**
Australian Pochard 68
Aythya (Anatidae)
    australis 68
    ferina 67, **69**
    fuligula 69, **71**

Barnacle Goose 53
Barn Owl 142, **143**
Bearded Tit 238
Bittern 44
Blackbird 202, **203**
Blackcap 223, **225**
Black Grouse 87
Black-headed Gull 124, **125**
Black-necked Grebe 43
Black Redstart 193
Black-tailed Godwit 118
Blue Tit 246, **247**
Botaurus (Ardeidae)
    stellaris 44
Brambling 272
Branta (Anatidae)
    canadensis 50, **51**
    leucopsis 53
Bucephala (Anatidae)
    clangula 71
Bullfinch 284, **287**
Bunting, Cirl 290
    Corn 292, **293**
    Reed 290, **291**
Buteo (Accipitridae)
    buteo 80, **81**
Buzzard 80, **81**
    Honey 75

Calidris (Scolopacidae)
    alpina 112, **113**
Canada Goose 50, **51**
Caprimulgus (Caprimulgidae)
    europaeus 151, **153**
Carduelis (Fringillidae)
    cannabina 278, **279**
    carduelis 275, **275**
    chloris 273, **273**
    flammea 281, **283**
    flavirostris 279, **281**
    spinus 277
Certhia (Certhiidae)
    familiaris 252, **253**
Cettia (Sylviidae)
    cetti 210
Cetti's Warbler 210
Chaffinch 271, **271**
Charadrius (Charadriidae)
    dubius 104, **105**
    hiaticula 106, **107**
Chiffchaff 228, **229**
Cinclus (Cinclidae)
    cinclus 184, **185**
Circus (Accipitridae)
    aeruginosus 76
    cyaneus 76
    pygargus 76
Cirl Bunting 290
Coal Tit 245, **245**

Coccothraustes (Fringillidae)
    coccothraustes 286, **287**
Collared Dove 136, **137**
Columba (Columbidae)
    livia 130, **131**
    oenas 132, **133**
    palumbus 134, **135**
Common Sandpiper 122, **123**
Common Tern 128, **129**
Coot 101, **101**
Cormorant 43
Corn Bunting 292, **293**
Corncrake 97
Corvus (Corvidae)
    corax 265
    corone 262, **263**
    frugilegus 260, **261**
    monedula 258, **259**
Coturnix (Phasianidae)
    coturnix 92, **93**
Crake, Spotted 96
Crex (Rallidae)
    crex 97
Crossbill 283, **285**
Crow 262, **263**
Cuckoo 140, **141**
Cuculus (Cuculidae)
    canorus 140, **141**
Curlew 118, **119**
Cygnus (Anatidae)
    olor 46, **47**

Delichon (Hirundinidae)
    urbica 172, **173**
Dendrocopos (Picidae)
    major 161, **163**
    minor 162, **163**
Dipper 184, **185**
Dove, Collared 136, **137**
    Stock 132, **133**
    Turtle 138, **139**
Duck, Mandarin 56, **57**
    Ruddy 73, **75**
    Tufted 69, **71**
Dunlin 112, **113**
Dunnock 188, **189**

Emberiza (Emberizidae)
    cirlus 290
    citrinella 288, **289**
    schoeniclus 290, **291**
Erithacus (Turdidae)
    rubecula 190, **191**

Falco (Falconidae)
    columbarius 82
    peregrinus 84
    subbuteo 84
    tinnunculus 81, **83**

Falcon, Peregrine 84
Feral pigeon 130, **131**
Ficedula (Muscicapidae)
    hypoleuca 236, **239**
Fieldfare 205
Firecrest 234, **235**
Flycatcher, Pied 236, **239**
    Spotted 234, **237**
Fringilla (Fringillidae)
    coelebs 271, **271**
    montifringilla 272
Fulica (Rallidae)
    atra 101, **101**

Gadwall 58, **59**
Gallinago (Scolopacidae)
    gallinago 114, **115**
Gallinula (Rallidae)
    chloropus 98, **99**
Garden Warbler 222, **223**
Garganey 64, **65**
Garrulus (Corvidae)
    glandarius 254, **255**
Godwit, Black-tailed 188
Goldcrest 232, **233**
Goldeneye 71
Golden Plover 108, **109**
Goldfinch 275, **275**
Goose, Barnacle 53
    Canada 50, **51**
    Greylag 49, **49**
Goshawk 77
Grasshopper Warbler 210, **211**
Great Crested Grebe 39, **41**
Great Spotted Woodpecker 161, **163**
Great Tit 248, **249**
Grebe, Black-necked 43
    Great Crested 39, **41**
    Little 38, **39**
    Slavonian 42
Greenfinch 273, **273**
Green Woodpecker 159, **159**
Grey Heron 44, **45**
Greylag Goose 49, **49**
Grey Partridge 90, **91**
Grey Wagtail 180, **181**
Grouse, Black 87
    Red 85, **85**
Gull, Black-headed 124, **125**
    Herring 128, **129**
    Lesser Black-backed 127, **127**
    Little 124

Haematopus (Haematopodidae)
    ostralegus 102, **103**
Harrier, Hen 76
    Marsh 75
    Montagu's 76

Hawfinch   286, **287**
Hen Harrier   76
Heron, Grey   44, **45**
Herring Gull   128, **129**
Hirundo   (Hirundinidae)
    rustica   170, **171**
Hobby   84
Honey Buzzard   75
Hoopoe   158
House Martin   172, **173**
House Sparrow   266, **269**

Jackdaw   258, **259**
Jack Snipe   114
Jay   254, **255**
Jynx   (Picidae)
    torquilla   158

Kestrel   81, **83**
Kingfisher   156, **157**
Kite, Red   75

Lagopus   (Tetraonidae)
    lagopus   85, **85**
Lanius   (Laniidae)
    collurio   254
Lapwing   110, **111**
Larus   (Laridae)
    argentatus   128, **129**
    fuscus   127, **127**
    minutus   124
    ridibundus   124, **125**
Lesser Black-backed Gull   127, **127**
Lesser Spotted Woodpecker   162, **163**
Lesser Whitethroat   217, **219**
Limosa   (Scolopacidae)
    limosa   118
Linnet   278, **279**
Little Grebe   38, **39**
Little Gull   124
Little Owl   145, **145**
Little Ringed Plover   104, **105**
Little Tern   130
Locustella   (Sylviidae)
    naevia   210, **211**
Long-eared Owl   149
Long-tailed Tit   239, **241**
Loxia   (Fringillidae)
    curvirostra   283, **285**
Lullula   (Alaudidae)
    arborea   165
Luscinia   (Turdidae)
    megarhynchos   192
Lymnocryptes   (Scolopacidae)
    minimus   114

Magpie   256, **257**
Mallard   61, **63**

Mandarin Duck   56, **57**
Marsh Harrier   76
Marsh Tit   241, **243**
Martin, House   172, **173**
    Sand   166, **169**
Meadow Pipit   176, **177**
Merganser, Red-breasted   72
Mergus   (Anatidae)
    albellus   72
    serrator   72
Merlin   82
Miliaria   (Emberizidae)
    calandra   292, **293**
Milvus   (Accipitridae)
    milvus   75
Mistle Thrush   208, **209**
Montagu's Harrier   76
Moorhen   98, **99**
Motacilla   (Motacillidae)
    alba   182, **183**
    cinerea   180, **181**
    flava   178, **179**
Muscicapa   (Muscicapidae)
    striata   234, **237**
Mute Swan   46, **47**

Netta   (Anatidae)
    rufina   67
Nightingale   192
Nightjar   151, **153**
Numenius   (Scolopacidae)
    arquata   118, **119**
Nuthatch   250, **251**

Oenanthe   (Turdidae)
    oenanthe   199, **199**
Ouzel, Ring   200, **201**
Owl, Barn   142, **143**
    Little   145, **145**
    Long-eared   149
    Short-eared   150, **151**
    Tawny   147, **147**
Oxyura   (Anatidae)
    jamaicensis   73, **75**
Oystercatcher   102, **103**

Panurus   (Timaliidae)
    biarmicus   238
Parakeet, Ring-necked   140
Partridge, Grey   90, **91**
    Red-legged   88, **89**
Parus   (Paridae)
    ater   245, **245**
    caeruleus   246, **247**
    major   248, **249**
    montanus   243, **244**
    palustris   241, **243**

Passer   (Passeridae)
    domesticus   266, **269**
    montanus   268, **269**
Perdix   (Phasianidae)
    perdix   90, **91**
Peregrine Falcon   84
Pernis   (Accipitridae)
    apivorus   75
Phalacrocorax   (Phalacrocoracidae)
    carbo   43
Phasianus   (Phasianidae)
    colchicus   93, **95**
Pheasant   93, **95**
Philomachus   (Scolopacidae)
    pugnax   114
Phoenicurus   (Turdidae)
    ochruros   193
    phoenicurus   193, **195**
Phylloscopus   (Sylviidae)
    collybita   228, **229**
    sibilatrix   225, **227**
    trochilus   230, **231**
Pica   (Corvidae)
    pica   256, **257**
Picus   (Picidae)
    viridis   159, **159**
Pied Flycatcher   236, **239**
Pied Wagtail   182, **183**
Pigeon, feral   130, **131**
    Wood   134, **135**
Pintail   63, **65**
Pipit, Meadow   176, **177**
    Rock   178
    Tree   174, **175**
Plover, Golden   108, **109**
    Little Ringed   104, **105**
    Ringed   106, **107**
Pluvialis   (Charadriidae)
    apricaria   108, **109**
Pochard   67, **69**
    Australian   68
    Red-crested   67
Podiceps   (Podicipedidae)
    auritus   42
    cristatus   39, **41**
    nigricollis   43
Porzana   (Rallidae)
    porzana   96
Prunella   (Prunellidae)
    modularis   188, **189**
Psittacula   (Psittacidae)
    krameri   140
Pyrrhula   (Fringillidae)
    pyrrhula   274, **287**

Quail   92, **93**

Rail, Water   94, **97**

Rallus   (Rallidae)
    aquaticus   94, **97**
Raven   265
Red-backed Shrike   254
Red-breasted Merganser   72
Red-crested Pochard   67
Red Grouse   85, **85**
Red Kite   75
Red-legged Partridge   88, **89**
Redpoll   281, **283**
Redshank   120, **121**
Redstart   193, **195**
    Black   193
Redwing   207
Reed Bunting   290, **291**
Reed Warbler   215, **217**
Regulus   (Sylviidae)
    ignicapillus   234, **235**
    regulus   232, **233**
Ringed Plover   106, **107**
    Little   104, **105**
Ring-necked Parakeet   140
Ring Ouzel   200, **201**
Riparia   (Hirundinidae)
    riparia   166, **169**
Robin   190, **191**
Rock Pipit   178
Rook   260, **261**
Ruddy Duck   73, **75**
Ruff   114

Sand Martin   166, **169**
Sandpiper, Common   122, **123**
Saxicola   (Turdidae)
    rubetra   195, **197**
    torquata   196, **197**
Scolopax   (Scolopacidae)
    rusticola   116, **117**
Sedge Warbler   213, **213**
Shelduck   54, **55**
Short-eared Owl   150, **151**
Shoveler   66, **67**
Shrike, Red-backed   254
Siskin   277
Sitta   (Sittidae)
    europaea   250, **251**
Skylark   165, **167**
Slavonian Grebe   42
Smew   72
Snipe   114, **115**
    Jack   114
Song Thrush   205, **207**
Sparrow, House   266, **269**
    Tree   268, **269**
Sparrowhawk   78, **79**
Spotted Crake   96
Spotted Flycatcher   234, **237**
Starling   265, **267**

Sterna (Sternidae)
    albifrons 130
    hirundo 128, **129**
Stock Dove 132, **133**
Stonechat 196, **197**
Streptopelia (Columbidae)
    decaocto 136, **137**
    turtur 138, **139**
Strix (Strigidae)
    aluco 147, **147**
Sturnus (Sturnidae)
    vulgaris 265, **267**
Swallow 170, **171**
Swan, Mute 46, **47**
Swift 153, **155**
Sylvia (Sylviidae)
    atricapilla 223, **225**
    borin 222, **223**
    communis 220, **221**
    curruca 217, **219**

Tachybaptus (Podicipedidae)
    ruficollis 38, **39**
Tadorna (Anatidae)
    tadorna 54, **55**
Tawny Owl 147, **147**
Teal 60, **61**
Tern, Common 128, **129**
    Little 130
Tetrao (Tetraonidae)
    tetrix 87
Thrush, Mistle 208, **209**
    Song 205, **207**
Tit, Bearded 238
   Blue 246, **247**
   Coal 245, **245**
   Great 248, **249**
   Long-tailed 239, **241**
   Marsh 241, **243**
   Willow **243**, 244
Treecreeper 252, **253**
Tree Pipit 174, **175**
Tree Sparrow 268, **269**
Tringa (Scolopacidae)
    totanus 120, **121**
Troglodytes (Troglodytidae)
    troglodytes 186, **187**
Tufted Duck 69, **71**

Turdus (Turdidae)
    iliacus 207
    merula 202, **203**
    philomelos 205, **207**
    pilaris 205
    torquatus 200, **201**
    viscivorus 208, **209**
Turtle Dove 138, **139**
Twite 279, **281**
Tyto (Tytonidae)
    alba 142, **143**

Upupa (Upupidae)
    epops 158

Vanellus (Charadriidae)
    vanellus 110, **111**

Wagtail, Grey 180, **181**
   Pied 182, **183**
   Yellow 178, **179**
Warbler, Cetti's 210
   Garden 222, **223**
   Grasshopper 210, **211**
   Reed 215, **217**
   Sedge 213, **213**
   Willow 230, **231**
   Wood 225, **227**
Water Rail 94, **97**
Wheatear 199, **199**
Whinchat 195, **197**
Whitethroat 220, **221**
   Lesser 217, **219**
Wigeon 57, **59**
Willow Tit **243**, 244
Willow Warbler 230, **231**
Woodcock 116, **117**
Woodlark 165
Woodpecker, Great Spotted 161, **163**
   Green 159, **159**
   Lesser Spotted 162, **163**
Woodpigeon 134, **135**
Wood Warbler 225, **227**
Wren 186, **187**
Wryneck 158

Yellowhammer 288, **289**
Yellow Wagtail 178, **179**

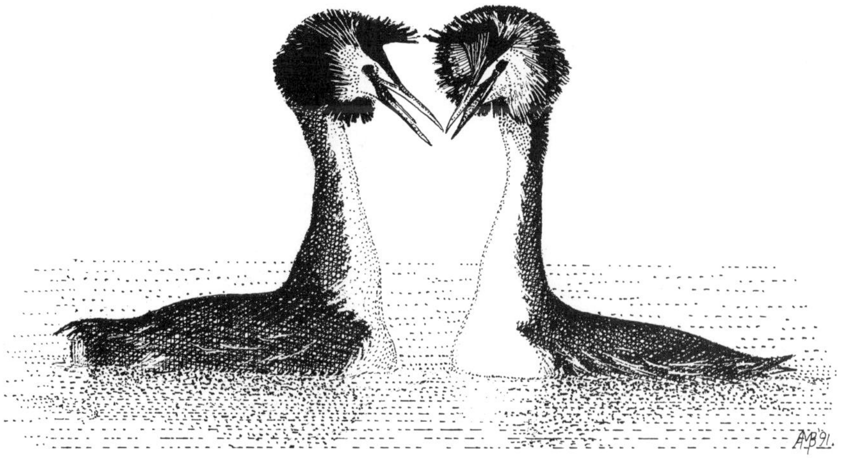